Lecture Notes in Networks and Systems

Volume 212

The series "Lecture Notes in Networks and Systems" publishes the latest developments in Networks and Systems—quickly, informally and with high quality. Original research reported in proceedings and post-proceedings represents the core of LNNS.

Volumes published in LNNS embrace all aspects and subfields of, as well as new challenges in, Networks and Systems.

The series contains proceedings and edited volumes in systems and networks, spanning the areas of Cyber-Physical Systems, Autonomous Systems, Sensor Networks, Control Systems, Energy Systems, Automotive Systems, Biological Systems, Vehicular Networking and Connected Vehicles, Aerospace Systems, Automation, Manufacturing, Smart Grids, Nonlinear Systems, Power Systems, Robotics, Social Systems, Economic Systems and other. Of particular value to both the contributors and the readership are the short publication timeframe and the world-wide distribution and exposure which enable both a wide and rapid dissemination of research output.

The series covers the theory, applications, and perspectives on the state of the art and future developments relevant to systems and networks, decision making, control, complex processes and related areas, as embedded in the fields of interdisciplinary and applied sciences, engineering, computer science, physics, economics, social, and life sciences, as well as the paradigms and methodologies behind them.

Indexed by SCOPUS, INSPEC, WTI Frankfurt eG, zbMATH, SCImago.

All books published in the series are submitted for consideration in Web of Science.

More information about this series at http://www.springer.com/series/15179

Petro Vorobiyenko · Mykhailo Ilchenko ·
Iryna Strelkovska
Editors

Current Trends in Communication and Information Technologies

Springer

problem of the deployment of 5G technology in EU countries, the problem of constructing an adaptation model for optimizing service platforms in infocommunications, the results of a full-factor experiment and analysis of experimental data to increase the performance between network screens in information security systems in telecommunications networks as well as factors of providing sufficient resources of modern communication systems with the 5G ideology.

Part IV Research of public and special applications of information technologies includes various aspects of scientific and educational applications. The presented studies of this Part cover, on the one hand, special applications of information technologies, such as the conceptual principles of combining the bases of transparency and confidentiality of personal life and the need to track personal contacts with possible technical solution using blockchain technology during coronavirus pandemic 2019 (COVID-19), conceptual approach to technical, including cyberphysical, systems quality evaluation at the stage of exploitation in three aspects: functional, technical and process. This group of problems also includes methods that can be used to solve the problem of volume visualization, using the capabilities of standard GPU, methods of sustainability in the distributed SCADA system based on the analysis of changes in the information flow states.

On the other hand, it proposed public and special applications of information technologies, such as analysis of current state of distributed virtual learning systems, their key tasks, advantages and disadvantages, as well as problem of inconsistency while choosing standards for information resources metadata creation in the educational intelligent analytical system and the various questions of creation of integrated development environment for computer training systems.

For the convenience of the readers, we briefly summarize contents of the chapters accepted.

Part I Research of Telecommunication System Characteristics at the Physical Level

The first chapter presented by **I. Strelkovskaya, I. Solovskaya, J. Strelkovska Spline-Approximation and Spline-Extrapolation Methods in Telecommunication Problems** determines the place of spline methods in the problems of analyzing the characteristics of telecommunication systems. These include the tasks of restoring random signals and self-similar traffic, managing network objects and the network as a whole, operating procedures for telecommunication objects and networks. When solving such problems, the spline-approximation method was used and a spline-extrapolation method was developed based on various types of spline functions, namely: linear, quadratic, quadratic *B*-splines, cubic, cubic *B*-splines. As the selection criterion of a particular type of spline while using spline-approximation method and spline-extrapolation method, the error of reconstruction or prediction is selected, the accuracy of which can be increased by using one or another type of spline, depending on the restored or predicted object. In some cases, the expediency of using wavelet approximation and wavelet extrapolation is shown, depending on

Preface

This volume is a collection of the most important research results in the fields of information and telecommunication technologies provided by different groups of researchers from Ukraine in collaboration with scientists from different countries. The authors of the chapters from this collection present in-depth extended research results in their scientific fields.

The volume consists of **4 Parts.**

Part I Research of telecommunication system characteristics at the physical level contains a discussion of various aspects of the signal transmission quality indicators analysis for solving practically important issues in telecommunication systems. This applies to spline methods in the tasks of analyzing signal characteristics, to studying the features of information transmission through fiber-optic systems, to harmonizing reception quality indicators in the tasks of the classical theory of noise immunity and in the regulations of international standards, to increasing the efficiency of phase-locked loop systems, as well as to methods of signal detection and recognition in cognitive radio networks.

Part II Research of telecommunication traffic and its performance indicators introduces original chapters dealing with many aspects of research and forecasting of services characteristics of telecommunication systems. This Part is about the problem of constructing a stream model of single-path or multipath routing, with the use of the Kolmogorov—Wiener equation for stationary forecasting of telecommunication traffic, and on a new trend in the field of ICT, called the trend of big data processing. In this Part, the authors present a scientific and technical principles of designing and planning a communication system for automated control systems with high requirements for security, survivability and reliability and the methodology and technological solutions based on a model of a realistic modern switching (routing) device.

Part III Research of information systems technological parameters deals with actual chapters, which show some effective technological solutions that can be used for the implementation of novel systems. The presented studies of this Part cover a wide range of information and telecommunication technologies, including relevant approach to the problem of using the terahertz frequency range, the

Editors
Petro Vorobiyenko
O.S. Popov Odesa National Academy
of Telecommunications
Odesa, Ukraine

Mykhailo Ilchenko
National Technical University of Ukraine
“Igor Sikorsky Kyiv Polytechnic Institute”
Kyiv, Ukraine

Iryna Strelkovska
O.S. Popov Odesa National Academy
of Telecommunications
Odesa, Ukraine

ISSN 2367-3370 ISSN 2367-3389 (electronic)
Lecture Notes in Networks and Systems
ISBN 978-3-030-76342-8 ISBN 978-3-030-76343-5 (eBook)
https://doi.org/10.1007/978-3-030-76343-5

This Springer imprint is published by the registered company Springer Nature Switzerland AG
The registered company address is: Gewerbestrasse 11, 6330 Cham, Switzerland

the problems being solved. The results obtained allow to provide the required characteristics and scalability of the solutions obtained.

In the chapter *Evaluation of Quality Parameters of an Intensity-Modulated Optical Transmission System* by **V. Pedyash, A. Mazur, D. Rozenvasser**, the authors investigated the features of fiber-optic transmission systems that are used for high-speed traffic transmission in modern telecommunications networks including NGN and IMS. Optical channels of optical transport hierarchy (OTH) communication systems have a bandwidth of 2.5 to 100 Gbps, so the study of their characteristics is an urgent task. The problem of studying the quality parameters (Q-factor) of optical OTH signals of the OTU1 and OTU2 types is solved in the current section. A simulation model of an intensity-modulated optical transmission system has been developed to solve this problem. Optical fiber simulation model allows to introduce linear and nonlinear distortions of the propagation medium into the signal, and it is based on the split-step Fourier method. Comparison of the results of modeling and analytical calculations was carried out to check the accuracy of the proposed simulation model. The influence of the signal power of the transmitter and the quantity of linear path optical sections on the Q-factor of the signal in the receiver are calculated. It is proposed to optimize the transmitter signal power by the criterion of maximizing the Q-factor to increase the length of the regeneration section.

The chapter *Application of the Classical Noise Immunity Theory for Prediction of the Parameters of Perspective Multiservice Telecommunications in Accordance with Modern Digital Standards* by **L. Uryvsky, A. Moshynska,V. Solianikova,B. Shmigel** focuses on the problem of synthesizing a unified methodology for assessing communication quality indicators in multiservice communication systems. In the classical theory of noise immunity, an indicator of reliability is the probability of an information bit error. In modern international standards and recommendations, there is an indicators hierarchy of the communication channel quality, based on measurements, which is not related to the classical theory of noise immunity. This technique differs from the existing methods for assessing communication quality indicators in that it allows to display the results of instrumental assessment according to the current international standards in the indicators structure of the noise immunity classical theory. The advantage of the proposed methodology and technique is that the fundamental theory of noise immunity makes it possible to calculate the corresponding quality indicators of a digital channel at the design stage of a telecommunication system, while ITU-T standards determine the quality indicators of information transmission in multiservice systems only on the basis of measurements of an already operating digital communication system.

The chapter *Improving Efficiency of the Phase-Locked Loop for Reference Oscillator of the Multichannel System for Time Synchronization Signals Telemonitoring* by **V. Koval, V. Lysenko, D. Kalian, O. Osinskyi, O. Samkov** devotes a multichannel automated telemonitoring system for time synchronization signals. It simultaneously measures the parameters of several synchronization signals relative to the reference oscillator signal, processes the received data and transmits them via an IP network to a computer of a centralized control system. In order to improve the efficiency of the phase-locked loop, the optimization was performed in regards to the speed of the reference oscillator control process. The structure of the phase-locked loop subsystem (PLLS) was synthesized in three steps: the first is analysis; the second is implementation; the third is simulation (using the Simulink program of the MATLAB package). The structure of the optimal PLLS with adaptive properties was designed, in which generation of the control signal is done in accordance with the speed-optimal control strategy. To implement optimal transient processes in PLLS, two hierarchical levels of control and an adaptive digital phase discriminator (ADPD) with a controlled form of the discriminating characteristic were used. Simulation modeling of the synthesized PLLS circuit with unlimited bandwidth of the LPF and an original adaptive digital phase discriminator has been performed. The simulation results showed a decrease of the duration of the transient response from 2.7 to 4 times, depending on the initial conditions of the phase-locked loop of the reference oscillator.

In the chapter *Methods of Signal Detection and Recognition in Cognitive Radio Networks* by **V. Bezruk, S. Ivanenko, O. Fedorov** the authors present techniques of signal detection and recognition to perform spectrum sensing in cognitive radio networks. The main focus is on extensive use of signal recognition techniques, which is not the case yet in the context of cognitive radio networks. In particular, the use of unconventional methods of signal recognition allows us to not only divide received signals into classes of those which belong to primary or secondary users but, additionally, register appearance of new, uncategorized, sources of radiation. As a consequence, all considered in the paper procedures may be employed by the local frequency control authorities to monitor radio frequency resource utilization. Performance analysis of presented in the paper algorithms for signal detection and recognition is performed by the method of Monte Carlo.

Part II Research of Telecommunication Traffic and Its Performance Indicators

The chapter *Research of the QoE Fast ReRouting Processes with Differentiated R-Factor Maximization for VoIP-Flows Using the Tensor Model of the Corporate Telecommunication Network* by **O. Lemeshko, O. Yeremenko, M. Yevdokymenko, T. Radivilova** focuses on the problem of construction of the flow-based model of the QoE Fast ReRouting. The model is based on the implementation of the single-path or multipath routing, the conditions of flow conservation. The model was supplemented by the conditions of structural network elements protection (node, link and route). The peculiarity of these conditions is the consideration of possible packet losses due to congestion of router interfaces. In order to obtain analytical expressions for the calculation of the R-factor for each of the VoIP-flows, a tensor generalization of the mathematical model of routing has been performed. Based on the tensor description of the network, it was possible to obtain expressions for calculating the average multipath end-to-end delay and packet loss probability, which allowed to formulate in analytical form the R-factor calculation expressions for each of the VoIP-flows. The novelty of the proposed model is the formulation of the problem of QoE Fast ReRouting in the optimization form when the optimality criterion was the maximum of the additive form, represented by the sum of weighted according to the IP-priority values of R-factor for each of the VoIP-flows. The results of the study of the proposed model confirmed its efficiency and adequacy, which was especially evident in the conditions of complex network topologies, high network congestion and flow differentiation relative to the IP-priority values of packets.

The chapter *Kolmogorov–Wiener Filter Weight Function for Stationary Traffic Forecasting: Polynomial and Trigonometric Solutions* by **V. Gorev, A. Gusev, V. Korniienko, M. Aleksieiev** is devoted to obtaining the Kolmogorov–Wiener filter weight function for stationary telecommunication traffic forecasting. The aim of the chapter is to obtain an approximate solution to the Kolmogorov–Wiener filter weight function for stationary traffic forecasting in the framework of the model of a power-law structure function and the model of fractional Gaussian noise and to investigate the validity of the obtained results. Two models of stationary traffic are considered: a model where traffic is considered to be a process with a power-law structure function and a model where traffic is considered to be fractional Gaussian noise. In the case of a large amount of data, traffic may be described as a continuous random process, and the Kolmogorov–Wiener filter weight function is a solution to the corresponding Wiener–Hopf integral equation. The scientific novelty is the fact that for the first time, the corresponding weight function is sought as a truncated trigonometric Fourier series. It is shown that trigonometric solutions are better than the polynomial ones for the model where the telecommunication traffic is considered as a process with a power-law structure function.

L. Globa, Z. Savchuk, O. Vasylenko, E. Siemens (QoS of Data Networks Analyzing Based on the Fuzzy Knowledge Base) discuss an approach to analysis of new trend in the ICT field that has been called the big data processing trend. There are a lot of up-to-day intelligent technics and systems used for overcoming this trend but the computational and data processing complexity under real-time requirements continues to be one of the important disadvantages for many engineering fields. The paper deals with an approach of big data stream structuring into fuzzy logic rules for fuzzy knowledge base development that has no large data processing complexity. To guarantee the correctness of the fuzzy knowledge base, the metagraph theory apparatus is used based on control conflicting and duplicate rules under consideration of their logical inter-connections. Usage of the fuzzy knowledge base during big data stream processing helps to decrease data processing time for decision-making systems in real engineering fields.

I. Strelkovskaya, R. Zolotukhin, A. Makoganiuk (Modeling of Telecommunication Components of Automated Control Systems in Low-Bandwidth Radio Networks) present the scientific and technical principles of designing and planning a communication system for automated control systems with high requirements for security, survivability and reliability, is the reasonable choice of the necessary telecommunications equipment to build communication networks with the required functional characteristics to provide access to services. The paper presents simulations of the telecommunication system at the lower level of management aimed at maintaining communication during the active movement of users. The work proposes the analysis of STANAG-4677 and ADatP-36 standards. As well, it offers the concept of their use to design telecommunication components of the automated control system for UHF/VHF radio communication networks. The software implementations of the JDSS, NFFI, FFI protocols have been developed using .NET framework and C# programming language. The data has been obtained on the average bitrate of traffic for the standards, possible packet lengths and average length, the time interval between incoming packets. The distribution law of time intervals between requests for service and the density of this distribution has been determined.

In the chapter *Modeling Unconditional Forwarding Decision within Switching Lattice* by **T. Shmeleva, I. Stetsenko**, the authors present the methodology and technological solutions based on a detailed model of a realistic modern switching (routing) device has been built that combines cut-through and store-and-forward switching techniques. The model is based on a non-deterministic unconditional forwarding decision made with regard to the input packet. A rectangular (square) switching lattice model is composed in the form of an infinite Petri net using dedicated transitions to connect devices. A parametric representation of p-invariants is obtained as a solution of an infinite Diophantine linear system of equations. P-invariance has been proven for a lattice an any size that allows the conclusion about boundedness and conservativeness which are properties of an ideal networking protocol model. Some comparisons with the alternative technique of object Petri nets and the corresponding simulation systems have been obtained which acknowledge mutual complementary of the two approaches.

Part III Research of Information Systems Technological Parameters

M. Ilchenko, T. Narytnyk, G. Avdeyenko (*Wireless Communication Systems of Terahertz Frequency Range)* present a novel approach for the problem of using the terahertz frequency range. The transition to the terahertz range solves both the issue of increasing the data rate transmission by using of wider frequency bands and the issue of electromagnetic compatibility ensuring between wireless communication systems, as well as elimination of intentional interferences. The aim of this chapter is to inform about the results of investigation and development in the terahertz range which completed by the authors in order to determine the possibilities and prerequisites for solving the fundamental problem of wireless communications systems development using world and domestic experience in microwave telecommunications. It is shown that in the terahertz range, the main influence on a radio link energy potential exerted by the electromagnetic waves attenuation in hydrometeors and gases. The terahertz frequency range sections that are most suitable for use in radio links are determined. The results of functional design, simulation and experimental research of the receiving and transmitting devices of terahertz wireless communication system with a gigabit data rate in the 130-134 GHz frequency range are presented. The concept of creating a software-defined radio technology based on Wi-Fi which could be used in order to ensure high data rates, reliability and security of transport distribution networks of the 5G mobile systems in the terahertz range was proposed.

The chapter *5G Slicing and Handover Scenarios: Compulsoriness and Machine Learning* by **A. Luntovskyy, B. Shubyn, T. Maksymyuk, M. Klymash** focuses on the problem of the deployment of 5G in EU countries. The most important requirements for 5G networks are examined like slicing, wide network inter-operability (with Wi-Fi 6, IoT devices, LoRa WAN, etc.). The operation in 5G micro-cells of small private providers is favored. The authors investigated the appropriate 5G blockchain (BC) slices architectures, recent BC types, BC applications, platforms as well resource consumption. The so-called smart contracting for 5G networks and DAO technology can be taken into account for use as 5G blockchained slices. Intelligent 5G handover and roaming within the 5G networks can be provided using standards of machine learning. Recurrent neural networks as well as their modeling and training processes are represented. Modeling with a neural network and prediction of the user number in a micro-cell are investigated. Recurrent neural networks, based on GRU cell, can optimize the handover in 5G cellular structures.

The chapter *Method for Constructing an Adaptive Model for Optimizing Service Platforms of Information and Communication Networks* by **L. Nikityuk, R. Tsaryov** considered some aspects related to the problems of constructing an adaptation model for optimizing service platforms in infocommunications. A feature of modern service platforms is that the requirements for their technical characteristics change dynamically depending on the conditions of both external and internal environment. Take to account this fact, that the process of optimization service platform should be continuous that is adapted to the changing situations, the proposed method for constructing a mathematical model allows solving the problems of structural and parametric optimization of the service platform during the life cycle from the moment of its design and further during maintenance at the stages of reconfiguration and reconstruction, taking into account the influence unforeseen factors of the environment and object itself. Solving the problems of reconfiguration and reconstruction is a cyclical process during all-time maintenance until the service platform will be disposal. The proposed method for constructing an adaptive mathematical model is based on models of the morphological and functional description of the service platform, which allow to formalize the problems of finding the optimal values of its parameters. The effectiveness of the proposed method is demonstrated on specific examples of constructing adaptive models for optimizing service platforms for IPTV, telemedicine, etc.

In the chapter *The Research of Ways to Increase the Efficiency of Functioning between Firewalls in the Protection of Information Web-portals in Telecommunications Networks* by **Z. Dudar, I. Shubin,V. Skovorodnikova, S. Litvin**, the authors discuss the results of a full-factor experiment and analysis of experimental data to increase the performance between network screens in information security systems in telecommunications networks. A number of factors that significantly affect the functioning between firewalls are considered, namely: fragmentation of network packets, security policy, network structure and auxiliary functions between firewalls. Empirical dependences characterizing the influence of the adopted security policy on the performance between network screens are obtained. The possibility of increasing the level of information security of the network by redistribution of security rules is proved. A technique for studying the performance of firewalls is described, which involves the primary processing of experimental data, and their secondary processing to ensure satisfactory accuracy of the results.

L. Globa, V. Prokopets (The Approach to Network Planning Process Improvement) discuss approach to the analysis of factors providing modern communication systems with the 5G ideology sufficient resources. Current trends like energy efficiency, planning with traffic uncertainty considerations require a significant change in existing network planning technologies. The main idea of this research is to provide a pipeline that will process the network planning process to focus on additional designing stage as marketing forecast, B2B, B2C, rollout plan, target cells and equipment in stock. The pre-planning algorithms for the definition of the required outputs, the algorithms of equipment-based optimization and problem detection (solution type selection) are suggested. The suggested approach takes into account all new key features and technologies and focuses on improving the pre-planning stage.

Part IV Research of Public and Special Applications of Information Technologies

In the chapter *Fundamentals of Functioning of Global Contact Monitoring Systems in the Context of COVID-19 Spread Prevention* by **A. Kuznetsov, N. Poluyanenko, A. Kiian**, the authors pay attention to the fact that coronavirus disease 2019 (COVID-19) has a clear potential for a prolonged global pandemic, high mortality and overload of health systems. The chapter considers the conceptual principles of combining the bases of transparency and confidentiality of personal life and the need to track personal contacts, substantiates the need for such a combination and provides a possible technical solution using blockchain technology. The practical development and deployment of the system based on such protocols will allow each citizen to track the personal history of probable contact with infected persons or other potential risks (including deciding on the need for self-isolation or additional examination). In this case, the decentralized system guarantees the privacy of such information both for each participant (due to the inability to determine the personal data of a potentially infected person) and for the system as a whole.

The chapter **A Three-Aspects Approach for Technical Systems Quality Evaluation** by **S. Volkov, L. Kolomiets, O. Hrabovskyi** presented conceptual approach to technical, including cyberphysical, systems quality evaluation at the stage of exploitation in three aspects: functional, technical and process. The dominance of the reductionist approach to the evaluation of technical systems, which is impractical in terms of management decisions at the current level of complexity of systems, is shown. A general approach, which allows considering and researching functional, technical and process approaches to evaluation, as a holistic evaluation system based on their interaction, within a single information structure based on project specifications and system information links, is offered. The presentation of the informational structure in the form of the sensory infrastructure is considered, and the model and the scale of the quality state evaluation, based on relative value of the amount of information, are given. A model, that combines technical, functional and process aspects into the quality state evaluation unified system, is developed.

In the chapter *A Method for Diagnosing SCADA Failures Based on Predicate Logic within the Expert System* by **O. Syrotkina, M. Aleksieiev, B. Moroz, I. Udovyk, A. Martynenko**, the authors address the issue of detecting and localizing failures in the distributed SCADA system based on the analysis of changes in the information flow states when passing through structural modules and levels of SCADA hierarchy. The structural and logical model presented describes SCADA of any topology. The configuration diagram of SCADA topology is presented using the parameters of the structural and logical model for diagnosing failures. This model is used by the predicate system within the expert system knowledge base. The model calculates the instability coefficients to evaluate the operability of each system's structural module in real time based on the analysis of changes in the state of SCADA information flows within the structural and logical model developed. The proposed method for detecting and localizing failures in the SCADA system can be applied both for current diagnostics and for predicting the operability of the structural modules of the SCADA system. The system operability prediction will be formed on the basis of statistical processing and analysis of the retrospective array containing the instability coefficients of the SCADA structural modules' operation.

The chapter *Comparison of Volume Rendering Methods Using GPU and Specialized Volumetric Accelerator* by **S. Vyatkin, O. Romanyuk, R. Chekhmestruk, S. Romanyuk, O. Romanyuk** covers methods that can be used to solve the problem of volume visualization, using the capabilities of standard GPU. The aim of the chapter is to research the features of using known and original volumetric visualization methods and evaluation of the quality images and performance. An overview of alternative implementations of the volume visualization problem and a comparison of the proposed solution with a system based on a highly specialized hardware accelerator for volume visualization is given. Obviously, we can hardly hope in the near future for a purely software implementation of volume visualization in real time, even in a multiprocessor system, without the cost of creating processing accelerators—optimization algorithms and hardware.

In the chapter **Individual Training Technology in Distributed Virtual University** by **Z. Dudar, I. Shubin, A. Kozyriev**, the authors are concerned with some issues related to the current state of distributed virtual learning systems, their key tasks, advantages and disadvantages. The analysis of building local informational and educational environments allowed formulating the basic principles of professional distance education: universality, invariance; scalability; openness; processability of creation, storage and use of resources; continuity, etc. The method for building an individual learning path is analyzed. The logical model for structuring the content of text educational materials is improved by introducing predicates of equivalence of document content and equivalence of concepts which allows obtaining a logically justified sequence of representation of educational materials on an individual learning path.

The chapter **An Approach to Designing the Educational Intelligent Analytical System** by **L. Globa, R. Novogrudska, M. Popova** is connected with some aspects related to a problem of inconsistency while choosing standards for information resources metadata creation. The urgent task is not the implementation of uniform requirements for the description of information resources reflecting the current state of science and technology, but the development of software and information tools that ensure the integration of existing disparate resources and data as well as their metadescriptions developed in accordance with different standards and in different formats. The chapter deals with the approach to design the built-inlanguage-invariant tools of semantic analysis and program modules for the dynamic structuring of various informational resources. Such software tools make it possible to link informational resources involved in educational and research process, as well as allow solving a number of important tasks, such as eliminating information inequality, organizing educational, scientific and research processes, promoting information and knowledge, intensifying the communication processes of the scientific community, popularizing science among young people, creating conditions for the formation of high-quality scientific personnel.

The chapter **Principles of Creating an Integrated Development Environment for Educational Computer Systems** by **Z. Dudar, I. Shubin, A. Kozyriev** considered the various questions of creation of integrated development environment for computer training systems. The information technologies that can be used for creation of the integrated development environment are described. The different didactic aspects of realization of such systems are analyzed. The ways to improve the efficiency and quality of learning process with computer training systems for distance education are pointed.

We would like to sincerely thank the authors of this collection, because without their hard work of preparing good chapters, this volume would not have been successfully prepared.

And last but not least, we wish to thank…for their dedication and help to implement and finish this large publication project on time maintaining the highest publication standards.

And last but not least, we wish to thank Series editor, Prof. Janusz Kacirzyk, Polish Academy of Sciences, Dr. Thomas Ditzinger, Executive Editor, Interdisciplinary and Applied Sciences & Engineering, Ms. Varsha Prabakaran, Project Coordinator, Books Production, Mr. Holger Schaepe, Applied Sciences and

Engineering Editorial Assistant, Ms. Mirnalini Paramasivam, Springer Project Coordinator (Books), Mr. Ravi Vengadachalam from Springer Nature **for their dedication and help to implement and finish this large publication project on time maintaining the highest publication standards.**

June 2021

The Editors
Series editor, Prof. Janusz Kacprzyk
Polish Academy of Sciences

Contents

Research of Telecommunication System Characteristics at the Physical Level

Spline-Approximation and Spline-Extrapolation Methods in Telecommunication Problems

I. Strelkovskaya[1(✉)], I. Solovskaya[1(✉)], and J. Strelkovska[2(✉)]

[1] O.S. Popov Odesa National Academy of Telecommunications, Odesa, Ukraine
{strelkovskaya,i.solovskaya}@onat.edu.ua
[2] National University «Odesa Law Academy», Odesa, Ukraine
4800632@gmail.com

Abstract. One of the main tasks that need to be solved in telecommunications are the problems of analysis and synthesis of telecommunication systems. These include the tasks of restoring random signals and self-similar traffic, managing network objects and the network as a whole, operating procedures for telecommunication objects and networks. The problems of forecasting are also considered, in particular, predicting the characteristics of network traffic, maintaining the quality characteristics of QoS during its service, and forming the requirements for network buffer devices. When solving such problems, the spline-approximation method was used and a spline-extrapolation method was developed based on various types of spline-functions, namely: linear, quadratic, quadratic B-splines, cubic, cubic B-splines. As the selection criterion of a particular type of spline while using spline-approximation method and spline-extrapolation method, the error of reconstruction or prediction is selected, the accuracy of which can be increased by using one or another type of spline, depending on the restored or predicted object. In some cases, the expediency of using wavelet-approximation and wavelet-extrapolation is shown, depending on the problems being solved. The results obtained allow to provide the required characteristics and scalability of the solutions obtained.

Keywords: Approximation · Interpolation · Extrapolation · Restoration · Prediction · Spline-approximation · Spline-extrapolation · Spline-function

1 Introduction

The rapid development of modern telecommunications systems and networks today leads to rapid change of technological solutions, continuous improvement of protocol solutions, mechanisms to maintain the necessary quality characteristics of QoS/QoE and algorithms for the operation of the systems in the direction of Future Networks [1–6]. Such technological changes require a radical revision of known solutions and the search for new methods for solving problems of analysis and synthesis of telecommunications. The creation of the theory of telecommunication systems and networks is not yet complete, however, engineering intuition today often precedes science and allows to find the necessary solutions to telecommunications problems at the intersection of different

P. Vorobiyenko et al. (Eds.): IPF 2020, LNNS 212, pp. 3–20, 2021.
https://doi.org/10.1007/978-3-030-76343-5_1

sciences, thus moving from purely technological solutions to solutions based on systems theory.

The main tasks that need to be solved in modern telecommunications networks include the following:

- tasks of recovery and assessment of states (data, signals, traffic);
- signal and image processing tasks, including data filtering and compression, signal detection and measurement tasks;
- tasks of improving the quality characteristics of QoS (Quality of Service) of modern telecommunications networks, which regulate the dependencies between the parameters of interaction in the network;
- tasks of forecasting the characteristics of the traffic of various applications, which is served in the telecommunications network.

To solve such problems it is necessary to search for effective tools for system analysis and synthesis of modern telecommunications systems and networks based on new mathematical methods. Existing solutions are based on a number of already known mathematical methods (functional analysis, integral and differential calculus, group theory, operator theory, spectral analysis and others) [7–9]. An important issue is the search for other mathematical methods, alternative to the known ones, which will greatly simplify the decision process and at the same time provide new solution possibilities. Depending on the nature of the considered problems, obtaining results is possible with the help of interpolation, extrapolation and approximation estimates, because the efficiency of solving problems of analysis and synthesis of telecommunication networks directly depends on the accuracy of the obtained results. The use of methods of interpolation, approximation and extrapolation of real processes and modes of network elements, as well as network functions will allow to obtain a solution of a large class of problems using spline-functions, which will more easily solve some telecommunications problems. Among the main advantages of using spline-functions are [10–13]:

1) splines are more resistant to local perturbations, ie the behavior of the spline in the vicinity of the point does not affect the behavior of the spline as a whole, as, for example, this occurs in polynomial interpolation;
2) good convergence of spline-interpolation in contrast to polynomial, in particular, for functions with irregular smoothness properties is an indisputable priority for spline-interpolation;
3) splines have useful extreme properties, an important practical advantage is a fairly simple implementation of spline-functions on personal computers.

In [14–18], the authors obtained solutions to the problem of speech signal recovery using discrete samples, using a comparison of recovery results using the Kotelnikov series and spline-approximation based on cubic splines. It is established that the use of spline-approximation for a speech signal with frequency oscillations allows to obtain better results with the least error. The results of recursive assessments of the state of network elements are considered in [19], where the use of Kalman-Bussey filter and spline-approximation suggests that the results of restoring the state of network elements of RTP (Real Time Protocol)/RTCP (Real Time Control Protocol) with the help of cubic splines can be significantly improved. The advantage of spline approximation in solving traffic recovery problems is proved by different types of spline-functions (linear, quadratic, cubic, B-splines, etc.) in [14–18]. At the present stage of development of telecommunications, an important issue is to solve problems of forecasting the characteristics of traffic served in the telecommunications network. To do this, in [20–31] the use of spline-extrapolation based on various spline-functions is proposed, which satisfy the accuracy of the forecast. The forecasting results will allow to predict the required size of buffer devices and characteristics of network objects, thus preventing network congestion and exceeding the normative values of QoS quality characteristics.

The methods of approximation theory, which include methods of interpolation, extrapolation and approximation, are increasingly used in solving various problems of the theory and practice of telecommunications [32]. Consider these methods in detail.

The purpose of this work is to find new methods for solving problems of analysis and synthesis in telecommunications to improve the quality of telecommunications systems.

2 Spline-Interpolation

Interpolation is a way of approximating any quantity by a known set of values of that or another quantity which are related to it. This interpolation is the basis for the implementation of many approximate methods of solution. When interpolating the function $f(x)$ on the segment $[a, b]$ by its values in the nodes x_k of the grid $\Delta_n = \{a \leq x_0 < x_1 < \ldots < x_n \leq b\}$ determines the construction of another function $L_n(x_k) = L_n(f; x_k)$ such that $L_n(x_k) = f(x_k)$, $k = 0, 1, \ldots, n$. As a general statement of the problem of interpolation function $f(x)$ is to construct $L_n(x)$ not only from the condition of coincidence of the function $L_n(x)$ and $f(x)$ on the grid Δ_n, but also claims overlap in some nodes derivatives to some order or some other relations connecting $f(x)$ and $L_n(x)$. Usually, $L_n(x)$ is given as: $L_n(x) = \sum_{i=0}^{n} a_i\phi_i(x)$, where the formula $\{\phi_i(x)\}_{i=0}^{n}$ is some system of linearly independent functions [32].

In practice, spline-interpolation based on various spline-functions has found its use in solving problems of analysis and synthesis of telecommunication networks and systems. Based on the obtained results, an approximation can be performed.

The approximation allows to investigate the numerical characteristics and qualitative properties of the analyzed object, reducing the task to the study of simpler or more convenient objects (for example, those whose characteristics are easy to calculate or whose properties are already known).

Then, the obtained results allow using spline-approximation to solve a significant class of problems of modern telecommunications, among which we note the following [14–19]:

- task of restoring self-similar random signals and traffic solutions which yield the required value interpolation between the nodes with the required accuracy [14–18];
- tasks of restoring the state of network objects and characteristics in real time, the solution of which will provide recursive estimates of the delay time of the RTP/RTCP protocol, RED buffer loading capacity and others in order to reduce the recovery error [19];

- tasks of management of network objects and a network as a whole which are based on results of monitoring of a network and processing of the received data;
- tasks of support of procedures of functioning of objects and a network as a whole for the purpose of increase of characteristics of quality of QoS functioning.

Consider the solution of some problems of telecommunications, based on the use of spline-approximation [14–19].

2.1 The Task of Restoring the Signal *f(x)* Using a Cubic Spline

The solution of the problem of recovery of random signals provides an opportunity to improve the quality of recovery by spline-approximation in comparison with the Kotelnikov series used in practice [14–18].

Consider the recovery of random signals using spline-approximation based on a cubic spline [14–18]. Let the values of some function $f(x)$ and its derivative $f'(x)\, f_i = f(x_i),\ f_i' = f'(x_i)\quad i = 0, 1, \ldots, n$ be given in the nodes of the grid $\Delta\colon a = x_0 < x_1 < \ldots < x_n = b$.

A cubic spline is a piecewise continuous function $S_3(f;x) = S_3(x)$, which satisfies two conditions [11–13]:

1) at each of the segments $[x_i; x_{i+1}]$, $i = 0, \ldots, n - 1$,

$$S_3(x) = a_{io} + a_{i1}(x - x_i) + a_{i2}(x - x_i)^2 + a_{i3}(x - x_i)^3, i = 0, \ldots, n - 1, \quad (1)$$

2) $S'(f; a) = f'(a),\ S'(f; b) = f'(b).$

Cubic spline $S_3(x)$ has the form of [11–13]:

$$S_3(x) = f_i(1 - t)^2(1 + 2t) + f_{i+1}t^2(3 - 2t) + m_i h_i t(1 - t)^2 - m_{i+1}h_i t^2(1 - t), \quad (2)$$

while $t = (x - x_i)/h_i$, $S_3(x_i) = f_i$, $S_3(x_{i+1}) = f_{i+1}$, $m_i = S'(f; x_i)$, $hi = h = (b - a)/n$.

The results of the comparison of the recovery of the signal f (x) using the Kotelnikov series and the cubic spline are shown in Fig. 1

2.2 The Task of Restoring the State of Network Elements RTP/RTCP

When estimating the state of packet network network objects, namely, packet delay time in RTP (Real Time Protocol/Real Time Control Protocol) and load volume of RED buffer devices (Random early detection) using Kalman-Bussey filter and spline-approximation, a significant improvement of the a posteriori variance of the estimation error was obtained in 4–6 times due to the use of cubic spline-functions (Fig. 2) [19].

Condition assessment is performed using a Kalman-Bussy filter (KBF) according to the expression [19]:

$$\widehat{x}(k) = \Phi(k, k - 1)\widehat{x}(k - 1) + K(k)[y(k) - \Phi(k, k - 1)\widehat{x}(k - 1)], \quad (3)$$

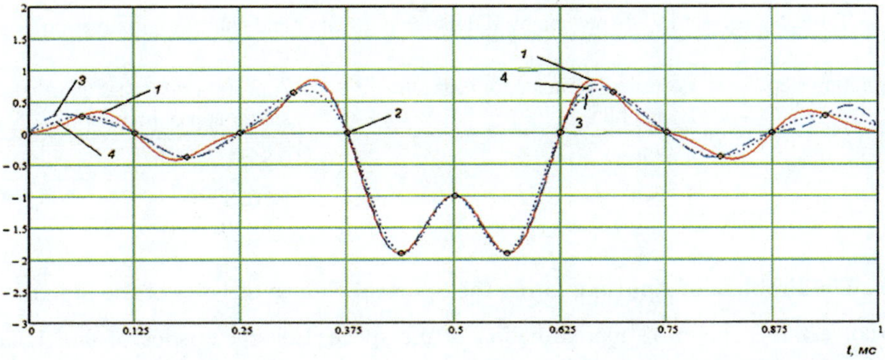

Fig. 1. The output signal *f(x)* (*line 1*), the value of the function *f(x)* in the nodes of the grid (*line 2*), the restored signal using the Kotelnikov series (*line 3*), the restored signal using the cubic spline (*line 4*).

where $\hat{x}(k)$ – the current estimate, which is obtained in the k-step, $y(k)$ – the observation process, $K(k)$– the gain of the Kalman-Bussey filter, which is recursive by the formula $K(k) = D_{\tilde{x}}(k)H^T(k)D_n^{-1}(k)$, T – sign of transposition, $D_{\tilde{x}}(k+1|k)$ – a priori variance of the estimation error, $D_{\tilde{x}}(k+1|k) = \Phi(k+1,k)D_{\tilde{x}}(k)\Phi^T(k+1,k) + \Gamma(k)D_{\omega}(k)\Gamma^T(k)$, $D_{\tilde{x}}(k)$ – a posteriori variance of the estimation error $D_{\tilde{x}}(k+1|k) = [I - K(k)H(k)]D_{\tilde{x}}(k|k-1)$.

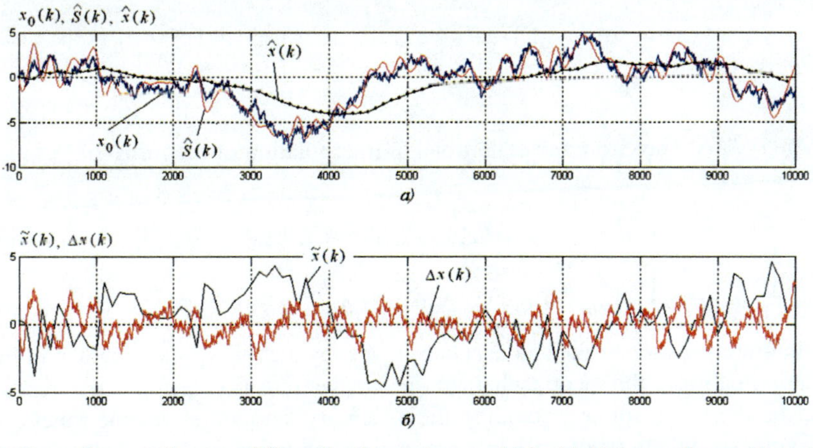

Fig. 2. Dependence at the step of discretization $\Delta t/\tau_{\text{kor}} = 1{,}10$, a) initial sampling $x_0(k)$, estimates of spline-approximation $S(k)$ (*thin line*); scores obtained by the KBF procedure (*line with points*); b) estimation errors $\tilde{x}(k) = \hat{x}(k) - x_0(k)$ and $\Delta x(k) = \hat{S}(k) - x_0(k)$.

The evaluation results are summarized in Table 1.

The obtained results prove that the value of the a posteriori dispersion in the established state depends both on the number of steps on the correlation interval and on the value of the level of the evaluated process.

Table 1. A posteriori dispersion by Kalman-Bussey filter and spline-approximation

Sampling step	Posteriori variance for the Kalman Byusy filter	A posteriori dispersion in spline-approximation
1/10	4,6756	0,8551
1/2	7,7644	1,7638

2.3 The Problem of Approximating the Spectral Characteristics of the Signal

To solve the problem of approximation of the spectral characteristics of the signal, consider the spectral density of the signal of the form (4) [33]:

$$|\,G(j\omega)\,| = \begin{cases} UT, & |\omega| < \omega, \\ G_{\Delta 1}(\omega), & \omega_A \le |\omega| \le \omega_C, \\ G_{\Delta 2}(\omega), & \omega_C < |\omega| \le \omega_B, \\ 0, & \omega_B < |\omega|, \end{cases} , \tag{4}$$

where $U = g(0)$; $\omega_A = (1-\alpha)\,\omega_C$; $\omega_B = (1+\alpha)\,\omega_C$; $\omega_C = {}^{\pi}/_{T}$ $\alpha = (\omega_C - \omega_A)/\omega_C = (\omega_B - \omega_C)/\omega_C$ – rounding coefficient of spectral density $(0 \le \alpha \le 1)$, which determines the width of the transition area $[\omega_A, \omega_B]$; $2\,\Delta\omega = 2\alpha\,\omega_C$ – the width of the transition area, $\alpha = \Delta\omega/\omega_C$.

Given the odd symmetry of the spectral density of the signal (4) relative to point C, the function $G_{\Delta 1}(\omega)$ та $G_{\Delta 2}(\omega)$ related by equations [33].

$$G_{\Delta 1}(\omega) = UT - G_{\Delta 2}(2\omega_C - \omega)\,, \quad \omega_A \le |\,\omega\,| \le \omega_C,$$
$$G_{\Delta 2}(\omega) = y(\omega) = a + b\,(\omega - \omega_C) + c\,(\omega - \omega_C)^2 + d\,(\omega - \omega_C)^3, \quad \omega_C \le \omega \le \omega_B, \tag{5}$$

where *a, b, c, d* are from the conditions of interpolation of the form (6)

$$\begin{cases} y(\omega_C) = a = y_C; \\ y(\omega_B) = a + b\Delta\omega + c\Delta\omega^2 + d\Delta\omega^3 = y_B; \\ y'(\omega_C) = b = y'_C; \\ y'(\omega_B) = b + 2c\Delta\omega + 3d\Delta\omega^2 = y'_B. \end{cases} \tag{6}$$

Spline-interpolation of the signal spectrum *f(x)* by a cubic spline is shown in Fig. 3.

Approximation of the spectral characteristic in the transition region makes it possible to describe the cubic spline, providing the necessary smoothness of the function and signal synthesis, which practically leads to rapid attenuation of the signal function and energy concentration in individual petals [33].

Thus, considering the above, it can be noted that in solving problems of analysis and synthesis in telecommunications, the use of spline-approximation is proposed, which makes it much easier to obtain solutions to a class of problems such as recovery and assessment (data, signals, traffic), processing problems signals and images, including the tasks of filtering and compressing data, detecting and measuring signals and improving the quality characteristics of QoS operation of telecommunications networks with the necessary error.

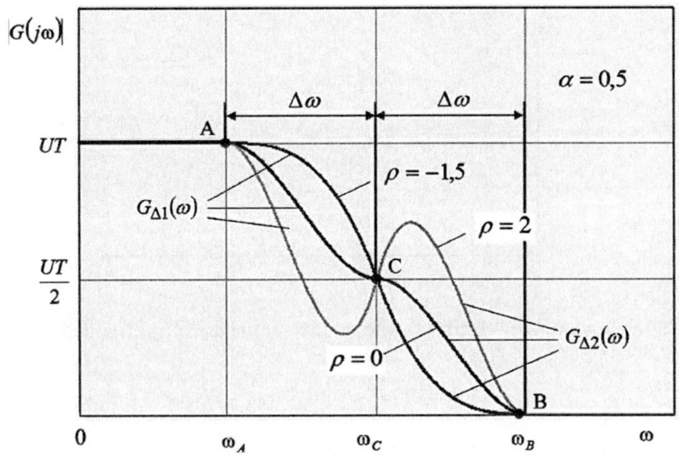

Fig. 3. Spline-interpolation of the signal spectrum *f(x)* by a cubic spline

3 Spline-Extrapolation

Extrapolation is a continuation of a function beyond its domain, in which the extended function belongs to a given class. The function is extrapolated using formulas that use information about the behavior of the function in the interpolation nodes that belong to its domain.

If the given values of the function f: $[a, b] \rightarrow R$ in the nodes $x_k \in [a, b]$, $k = 0, 1, \ldots, n$, then the interpolation polynomial $L_n(x)$ is defined on the entire time axis R is an extrapolation of the function f on the segment $[a, b]$ in the class of polynomials of degree not higher than n.

It should be noted that the definition of *interpolation of functions* is used as a contrast to the concept of extrapolation of functions, when the value of functions in the areas of their definition is constructively restored [32].

In the works of the authors [20–31] the authors developed a method of spline-extrapolation, which, in contrast to previously known methods of extrapolation using polynomials of Lagrange, Chebyshev and others, allowed to increase the accuracy of the forecast and ensure scalability.

Consider spline-extrapolation on the example of self-similar traffic, when it is necessary to predict the characteristics of traffic outside the segment $[a; b]$ [32], namely, to the right of point b. Let, for certainty, it be a point x_c, $x_c > x_N = b$, $x_c - b = h$, where h is the step of partitioning the segment $[a; b]$. We consider a uniform partition of the segment $[a; b]$, i.e., $h_i = h = \frac{b-a}{N}$, $i = 0, 1, \ldots, N - 1$. Construct a spline-function (linear, quadratic or cubic) on the segment $[b; x_c]$. Consider two options (Fig. 4 and Fig. 5). In the first case, we assume that $f(x_c) = f(x_1)$ and construct a spline- function, respectively, on the segment $[b; x_c]$ (Fig. 4).

In the second case (Fig. 5) we set $f(x_k) = f(x_c)$, where $x_k = x_0 + kh$, where k is natural, if $kh \neq x_c - b$, then as $f(x_c)$ we take the value of the function $f(x)$, which is closest to the point x_k [22, 23].

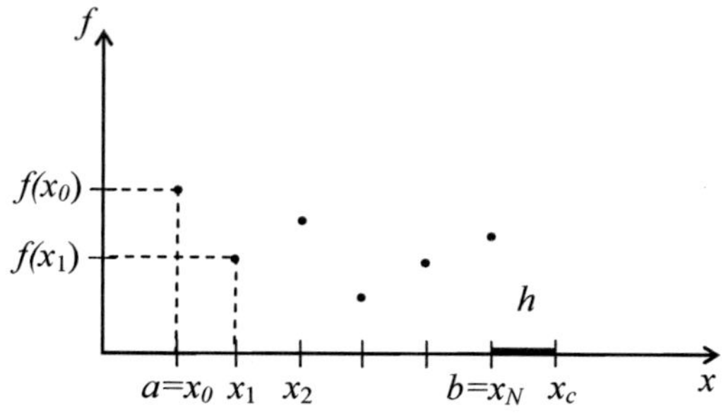

Fig. 4. Extrapolations of self-similar traffic on the segment $[a; b]$ provided $f(x_c) = f(x_1)$

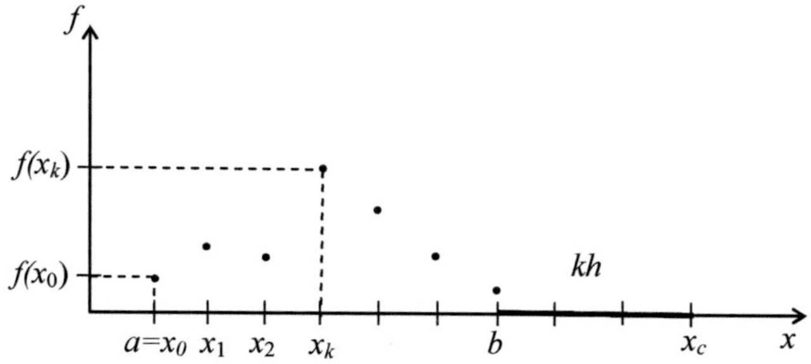

Fig. 5. Extrapolations of self-similar traffic on the segment $[a; b]$ provided $f(x_{kh}) = f(x_c)$

The developed method of spline-extrapolation allows to solve a significant class of problems of analysis and synthesis of telecommunications, which are shown in [20–31]:

- tasks of forecasting the traffic characteristics of network objects, the solution of which in real time for different types of traffic (multimedia data, video, telemetry) will provide opportunities to maintain the quality characteristics of QoS [29–31];
- tasks of maintaining the quality characteristics of QoS, namely, the characteristics of the delay time and the probability of loss and distortion of packets in the maintenance of various types of traffic and the formation of requirements for network buffer devices [33];
- task of choosing the optimal configuration of network objects whose decisions are based on the results of forecasting traffic characteristics and network facilities as a whole to enhance the quality characteristics of functioning telecommunications network.

3.1 The Problem of Predicting the Characteristics of Video Traffic

The use of the spline-extrapolation method in solving traffic forecasting problems, according to [30, 31], allows to increase the accuracy of the predicted traffic estimates (data, speech and video) by selecting the spline-function. Consider an example of predicting the characteristics of real-time video traffic, which is characterized by frequent and significant "bursts" of packet intensity.

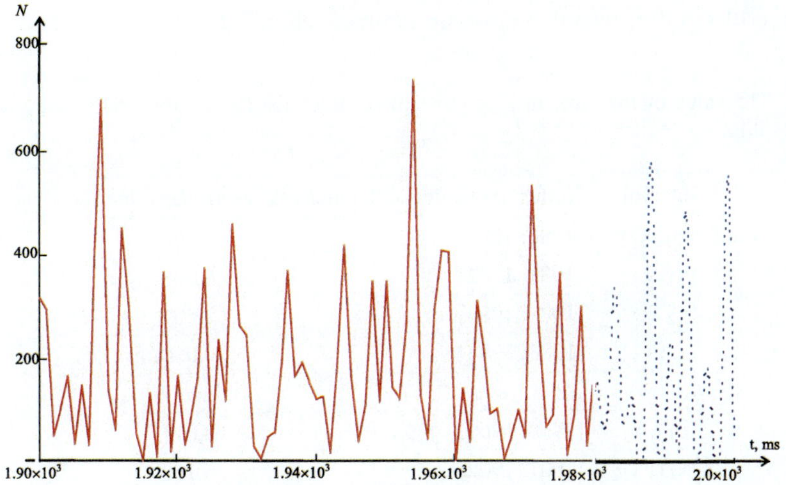

Fig. 6. Video traffic obtained using the Open Source resource Net Flow Analyzer, which is considered on the segment [1900; 1980] ms (red line) and its extrapolation on the segment [1980; 2000] (blue dotted line)

Such traffic is particularly critical to service with specified QoS quality characteristics, so it will allow the participation of traffic characteristics for a given period of time and provide the necessary performance of network objects to prevent congestion and support the necessary quality characteristics of QoS [30, 31]. Using real-time video traffic obtained using the Open Source resource of the Net Flow Analyzer [34], using the spline-extrapolation method based on the cubic Hermite spline, the prediction results are obtained, which are shown in Fig. 6 and Fig. 7.

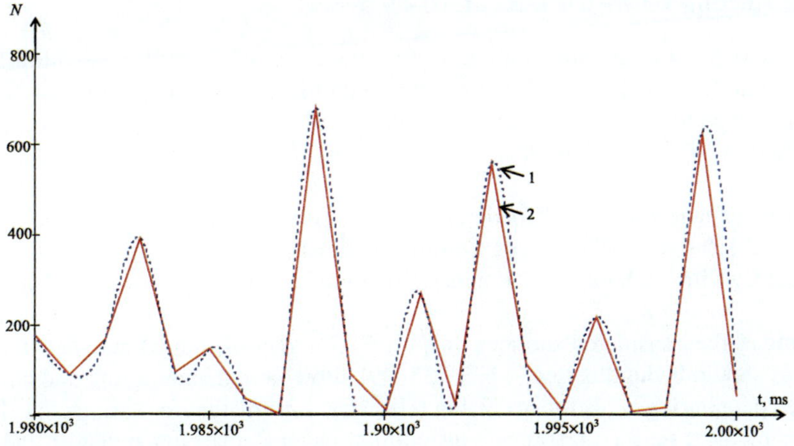

Fig. 7. Comparison of video traffic forecasting results using spline-extrapolation based on a cubic Hermite spline (*line 1*) and real traffic (*line 2*)

The results of the prediction error are summarized in Table 2.

Table 2. The value of the error in predicting the characteristics of video traffic using a cubic Hermite spline

Interval	Numerical values of the interval, ms	Error value
$[x_0; x_1]$	[1980; 1981]	0,060
$[x_1; x_2]$	[1981; 1982]	0,003
$[x_2; x_3]$	[1982; 1983]	2,501
$[x_3; x_4]$	[1983; 1984]	0,050
...	...	...
$[x_{10}; x_{11}]$	[1900; 1901]	2,017
$[x_{11}; x_{12}]$	[1901; 1902]	5,150
$[x_{12}; x_{13}]$	[1902; 1903]	6,850
$[x_{13}; x_{14}]$	[1904; 1905]	2,009
...	...	...

Thus, the use of spline-extrapolation to solve problems of forecasting traffic characteristics allows to increase the accuracy of the forecast, ensuring its scalability and use for different types of traffic and the most critical video traffic, thereby preventing network congestion and exceeding the normative values of QoS. It is advisable to use the methods of wavelet approximation and wavelet extrapolation to reduce the error and increase the accuracy of the results obtained in solving some problems of telecommunications.

3.2 Predicting Telemetric Data of IoT-Devices

The development of mobile communication networks of the fifth generation 5G/NR (New Radio) is associated with the rapid development of the range of high-speed services provided to users and the range of Internet of Things [35–39] services classified, according to [40, 41], into groups:

- eMBB (enhanced Mobile Broadband),
- mMTC (Massive Machine-Type Communications),
- URLLC (Ultra-Reliable Low Latency Communication).

Among the considered services of the IoT (Internet of things) network, telemetry services should be highlighted, which [35–39] allow remote monitoring and collecting detailed data into systems for collecting telemetric information for various IoT devices, such as medical sensors, resource consumption meters, smart home device, individual bracelets, which limit the right to move (convicted, under investigation, house arrest, etc.).

The development of telemetry services requires the operator, first of all, to ensure the required values of the QoS characteristics. It should be noted that most telemetry services do not require high bandwidth, but are quite critical to the characteristics of packet delay time and the probability of packet loss [35–39]. The most critical to the delay time is the telematic data of medical services, which include medical sensors for monitoring the patient's e-healthy and mHealth conditions, including transmural states, as well as ePrescription services, which are less critical to delays, and Medscape and MDLinx. Accordingly, telematic data from industrial automation, remote control using drones, surveillance cameras, Smart Grid networks and emergency notification services also belong to the group of low latency services URLLC. Agricultural services (telematic data on soil conditions, humidity, on-line observations of weather conditions) and logistics services are less critical to the delay time values [35–39].

Traffic of IoT-devices is usually formed by a deterministic data flow, which is a sequence of homogeneous events occurring at strictly defined intervals [20–31]. When servicing telemetry traffic, deterministic sensor polling intervals are often used, which leads to the formation of such data packet streams. In an IoT network, the characteristics of IoT traffic are determined by IoT-devices operating according to a given schedule and functional features that determine the functioning of the system in response to data requests to sensors. Then, considering that a telemetry traffic bursts intensity in well-defined time intervals, this could lead to a sharp increase in packet delay time, which causes an overload of network nodes, and buffer devices and thus has a significant impact on QoS characteristics.

Consider the solution of providing QoS quality characteristics serving traffic IoT-telemetry device using the prediction. The prediction of the characteristics of telemetry traffic of IoT devices will allow the fate of its characteristics for a given period of time and provide for the required performance of IoT devices, and will also allow using the results obtained for various IoT devices in order to prevent congestion and maintain the necessary QoS characteristics [40].

Traffic forecasting issues are considered in works [20–31, 40]. In [41], a classification of the main types and models of traffic in IoT networks is proposed and requirements for each of the service groups are formulated, including the values of the QoS characteristics (average latency, packet loss probability, battery life, etc.). In predicting the traffic characteristics of IoT devices, from the point of view of the authors of the article [42], the use of ARIMA models is considered to be quite effective, although obtaining the required trend of a specific device traffic in real time is often quite difficult. The use of neural networks for forecasting, proposed in [43, 44], allows to obtain the required result, but from the point of view of using this method for an IoT-device, the authors see it as inappropriate. As an alternative to the above methods, the authors in [20–31] proposed the use of spline- and wavelet-extrapolation methods for short-term forecasting based on various spline- and wavelet-functions. The use of cubic splines showed clear advantage of their use, especially for real-time predictions with a given accuracy [23–25, 28–31].

In this work is proposed to solve the problem of real time traffic characteristics predict the IoT-telemetry devices using the method of spline-extrapolation based on the cubic Hermite spline.

Consider a cubic Hermite spline. Let the grid nodes Δ: $a = x_0 < x_1 < \ldots < x_n = b$ be given the values of some function $f(x)$ and its derivative $f'(x)$ [11–13]:

$$f_i = f(x_i), f_i' = f'(x_i), i = 0, 1, \ldots, n. \quad (7)$$

By cubic Hermite spline we mean a function $S_3(f; x) = S_3(x)$, satisfying two conditions [11–13]:

1) at each of the segments $[x_i; x_{i+1}]$, $i = 0, \ldots, n - 1$,

$$S_3(x) = a_{io} + a_{i1}(x - x_i) + a_{i2}(x - x_i)^2 + a_{i3}(x - x_i)^3, \quad (8)$$

2) $S_3(x_i) = f_i, S_3'(x_i) = f_i',\ i = 0, \ldots, n.$

Obviously, $S_3(x) \in C^1[a, b]$, where $C^1\ [a, b]$ – the class of functions $f(x)$ having a continuous first derivative on the interval $[a, b]$. To determine the coefficients $a_{i\alpha}$, $\alpha = 0, 1, 2, 3$, $i = 0, 1, \ldots, n - 1$ we obtain a system of equations:

$$\begin{cases} S_3(x_i) = f_i, S_3(x_{i+1}) = f_{i+1}, \\ S_3'(x_i) = f_i', S_3'(x_{i+1}) = f_{i+1}'. \end{cases} \quad (9)$$

The cubic Hermite spline has the form [11–13]:

$$S_3(x) = \varphi_1(t) f_i + \varphi_2(t) f_{i+1} + \varphi_3(t)\, h_i f_i' + \varphi_4(t)\, h_i f_{i+1}', \quad (10)$$

moreover,

$$\varphi_1(t) = (1 - t)^2(1 + 2t), \varphi_2(t) = t^2(3 - 2t), \varphi_3(t) = t(1 - t)^2,$$

$$\varphi_4(t) = -t^2(1 - t), h_i = x_{i+1} - x_i, \mathrm{t} = (x - x_i)/h_i, i = 0, \ldots, n - 1.$$

To estimate the error of extrapolation by a cubic Hermite spline, we use estimates for the remainder of the interpolation $R(x) = S_3(f; x) - f(x)$ depending on the smoothness of the function $f(x)$ [11–13].

Theorem 1. If $S_3(x)$ interpolates the function f(x) on the grid Δ, then the estimates

$$\left\| S_3^{(r)}(x) - f^{(r)}(x) \right\|_\infty \le R_r,\ r = 0, 1, 2, 3. \quad (11)$$

where R_r is values for various classes of functions.

For the experiment, we use the telemetry traffic of the IoT device of the 5G network, obtained using the Open Source resource Net Flow Analyzer [34], which is shown in Fig. 8.

In Fig. 8 is shown the telemetry traffic for an IoT device, where N is the number of packets (thousand packets), t is the time of packet arrival (s). A fragment of the considered telemetry traffic of an IoT device on the interval [1; 990] s, has the presence of "bursts" of packet intensity at the moments of telemetry information transmission. When forecasting, it is necessary to take into account the structure of such traffic, when it is

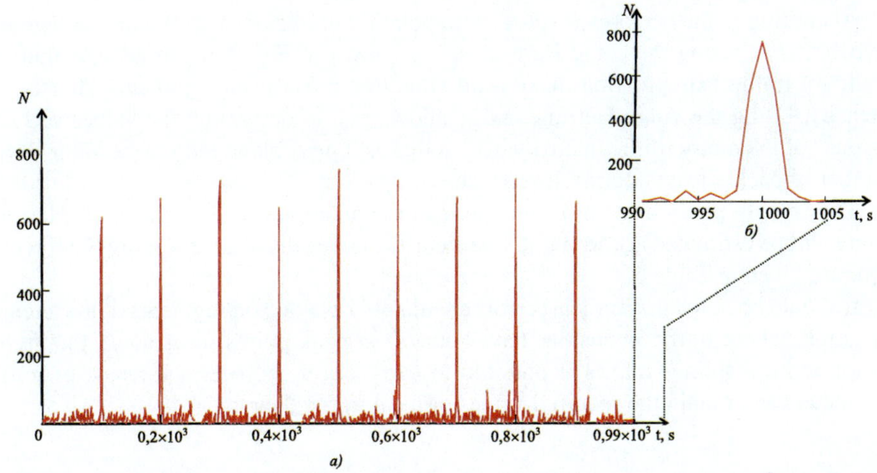

Fig. 8. Real-time traffic IoT-telemetry device, *a*) viewed in the interval [1, 990] s, *b*) prediction interval [990, 1005] s

important to determine exactly the value of the traffic "burst" as accurately as possible. To conduct an experiment using the spline-extrapolation method, let us consider a fragment of the telemetry traffic of an IoT device on a given interval [990; 1005] ms.

For the telemetry traffic fragment under consideration, we perform spline-extrapolation on the segment [990; 1005] ms using the cubic Hermite spline $S_3(x)$ of the form (8). The spline-extrapolation results are shown in Fig. 9.

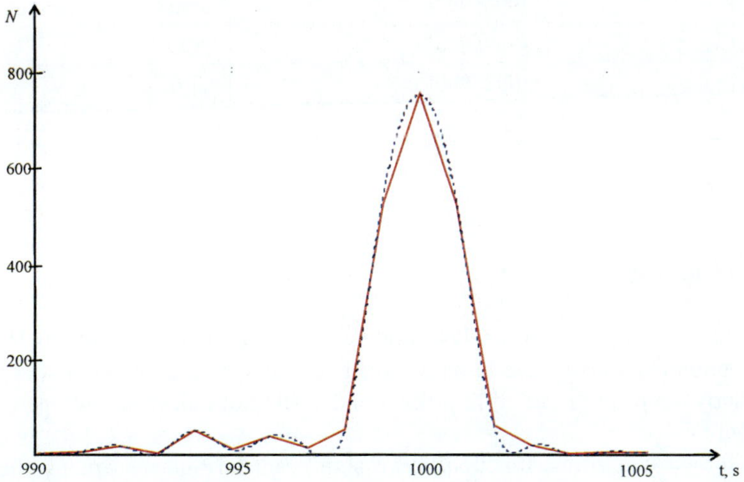

Fig. 9. Comparison of the results of telemetry traffic prediction using spline-extrapolation based on the cubic Hermite spline and real traffic telemetry.

According to the performed spline-extrapolation of telemetry traffic on the segment [990;1005] ms using the cubic Hermite spline shown in Fig. 9, it can be seen that the results of traffic extrapolation make it possible to recover even significant "bursts" of intensity. Using the cubic Hermite spline allows you to determine the values of local "bursts" of telemetry traffic intensity, which makes it possible to reduce the value of the number of packet losses during its service.

The error in predicting traffic with spline-extrapolation based on the cubic Hermite spline will be estimated according to Theorem 1. The results of determining the forecast error are given in Table 3.

It should be noted that forecast errors are mainly obtained on segments of the telemetry graph, where traffic intensities have "bursts" at peak points. In general, the use of cubic Hermite splines makes it possible to carry out a short-term forecast of traffic parameters and obtain the required values of local traffic "bursts".

Table 3. Error in predicting telemetry traffic based on spline-extrapolation method using cubic Hermite spline

Considered segment of extrapolation	Numerical values of the segment, s	Extrapolation error value
$[x_0; x_1]$	[990; 991]	0,005
$[x_1; x_2]$	[991; 992]	0,003
$[x_2; x_3]$	[992; 993]	1,050
...	...	...
$[x_{10}; x_{11}]$	[995; 996]	2,017
$[x_{11}; x_{12}]$	[996; 997]	3,600
$[x_{12}; x_{13}]$	[997; 998]	1,800
$[x_{13}; x_{14}]$	[998; 999]	8,009
...	...	...

4 Conclusions

1. A number of problems of analysis and synthesis that require solutions in modern telecommunications on the basis of effective mathematical methods that will significantly simplify the solution process and at the same time provide new solution capabilities.
2. It is shown that spline-functions have significant advantages among the known approximation methods. Splines are more resistant to local perturbations, the behavior of the spline around the point does not affect the behavior of the spline as

a whole, as, for example, is the case with polynomial interpolation; good convergence of spline-interpolation in contrast to polynomial, in particular, for functions with irregular smoothness properties is an indisputable priority for spline-interpolation; useful extreme properties, an important practical advantage is a fairly simple implementation of spline-functions on personal computers.
3. The use of spline-approximation is proposed, which makes it much easier to obtain solutions to a class of problems, such as recovery and evaluation of states (data, signals, traffic), signal and image processing, including data filtering and compression, signal detection and measurement and improving the quality characteristics of QoS operation of telecommunications networks with the necessary error.
4. A method of spline-extrapolation has been developed, which allows to obtain solutions to problems of forecasting traffic characteristics and increase the accuracy of the forecast, ensuring its scalability and use for different applications.
5. To reduce the error and increase the accuracy of the results obtained in solving some problems of telecommunications, the use of wavelet approximation and wavelet extrapolation is proposed.
6. To predict the characteristics of telemetry traffic of IoT devices, the spline-extrapolation method using the cubic Hermite spline was used. The use of a cubic Hermite spline in spline-extrapolation shows sufficient simplicity in implementation, has low resource consumption and sufficient forecast accuracy.
7. The obtained results of predicting the telemetry traffic of IoT devices allow to improve the accuracy of the forecast, ensure its scalability and use for various IoT applications, thereby avoiding network congestion and exceeding the standard values of QoS characteristics.

References

1. Roslyakov, A.V., Vanyashin, S.V.: Future networks. VSUTI, Samara (2015)
2. Donovan, J.: Building the network of the future: getting smarter, faster, and more flexible with a software centric approach. In: John, D., Krish, P. (eds.) Chapman and Hall/CRC (2017)
3. Vorobiyenko, P.: Industry 4.0 and information communication technologies. Paper Presented at the International Conference on 2nd Information and Telecommunication Technologies and Radio Electronics (UkrMiCo 2017), Odessa National Academy of Telecommunication named after O.S. Popov, Odessa, Ukraine, September 2017, pp. 1–4 (2017). https://doi.org/10.1109/UkrMiCo.2017.8095359
4. Ilchenko, M., Uryvsky, L., Moshynska, A.: Developing of telecommunication strategies based on the scenarios of the information community. In: Springer Science, Business Media, pp. 905–913 (2017). https://doi.org/10.1007/s10559-017-9992-9
5. Ilchenko, M., Uryvsky, L., Osypchuk, S.: The main directions of improving infocommunications in the global tendencies. In: Advances in Information and Communication Technologies. Lecture Notes in Networks and Systems, Monograph, pp. 3–22. Springer, Cham (2020). https://doi.org/10.1007/978-3-030-58359-0_1
6. Vorobiyenko, P.: The role of infocommunications in the world-wide human society development. Paper Presented at the IEEE 11th International Conference on Advanced Trends in Radioelectronics, Telecommunications and Computer Engineering (TCSET), Lviv Polytechnic National University, Slavske, Ukraine, February 2012, pp. 25–26 (2012)

7. Lemeshko, O., Nevzorova, O., Hailan, A.: The increasing convergence of coordination procedure in the implementation of multipath hierarchical routing. In: 2014 First International Scientific-Practical Conference Problems of Infocommunications Science and Technology, pp. 45–48. IEEE (2014)
8. Lemeshko, O., Semenyaka, M., Simonenko, O.: Researching and designing of the dynamic adaptive queue balancing method on telecommunication network routers. In: 2013 12th International Conference on the Experience of Designing and Application of CAD Systems in Microelectronics (CADSM), pp. 204–207. IEEE (2013)
9. Lemeshko, O., Yeremenko, O., Tariki, N.: Solution for the default gateway protection within fault-tolerant routing in an IP network. Int. J. Electr. Comput. Eng. Syst. **8**(1), 19–26 (2017)
10. Strelkovska, I.: Theory and methods of spline-approximation in telecommunications: dissertation. For the degree of Doctor of Technical Sciences. O.S. Popov Odessa National Academy of Communications (2010)
11. Zavyalov, Yu.S., Kvasov, B.I., Miroshnichenko, V.L.: Methods of Spline Functions. Science, Moscow (1980)
12. Stechkin, S.B., Satbotin, Yu.N.: Splines in Computational Mathematics. Science, Moscow (1976)
13. Ahlberg, J., Nilsson, E., Walt, J., Stechkin, S.B. (eds.): Theory of Splines and its Applications. World. Moscow
14. Strelkovskaya, I.V., Bukhan, D.Yu.: Comparative analysis of signal recovery using the sampling theorem and spline interpolation. In: Proceedings of the O.S. Popov ONAT, vol. 1, pp. 149–152 (2007)
15. Strelkovskaya, I.V., Bukhan, D.Yu.: Restoration of continuous signals on the basis of a number of Nyquist and cubic splines. Radio Eng. All-Ukr. Int. Sci. Tech. **151**, 181–185 (2007)
16. Strelkovskaya, I., Lysiuk, O., Zolotukhin, R.: Comparative analysis of reconstruction of continuous signals by the Kotelnikov series and spline functions. Eastern-Eur. J. Enterp. Technol. **2/9**(62), 12–15 (2013)
17. Strelkovskaya, I., Lysiuk, O., Paskalenko, S.: Application of different kinds of approximation in signals restoration. In: Problems of Infocommunications Science and Technology, Kharkiv, Ukraine, October 2015, pp. 177–180 (2015). https://doi.org/10.1109/INFOCOMMST.2015.7357306
18. Strelkovskaya, I., Solovskaya, I., Severin, N., Paskalenko, S.: Spline approximation based restoration for self-similar traffic. Eastern-Eur. J. Enterp. Technol. **3/4**(87), 45–50 (2017). https://doi.org/10.15587/1729-4061.2017.102999
19. Strelkovskaya, I.V., Bukhan, D.Yu.: Restoration of continuous signals based on the Kalman-Bucy filter and cubic splines. Radio Eng. All-Ukr. Int. Sci. Technol. **156**, 61–63 (2009)
20. Strelkovskaya, I., Solovskaya, I., Makoganiuk, A.: Predicting characteristics of self-similar traffic. Paper Presented at the Third International Conference on Information and Telecommunication Technologies and Radio Electronics (UkrMiCo 2018), O.S. Popov Odessa National Academy of Telecommunication, Odessa, Ukraine, September 2018 (2018). https://doi.org/10.1109/UkrMiCo43733.2018.9047536
21. Strelkovskaya, I., Solovskaya, I., Makoganiuk, A., Severin, N.: Predicting the characteristics of self-similar traffic using spline-extrapolation. Bull. Univ. "Ukraine" **1**(22), 87–94 (2019)
22. Strelkovskaya, I., Solovskaya, I., Makoganiuk, A.: Spline-extrapolation method in traffic forecasting in 5G networks. J. Telecommun. Inf. Technol. **3**, 8–16 (2019). https://doi.org/10.26636/jtit.2019.134719
23. Strelkovskaya, I., Solovskaya, I.: Using spline-extrapolation in the research of self-similar traffic characteristics . J. Electr. Eng. **70**(4), 310–316 (2019). https://doi.org/10.2478/jee-2019-0061
24. Strelkovskaya, I.V., Solovskaya, I.N., Paskalenko, V.M.: Spline-extrapolation method for restoring self-similar traffic. Digit. Technol. Collect. Sci. Works **25**, 42–51 (2019)

25. Strelkovskaya, I., Solovskaya, I., Makoganiuk, A.: Different extrapolation methods in Problems of Forecasting. In: Ilchenko, M., Uryvsky, L., Globa, L. (eds.) Advances in Information and Communication Technology and Systems, MCT 2019. Lecture Notes in Networks and Systems, vol 152. Springer, Cham (2019). https://doi.org/10.1007/978-3-030-58359-0_12
26. Strelkovskaya, I., Solovskaya, I., Makoganiuk, A.: Predicting self-similar traffic using cubic B-splines. Paper Presented at the 3nd IEEE International Conference Advanced Information and Communication Technologies (AICT-2019), Lviv, Ukraine, July 2019, pp. 153–156 (2019). https://doi.org/10.1109/AIACT.2019.8847761
27. Strelkovskaya, I., Solovskaya, I., Tolmak, V.: Use of spline-extrapolation to increase the quality indicators of telecommunication systems. In: Proceedings of the O.S. Popov ONAT, vol. 2, pp. 77–85 (2019). https://doi.org/10.33243/2518-7139-2019-1-2-77-85
28. Strelkovskaya, I., Solovskaya, I., Makoganiuk, A., Tsyra, O.: Comparative analysis of the methods of wavelet- and spline-extrapolation in problems of predicting self-similar traffic. Paper Presented at the Fourth International Conference on Information and Telecommunication Technologies and Radio Electronics (UkrMiCo 2019), Odessa National Academy of Telecommunication named after O.S. Popov, Odessa, Ukraine, September 2019, pp. 1–6 (2019). https://doi.org/10.1109/UkrMiCo47782.2019.9165432
29. Strelkovskaya, I., Solovskaya, I., Makoganiuk, A.: Forecasting 5G network multimedia traffic characteristics. Paper Presented at the IEEE 15th International Conference on Advanced Trends in Radioelectronics, Telecommunications and Computer Engineering (TCSET), Lviv Polytechnic National University, Slavske, Ukraine, February 2020, pp. 982–987 (2020). https://doi.org/10.1109/TCSET49122.2020.235585
30. Strelkovskaya, I., Solovskaya, I., Makoganiuk, A., Rodionova, T.: Multimedia traffic prediction based on wavelet- and spline-extrapolation. Paper Presented at the 8th Black Sea Conference on Communications and Networking, Odessa, Ukraine, May 2020, pp. 1–4 (2020). https://doi.org/10.1109/BlackSeaCom48709.2020.9234998
31. Strelkovskaya, I., Solovskaya, I., Strelkovska, J.: Spline-extrapolation of video traffic of IoT-devices based on various cubic splines. Paper Presented at the 2020 International Scientific-Practical Conference Proceedings of Problems of Infocommunications Science and Technology (PICS&T 2020), Kharkiv, Ukraine, Kharkiv National University of Radio Electronics, October 2020 (2020)
32. Popovsky, V.V., Strelkovskaya, I.V.: Accuracy of procedures for filtering, extrapolating and interpolating random processes. Telecommun. Prob. **1**(3), 3–10 (2011)
33. Strelkovskaya, I., Solovskaya, I., Makoganiuk, A.: Different approaches to studying the extreme properties of signal functions synthesized with splines. In: Lecture Notes on Data Engineering and Communications Technologies (LNDECT), vol. 48, pp. 17–33 (2020). https://doi.org/10.1007/978-3-030-43070-2_2
34. https://www.manageengine.com/products/netflow/download-free.html
35. Uryvsky, L., Osypchuk, S., Moshynska, A., Kyrashchuk, V.: IoT solutions research and development for wide range applications. Sci. Europe Praha, Czech Republic **1**(36), 40–54 (2019). https://europe-science.com/wp-content/uploads/2019/03/VOL-1-No-36-2019.pdf
36. Uryvsky, L., Osypchuk, S., Moshynska, A.: Internet of Things Solutions Research and Development for Widespread Usage and Applications. The Actual Problems of the World Today, pp. 254–266. Monograph. SCIEMCEE, London (2019)
37. Globa, L., Kurdecha, V., Ishchenko, I., et al.: The intellectual IoT-system for monitoring the base station quality of service. Paper Presented at the IEEE International Black Sea Conference on Communications and Networking (2018). https://doi.org/10.1109/BlackSeaCom.2018.8433715

38. Globa, L., Ishchenko, I., Gvozdetska, N., et al.: An approach to the Internet of Things system with nomadic units developing. In: IEEE International Black Sea Conference on Communications and Networking (2017). https://doi.org/10.1109/BlackSeaCom.2016.7901587
39. Kot, T., Reverchuk, A., Globa, L., Schill, L.: A novel approach to increase efficiency of OSS/BSS workflow planning and design. Paper Presented at the International Conference on Business Information Systems, pp. 142–152. Springer, Heidelberg (2012). https://doi.org/10.1007/978-3-642-30359-3_13
40. Strelkovskaya, I., Solovskaya, I., Strelkovska, J.: Predicting telemetric data of IoT devices. Paper Presented at the X International Scientific and Practical Conference "Infocommunications – Present and Future", Odesa, November 2020, pp. 386–391 (2020)
41. Mocnej, J., Pekar, A., Seah, K.G.W., Zolotova, I.: Network Traffic Characteristics of the IoT Application Use Cases School of Engineering and Computer Science, Victoria University of Wellington (2018)
42. Sunil, K.P.V.: An ARIMA based approach for traffic prediction. Paper Presented at the International Conference on Information Systems and Technology (ICIST), August 2017, pp. 7–10 (2017)
43. Abdellah Ali, R., Omar Abdul Kareem, M., Paramonov, A., Koucheryavy, A.: IoT traffic prediction using multi-step ahead prediction with neural network. Paper Presented at the 11th International Congress on Ultra Modern Telecommunications and Control Systems and Workshops, Dublin, Ireland, October 2019 (2019). https://doi.org/10.1109/ICUMT48472.2019.8970675
44. Tang, F., Fadullah, Z.Md., Mao, B., Kato, N.: An intelligent traffic load prediction-based adaptive channel assignment algorithm in SDN-IoT: a deep learning approach. Paper Presented at the IEEE Internet of Things Journal, vol. 5, issue 6 (2018). https://doi.org/10.1109/JIOT.2018.2838574

Evaluation of Quality Parameters of an Intensity-Modulated Optical Transmission System

V. Pedyash(✉), A. Mazur(✉), and D. Rozenvasser(✉)

O. S. Popov Odessa National Academy of Telecommunications,
Kuznechnaya str., 1, Odessa 65029, Ukraine
{pedyash,rozenvasser}@onat.edu.ua, anna2102@i.ua

Abstract. Fiber optic transmission systems are used for high-speed traffic transmission in modern telecommunication NGN and IMS networks. Optical channels of optical transport hierarchy (OTH) communication systems have a bandwidth of 2.5 to 100 Gbps, so the study of their characteristics is an urgent task. The problem of studying the quality parameters (Q-factor) of optical OTH signals of the OTU1 and OTU2 types is solved in the current section. A simulation model of an intensity modulated optical transmission system has been developed to solve this problem. This model contains a transmitter, an optical transmission path with optical amplifiers and an optical signal receiver. Optical fiber simulation model allows to introduce linear and nonlinear distortions of the propagation medium into the signal and it is based on the Split-Step Fourier Method. The noise of the optical amplifier was generated by using the inverse discrete Fourier transform to increase the performance of the model. Comparison of the results of modeling and analytical calculations was carried out to check the accuracy of the proposed simulation model. The influence of the signal power of the transmitter and the quantity of linear path optical sections on the Q-factor of the signal in the receiver is calculated. It is proposed to optimize the transmitter signal power by the criterion of maximizing the Q-factor to increase the length of the 3R regeneration section.

Keywords: Fiber · OTH · Communication system · Quality · Modeling

1 Introduction

Optical fiber transmission systems (OFTS) form the backbone of a modern telecommunications network. The transmission speed of digital transmission systems over optical fiber has increased from several megabits per second to several hundred gigabits per second.

The development of an Erbium Doped Fiber Optic Amplifier (EDFA) made it possible to amplify the signal (1R regeneration) without converting it into an electrical form. This reduced the cost of the transmission system and allowed the construction of

P. Vorobiyenko et al. (Eds.): IPF 2020, LNNS 212, pp. 21–37, 2021.
https://doi.org/10.1007/978-3-030-76343-5_2

a wavelength division multiplexing (WDM) transmission system. Today, on the backbone section of the network, dense wavelength multiplexing (DWDM) systems with a frequency grid of 100 GHz and below are used. An important parameter of the optical transmission system is the type of modulation of the optical signal. The transmission speed, signal immunity and the total cost of the system depend on it. Today, the most common is signal intensity modulation (IM), which is used when the digital stream has a speed of 10 Gbps or lower.

DWDM systems have evolved to optical transport hierarchy (OTH) transmission systems. In these systems, user information is multiplexed into optical channel transport blocks (OTUk) of four hierarchy levels (k = 1, 2, 3, 4). In OTH transponders of the OTU1 (2.7 Gbps) and OTU2 (10.7 Gbps) levels, the intensity modulation of the laser diode is used.

The optical fiber line path of OTH systems contains the propagation medium (optical fiber) and optical amplifiers. The propagation medium introduces linear and nonlinear distortions into the information signal [1]. The optical amplifier also introduces the additive noise of amplified spontaneous emission [2].

Calculation of signal immunity in the receiver of the transmission system is one of the urgent tasks. To solve the problem of studying the noise immunity of transmission system signals, it is possible to use two modeling methods: analytical (mathematical) and simulation modeling.

The first method involves obtaining analytical expressions that correspond to the signal processing processes in the functional blocks of the transmission system and their statistical parameters. Models built by this method are not flexible, since a small change in the system configuration requires the development of a new or significant revision of the existing model. Today, accurate simulation of linear and non-linear distortion of the propagation medium is performed only by numerical methods. On their basis, it is possible to build full-fledged simulation models of transmission systems. These models allow to obtain reliable data on the quality of the signal at various points in the transmission system.

This article describes the mathematical and simulation model of the intensity modulated optical transmission system in the MatLab program.

2 Analytical Model of Optical Transmission System

In this subsection, we analyze the noise immunity of the generalized model of the optical transmission system (Fig. 1). The block diagram contains a transceiver, an optical fiber transmission path (OFTP) and a signal receiver.

We begin our analysis of the optical transmission system using the analytical and simulation methods and without optical signal modulation. Further, this model can be expanded to a transmission system with any modulation.

The transmitter signal in complex form can be written as an expression

$$\dot{E}_{\text{tr}}(t) = \sqrt{P_{\text{tr}}(t)}\, e^{j\varphi_{\text{tr}}(t)} e^{j\omega_0 t}, \tag{1}$$

where $P_{\text{tr}}(t)$ and $\varphi_{\text{tr}}(t)$ - average power and phase of the transmitter signal,

ω_0 and φ_0 - frequency and initial phase of the laser signal.

The optical transmission path consists of N_{sect} identical sections (Fig. 1), each of which contains a length L_{of} of fiber and an optical erbium-doped fiber amplifier (EDFA).

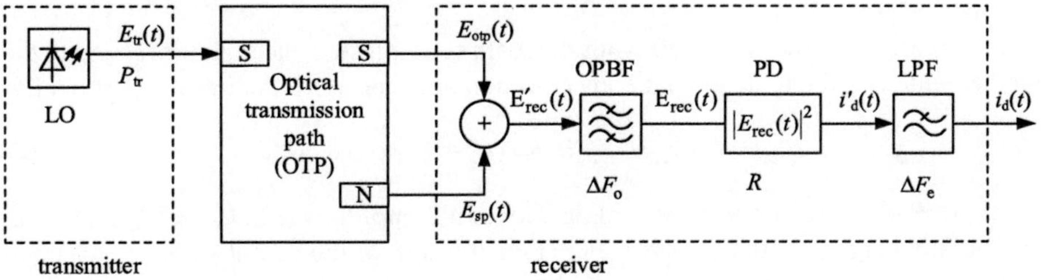

Fig. 1. Generalized model of a channel of a transmission system with wavelength multiplexing

The mathematical model of the optical transmission path is shown at the Fig. 2. It assumes linear amplification of the signal and the generation of additive noise in the optical amplifier. In further calculations, we will assume that the signal attenuation in the fiber is completely eliminated in the optical amplifier ($G = 10^{0,1\alpha L_{\text{of}}}$). At the amplifier output, there is also spontaneous emission noise $E_{\text{sp1}}(t)$ in addition to the useful signal.

The noise is formed in two polarization planes and the average noise power of one amplifier in one polarization plane is determined by the expression [2]

$$P_{\text{sp1}} = PSD_{\text{sp1}}\, \Delta F_o = n_{\text{sp}}(G-1)\, h\nu\, \Delta F_o, \tag{2}$$

where PSD_{sp1} - the spontaneous emission noise power spectral density;

n_{sp} - the spontaneous emission coefficient of an optical amplifier;

G - the amplifier gain;

h - the Planck's constant ($h = 6{,}626 \cdot 10^{-34}$ J * s);

ν - the optical signal frequency (for bandpass signal coincides with laser frequency $\nu = \omega_0/2\pi$);

ΔF_{o} - bandwidth of the optical bandpass filter (OBPF) in a receiver of a transmission system.

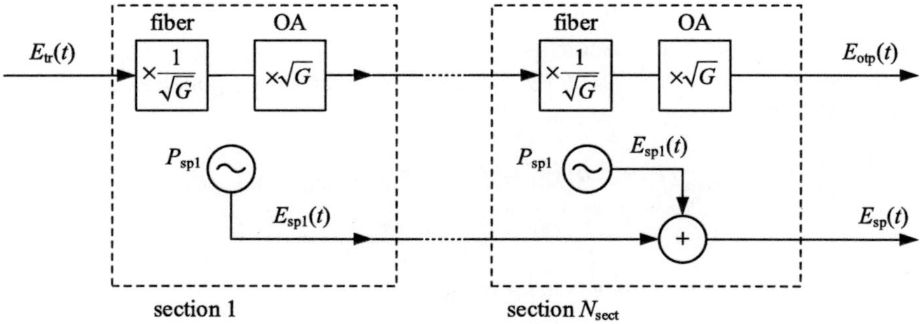

Fig. 2. Block diagram of a model for an optical transmission path

The spontaneous emission factor can be calculated using the noise factor *NF* of the optical amplifier

$$n_{\mathrm{sp}} = \frac{NF}{2} = \frac{10^{0,1nf}}{2}. \tag{3}$$

In the case when the optical transmission path contains several amplifiers N_{oa} with the same parameters, the total spontaneous emission noise power increases proportionally:

$$P_{\mathrm{sp}} = PSD_{\mathrm{sp}}\,\Delta F_o = n_{\mathrm{sp}}(G-1)\,h\nu\,\Delta F_o\,N_{\mathrm{oa}}, \tag{4}$$

where PSD_{sp} - total noise spectral density of all amplifiers ($PSD_{\mathrm{sp}} = N_{\mathrm{oa}}\,PSD_{\mathrm{sp1}}$).

The input signal of the receiver (Fig. 1) is the sum of the useful signal $E_{\mathrm{otp}}(t)$ and the total spontaneous emission noise $E(t)$:

$$E'_{\mathrm{rec}}(t) = E_{\mathrm{otp}}(t) + E_{\mathrm{sp}}(t) = E_{\mathrm{tr}}(t) + E_{\mathrm{sp}}(t). \tag{5}$$

In this expression, the equality of signals $E_{\mathrm{otp}}(t) = E_{\mathrm{tr}}(t)$ is accepted, since in this version of the transmission system model there are no signal distortions in the propagation medium and full compensation of the signal attenuation in the amplifier is assumed.

The band-pass optical filter does not introduce significant distortions into the signal spectrum, since in this version of the system model there is no signal modulation in the transmitter. Therefore, in the further text of the article, the equality of the input and output signal of the filter is also accepted $E_{\mathrm{rec}}(t) = E'_{\mathrm{rec}}(t)$.

The electric field of the spontaneous emission noise has a uniform spectral power density (Fig. 3) and it is described by the expression [3]

$$\dot{E}_{\mathrm{sp}}(t) = \sum_{k=-M}^{M} \sqrt{PSD_{\mathrm{sp}}\delta\nu}\,\exp\big[j((\omega_0 + 2\pi k\delta\nu)t + \Phi_k)\big], \tag{6}$$

where Φ_k - random phase of noise component *k* ($\Phi_k \in [0,\ 2\pi]$).

The photodiode (PD) is a square-law detector, so the current at its output has the form

$$i'_{\mathrm{d}}(t) = R\,|E_{\mathrm{rec}}(t)|^2, \tag{7}$$

there *R* - the photodiode responsivity.

The electric current consists of the sum of three separate components at the output of the photodiode:

$$\begin{aligned} i'_{\mathrm{d}}(t) = R\,|E_{\mathrm{rec}}(t)|^2 &= R\Big(|E_{\mathrm{tr}}(t)|^2 + 2\,|E_{\mathrm{tr}}(t)|\,\big|E_{\mathrm{sp}}(t)\big| + \big|E_{\mathrm{sp}}(t)\big|^2\Big) \\ &= i'_{\text{s-s}}(t) + i'_{\text{s-sp}}(t) + i'_{\text{sp-sp}}(t). \end{aligned} \tag{8}$$

These terms are beating between components:

1) signal $i'_{\text{s-s}}(t) = R\,P_{\mathrm{tr}}$;

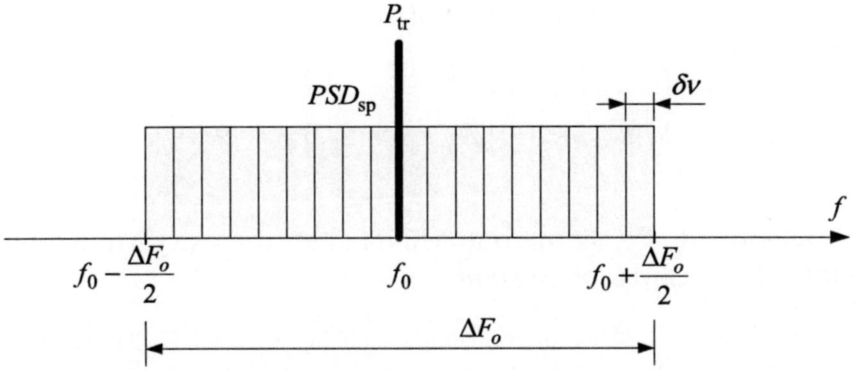

Fig. 3. Model of the signal spectrum and noise at the output of the optical filter

2) signal and spontaneous emission $i'_{\text{s-sp}}(t)$;
3) spontaneous emission $i'_{\text{sp-sp}}(t)$.

The presentation of signal and noise in a complex form allows us to apply elements of the theory of vector analysis. The signal vector at the receiver input $\overline{E}_{\text{rec}}$ is the sum of the vectors of the useful signal $\overline{E}_{\text{tr}}$ of the transmitter and the spontaneous emission noise $\overline{E}_{\text{sp}}$ (Fig. 4).

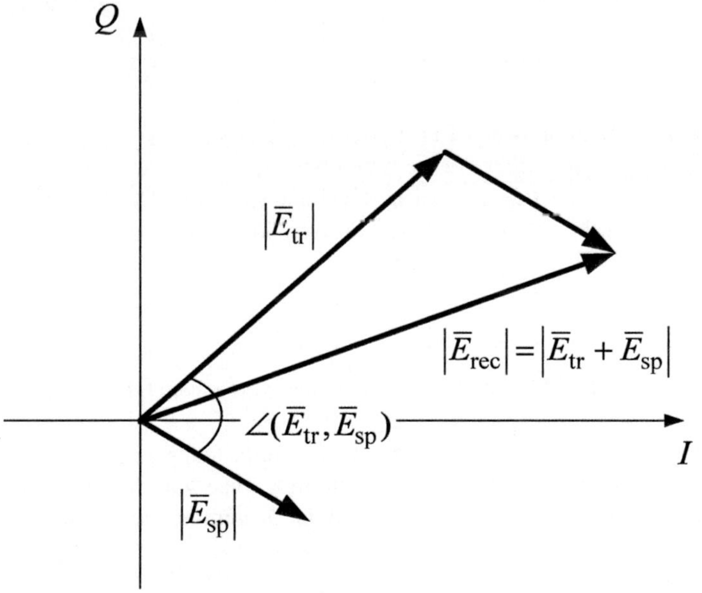

Fig. 4. The vector sum of the signal at the input of the optical receiver

In scalar form, the current at the output of the photodiode has the expression:

$$\begin{aligned} i'_{\mathrm{d}} &= R\left|\overline{E}^2_{\mathrm{rec}}\right| = R\left|\overline{E}^2_{\mathrm{tr}} + 2\overline{E}_{\mathrm{tr}} \cdot \overline{E}_{\mathrm{sp}} + \overline{E}^2_{\mathrm{sp}}\right| \\ &= R\Big[\left|\overline{E}_{\mathrm{tr}}\right|^2 + 2\left|\overline{E}_{\mathrm{tr}}\right|\left|\overline{E}_{\mathrm{sp}}\right| \cos \angle(\overline{E}_{\mathrm{tr}}, \overline{E}_{\mathrm{sp}}) + \left|\overline{E}_{\mathrm{sp}}\right|^2\Big]. \end{aligned} \quad (9)$$

3 Development of a Simulation Model of an Intensity Modulated Optical Transmission System

In this subsection, the problem of evaluating the quality of OTU1/OTU2 OTH signals transmission using intensity modulation of the optical signal in the transmitter is solved. It also includes the task of optimizing the main parameters of the system (functional blocks and signal). To solve this problem, it is necessary to develop a simulation model of the transmission system, which will take into account the signal formation algorithm in the transmitter, its distortion in the propagation medium and detection in the receiver.

In the simulation model, the continuous signal is replaced by its samples, which are generated with a sampling interval $T_{\mathrm{s}} = 1/F_{\mathrm{s}}$, where F_{s} is the sampling frequency. In this case, the continuous time t in the previous expressions is replaced by a discrete value $t_k = k\,T_{\mathrm{s}}$. It is necessary to use the function of the inverse fast Fourier transform to accelerate the generation of samples of the spontaneous emission interference according to expression (6) in the passband $(f_{\mathrm{min}}; f_{\mathrm{max}})$ of the optical bandpass filter.

It is implemented according to the following expression in the MatLab system [4]

$$\dot{x}(k) = ifft(\dot{A}(n)) = \frac{1}{N_{FFT}} \sum_{n=0}^{N_{FFT}} \dot{A}(n) \exp\left(j\frac{2\pi nk}{N_{FFT}}\right), \quad (10)$$

where N_{FFT} - the dimension of Fast Fourier Transform;
k - the signal sample number in the time domain;
n - the complex amplitude number in the signal spectrum ($k,\ n \in [0; N_{FFT} - 1]$).

The spectrum of the discrete Fourier transform contains N_{FFT} complex amplitudes $\dot{A}(n)$ with frequencies

$$f_n = F_{\mathrm{s}}\frac{n}{N_{FFT}}. \quad (11)$$

They are uniformly distributed in the frequency range from 0 to $F_{\mathrm{s}}(N_{FFT} - 1)/N_{FFT}$ with an interval $\Delta f_{FFT} = F_{\mathrm{s}}/N_{FFT}$.

It is necessary to find the sum of the complex amplitudes with positive frequencies $(f_{\mathrm{min}}; f_{\mathrm{max}})$ to synthesize the narrowband complex noise with a uniform power spectral density. These terms correspond to the range of numbers $n \in [n_{\mathrm{min}}; n_{\mathrm{max}}]$ in the expression (10):

$$n_{\mathrm{min}} = \frac{f_{\mathrm{max}} N_{FFT}}{F_{\mathrm{s}}}, \quad (12)$$

$$n_{\mathrm{max}} = \frac{f_{\mathrm{max}} N_{FFT}}{F_{\mathrm{s}}}. \quad (13)$$

The modulus of the complex amplitudes should be multiplied by a factor Δf_{FFT} since the discrete Fourier transform is applied:

$$\dot{A}(n) = \sqrt{PSD_{\text{sp}}\Delta f_{FFT}}\, \exp(j\,\Phi_n), \tag{14}$$

where Φ_n is the random phase of the spectrum component *n*, which has a uniform probability distribution density ($\Phi_n \in [0; 2\pi]$).

The IFFT expression in MatLab contains an additional factor $1/N_{FFT}$, so the *ifft*() function should be multiplied by the inverse coefficient N_{FFT}:

$$\dot{E}_{\text{sp}}(k) = N_{FFT}\; \textit{ifft}(\dot{A}(n)). \tag{15}$$

In the MatLab program according to Fig. 1, a model of a transmission system without modulation of the transmitter signal was created to check the validity of the proposed expressions.

Optical bandpass filters OPBF-1 and OPBF -2 are identical in implementation, have a bandwidth ΔF_{o} centered on the frequency $\omega_0/2\pi$. The signal samples at the output of the OPBF models is a discrete convolution of the input signal samples and the impulse response $h(k)$ of the filter [4]:

$$y(k) = \sum_{m=-\infty}^{k} x(m)h(k-m) = x(k) * h(k). \tag{16}$$

The impulse response samples $h(k)$ were calculated using the standard MatLab function fir1(). The transmission coefficient of the designed digital filter in the passband $(\omega_0/2\pi - \Delta F_{\text{o}}/2;\, \omega_0/2\pi + \Delta F_{\text{o}}/2)$ is close to 1.

Since thc transmitter signal is unmodulated in this version of the transmission system model, the following approximations can be used

$$\dot{E}_{\text{tr}}(k) = \dot{E}_{\text{tr}}(k) * h_{\text{pbf}}(k) \approx \dot{E}'_{\text{tr}}(k), \tag{17}$$

$$\dot{E}_{\text{sp}}(k) = \dot{E}_{\text{sp}}(k) * h_{\text{pbf}}(k) \approx \dot{E}'_{\text{sp}}(k). \tag{18}$$

The impulse response of the low-pass filter model with a cutoff frequency ΔF_{e} was also calculated using the fir1() function. The verification of the reliability of the described models of the transmission system was carried out using the example of the initial data in Table 1.

The results obtained by analytical calculation according to the expressions [3] were taken as the actual power values:

$$P_{\text{s-sp}} = PSD_{\text{s-sp}}\,\Delta F_{\text{e}} = 4R^2 P_{\text{tr}} PSD_{\text{sp}} \Delta F_{\text{e}}, \tag{19}$$

$$P_{\text{sp-sp}} = R^2\, PSD_{\text{sp}}^2 \Delta F_o (\Delta F_o + 2\Delta F_{\text{e}}). \tag{20}$$

The calculation result was compared with the results of simulation models with real and complex signal representation. Simulation was also carried out in the Optiwave

Table 1. Transmission system model parameters

Functional block	Parameter	Value
Transmitter	Carrier frequency ω_0	100 GHz
	P_{tr}	$1 \cdot 10^{-3}$ W
	Sampling frequency F_s	640 GHz
	Samples quantity	$64 \cdot 10^5$
Optical transmission path	Optical section length L_{of}	100 km
	Attenuation coefficient α	0,2 dB/km
	N_{sect}	10
	Noise figure nf	6 dB
	PSD_{sp}	$2,521 \cdot 10^{-16}$ W/Hz
Receiver	ΔF_o	100 GHz
	Phodiode responsivity R	1 A/W
	ΔF_e	7,5 GHz

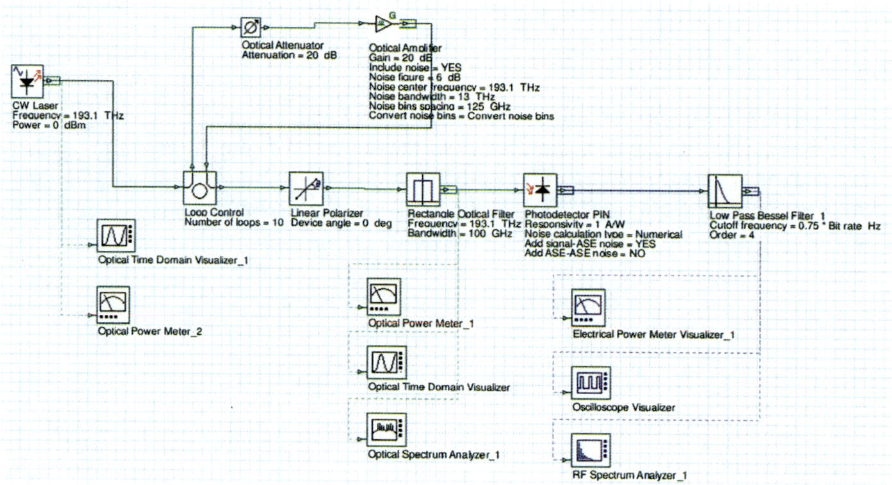

Fig. 5. Functional diagram of the simulation model in the Optiwave Optisystem

Optisystem 7 program according to the functional diagram of the virtual layout of the system (Fig. 5). All calculation and simulation results are summarized in Table 2.

The reliability of the results obtained was carried out by calculating the relative error δ of the signal power for each component of the output current [5]:

$$\delta = \frac{\Delta P}{P_{eT}} \cdot 100\% = \frac{P_{sim} - P_{ref}}{P_{ref}} \cdot 100\%, \tag{21}$$

where P_{ref} and P_{sim} - the reference and simulated power values, respectively.

Table 2. Results of calculation and simulation of a transmission system without modulation of an optical signal

Noise term	Noise power for model type, W (relative error δ)			
	Analytical model	Simulation model (real signal)	Simulation model (complex signal)	Simulation model (Optiwave Optisystem)
P_{s-s}	$1 \cdot 10^{-6}$	$1 \cdot 10^{-6}$	$1 \cdot 10^{-6}$	$1 \cdot 10^{-6}$
P_{s-sp}	$7{,}564 \cdot 10^{-9}$	$7{,}341 \cdot 10^{-9}$ (−2,95%)	$7{,}322 \cdot 10^{-9}$ (−3,20%)	$8{,}37 \cdot 10^{-9}$ (10,66%)
P_{s-sp}	$7{,}311 \cdot 10^{-10}$	$7{,}177 \cdot 10^{-10}$ (−1,83%)	$7{,}17 \cdot 10^{-10}$ (−1,93%)	$7{,}34 \cdot 10^{-10}$ (0,4%)

Comparing the results from Table 2, we can conclude that the error in the power values $P_{\text{s-sp}}$ and $P_{\text{s-sp}}$ from the MatLab model does not exceed 2–3%. The noise power calculated using the Optiwave Optisystem virtual layout has an error of 10.66%.

Now let us consider the operation of the model of the transmission system with intensity modulation according to the block diagram in Fig. 6. This model is developed using the theory of complex numbers, since only this format allows for modeling the fiber using the Split Step Fourier Method.

A pseudo-random sequence generator (PRBS) generates a sequence of information symbols $a_n = \{0, 1\}$ that control an optical modulator (OM). Binary symbols 0 and 1 are transmitted by pulses of an optical signal with a power of P_0 and P_1, respectively. Taking into account the same probability of the formation of binary symbols in the information source ($P(0) = P(1) = 0{,}5$), the average power of the transmitter signal can be calculated by the expression:

$$P_{\text{tr}} = P_0 \cdot P(0) + P_1 \cdot P(1) = 0{,}5(P_0 + P_1). \tag{22}$$

In this work, to reduce the quantity of variable parameters that affect the quality indicators of the system, a zero value of the power of the zero symbol is used ($P_0 = 0$). In this case, the average power of the transmitter signal, according to formula (22), will be half the peak power of a binary unit pulse ($P_{\text{tr}} = 0{,}5P_1$).

Given the above statistical parameters, the transmitter output signal can be written as

$$\dot{E}_{\text{tr}}(t) = \sqrt{P_{\text{tr}}}\sqrt{2}\sum_{n=1}^{N_{\text{symb}}} a_n \exp\left[j\omega_0(t - nT_{\text{T}})\right], \tag{23}$$

where N_{symb} - the number of information symbols in the test sequence of the mathematical model of the system;

n - the number of the current binary symbol.

To compensate for chromatic dispersion, two main methods are used in optical transmission systems in practice:

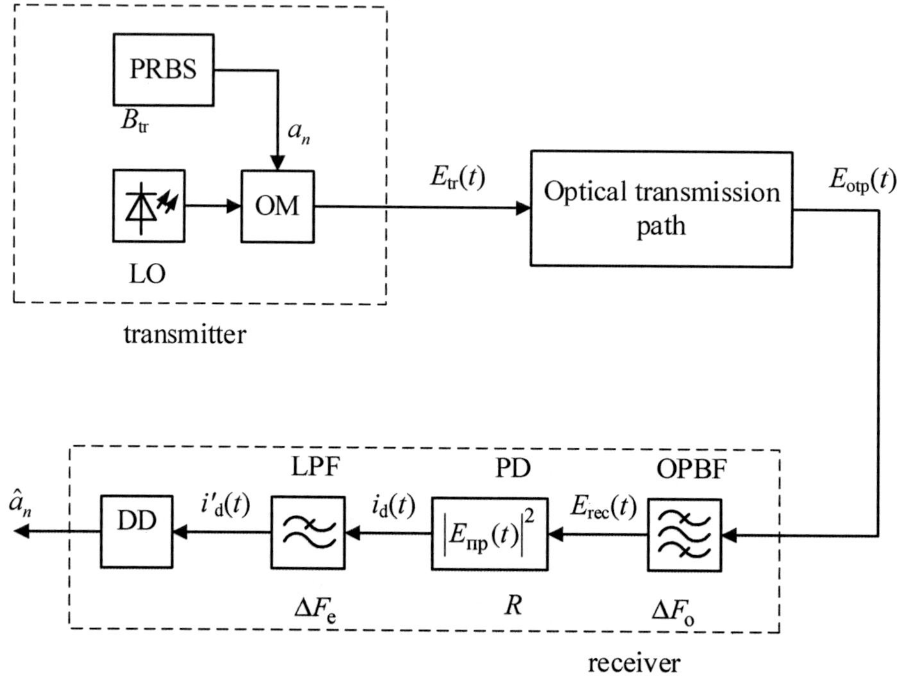

Fig. 6. Functional diagram of the model of the transmission system with intensity modulation

1) separate dispersion compensation units (DCU) as part of each section or at the output of the transmission path;
2) algorithms for electronic compensation of dispersion in the transponder receiver.

In this section, a study of the qualitative parameters of the transmission system with separate dispersion compensation devices as part of each section is carried out.

The next block of the functional diagram of the model (Fig. 2) is an optical band-pass filter (OPBF). In the previous section, a model based on an FIR filter was used to verify the accuracy of the spontaneous emission noise generation and the optical signal detection process. Its order affects the width of the frequency response transition regions and the quality characteristics of the system. Therefore, in this section of the work, an ideal optical bandpass filter was used to reduce the number of variable parameters that affect the quality characteristics of the system. It is easy to implement by filtering the signal based on IFFT and FFT operations in the frequency domain:

$$E_{\text{rec}}(t) = F_{-1}\left\{K_{\text{opbf}}(\omega)F\left\{E_{\text{otp}}(t)\right\}\right\}. \tag{24}$$

The filter transmission coefficient $K_{\text{opbf}}(\omega)$ has a unit value in the operating frequency band $(\omega_{\min}; \omega_{\max})$ of the optical channel:

$$K_{\text{opbf}}(\omega) = \begin{cases} 1, & \omega_{\min} \leq |\omega| \leq \omega_{\max} \\ 0, & |\omega| < \omega_{\min},\ |\omega| > \omega_{\max}. \end{cases} \tag{25}$$

The photodiode model is implemented according to expressions (7) and (8). A low-pass filter at the output of the photodiode has a significant effect on the appearance of the eye diagram of the received signal. In the model without modulation of the optical signal, which was used in the previous subsection, a high-order FIR filter was applied.

In real optical transmission systems with signal intensity modulation, a Bessel-Thompson filter of the 4th order is used as a low-pass filter.

Its advantage is the maximally flat phase response in the passband of the filter [4]. The transfer characteristic of this filter is written by the following expression:

$$H_4(\omega) = \frac{B_0}{\sum\limits_{k=0}^{4} B_k y^k} = \frac{B_0}{B_0 + B_1 y + B_2 y^2 + B_3 y^3 + B_4 y^4}, \tag{26}$$

where B_k - are the coefficients of the Bessel-Thompson filter ($B_0 = B_1 = 105$, $B_2 = 45$, $B_3 = 10$, $B_4 = 1$ [6]);

$y = 2{,}114 j\omega/\omega_{\text{cut}}$ - the imaginary frequency variable;

$\omega_{\text{cut}} = 2\pi \cdot 0{,}75 \cdot f_{\text{clk}} = 1{,}5\pi f_{\text{clk}}$ - the LPF cutoff frequency ($|H(\omega_{\text{cut}})| = 1/\sqrt{2}$);

f_{clk} - the clock frequency of an information signal ($f_{\text{clk}} = B_{\text{tr}}$).

The frequency response of this filter normalized to the cutoff frequency is shown in Fig. 7.

One of the main parameters of digital channels and paths of optical transmission systems is the bit error rate (BER). This factor depends on the signal-to-noise ratio at the receiver input. During the operation of the system, the parameters of the eye diagram of the signal at the input of the receiver's decision device (DD) are studied. The bit error probability depends on the intermediate quality parameter Q-factor, which is calculated according to the expression:

$$Q = \frac{\mu_1 - \mu_0}{\sigma_1 + \sigma_0}, \tag{27}$$

where μ_1 and μ_0 - average values (mathematical expectation) of the samples of symbols 1 and 0, respectively;

σ_1 and σ_0 - average deviation of samples of symbols 1 and 0.

The optimal value of the signal sampling offset time ΔT in the decision device should be determined in the process of calculating the Q-factor. The offset time ΔT is within the range ($0 \leq \Delta T < T_{\text{clk}}$) and the maximum value of the variable Q corresponds to its optimal value. Taking this limitation into account, the determination of signal parameters μ and σ in expression (27) should be performed according to the following expressions:

$$\begin{aligned}
\mu_1 &= M\left[i'_{\text{d}}(nT_{\text{T}} + \Delta T)\big|_{a_n=1} \right], \\
\mu_0 &= M\left[i'_{d}(nT_{\text{T}} + \Delta T)\big|_{a_n=0} \right], \\
\sigma_1 &= \sqrt{D\left[i'_{\text{d}}(nT_{\text{T}} + \Delta T)\big|_{a_n=1} \right]}, \\
\sigma_0 &= \sqrt{D\left[i'_{\text{d}}(nT_{\text{T}} + \Delta T)\big|_{a_n=0} \right]},
\end{aligned} \tag{28}$$

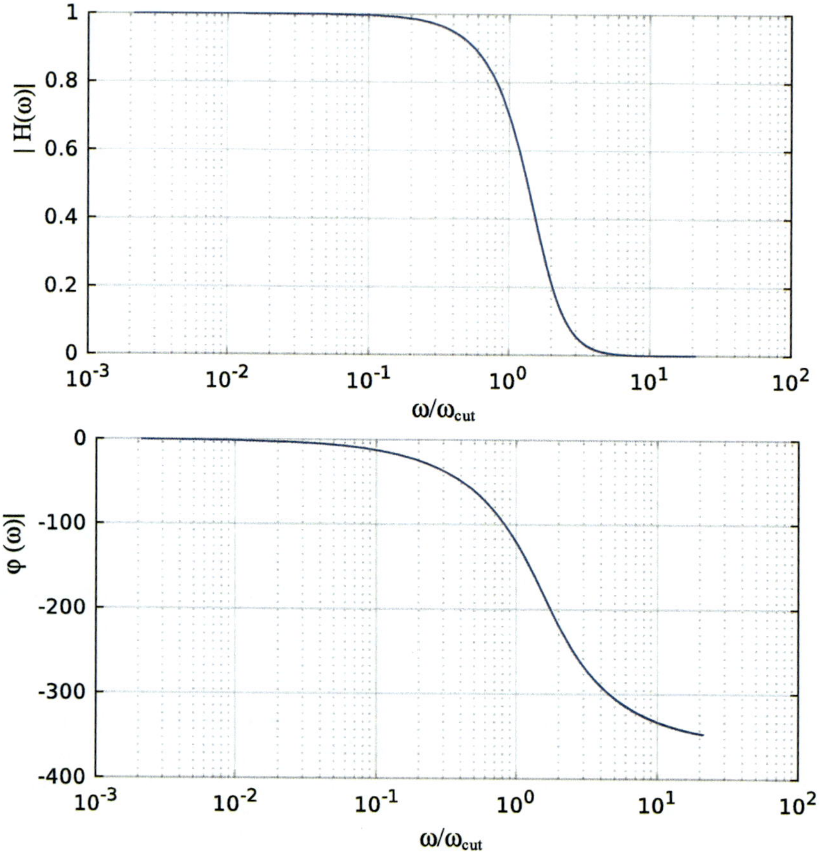

Fig. 7. Normalized frequency response of the 4th order Bessel LPF

where M [] and D[] - are operations for calculating the mathematical expectation and variance of a discrete random variable, respectively;

a_n - are binary symbols at the PRBS output ($1 \leq n \leq N_{\text{symb}}$).

4 Validation of the Simulation Model and Optimization of the Optical Signal Parameters

According to the above functional diagram (Fig. 6), a simulation model of one OFTS channel with IM was built in the MatLab environment. Its quality characteristics depend on the parameters of the transmitter, OTP and receiver. In the further text of the work, the effect of the transmitter power and the quantity of OTP sections on the total noise power $P_{n\sum} = P_{\text{s-sp}} + P_{\text{sp-sp}}$ and Q-factor of the OFTS channel is considered. Other parameters remained unchanged, and their numerical values corresponded to those given in table. 1. The simulation was carried out for standard digital OTH channels of the OTU1 ($B_{\text{tr}} = 2.666$ Gbps) and OTU2 ($B_{\text{tr}} = 10.709$ Gbps) types, in which it is advisable to use the intensity modulation of the optical signal.

To check the reliability of the simulation results, the Q-factor was also calculated according to the modified version of the expression (27).

Simulation showed (Fig. 8) that for a fixed number of OTP sections, the dependence of the Q-factor on the transmitter signal power has an extremum in the form of a maximum.

Its origin can be explained by the fact that the signal-to-noise ratio of the receiver is influenced by two factors: interference noise of spontaneous emission $P_{\mathrm{n}\sum}$ and nonlinear signal distortions in the propagation medium. At low values of the transmitter signal power, nonlinear distortions practically do not appear and the OFTS channel operates in a linear mode. In such a case, an increase in the transmitter signal power P_{tr} results in a proportional increase in the signal-to-noise ratio and Q-factor. Further increase in transmitter power requires to the manifestation of nonlinear distortion. They lead to a change in the temporal waveform and increasing of its spectrum, which, in combination with spectrum limitation in the OBPF, leads to intersymbol interference of the digital signal. Therefore, for each value N_{sec} dependence $Q(N_{\mathrm{sec}}, P_{\mathrm{tr}})$ has an extremum corresponding to the optimal power value $P_{\mathrm{tr\ opt}}$. Dependence $P_{\mathrm{tr\ opt}}(N_{\mathrm{sec}})$ at the bottom of Fig. 8 shows that increasing the OTP length leads to a decreasing in the required transmitter signal power. This is due to the accumulation of signal distortions with an increasing in the length of the propagation medium. For a digital path of the OTU1 type, the optimal power value is higher (14 mW for one OTP section versus 2 mW for OTU2), since self-modulation distortions are stronger for a shorter clock interval (higher transmission rate).

To check the reliability of the proposed model of the FOTS channel with MI, a simulation model with a similar diagram and parameters of functional blocks was developed in the Optiwave Optisystem environment (Fig. 9). It was later used as a source of valid data to check the accuracy of the MatLab model. For the optimal power value $P_{\mathrm{tr\ opt}}(N_{\mathrm{sec}})$ calculated in MatLab, modeling was carried out in the Optiwave Optisystem.

After that, the relative simulation error was calculated by comparing the power of the total noise $P_{\mathrm{n}\sum}$ and the Q-factor of the signal:

$$\delta Q = \frac{\Delta Q}{Q_{\mathrm{o}}} = \frac{Q_{\mathrm{M}} - Q_{\mathrm{o}}}{Q_{\mathrm{o}}} \cdot 100\%, \tag{29}$$

where Q_{M} and Q_{o} are the numerical values of the Q-factor obtained by modeling in MatLab and Optiwave Optisystem at the same values P_{tr} and N_{sec}.

Comparison of power graphs $P_{\mathrm{n}\sum}(N_{\mathrm{sec}})$ in Fig. 10 shows an almost complete coincidence of the results obtained for both considered variants of the OFTS channel of the OTU1 and OTU2 type. The relative Q-factor error for one section of the optical transmission path is equal to 22.67% for OTU1 and 17.23% for OTU2. When the number of sections is more than 5, the relative modeling error does not exceed 4–5% for both types of optical channel that were studied. The discrepancy in the results for short-length OTP is explained by some differences in the implementation of certain functional blocks by the models, for example, Butterworth LPF. The indicated error is not critical, since the main task of modeling is to check the reliability of expressions and optimize the parameters of the OFTS signal.

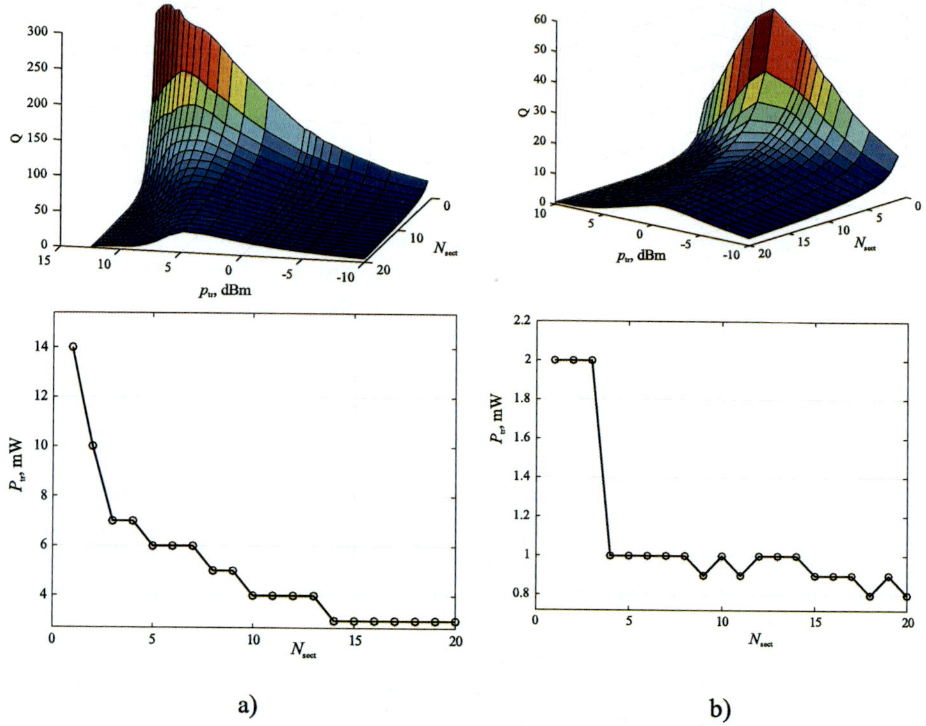

Fig. 8. Influence of the power and the number of OTP sections on the output parameters of the OFTS OTH channel of the OTU1 (a) and OTU2 (b) types

As stated earlier, the Q-factor is an intermediate parameter for determining the bit error probability (BER). For intensity modulation, the relationship between them is determined by the following expression [2, 7]:

$$BER = \frac{1}{2}erfc\left(\frac{Q}{\sqrt{2}}\right). \tag{30}$$

The considered dependence $BER(Q)$ is exponential (Fig. 11), therefore the value $Q \approx 6$ corresponds to the optical channel with standard quality ($BER = 10^{-10}$), and for the channel with improved performance ($BER = 10^{-12}$) the parameter Q increases to 7.

For such values of the Q-factor, according to the Fig. 10, a relatively small simulation error of up to 5% is already achieved. It should also be noted that the use of numerical modeling data obtained by commercial software such as Optiwave Optisystem is conditional.

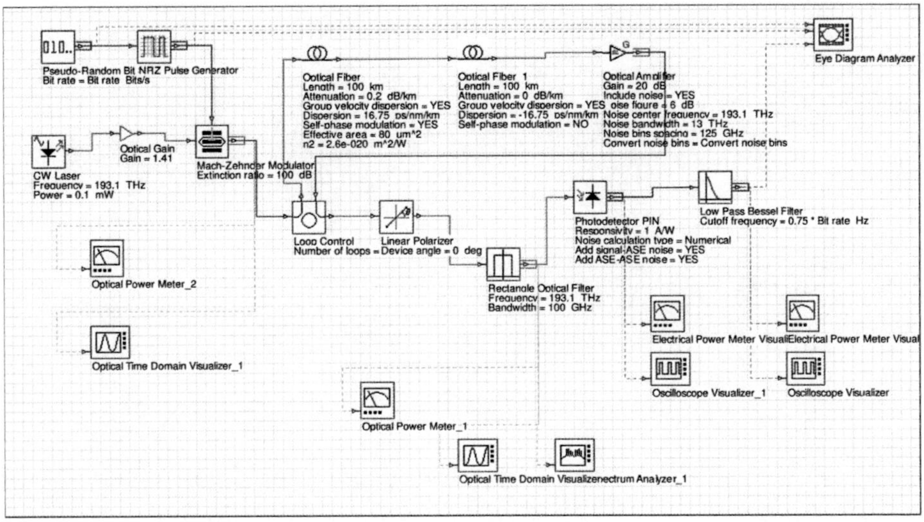

Fig. 9. Functional diagram of the OFTS with IM channel simulation model in Optiwave Optisystem

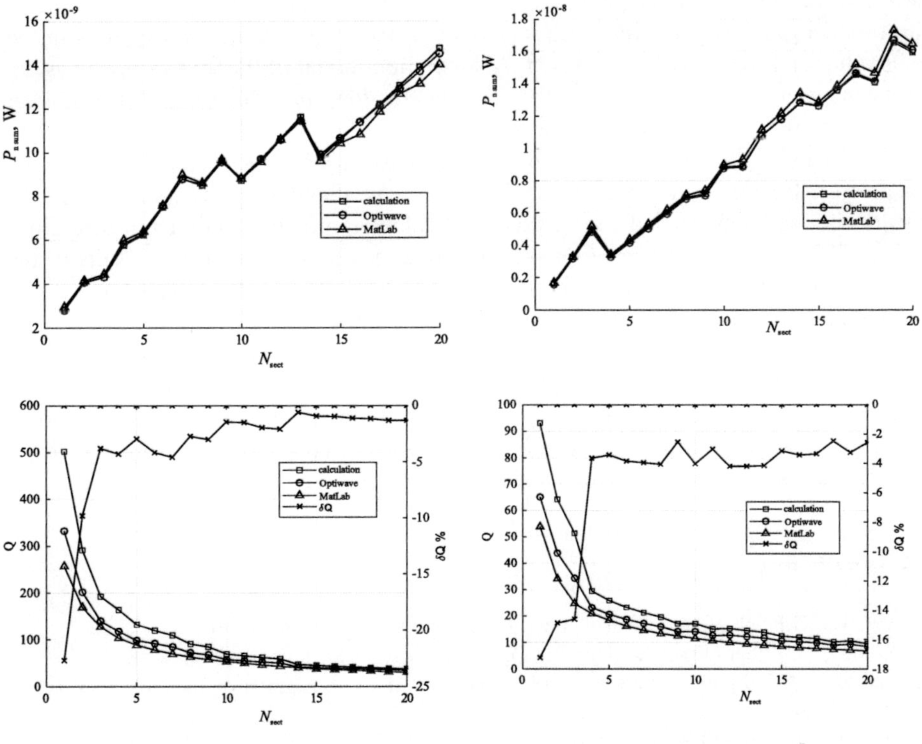

a) b)

Fig. 10. Quality parameters of optical channel of OTU1 (a) and OTU2 (b) type

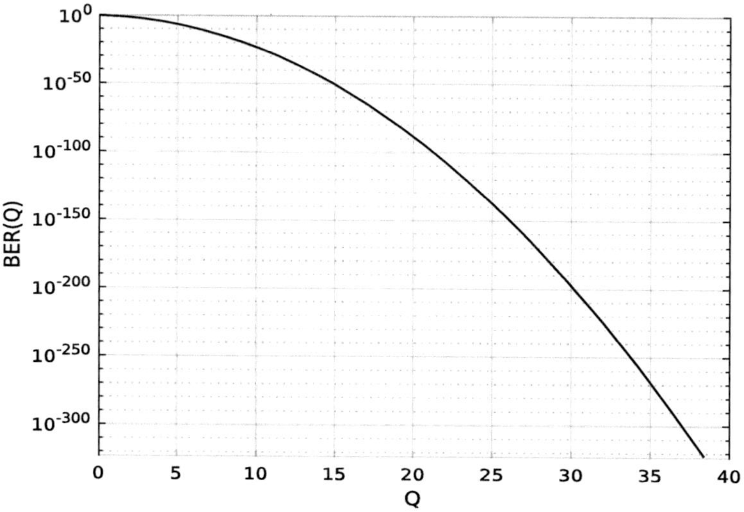

Fig. 11. Influence of Q-factor signal at a bit error probability

5 Conclusion

In this work, the qualitative parameters of the optical transmission system with intensity modulation was performed. An analytical model of signal transformations in the transmitter and receiver was proposed. A simulation model of the transmission system has been developed in the MatLab program, which allows introducing distortions of the propagation medium and amplifier noise.

A variant of the formation of the additive noise of an optical amplifier based on the discrete Fourier transform was proposed. The verification of the reliability of the results obtained was carried out by a control calculation using analytical expressions and modeling in the Optiwave Optisystem program. Comparison of the results of the model without signal modulation in MatLab showed that the simulation error is 2–3% compared to the calculation using analytical expressions.

Comparison of the system simulation with intensity modulation showed that in the case of an optical path from one optical section, an error of the Q-factor of the order of 20% is observed. This is caused by a difference in the internal construction of some elements of the transmission system block diagram. When the number of optical sections is more than 4, the simulation error does not exceed 4–5%.

References

1. Agrawal, G.P.: Nonlinear Fiber Optics. Academic Press, London (2019)
2. Azadeh, M.: Fiber Optics Engineering Optical Networks. Springer, Boston (2009)
3. Olsson, N.A.: Lightwave systems with optical amplifiers. J. Lightwave Technol. **7**(7), 1071–1082 (1989)
4. Quinquis, A.: Digital Signal Processing Using MATLAB. Wiley, London (2008)

5. DeGroot, M.H., Schervish, M.J.: Probability and Statistics, 4th edn. Addison-Wesley, Boston (2012)
6. Bottacchi, S.: Noise and Signal Interference in Optical Fiber Transmission Systems. Wiley, Chichester (2008)
7. Gumaste, A., Antony, T.: DWDM Network Designs and Engineering Solutions. Cisco, London (2003)

Application of the Classical Noise Immunity Theory for Prediction of the Parameters of Perspective Multiservice Telecommunications in Accordance with Modern Digital Standards

L. Uryvsky(✉), A. Moshynska(✉), V. Solianikova(✉), and B. Shmigel(✉)

Telecommunication Systems Department, Igor Sikorsky Kyiv Polytechnic Institute,
Industrialnyi Lane, 2 (Campus 30), Kyiv 03056, Ukraine
leonid_uic@ukr.net

Abstract. In the classical theory of noise immunity, an indicator of reliability is the probability of an information bit error. In modern international standards and recommendations, there is an indicators hierarchy of the communication channel quality, based on measurements, which is not related to the classical theory of noise immunity.

A unified methodology for assessing communication quality indicators in multiservice communication systems is proposed. This technique differs from the existing methods for assessing communication quality indicators in that it allows to display the results of instrumental assessment according to the current international standards in the indicators structure of the noise immunity classical theory. The advantage of the proposed technique is that the fundamental theory of noise immunity makes it possible to calculate the corresponding quality indicators of a digital channel at the design stage of a telecommunication system, while ITU-T standards determine the quality indicators of information transmission in multiservice systems only on the basis of measurements of an already operating digital communication system.

Keywords: Multiservice communication systems · Indicators of communication quality · Forecasting the quality of digital communication channels · International ITU-T standards

1 Introduction

The category "quality of digital signal transmission" is one of the important characteristics for assessing the efficiency of information transmission in a telecommunication system [1–5].

In the fundamental theory, the quality of signal transmission is determined by the noise immunity of the channel with continuous transmission of symbols and the reliability of information transmission. The reliability indicators are the probability of a symbol error in the communication channel p_s (SER – Symbol Error Ratio) and the probability of an information bit error p_b (BER – Bit Error Ratio) [6–9].

P. Vorobiyenko et al. (Eds.): IPF 2020, LNNS 212, pp. 38–59, 2021.
https://doi.org/10.1007/978-3-030-76343-5_3

In modern standards and recommendations of the ITU-T Telecommunication Union [1], there is a hierarchy of communication quality indicators, which is not related to the classical theory of noise immunity, since the data stream is packetized, therefore, accordingly, the quality of the digital signal transmission is associated with the transmission of packets, but not a continuous stream.

For digital paths operating at primary rate or higher, ITU-T G.826 is based on the usage of blocks into which bit information is divided, and digital connections operating at a lower rate are estimated using bit errors and bit error rates. It does not allow evaluating the channel during design, before the start of operation.

This section proposes a unified methodology for assessing communication quality indicators in multiservice communication systems, which differs from existing methods for assessing communication quality indicators that allows displaying the results of an instrumental assessment of the digital communication channels quality according to the current international standards in the structure of noise immunity classical theory indicators.

Purpose of the study is to create a unified methodology for assessing the quality of communication, with the help of which, at the design stage, the reliability indicators proposed in the classical theory of noise immunity are determined. These indicators establish their compliance with modern international recommendations and standards at the operation stage of a telecommunications system.

So, the knowledge offered in the theory of noise immunity and expressed by analytical regularities becomes applied in accordance with the modern recommendations of the International Telecommunication Union ITU-T.

The proposed methodology is synthesized on the basis of the proposed analytics of the noise immunity theory, while the indicators according to modern ITU standards are based on instrumental measurements.

The procedures for measuring the quality indicators of information transmission are possible only using an already created telecommunication system. The usage of the classical theory of noise immunity makes it possible to provide for the energy parameters compliance of the future communication system with the necessary quality indicators.

If the probability of one stream distortion is known, that is, the Bit Error Ratio (BER) consecutive stream p, then transmission quality indicators examples of a digital stream with a known BER value are the probabilities:

- Errored Block (EB) of n length;
- Errored Second (ES) at channel speed V [bit/s];
- Severely Errored Second (SES), containing more than 30% of blocks n length with errors at channel rate V [bit/s].

2 Systematization of Ratio Between Digital Signal Quality Parameters Based on Indicators of the Fundamental Theory of Noise Immunity

2.1 Analysis of the Hierarchy of Digital Signal Quality Standardized Parameters in Combination with Indicators of the Fundamental Theory of Noise Immunity

To systematize the relationships between various parameters of digital signal quality based on the indicators of the noise immunity fundamental theory, a universal hierarchical diagram of the digital signal quality standardized parameters relationship in combination with indicators of the noise immunity fundamental theory is proposed (Fig. 1).

At the head of the hierarchy is the key parameter of the classical theory of noise immunity – the signal-to-noise ratio h^2. For the given parameters of the digital communication channel, the signal-to-noise ratio is definitely related to the probability of the Bit Error Ratio (EBR) p_b.

Further, according to Fig. 1, due to the analytical relations proposed in this section, a connection is established between the parameters of the noise immunity classical theory and the information transmission quality indicators in digital communication channels used in modern telecommunication standards and recommendations.

The general scheme is divided into two parts.

The first part reflects the hierarchy of Errored Blocks.

The second part of the hierarchy reflects the relationship of errors associated with the presence of Errored Second.

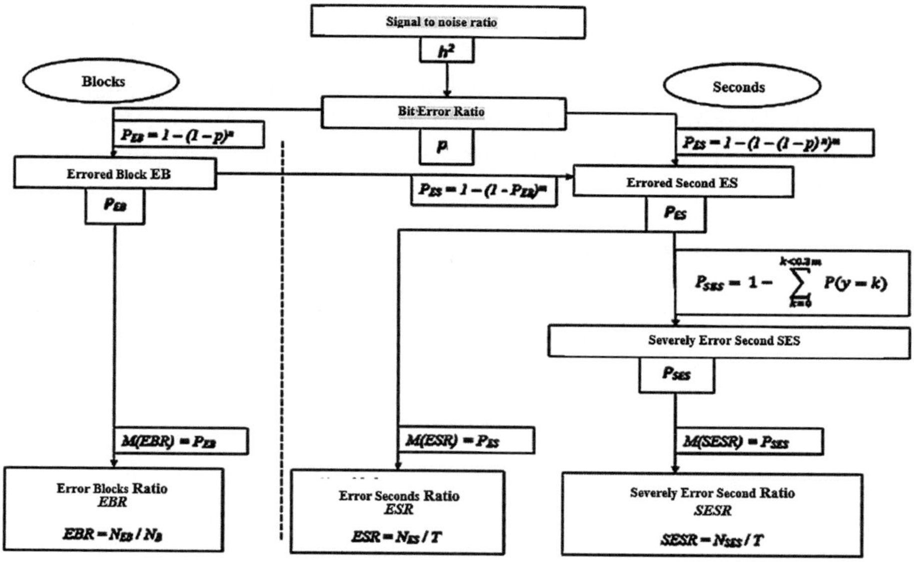

Fig. 1. Hierarchy of information transmission quality parameters

So, the block error hierarchy contains indicators: the probability of an Errored Blocks (EB) and the ratio of Errored Blocks Ratio (EBR).

By definition, the EBR is equal to the ratio of the Errored Blocks N_{EB} to the total number of transmitted blocks N_B:

$$EBR = N_{EB}/N_B \tag{1}$$

As can be seen from the hierarchy diagram, the probability of the EB occurrence P_{EB} is related to the BER p_b. The relationship between the P_{EB} and the EBR is established by understanding that, on average, the number of Errored Blocks in a given measurement interval is determined by the P_{EB}. And the average number of EB relative to the total number of transmitted blocks, by definition, gives the mathematical expectation of the EBR, which is.

$$M\,(EBR) = P_{EB}. \tag{2}$$

The error hierarchy associated with the presence of EB in seconds contains the metrics: the probability Errored Second (ES) and the probability of Severely Errored Second (SES), as well as the corresponding Errored Second Ratio (ESR) and Severely Errored Second Ratio (SESR) coefficients.

By definition, the Errored Second Ratio (ESR) is equal to number of ES – N_{ES} to the total number of measurement seconds T:

$$ESR = N_{ES}/\,T, \tag{3}$$

and the Severely Errored Second Ratio (SESR) is the ratio of the SES number – N_{SES} to the total number of measurement seconds T:

$$SESR = N_{SES}/T, \tag{4}$$

The proposed hierarchy is a graphical representation of the methodology for assessing the quality of digital signals, which is synthesized in this section.

2.2 Analysis of Digital Communication Channels Characteristics According to the ITU-T Recommendation

The methodologies for assessing the quality of digital communication channels proposed in the ITU-T G.821 and ITU-T G.826 standards consider the parameters and norms of end-to-end error characteristics for international digital paths and connections with constant bit rate, but do not specify the channel quality and do not give its precise characteristics in measuring classical reliability indicators.

There is a need to create a methodology for quality comparative assessment of digital communication channels for multiservice communication systems with known reliability characteristics for given parameters, in various operating conditions.

For digital paths operating at primary rate or higher, ITU-T G.826 is based on the usage of blocks into which bit information is divided, and digital connections operating at a lower rate are estimated using bit errors and bit error rates. It does not allow evaluating the channel during design, before the start of operation.

The G.826 standard provides a table of calculated parameters for evaluating digital communication channels with specified characteristics. The options are considered for various types of virtual containers with certain values of speed, number of blocks per time unit and BER value, according to this ESR, SESR, SEPI and BBER are calculated.

Table 1. Characteristics of digital communication channels according to ITU-T G.826 recommendation

Rate (kbit/s) parameters	1664 (VC-11)	2240 (VC-12)	6848 (VC-2)	48 960 (VC-3)	150 336 (VC-4)	601 344 (VC-4-4c)	2 405 376 (VC-4-16c)	9 621 504 (VC-4-64c)
Blocks/second	2000	2000	2000	8000	8000	8000	8000	8000
ESR (according to ITU-T Recommendation G.826, Appendix 2)	0,005	0,005	0,005	0,01	0,02	NA	NA	NA
% of time SESR	0,001	0,001	0,001	0,001	0,001	0,001	0,001	0,001
SEPI event/s (according to ITU-T Recommendation G.826, Appendix 3)	1×10^{-4}	1×10^{-4}	$0,8 \times 10^{-4}$	1×10^{-4}	1×10^{-4}	1×10^{-4}	1×10^{-4}	1×10^{-4}
BBER (according to ITU-T Recommendation G.826, Appendix 2)	$2,5 \times 10^{-5}$	$2,5 \times 10^{-5}$	$2,5 \times 10^{-5}$	$2,5 \times 10^{-5}$	5×10^{-5}	5×10^{-5}	5×10^{-5}	5×10^{-5}

Table 1 shows the limit parameters for virtual containers VC-11, VC-12, VC-2, VC-3, VC-4, VC-4-4c, VC-4-16c, and VC-4-64c with speeds of 1664, 2240, 6858, 48960, 150336, 601344, 2405376, 9621504 kbps, respectively. Each of these types of virtual containers corresponds to its own set of characteristics with certain types of digital and operational connection and transmission paths, and explains the different data transfer rates. There is a need to calculate the limit values for the quality parameters of digital communication channels. Let's perform the task setting in accordance with our needs. Consider a digital channel with specified characteristics and calculate the main parameters for channel estimation.

It is advisable to apply the methodology for assessing the digital communication channel in accordance with the requirements of ITU-T standards and carry out the calculation, particular, for the limit values:

- indicators of the Errored Second Ratio (ESR);
- time % versus Severely Errored Second Ratio (SESR);
- Background Block Error Ratio (BBER);

2.3 Research of Signal Conversion Mathematical Models in Multiservice Information Transmission Systems

To characterize the dependence, the formulas of J. Prokis will be used in order to calculate the probability of a symbol and bit error [6].

It is known that ensuring the efficiency of using limited energy and frequency resources is an urgent task in telecommunication systems. On the one hand, there is a desire to use the spectrum more efficiently, and on the other hand, the greater the keying frequency, the more the signal energy deteriorates [5–9]. Therefore, studies to improve the reliability will be conducted according to the methodology described below.

The initial data for the technique are the following parameters:

n – is the length of the block of the error-correcting code;

h^2 – is an energy parameter characterizing the ratio of the average signal energy at the receiver in one-sided spectral noise power:

$$h^2 = \frac{E}{N_0} = \frac{P_c}{V_c \cdot N_0}, \tag{5}$$

where $V_c = \frac{1}{T}$;

α – energy parameter equal to the ratio of the useful signal power P_c and the noise one-sided spectral power N_0.

$$\alpha = \frac{P_c}{V_c \cdot N_0}, \tag{6}$$

where V_c is the symbol rate in the communication channel.

Using these parameters, the available bit error probability in the channel can be determined. It depends on how the signal is generated and processed.

For QPSK and QAM-M keying, the symbol and bit error probability will be determined by the formulas:

$$p_{symbQPSK} = \frac{3}{4} - \frac{1}{2}\Phi\left(\sqrt{h^2}\right) - \frac{1}{4}\Phi^2\left(\sqrt{h^2}\right), \tag{7}$$

– symbol error ratio for the Quadrature Phase Shift Keying (QPSK) signal;

$$p_{bitQPSK} = \frac{1}{2}\left[1 - \Phi\left(\sqrt{2h^2}cos\frac{\pi}{4}\right)\right], \tag{8}$$

For Quadrature amplitude modulation (QAM) signals, the following formulas are valid:

$$p_{symbQAM} = 1 - \left(1 - \frac{2\left(1 - \frac{1}{\sqrt{M}}\right)}{\sqrt{2\pi}} \int_{\sqrt{\frac{3}{M-1}h^2}}^{\infty} exp\left(\frac{-u^2}{2}\right)du\right)^2, \tag{9}$$

– the symbol error ratio;

$$p_{bitQAM} = \frac{4\left(1 - \frac{1}{\sqrt{M}}\right)}{\sqrt{2\pi}} \int_{\sqrt{\frac{3}{M-1}h^2}}^{\infty} exp\left(\frac{-u^2}{2}\right)du, \tag{10}$$

– is the bit error probability, where M is the order of the modulation.

It was found that when transmitting information using multi-position signals, a symbolic error probability arises in the channel. The symbolic error probability is highly dependent on the energy parameters of the signal. That is, as the signal/noise parameter improves, the error probability decreases.

The symbol rate is completely determined by the type of modulation used, that is, each type of modulation provides a specific symbol rate. Thus, high data density is achieved due to the ability to support high-speed modulation types over long distances compared to other systems. Today, QAM is one of the most efficient modulation methods, which makes it possible to achieve the maximum possible data transmission rate [5, 7].

Using the formulas above, probability dependence graphs of symbolic and bit errors on the communication channel energy parameters were constructed.

Symbol error investigation graphs and the signal-to-noise ratio for each modulation are calculated and constructed in the MATHLAB program are presented in Fig. 2.

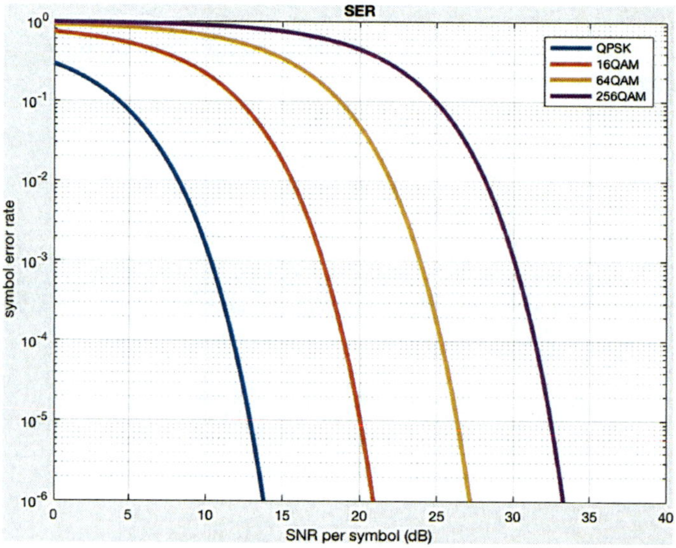

Fig. 2. The graph of the symbolic error probability dependence on h^2 for QPSK, QAM-16, QAM-64 and QAM-256

The bit error probability of multi-position signals for the QPSK signal is less than the symbol error probability, since not every symbol distortion leads to incorrect reception of a single bit of information.

The graphs of the bit error probability dependence on the energy parameter h^2 are shown in Fig. 3.

From the presented dependences in Figs. 2, 3, the following conclusions can be drawn:

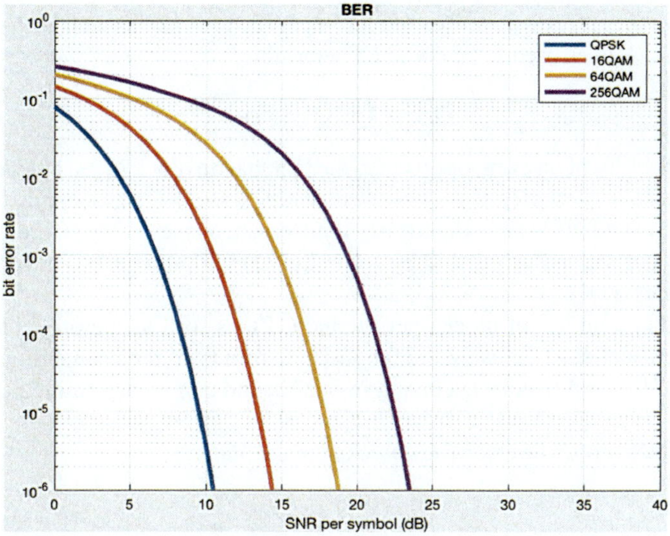

Fig. 3. The graph of the bit error probability dependence on h^2 for QPSK, QAM-16, QAM-64 and QAM-256

- to ensure the specified reliability, it is necessary to correct a larger number of errors at low energy in the communication channel;
- in channels with a large order, a greater number of errors occur in comparison with channels with a lower order, which is due to a smaller phase-amplitude distance between the positions of these signals.

3 Methodology Synthesis for Assessing the Quality of Telecommunication Channels Digital Signals

3.1 Digital Channel Parameters Determination of a Multiservice Telecommunications System in the Analytical Model Through Traditional Indicators of Noise Immunity

According to the hierarchy defined by the ITU-T standards and shown in Fig. 1, it should be established in accordance with the basic indicator of the classical theory of noise immunity - the Bit Error Ratio (BER) consecutive stream *p*.

1. The probability of an event opposite to none of the n bits in the block being corrupted is determined by:

$$P_{EB} = 1 - (1 - p)^n \tag{11}$$

2. With a known blocks bit rate V, M of n bits are transmitted in one second:

$$M = V/n \tag{12}$$

Therefore, with a known number of blocks per second (12), as well as the probability that the block is errored (11), could be determine:

$$P_{ES} = 1 - (1 - P_{EB})^{V/n} \tag{13}$$

– the probability of an event opposite to the fact that no block in a second was errored. It is necessary to clarify that a block that hits two adjacent seconds at once will affect both seconds when identifying their erroneousness. Therefore, the number of blocks per second calculated by formula (12) will always be rounded up to the nearest integer.

3. Knowing the number of blocks per second (12), as well as the probability that the block is erroneous (11), using the binomial distribution, the probability of a certain number of Errored Blocks occurring in one second can be determined.

Let y be the number of erroneous blocks per second,

$$y = 0,\ 1,\ 2 \ldots k, \tag{14}$$

then

$$P(y = k) = \frac{M!}{(M-k)!k!} P_{EB}^{M} (1 - P_{EB})^{M-k} \tag{15}$$

For a second with multiple errors to occur, it is necessary that 30% or more blocks per second are erroneous, that is

$$y \geq 0.3\,m, \tag{16}$$

Then, using (12) and (15) taking into account conditions (14) and (16)

$$P_{SES} = 1 - \sum\nolimits_{k=0}^{k<0.3m} P(y = k), \tag{17}$$

where the subtracted is the sum of the probabilities of events, each of which consists in the appearance of the Errored Blocks number in a given second, less than 30%.

It can be seen that formula (15) at large values of M becomes rather inconvenient for calculation for practical purposes, especially when used in formula (17). This inconvenience is due to the need to calculate the factorials of large numbers a large number of times. The situation is further aggravated when it becomes necessary to calculate the value using formula (17) for a specific set of input data.

To optimize the calculation of factorials, specific values of k in the expression with formula (17) containing factorials should be substituted:

$$\sum\nolimits_{k=0}^{k<0.3M} \frac{M!}{(M-k)!k!} = \frac{M!}{M!0!} + \frac{M!}{(M-1)!1!} + \frac{M!}{(M-2)!2!} + \ldots + \frac{M!}{(M-0.3M)!(0.3M)!} \tag{18}$$

If M! is taken out of brackets, then the denominators are transformed:

$$\sum\nolimits_{k=0}^{k<0.3M} \frac{M!}{(M-k)!k!} = M!\left(\frac{1}{M!} + \frac{M}{M!1!} + \frac{(M-1)M}{M!2!} + \ldots + \frac{(M-0.3M+1)\ldots(M-1)M}{M!(0.3M)!}\right) \tag{19}$$

Abbreviate M! and an optimized expression is obtained for calculating the part of the formula containing factorials:

$$\sum_{k=0}^{k<0.3M} \frac{M!}{(M-k)!k!} = 1 + \frac{M}{1!} + \frac{(M-1)M}{2!} + \ldots + \frac{(M-0.3M+1)\ldots(M-1)M}{(0.3M)!} \quad (20)$$

The resulting formula for the probability of a second with multiple errors P_{SES}, optimized for practical calculations:

$$P_{SES} = 1 - \left(1 + \frac{M}{1!} + \frac{(M-1)M}{2!} + \ldots + \frac{(M-0.3M+1)\ldots(M-1)M}{(0.3M)!}\right) \times \sum_{k=0}^{k<0.3m} P_{EB}^{M}(1-P_{EB})^{M-k} \quad (21)$$

Formula (21) is optimal from the point of view of the need factorials to calculation. Calculation of three factorials in the expression $\frac{M!}{(M-k)!k!}$ replaced by calculating the factorial in the denominator $(0.3M)!$ and calculating the numerator $(M-0.3M+1)\ldots(M-1)M$, is the product of the last $M-0.3M$ factorial factors of M.

Thus, a new method is proposed for assessing the quality indicators of information transmission. By specifying the initial data, namely the source rate V, the block size n and the BER p, the probability of errors in one block and the probability of one or more Errored Blocks occurring in the transmitted sequence of symbols can be determined, giving theoretical indicators of noise immunity in standardized reliability indicators in telecommunications systems.

Therefore, to solve this problem, formulas (11), (13), (21) and the software package for numerical analysis MATLAB should be used to calculate the specified parameters.

The main parameters for the calculation are the flow rates, in accordance with the types of path, the block length and the probability of error in a continuous stream BER. The maximum block value for TCP-IP protocols is 65535. With regard to the specified characteristics: block size, speed and false bit value, the calculated characteristics are selected according to the recommendations of ESR, SESR, SEPI and BBER and the BER limits are determined.

With an increase in the transmission rate in the channel, the reliability of the reception will drop, so it's needed to adjust the parameters to obtain indicators of the quality of the channel.

With an increase in the BER value, the quality indicators of the channel drop sharply, the transmission reliability in such a channel is extremely low for its productive use.

By changing the specified parameters, it is possible to calculate and fill in the value in the table of digital communication channels limit characteristics, namely, the values of ESR, SESR, SEPI and BBER indicators.

3.2 Comparison of Digital Signal Transmission Quality Indicators of Channels in Multiservice Telecommunication Systems

The analysis of the digital communication channel quality indicators behavior is based on analytical relations (1–4) and (11), (13), (21) for various values of parameters such as the probability of a bit error, transmission rate and block length, and is also confirmed by the results of the simulation modeling [10].

Let turn to individual indicators of the digital hierarchy and assess the sensitivity of these indicators to changes in the classical indicators of reliability and noise immunity.

3.2.1 Errored Block

Figure 4 illustrates the content of the category an Errored Block (EB) is a block in which one or more bits are in error. In turn, the EBR is equal to the ratio of the number of Errored Block N_{EB} to the total transmitted N_B. The Bit Error Ratio is determined by formula (11).

According to G.826 terminology, a block is a set of consecutive bits associated with a path; each bit belongs to one and only one block. An Errored Block is a block in which at least one bit was received with an error (Fig. 4).

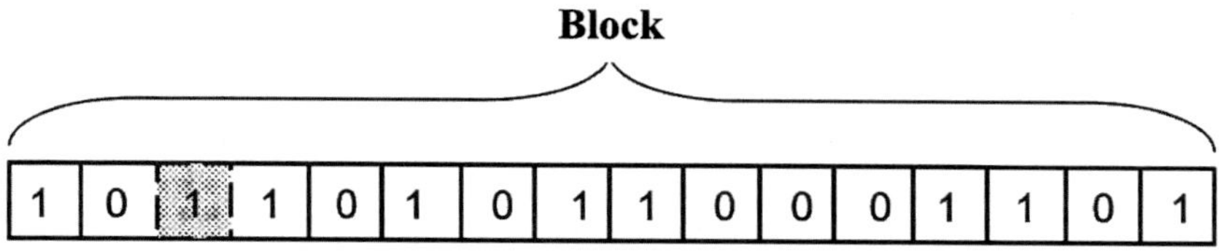

Fig. 4. Errored block

In telecommunication channels, to which the G.826 recommendation applies, the block length reaches tens of thousands of bits per block. As can be seen from the definition and diagram, for a block of any length, one single erroneous bit is sufficient to be considered an erroneous block. So, the block length is a significant factor in the study of quality indicators, because with the same probability of error per bit, the probability of an Errored Block will differ significantly for channels with different block lengths [8–10].

Figure 5 shows the characteristic of the EBR block error rate on the axis of the bit error probability p_b for a block length n = 1000 bits/block and a transmission rate V = 100000 bit/s and with a change in the reliability of bit transmission in a continuous bit stream from 10^{-5} to $3.5 * 10^{-3}$.

It can be seen that a change in the P_{bit} probability by 1 order of magnitude (from 10^{-4} to10^{-3}) changes the EBR coefficient from 0.1 to 0.6, while the normative values of the EBR coefficient have values of the 10^{-3} order (dashed line).

The plot starts at the origin, because at zero bit error probability, the block cannot be hit. The graph asymptotically approaches the value 1.

It was found that with block length increasing EBR $n = 1000 \ldots 1400$ bits/block grows faster on the p_b axis. Indeed, as the length of a block increases, increases the likelihood that at least one bit in it will be erroneous, that leads to EB appearance. This trend corresponds to the probability of an event opposite to the fact that none of the n bits in the block was distorted (6), from which it can be seen that with an increase in the block length, the probability of ES occurrence increases [10, 11].

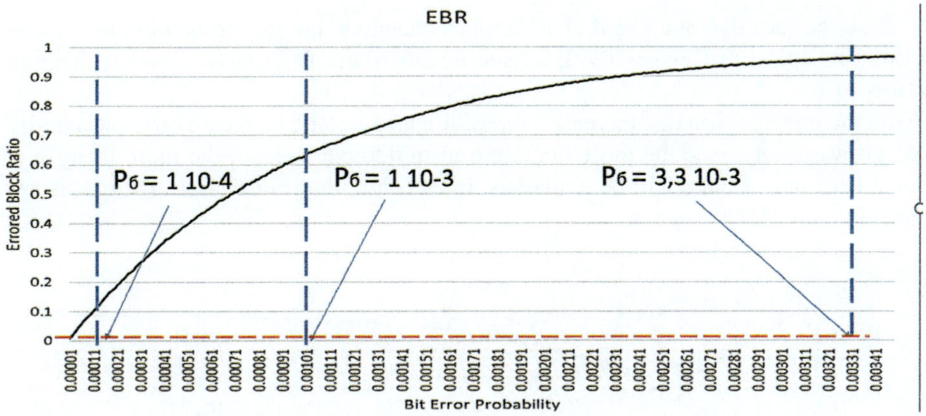

Fig. 5. Dependence of EBR on the bit error probability

More interesting for research from a practical point of view are the quality indicators associated with seconds of channel operation, since norms and restrictions are established in international standards for them.

3.2.2 Errored Seconds

Figure 6 illustrates the content of the **Errored Second (ES)** category: A second period with one or more false blocks or at least one defective. The probability of an Errored Second is determined by the formula (13).

The first type of error associated with the time of information transmission is an **Errored Second** - a second containing at least one block with errors (Fig. 6).

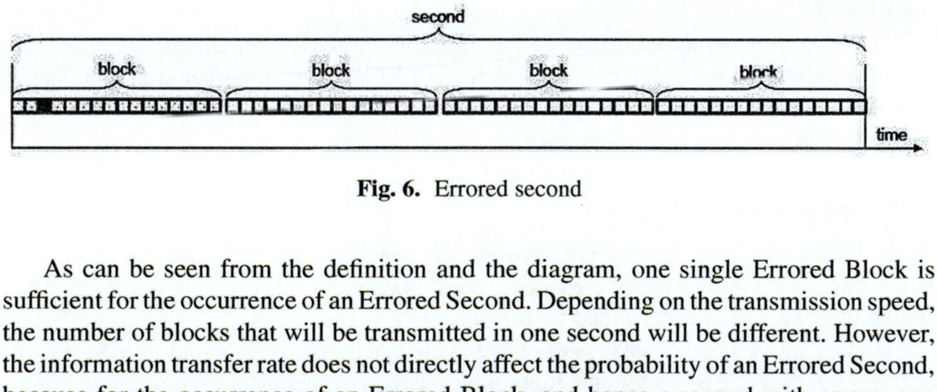

Fig. 6. Errored second

As can be seen from the definition and the diagram, one single Errored Block is sufficient for the occurrence of an Errored Second. Depending on the transmission speed, the number of blocks that will be transmitted in one second will be different. However, the information transfer rate does not directly affect the probability of an Errored Second, because for the occurrence of an Errored Block, and hence a second with errors, one single errored bit is enough [1, 9, 10].

Figure 7 shows the ESR characteristics at a transmission rate of V = 100 kbit/s and 1 Mbit/s with a change in the reliability of bit transmission in a continuous bit stream from 10^{-7} to $5{,}5 * 10^{-5}$ for two values of the symbol rate in the channel (0,1 Mbps and 1 Mbps).

It can be seen that at a speed of 1 Mbit/s, a change in the P_{bit} probability by 1 order (from 10^{-7} to 10^{-6}) changes the ESR coefficient from 10^{-4} (dashed line) to 0.6 and further to 1.

As the transmission rate increases, the ESR characteristic on the bit error probability axis grows faster, since the more bits are transmitted per second, the more likely it is that at least one of them will be erroneous. In this case, the ESR characteristic does not depend on the block length n.

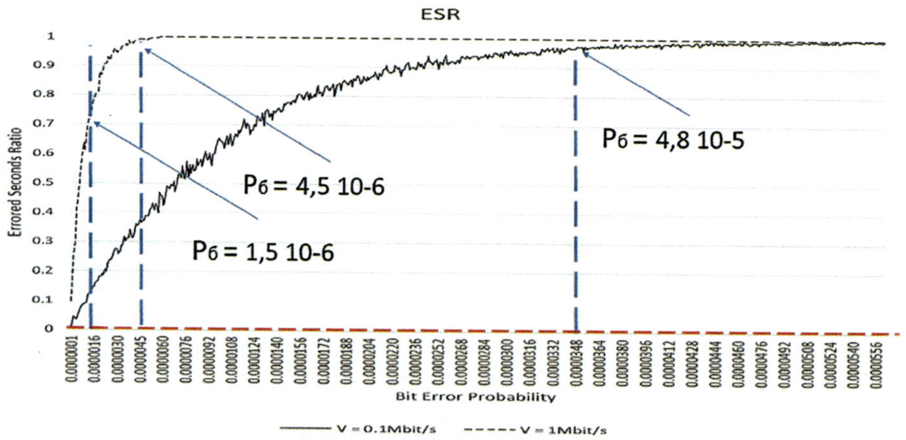

Fig. 7. Correlation of ESR characteristics at different ratio transmission values

So, the characteristic of Errored Second rate on the bit error probability axis behaves like the characteristic of the Errored Block rate, but it grows much faster. It grows faster with increasing symbol rate in the communication channel and does not depend on the block length.

The most interesting parameter of the digital communication channels quality for the study of the Severely Errored Second Ratio (SESR), which is tied simultaneously to blocks and seconds. A Severely Errored Second is logged when 0.3 or more blocks transmitted in that second contain errors.

3.2.3 Severely Errored Second

Figure 8 illustrates the content of the **Severely Errored Second (SES)** category: a second period containing ≥30% block errors or at least one defect. SES is a subset of ES. The probability of a SES is determined by formula (18).

An alternative type of error associated with the transmission time is a **Severely Errored Second (SES)** - a second containing 30% or erroneous blocks.

To find the probability of a second with multiple errors, it is convenient to use the **binomial distribution**. If n independent tests are sequentially carried out, in each of which some event A can occur with a constant probability p, then the total number of tests with the result A is a random variable (denote it by μ), and

$$P\{\mu = i|n, p\} = C_n^i p^i (1 - p)^{n-i} (i = 0, 1, 2, \ldots, n) \quad (22)$$

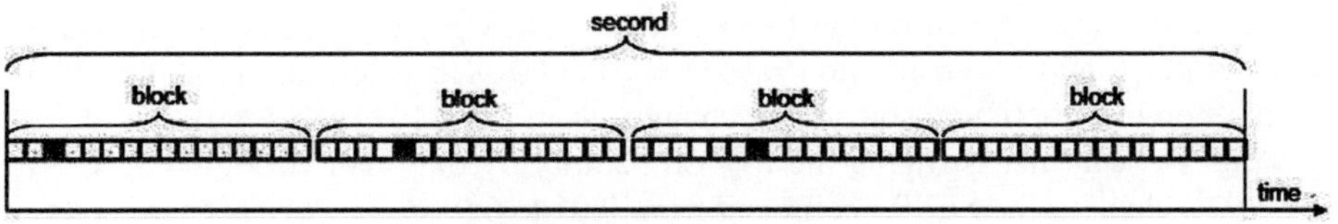

Fig. 8. Errored second

The transmission of some block is an action, it is a sequence of actions that consist in the transmission of the bits of this block. Assuming that the transmission of one bit does not affect the transmission of other bits, it can be argued that the transmission of one block does not affect the transmission of other blocks, since the division into blocks does not affect the transmission of the bits that make up these blocks. If the bit error probability is constant, then the probability of a block with errors is constant.

Knowing the probability of erroneous transmission of a block, one can abstract from the block as a sequence of bits and accept the sequence of blocks as a sequence of independent tests with a constant probability of some result. This gives the right to apply the binomial distribution of the number of false blocks among the total number of transmitted blocks [11].

Figure 9 graphically depicts the application of the binomial distribution to find the probability of Severely Errored Second occurrence P_{SES}. On it $P\,(y = k)$ - the probability that the number of false blocks in a given second will be equal to k.

Each column of the histogram corresponds to the probability $P\ (y\ =\ k)$ for each value k = 0,1, 2,…, m, where m is the total number of transmitted blocks per second. The probability of a second with multiple P_{SES} errors is equal to the sum of columns 0, 1, 2,…, 0.3m. The sum of all the columns gives one, since the each value occurrence of the Errored Blocks number from one to all transmitted is a complete group of events.

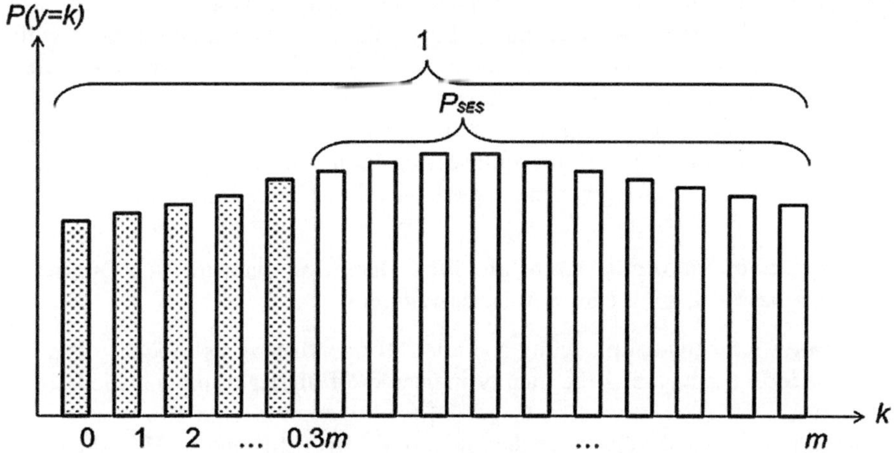

Fig. 9. Errored second

Figure 10 shows the SESR on the axis of the BER p_b for a block length of $n = 1000$ *bits/block* and a transmission rate $V = 100000$ *bit/s* when the reliability of bit transmission in a continuous bit stream changes from 1,5 * 10^{-4} to 5,5 * 10^{-4}.

It can be seen that changing the P_{bit} probability by only half (from 2,4 * 10^{-4} to 4,8 * 10^{-4}) changes the SESR coefficient from 0.01 to 0.9.

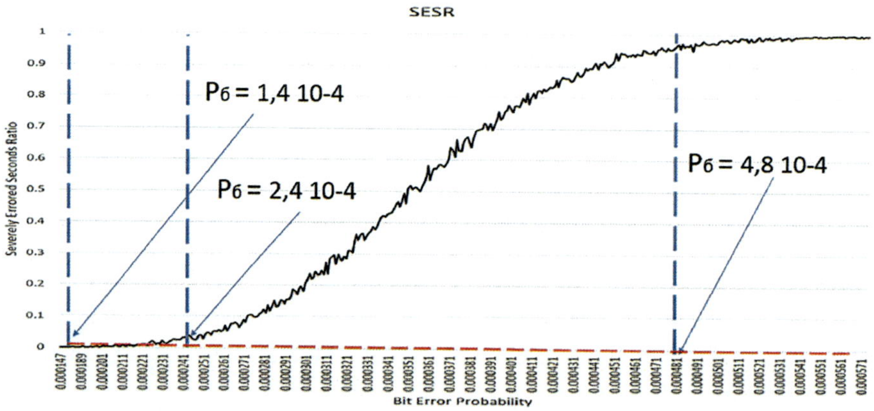

Fig. 10. Dependence of SESR on the BER

The behavior of the SESR on the p_b axis differs significantly from the behavior of the parameters that were considered before. This characteristic begins to grow sharply beyond a certain value of the BER, other than zero. Moreover, the first derivative also increases up to a certain point, after which it begins to subside. The graph is symmetrical about some point at which the first derivative begins to fall off.

So, the methods for assessing the quality of digital communication channels proposed in the ITU-T G.821 and ITU-T G.826 standards consider the parameters and norms of end-to-end error characteristics for international digital paths and connections with a constant bit rate, but does not specify the channel quality and does not give its exact characteristics in the measurement of classical reliability indicators.

Therefore, a method is proposed for the comparative assessment of the digital communication channels quality for multiservice communication systems with known reliability characteristics for given parameters, in various operating conditions.

3.3 Description of a Simulation Model for Calculating Transmission Quality Indicators for Digital Communication Signals

An alternative to analytical modeling and a tool for confirming its adequacy is a simulation model for calculating the quality indicators of digital communication signals transmission.

Such a model was developed to determine the EBR, ESR and SESR indicators at various values of the Bit Error Ratio p (BER), transmission rates V, block lengths n [10].

A simulation model is a computer program that, based on a random number generator, simulates a noisy digital communication channel. The main tool is the high-level programming language C# 6.0 and the Microsoft.Net Framework 4.5. C# is an object-oriented, typed-safe programming language for the.NET platform. The storage medium for the data that is the research results is Microsoft SQL Server 2016.

General sequence of actions for conducting research:

1) Setting the initial parameters;
2) Running the algorithm for execution. Execution results will be displayed in the console window and written to the database;
3) Selection of the necessary data from the database;
4) Plotting in MS Office Excel.

The simulation model is based on the Monte Carlo method.

The pseudo-random (onwards - random) number generator generates a sequence of random numbers, which is an image of bits transmitted sequence over a digital communication channel. The value of each random number determines whether the transmitted bit was received in error.

In order to determine if the transmitted bit was in error, the random number associated with this bit is compared with some threshold value.

If this random number is less than the threshold value, then the bit is considered to have been received in error. The threshold against which the random numbers are compared gives the BER, p_b.

If the random number generator generates numbers in the range from 0 to R, then the probability that the generated random number v will be less than some value r will be $P(v < r) = r/R$.

Through relation (1), it is possible to choose the value of r and R in such a way as to reproduce the probability of a bit error p_b, necessary for modeling certain conditions in a digital communication channel, equating $p_b = P(v < r)$. r and R are chosen so that the interval $0 \ldots r$ is large enough to avoid anomalous behavior of the random number generator at extreme values close to zero, where the distribution of random numbers may differ from the uniform one.

The quality indicators of those studied using the simulation model in this work include:

- Errored Block Ratio (EBR), equal to the ratio of blocks with errors to the total number of blocks in a measurement session.
- Errored Second Ratio (ESR), equal to the ratio of seconds with block errors to the total number of seconds of measurement;
- Severely Errored Second Ratio (SESR), equal to the ratio of seconds with strong block errors to the total number of seconds of measurement.

These indicators are of the greatest practical importance, since they allow assessing the properties of noise immunity of a digital communication channel without reference to a specific number of affected bits and blocks.

The norms for these indicators are given in the ITU-T recommendations and can be applied to the design of telecommunication systems that meet international standards.

Therefore, they turned out to be the most interesting from a scientific point of view and were selected for research using simulation modeling in this work.

Since the parameters under study are associated with blocks and seconds of transmission, the simulation process (namely, the process of "transmitting" bits and establishing their error) is also divided into blocks and seconds.

Figure 11 shows the general algorithm of the computer program, which is used to implement the simulation model.

The input data is initiated first, namely:

- the bit error probability p_b,
- transmission rate V, bit/s,
- block length n, bits/block,
- simulation time T, s.

The bit error probability p_b specifies the maximum value of the random number R and the threshold value r with which the generated random number is compared to determine if the bit is errored.

The baud rate V determines the number of random numbers that must be generated and compared to the threshold within one second.

The simulation time T sets the number of such seconds and affects the error in calculating the investigated quality indicators.

The larger T, the more accurate the simulation result, the closer the value of the indicators obtained by modeling to those that would have been calculated analytically.

While working with this model, T was chosen equal to $T = 1000$ (s).

Further, for each second with T, it is determined whether the second was given by the second with ES errors and whether it was a second with multiple SES errors. To do this, for each block, "transmitted" in this second, the given block was determined by the block with errors EB.

To determine if a block is in error, each bit is checked for error. If the bit is wrong, then the current block is designated as EB.

Further, for each second with T, it is determined whether it was an Errored Second (ES) or it was a Severely Errored Second (SES). To do this, for each block that was "transmitted" in a given second, it is determined whether this block was an Errored Block (EB).

To determine if a block is in error, each bit is checked for error. If the bit is wrong, then the current block is designated as an EB.

To distinguish which block and second a certain bit belongs to, so-called counters are used (in the diagram in Fig. 11 i and j, respectively), which are equal to zero at the beginning of the simulation, increase by one with each bit and reset to zero when the block length n and the number of bits in the second V respectively.

When a block or a second ends, the counters of the EB, ES, SES are updated (in the diagram in Fig. 2 N_{EB}, N_{ES}, N_{SES}, respectively).

When the number of seconds reaches T, the values of the studied quality indicators are calculated. They are the result of an experiment.

The experiment result and input data are then written to the database. This concludes the experiment.

In order to investigate the behavior of quality indicators for various input data, a large number of experiments were carried out, the results of which were stored in a database. The BER p_b was chosen as the main variable, and the bit rate V and block length n were fixed during a series of experiments.

On the basis of experiments series, a selection was made from the database according to the input data that interested the researcher and the analysis was carried out. In order to investigate the effect of bit rate and block length on quality parameters, a sample of several experiments series with different values of one of these input parameters was made.

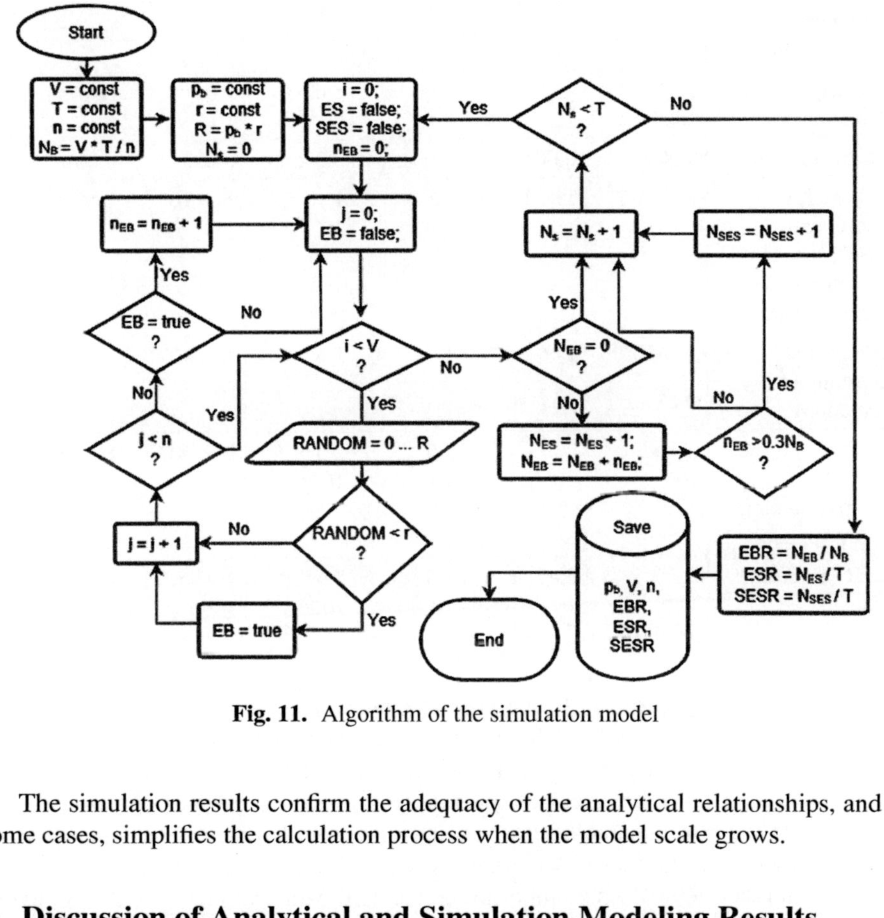

Fig. 11. Algorithm of the simulation model

The simulation results confirm the adequacy of the analytical relationships, and in some cases, simplifies the calculation process when the model scale grows.

4 Discussion of Analytical and Simulation Modeling Results

The results of analytical and simulation modeling obtained using the proposed unified technique are shown in Table 2.

Regarding the data in Table 2, it should be noted that ESR parameters, as a rule, lose value for channels with a high transmission rate, therefore they are not specified for channels with a bit rate of more than 160 Mbit/s. However, synchronous digital channels

Table 2. Calculated limiting characteristics of digital communication channels

Rate (kbit/s) parameters	1664 (VC-11)	2240 (VC-12)	6848 (VC-2)	48 960 (VC-3)	150 336 (VC-4)	601 344 (VC-4-4c)
Block length	832	1120	3424	6120	18792	75168
Blocks/second	2000	2000	2000	8000	8000	8000
BER	**6.02 x 10–6**	**4.44 x 10^{-6}**	**1.46 x 10^{-6}**	**0.82 x 10^{-6}**	**0.37 x 10^{-6}**	**NA**
ESR (according to ITU-T Recommendation G.826, Appendix 2)	5×10^{-3}	5×10^{-3}	5×10^{-3}	1×10^{-2}	2×10^{-2}	NA
ESR (according to calculation)	5.01×10^{-3}	4.98×10^{-3}	5.00×10^{-3}	5.00×10^{-3}	7.04×10^{-3}	NA
% of time SESR	0,001	0,001	0,001	0,005	0,005	0,001
SEPI event/s (according to ITU-T Recommendation G.826, Appendix 3)	1×10^{-4}	1×10^{-4}	$0{,}8 \times 10^{-4}$	1×10^{-4}	1×10^{-4}	1×10^{-4}
SEPI event/s (according to calculation)	1×10^{-4}	1×10^{-4}	1×10^{-4}	1×10^{-4}	1×10^{-4}	1×10^{-4}
BBER (according to ITU-T Recommendation G.826, Appendix 2)	$2{,}5 \times 10^{-5}$	$2{,}5 \times 10^{-5}$	$2{,}5 \times 10^{-5}$	$2{,}5 \times 10^{-5}$	5×10^{-5}	5×10^{-5}
BBER (according to calculations)	$2{,}5 \times 10^{-5}$	2,47 × 10–5	$2{,}5 \times 10^{-5}$	$2{,}49 \times 10^{-5}$	4.93 × 10–5	

are error free for long periods of time even at Gbits/s rates, and a high ESR indicates a degraded transmission system. Therefore, for ES maintenance purposes, monitoring must be performed within any measuring devices operating in error at these speeds.

So, Table 2 contains the calculated parameters for assessing digital communication channels with the given characteristics. The options considered are presented for various types of virtual containers in accordance with the G.826 standard with certain speed values, the number of blocks per unit time and the BER value, according to this, individual indicators are calculated, in particular, ESR, SESR, SEPI (Severely Errored Period Intensity) and BBER (Back-ground Block Error Ratio).

Two essential features of the data presented in Table 2 should be emphasized.

The first feature is that for fixed values of bit rate and block size, one value of error probability of continuous stream corresponds to all corresponding quality indicators of digital packet stream.

The second feature is that with an increase in the bit rate, the requirements for reliable symbols transmission grow.

In practice, this means that an increase in speed by 1 order of magnitude requires an increase in the signal power at the receiving point by 2 orders of magnitude.

5 Conclusions

ITU-TG.826 standards and recommendations have been investigated and, based on the proposed correlations, values of digital communication channels characteristics with given reliability indicators have been computed and marginal characteristics for qualitative work in multiservice networks have been defined. The values obtained by using the calculation are represented in the Table 2.

They allow obtaining borders for qualitative data transmission in digital communication channels with given reliability at the telecommunication systems design stage.

The digital communication channels quality assessing methods proposed in the ITU-T G.821 and ITU-T G.826 standards, take into account the parameters and norms of damage characteristics for international digital paths and connections with constant data rate.

Using **correlations** (1–4) and (11), (13), (21), it is possible to calculate corresponding digital channel quality indicators at the **telecommunication systems design stage**, while ITU-T standards define information transmission quality indicators in multiservice systems based on measurements of an already operational digital communication system.

Therefore, a new technique for information transmission quality indicators assessment has been suggested. It allows determining the quality indicators values from the G.826 recommendation of the International Telecommunication Union, bringing the theoretical noise immunity indicators to standardized reliability indicators in telecommunication systems.

The behavior of the EBR, ESR and SESR has been analyzed on the bit error probability axis.

Figures 5, 7, 10 illustrate a high sensitivity of specified indicators of standards to the bit transferring reliability changes in a continuous bit stream, that requires from telecommunication network operators guaranteed, first of all, energy conditions for digital traffic symbols transmission.

Characteristic of the EBR on the BER axis begins to grow sharply with appearance of the bit error probability and asymptotically approaches to one. This characteristic increases faster with increasing block length and does not depend on the rate of characters transmission in the communication channel.

Received and analyzed characteristic of the ESR on the BER axis, which behaves similarly to the characteristic of the EBR, but increase significantly faster. It increases faster with increasing characters transmission rate in the communication channel and does not depend on the length of the block.

Received and analyzed characteristic of the SESR on the BER axis, which abruptly goes from zero to one in the range around the point at which the EBR becomes 0.3.

It grows faster with increasing block length. With the increasing transmission rate, the transition range from zero to one decreases.

From the point of view of harmonization of indicators of the classical noise immunity theory and standardized indicators of digital packet transmission of information an important conclusion is that for fixed values of digital flow rate and block size to all corresponding quality indicators of the digital packet stream corresponds one value of the probability of the continuous stream error.

A significant conclusion from the usage of the proposed model for the digital multiservice telecommunications networks design specialists is the fact that an increase in speed by 1 order requires, to meet the requirements of international standards, an increase in signal power at the receiving point by 2 orders.

References

1. ITU-T Recommendation G.826. End-to-end error performance parameters and objectives for international, constant bit-rate digital paths and connections, [S. l.], pp. 6–9 (2002). https://www.itu.int/rec/T-REC-G.826
2. Ilchenko, M., Uryvsky, L., Osypchuk, S.: The main directions of improving information and communication technologies in the global trends. In: Ilchenko, M., Uryvsky, L., Osypchuk, S. (eds.) Advances in Information and Communication Technology and Systems, MCT 2019. Lecture Notes in Networks and Systems, vol. 152, pp. 3–22 Springer, Cham (2019). https://doi.org/10.1007/978-3-030-58359-0_1
3. Ilchenko, M., Uryvsky, L., Moshynska, A.: Developing of Telecommunication Strategies Based on the Scenarios of the Information Community, pp. 905–913. Springer, New York (2017)
4. Strelkovskaya, I., Solovskaya, I., Makoganiuk, A.: Different extrapolation methods in Problems of Forecasting. In: Ilchenko, M., Uryvsky, L., Globa, L. (eds.) Advances in Information and Communication Technology and Systems, MCT 2019. Lecture Notes in Networks and Systems, vol 152, pp.217–228. Springer, Cham (2019). https://doi.org/10.1007/978-3-030-58359-0_12
5. Moshynska, A.: Resource estimation of data transmission digital channels in satellite communication systems. In: The Scientific Heritage, Budapest, Hungary, vol. 1, no. 33, pp. 50–54 (2019). https://www.tsh-journal.com/wp-content/uploads/2020/09/VOL-1-No-33-33-2019.pdf
6. Proakis, J., Manolakis, D.: Digital Signal Processing, p. 1084. Prentice-Hall (2006)
7. Uryvsky, L., Moshynska, A., Osypchuk, S.: Selection of signal-code sequences in IEEE 802.11 equipment. In: The Scientific Heritage, Budapest, Hungary, vol. 1, no. 17, pp. 60–63 (2017). https://tsh-journal.com/wp-content/uploads/2017/12/VOL-1-No-17-17-2017.pdf
8. Uryvsky, L., Moshynska, A., Osypchuk, S., Shmihel, B.: Comparison of methods for determining noise immunity indicators of a multiservice transmission system. In: Advances in Information and Communication Technologies. Lecture Notes in Electrical Engineering, vol. 560, pp. 167–185. Springer, Cham (2019). https://doi.org/10.1007/978-3-030-16770-7_8
9. Uryvsky, L., Moshynska, A., Osypchuk, S.: Analysis of digital signal quality indicators based on analytic and stochastic modeling. In: International Conference Radio Electronics & Info Communications (UkrMiCo), Ukraine, Odessa, pp. 1–4. IEEE Xplore Digital Library (2017). https://ieeexplore.ieee.org/document/8095414
10. Uryvsky, L., Moshynska, A., Osypchuk, S., Heorgin, D.: The simulation model for calculating the indicators of digital communication signals quality transmission. In: The Scientific

Heritage, Budapest, Hungary, vol. 1, no. 34, pp. 21–31 (2019). https://tsh-journal.com/wp-content/uploads/2019/05/VOL-1-No-34-34-2019.pdf
11. Uryvsky, L., Moshynska, A., Osypchuk, S., Pieshkin, A.: The comparative assessment of corrective parameters for antinoise convolutional and block codes. In: International Conference Radio Electronics & Info Communications (UkrMiCo), Ukraine, Odessa, pp. 1–4. IEEE Xplore Digital Library (2018). https://ieeexplore.ieee.org/xpl/conhome/9040460/proceeding. https://ieeexplore.ieee.org/document/9047530

Improving Efficiency of the Phase-Locked Loop for Reference Oscillator of the Multichannel System for Time Synchronization Signals Telemonitoring

V. V. Koval[1(✉)], V. P. Lysenko[1(✉)], D. O. Kalian[1(✉)], O. L. Osinskyi[2(✉)], and O. V. Samkov[3(✉)]

[1] National University of Life and Environmental Sciences of Ukraine, Heroiv Oborony Street 15, Kyiv 03041, Ukraine
{v.koval,lysenko}@nubip.edu.ua

[2] The National Academy of Sciences of Ukraine, Volodymyrska Street 54, Kyiv 01030, Ukraine
osinskiy@usnan.org.ua

[3] Institute of Electrodynamics of the National Academy of Sciences of Ukraine, Peremoha Avenue 56, Kyiv 03057, Ukraine
samkov@ied.org.ua

Abstract. A multichannel automated telemonitoring system for time synchronization signals is analyzed. It simultaneously measures parameters of several synchronization signals relative to the reference oscillator signal, processes the received data and transmits it via IP network to a computer of the centralized control system. In order to improve efficiency of a phase-locked loop, an optimization was performed in regards to the speed of the reference oscillator control process. Taking into account a presence of nonlinearities in the mathematical model of the phase-locked loop subsystem of the reference oscillator (PLLS), which are caused by periodicity of the control action in the function of phase coordinate and "saturation" limitations, a structure of the subsystem was synthesized in three steps: analysis, implementation, and simulation (using Simulink program of the MATLAB package). The structure of the optimal PLLS with adaptive properties was designed, in which the control signal is generated in accordance with the speed-optimal control strategy.

Simulation modeling of the synthesized PLLS circuit with unlimited bandwidth of the LPF and original adaptive digital phase discriminator, protected by patent of Ukraine for an invention, has been performed. The simulation results showed a decrease in the duration of the transient response from 2.7 to 4 times, depending on the initial conditions of the phase-locked loop of the reference oscillator.

Keywords: Automatic frequency control · Time synchronization signal · Telemonitoring · Optimization on response time · Maximum principle · Simulation · Adaptive digital phase discriminator

P. Vorobiyenko et al. (Eds.): IPF 2020, LNNS 212, pp. 60–79, 2021.
https://doi.org/10.1007/978-3-030-76343-5_4

1 Introduction

An effective functioning of the integrated power supply systems, built in accordance with the SMART-Grid concept, requires an automated control of their operating modes in real time [1–3]. According to the current regulatory documents, the systems of automated control of the technological processes of 220–750 kV substations of the UPS of Ukraine should have high accuracy (± 1 μs) time synchronization. Continuous monitoring of the parameters of power supply systems with reference to the time synchronization signals (TSS) ensures their high energy efficiency, reliability and survivability [1, 2].

To provide power supply objects with the TSS, different mechanisms are used to propagate synchronization signals and means to monitor their parameters [2]. An influence of the quality indicators of TSS on the monitoring results of the power supply systems' parameters is obvious, and necessitates a use of the measured means to monitor the quality of synchronization signals. An increase in the productivity and reliability of the monitoring process of the TSS is ensured by using a computer-integrated multichannel automated telemonitoring system (MATS) [2, 4–7], a block diagram of which is shown in Fig. 1.

The multi-channel automated TSS telemonitoring system is designed to measure the parameters of several synchronization signals relative to the signal of the reference oscillator and process the received data in real time in order to transmit it via IP network to the centralized system computer. The automated monitoring of TSS parameters is carried out on the basis of the deviation measurements of time interval of the controlled signal relative to the signal of the reference oscillator [2, 4–7]. A generator, which is used as a reference oscillator, is controlled by voltage, which is formed as a result of the operation of the phase-locked loop system (PLLS) [8–12]. As a part of MAST, the PLLS system carries out the functions of a subsystem which task is to form a highly stable reference digital signal with a standard value of the duration of a clock interval (period) in the autonomous mode or due to synchronization from the sources of sample oscillations [8–15].

One of the operation features of PLLS subsystem of the voltage-controlled reference oscillator is a presence of nonlinearities (for example, phase detectors with "saturation" effects, periodicity of the characteristic as a function of phase error) [1–3, 5–15].

Analysis of the requirements for the accuracy of reference signal formation makes it necessary to improve the operation quality of the voltage-controlled PPLLS oscillator. One of the requirements for the operating characteristics of the subsystem is a minimization of the transient processes' duration [8–13], which determines the relevance of the researches aimed at optimization on response time and increase of the automatic frequency control efficiency of the reference oscillator of multichannel telemonitoring system of TSS.

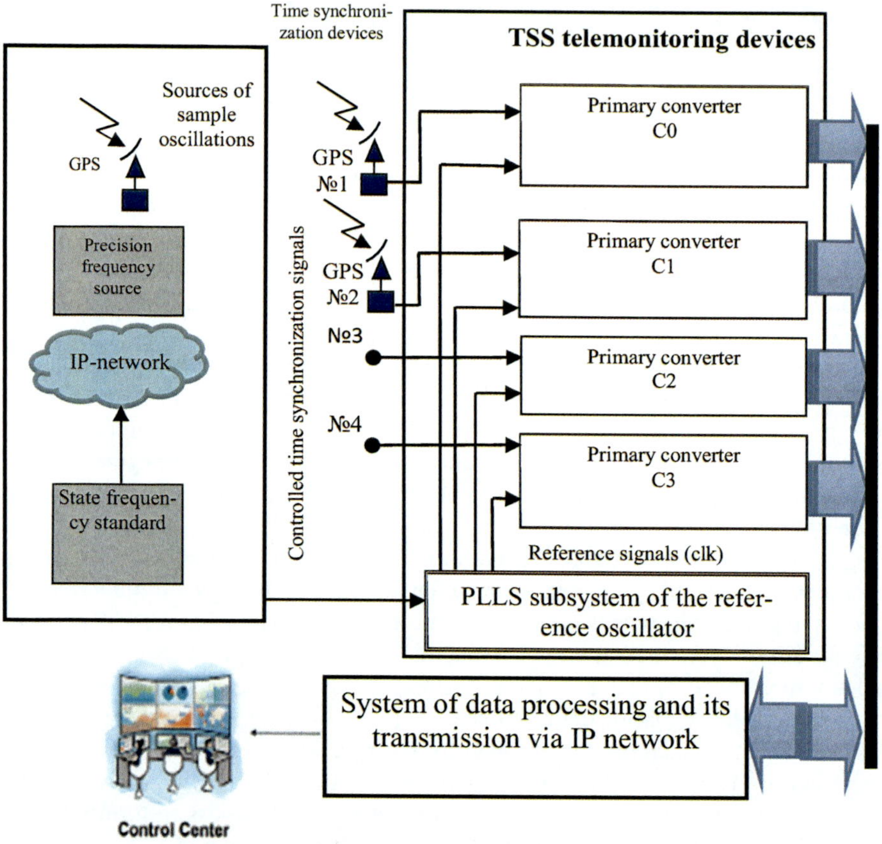

Fig. 1. Block diagram of a computer-integrated multi-channel automated system for telemonitoring the time synchronization signals

2 Optimization on Response Time of Automatic Frequency Control of the Reference Oscillator MAST of Time Synchronization Signals

We will carry out optimization on response time of the voltage-controlled reference oscillator PPLLS in order to improve the quality of reference signal and increase the efficiency of automatic frequency control of the reference oscillator of the multichannel telemonitoring system of TSS.

A diagram of the computer-integrated multichannel automated telemonitoring system for time synchronization, which is shown in Fig. 1, is an initial one for developing a functional diagram of PPLLS of the voltage-controlled reference oscillator.

In the process of developing a functional diagram of PPLLS, we will consider a source of sample signal (SSS) as an ideal one, that is, one which output signal has a stable frequency value and is without phase noise [12]. This approach can be implemented by using precision sources of time and frequency in the laboratory, which at the initial stage will simplify the optimization task and allow to investigate PPLLS, in which a low-pass filter has an infinite bandwidth. We will also assume that the disturbances acting on the subsystem are not significant, so we assume the absence of such an impact. A use of the 20-bit DAC and high-speed microcontroller allows to neglect digitization effects. Taking into account these conditions, a functional diagram of PPLLS of the reference voltage-controlled oscillator (RVCO), will look like the one presented in Fig. 2 [16].

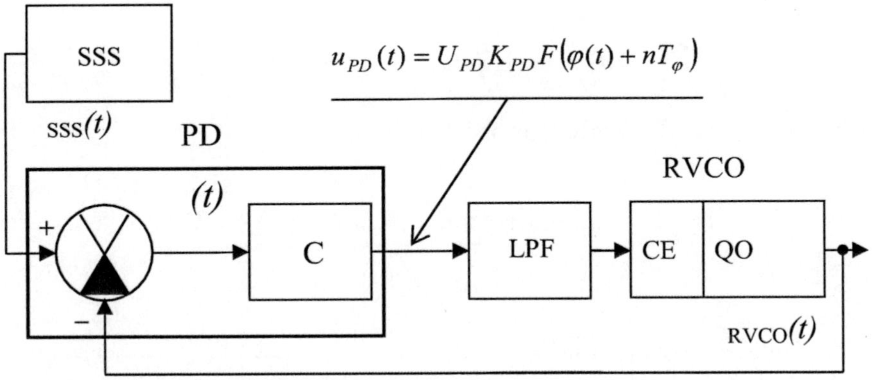

SSS – sample signal source;
PD – phase discriminator;
C – converter of phase difference (PLLS device error) into control action signal;
LPF – low pass filter (LPF);
CE – control element;
QO – voltage-controlled quartz oscillator;
RVCO – reference voltage controlled oscillator;
u_{PD} – PD output signal (control action of PLLS device);
$\varphi(t) = \varphi_{SSS}(t) - \varphi_{RVCO}(t)$ – PLLS device error;
${}_{SSS}(t)$ and ${}_{RVCO}(t)$ – the phases of SSS $u_{SSS}(t)$ and RVCO $u_{RVCO}(t)$.

Fig. 2. Functional diagram of PPLLS reference oscillator

In PPLLS of the voltage-controlled reference oscillator, a synchronizing signal fed to the first input of the phase discriminator (PD) (Fig. 2) and a signal generated by RVCO and fed to the second input of PD, are synchronized by controlling its frequency. A control action is formed on the basis of the phase difference determined by PD, due to the feedback channel (deviation control). With such a construction of the PLLS subsystem, an essence of RVCO control lies in the accurate automatic phase control of the signals arriving at the corresponding PD inputs, where they are compared and converted, in case of its implementation on digital logic microcircuits, into a digital code combination suitable for microcontroller processing. RVCO is controlled by a voltage generated by a digital-to-analog converter (DAC) based on a digital code combination received from the microcontroller.

A mathematical model of the PLLS subsystem of the oscillator is shown in Fig. 3 [16].

The phase discriminator, which performs a comparison function, is described by the algebraic equation:

$$\varphi(t) = \varphi_{LO}(t) - \varphi_{VCO}(t)$$

The specified phase difference (error of PPLLS of MATS) φ(t) generally is associated with a value of frequency difference ω(t) of RVCO signals $\omega_{\mathrm{RVCO}}(t)$ and SSS $\omega_{\mathrm{SSS}}(t)$ equation:

$$\varphi(t) = \varphi_0 + \int (\omega_{SSS} - \omega_{RVCO})dt,$$

where φ_0 – phase difference provided that $t = 0$.

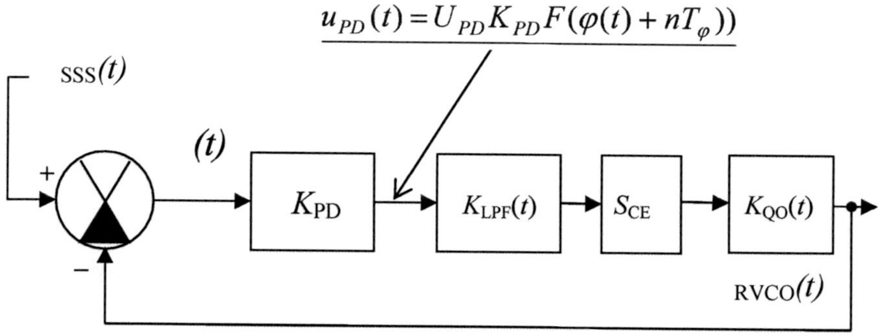

Fig. 3. Mathematical model of PPLLS reference oscillator

A mathematical model of the closed PLLS subsystem can be written as the following system of equations [16]:

$$\begin{cases} \frac{d\varphi(t)}{dt} = \Omega_{\mathrm{I}} - u_{\mathrm{PD}}(t)K_{LPF}(t)S_{CE}K_{QO}(t); \\ \varphi(t) = \varphi_{\mathrm{SSS}}(t) - \varphi_{\mathrm{RVCO}}(t); \\ u_{\mathrm{PD}}(t) = U_{\mathrm{PD}}K_{\mathrm{PD}}F(\varphi(t) + nT_{\varphi}); \\ \mathrm{U_{PDmin}} \le U_{PD} \le \mathrm{U}_{PD\mathrm{max}}; \\ -1 \le F(\varphi(t) + nT_{\varphi}) \le 1, \end{cases} \tag{1}$$

where $\varphi(t)$ – phase error of PPLLS;
t – time;
Ω_{I} – initial disorder of RVCO of PPLLS relative to SSS;
$u_{\mathrm{PD}}(t)$ – control action of PPLLS;
$K_{\mathrm{LPF}}(t)$ – transmission coefficient of the low-pass filter of PPLLS;
S_{CE} – slope of the characteristic of the control element of RVCO of PPLLS;
$K_{\mathrm{QO}}(t)$ – transmission ratio of the quartz oscillator of QO;
$\varphi_{SSS}(t)$ and $\varphi_{RVCO}(t)$ – phase of SSS and RVCO, respectively;
U_{PD} – the extreme value of the control signal of the PPLLS at the output of the PD;

K_{PD} – PD transmission ratio; $F(\varphi + \mathrm{nT}_{\varphi})$ – normalized periodic characteristic of the PD with a period coefficient n;

U_{PDmax} and U_{PDmin} – respectively, the maximum and minimum value of the signal at the output of the PD;

T_{φ} – period, periodic in the coordinate φ, discriminant characteristics of the PD of PPLLS.

Examples of a PD implementation with characteristic periodic in the coordinate φ are provided in publications [6, 8–13, 16, 17]. A mathematical model of PD with a rectangular normalized static discriminant characteristic periodic in the coordinate φ can be written as:

$$F(\varphi + nT_{\varphi}) = sign\ \sin(\varphi + nT_{\varphi})\forall\ \mathrm{n} = \overline{-\infty,\ +\infty} \tag{2}$$

The peculiarity of this work is the research and optimization on response time PLLS MAST subsystems, which take into account the control actions with periodicity caused by a periodicity of the discriminant characteristics of PD by phase coordinate. It should also be noted that a feature of the mathematical model of type (1) is a presence of two control components (φ and Tφ) in the automatic frequency control subsystem of RVCO MAST.

It is proposed to study the PLLS MAST subsystem in three stages. At the first stage, we focus only on an analytical research. At the second stage, we perform a synthesis of the optimal PLLS subsystem of the RVCO of MAST to define the requirements for its structure and make suggestions about its practical implementation. At the last third stage, we perform simulation by using Simulink software of the MATLAB package for experimental confirmation of the scientific results obtained at the previous research stages.

In addition to the specified periodicity along the coordinate φ it is necessary to take into account a real limitation on the extreme values of the control action of the "saturation" type:

$$U_{PD\,\min} \le u_{PD}(t) \le U_{PD\,\max}. \tag{3}$$

Thus, in the real PLLS device of MAST, in addition to time constraints on the control action interval, there is a limitation on the amplitude (maximum values) of the specified control action.

We will write down a mathematical model for the PLLS device of MAST and formulate a problem of an optimization on response time. In MAST, for the case of direct connection to a reference signal source, a first-order auto-tuning subsystem can be used, which will have a low-pass filter (LPF) with an infinite bandwidth, that is $K_{LPF}(p) = 1$. Taking into account LPF parameter with infinite bandwidth and the values of transfer coefficient of PD - $K_{PD} = 1$, and transfer coefficient of QO $K_{QO} = 1$, a mathematical model (1) will take the form:

$$\begin{cases} \frac{d\varphi(t)}{dt} = \Omega_{\mathrm{I}} - u_{\mathrm{PD}}(t) S_{\mathrm{CE}}; \\ \varphi(t) = \varphi_{\mathrm{SSS}}(t) - \varphi_{\mathrm{RVCO}}(t); \\ u_{\mathrm{PD}}(t) = U_{\mathrm{PD}} F\big(\varphi(t) + n\mathrm{T}_{\varphi}\big); \\ \mathrm{U}_{\mathrm{PDmin}} \le U_{PD} \le U_{PD\max}; \\ -1 \le F\big(\varphi(t) + n\mathrm{T}_{\varphi}\big) \le 1. \end{cases} \tag{4}$$

We formulate a problem of optimal on response time control of PLLS device of MAST, provided that the control action on the reference oscillator of the subsystem is a signal at the PD output $u_{\mathrm{PD}}(\varphi, t) = U_{\mathrm{PD}} F(\varphi(t) + n\mathrm{T}\varphi)$, which is formed on the basis of phase difference $\varphi(t)$.

It is logical that at the time $t = t_0$ (a beginning of the auto-tuning process), PLLS device of MAST is in a state determined by the initial conditions (4). It is necessary for the PLLS device of MAST to reach a state of stable equilibrium at t_C time at an admissible value of the signal of periodic control action $u_{\mathrm{PD}}(\varphi, t)$, subject to the condition of minimizing the integral optimality criterion, which is given in the form of the following functional

$$\Theta = \int_{t_0}^{t_k} 1 \cdot dt \Rightarrow \min. \tag{5}$$

First, we will solve the problem of optimal on response time control of PLLS device of MAST under the following conditions. At $t = t_0$ time (initial position), PLLS device of MAST is in a state determined by the initial conditions: $\varphi(t_0) = \varphi_0$, $\Omega_{\mathrm{I}}(t_0) = 0$. It is necessary to find such a control action $u_{\mathrm{PD}}(\varphi, t)$, which ensures the transition of PLLSD of MAST in the minimum time, that is, in the minimum value of the integral optimality criterion, from the initial state to the final position, which is specified by the following conditions $\varphi(t_{\mathrm{C}}) = 0$, $\Omega_{\mathrm{I}}(t_{\mathrm{C}}) = 0$.

To determine the optimal on response time control law in the studied PLLSD of MAST, we use the analytical results given in [18]. In particular, based on the methodology for applying the maximum principle developed by acad. L.S. Pontryagin [19–21], we perform optimization on response time of the PLLS of oscillator controlled by the MAST voltage. To do this, we compose the Hamilton function according to the first equation of system (4), which is a mathematical model of the PLLSD of MAST:

$$H = \Psi_1[\Omega_I - u_{PD}(t) S_{CE}] - 1, \tag{6}$$

and also the differential equation for the auxiliary variable ψ_1:

$$\frac{d\Psi_1}{dt} = \frac{dH}{d\varphi} = 0. \tag{7}$$

Solving the differential Eq. (7), we obtain an equation for the auxiliary variable Ψ_1:

$$\Psi_1 = const. \tag{8}$$

Then, it is necessary to find such a control law, namely, such a change in the value of $u_{PD}(t)$ voltage at the output of the PD PLLSD MAST, which gives as a result of a joint solution of Eqs. (6) and (7), a variable $\varphi(t)$, for which the value of H will be maximum in all its points. In this case, the Hamilton's function H in Eq. (6) should be the maximum of the control action, which is the voltage at the output of the PD PLLSD MAST. The found control law $u_{PD}(t)$ and the corresponding variable $\varphi(t)$ provide an optimal on response time mode in PLLSD MAST.

For the practical implementation of the obtained law in PLLSD MAST, we will first take into account only the actual existing restrictions on the amplitude of the control action of $u_{PD}(t)$ of the "saturation" type. Then, to ensure the maximum of the Hamilton's function (6) on $u_{PD}(t)$, it is necessary that the voltage at the PD output is determined from the equation [18]:

$$u_{PD}(t) = U_{PDextr} sign(S_{CE}\Psi_1), \tag{9}$$

where U_{PDextr} is equal to a limit value of the inequality in the system of Eqs. (4).

If we assume a symmetry of the discriminant characteristic of the PD relative to the abscissa axis, then the following relations are valid:

$$|U_{\mathrm{PDmax}}| = |U_{\mathrm{PDmin}}|,\ U_{\mathrm{PDextr}} = U_{\mathrm{PDmax}}, \tag{10}$$

that should be taken into account in the formula (9).

In accordance with the obtained optimal control law $u_{PD}(t)$ in the MAST subsystem, taking into account the second equation of the system of Eqs. (2), a formula for the normalized discriminant characteristic of PD PLLSD MAST is determined:

$$F(\varphi(t) + nT_{\varphi}) = sign(S_{CE} \cdot \Psi_1). \tag{11}$$

If we use the results of the analytical studies provided in [18], we can write the following analytical expression for the control signal of the optimal PLLSD MAST:

$$u_{PD}(t) = U_{extr} sign\, \varphi_0, \tag{12}$$

where φ_0 – an initial value of the phase error $\varphi(t)$. You can also determine a type of the normalized discriminant characteristics of PD:

$$F\big(\varphi(t) + n_{\varphi}\big) = sign\, \varphi_0 \tag{13}$$

and formulate requirements for its characteristics. In particular, the PD discriminant characteristic should be alternating, moreover, the discriminant characteristic of the PD is constant over the time interval of optimal transient processes, while the control action $u_{PD}(t)$ is extreme, and its sign is determined by a sign of the phase difference $\varphi(t) = \varphi_{SSS}(t) - \varphi_{RVCO}(t)$.

According to the optimal control law in PLLSD MAST obtained in mathematical form (9), we will formulate the requirements for its elements with the aim of practical implementation. Namely, the requirements for PD. First, a signal at the PD output for the time interval of the transient process, during which a phase difference in the PLLSD MAST is processed, should have a maximum $u_{PD}(t) = +\,U_{PDextr}$, or minimum $u_{PD}(t)$

$= - U_{PDextr}$ value, depending on the initial state of $\varphi(t_0) = \varphi_0$. By the theorem on *n*-intervals of the optimal on response time control law for the first-order PLLSD MAST, we have one interval of motion [22, 23]. A consequence is that the normalized discriminant characteristic of the PD $F(\varphi(t),\ \mathrm{nT}_\varphi),\ \forall\, \mathrm{n} = -\infty, +\infty$ should be a non-periodic function of time in the time interval of the transient process (in dynamics). Accordingly, the control signal $u_{PD}(t)$ should have a constant value equal to the maximum amplitude $u_{PD}(t) = +U_{PDextr}$ or minimum amplitude $u_{PD}(t) = -U_{PDextr}$, depending on the initial state $\varphi(t_0) = \varphi_0$.

Therefore, to implement an optimal on response time transient processes in the subsystem of phase auto-tuning MAST, it is necessary to develop a PD with a controlled form of normalized discriminant characteristics [17], namely:

$$F(\varphi(t), \mathrm{nT}_\varphi, \sigma),\ \forall\, \mathrm{n} = -\infty, +\infty, \tag{14}$$

where σ is a PD shape control signal.

The signal σ provides a change in the form of the discriminant characteristic of the PD in PLLS MAST on the time interval of the optimal on response time transient process Θ_{opt}. It should be noted that in statics, the discriminant characteristic of PD $F\big(\varphi(t), \mathrm{nT}_\varphi, \sigma\big),\ \forall\, \mathrm{n} = -\infty, +\infty$ PLLSD MAST may not change and can be periodic in the coordinate φ, with the period T_φ.

3 Development of a structure of the optimal on response time subsystem of the auto-tuning frequency of the reference oscillator MAST with adaptive digital phase discriminator

At the second stage, which is implementation, a structure and appropriate schemes were developed for the construction of adaptive digital phase discriminator (ADPD) [24], which will ensure a formation of discriminant characteristics of PD of a given type, essentially its adaptation, which should be carried out by the control channel (third input of the PD) to form an optimal control action of PLLSD MAST.

Figure 4 shows a generalized functional diagram of the optimal on response time subsystem of PLLS MAST, which has two hierarchical levels of the control and ADPD.

An additional ADPD input receives a control signal $\sigma(t)$ from the control unit, which changes a shape of the discriminant characteristic of the PD on a time interval of the optimal transient process depending on the magnitude and sign of the phase (in the general case, frequency) error of the PLLS MAST, which is optimal on response time. A control channel of the PD provides at the Θ_{opt} the fastest transition of a subsystem to the position of a stable equilibrium.

To ensure practical implementation of discriminant characteristics of PD of a given type, it is advisable to use modern technologies for digital signal processing. This approach allows to present ADPD, which will have digital inputs and outputs in the form of a generalized device diagram, shown in Fig. 5. To ensure an expansion of the functionality to control discriminant characteristics of the PD, a presence of *k*-inputs of the control channel is provided, which can receive data signals, presented in the form of the appropriate digital code combinations [17, 24].

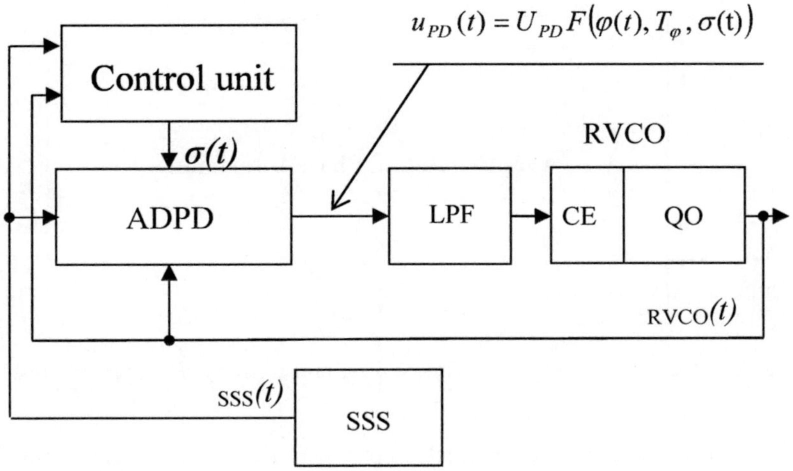

Fig. 4. Functional diagram of the optimal on response time subsystem PLLS MAST with two hierarchical control levels

According to the block diagram shown in Fig. 5, a schematic diagram of ADPD MAST in the Quartus Prime Lite Edition software environment was developed by using modern digital programmable logic schemes, for which an invention patent of Ukraine №113473 was obtained [24].

ADPD provides adaptively controlled formation of a code combination about the value of the phase shift angle (time difference) of two mutually independent periodic pulse sequences and increase of the reliability of the measurement results in digital form [17, 24]. Adaptive digital phase discriminator contains an input to which a signal of the first input pulse sequence (In.1) and an input to which a signal of the second input pulse sequence (In.2), as well as K-inputs of the presetting installation of the pulse counter (numbered *1*, *2…k*), that are the K digital inputs of the preliminary setting of the code (optimal switching $\sigma(t)$) of the device and *n*-outputs, that is, digital outputs of the device, on which a code combination is formed that corresponds to the time inconsistency between the first and the second input pulse sequences [17, 24].

Let us consider the basic principles of ADPD as a part of the PLLS subsystem of MAST (Fig. 4). The first input pulse sequence is fed to the In.1 input, and the second one - to the In.2 input of ADPD (Fig. 5). A code combination that will be formed at the outputs *1…n* of ADPD, corresponds to the value of the phase shift between the input pulse sequences and is a digital representation of the measured phase difference [24].

The K-inputs (numbered *1, 2 … k*), which are K digital inputs of the preset code setting, are supplied with a digital code combination that corresponds to the required discrete value of the initial state of the ADPD of PLLSD. The obtained possibility of changing an initial state of the ADPD allows to provide adaptive controlled formation of a code combination about the value of the phase shift angle of two pulse sequences. The digital code combination at the ADPD output after digital processing in the micro-controller and digital-to-analog conversion forms a value of the voltage supplied to the input of the RVCO of PLLSD.

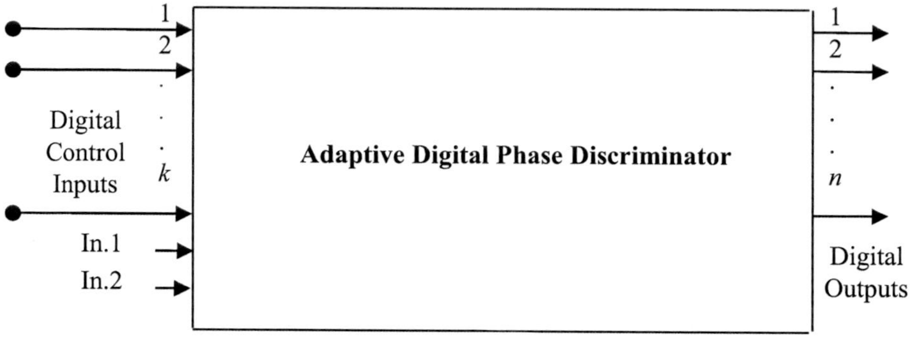

Fig. 5. Block diagram of an adaptive digital phase discriminator. (Reproduced from Kalian. 2018)

A technical solution of the invention [24] provides adaptive controlled formation of the code combination and increases reliability of the digital presentation of the results of phase difference measurements.

The use of ADPD in PLLSD MAST provides control of the form of PD characteristic on the time interval of optimal transient process Θ_{opt} depending on the value and sign of the phase (frequency) error. A control channel of the PD provides the fastest movement of a reflecting point to a stable equilibrium position and, as a result, an optimal on response time control of the RVCO of PLLSD MAST is realized.

4 Experimental studies of the optimal on response time subsystem of frequency auto-tuning of the reference oscillator MAST using simulation

To confirm the obtained analytical and implementation results, a simulation of the created scheme of PLLSD of RVCO with ADPD on a personal computer was performed. For comparison, Fig. 6 presents simulation models of the created scheme of PLLSD with ADPD (which is optimized) and also of a not optimized scheme (PLLSD with PD), which we will call traditional [16]. Simulation schemes were developed by using the Simulink/MATLAB software package [17]. In the Fig. 6 the reference voltage-controlled oscillator PLLSD (RVCO) is presented in the form of an integrator with the inscription “DCG”; PD is presented with the inscription “PD”; an adaptive digital phase discriminator with digital-to-analog converter – “ADPD”, microcontroller – “Control unit”.

Figure 7 shows oscillograms of a transient process, which are obtained at the output of the optimal on response time (scheme above) and non-optimized PLLSD MAST (lower scheme).

Figure 7 allows us to make the following conclusions. In the traditional subsystem (Fig. 7, bottom diagram), a transient process has a shape of a curve that goes to the point of a stable equilibrium. Thus, in the traditional PLLSD MAST, a phase dislocation of the subsystem decreases asymptotically to zero, while duration of the transient process will be $\Theta \approx 4$ s (Fig. 7, traditional) [17]. In the optimal on response time PLLS MAST

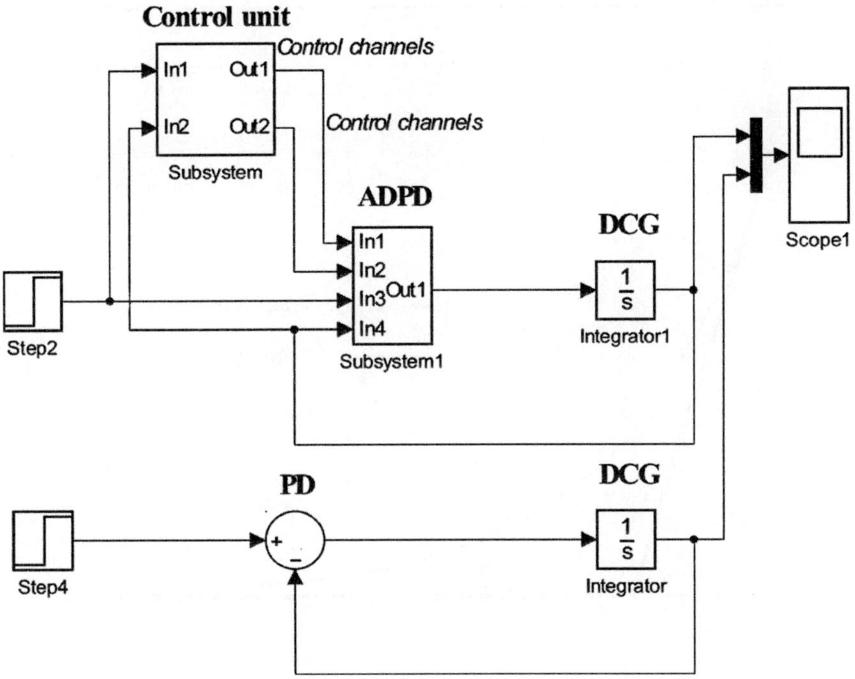

Fig. 6. Simulation models of optimal on response time and traditional PLLSD of the reference oscillator MAST. (Reproduced from Kalian. 2018)

(Fig. 7, top diagram), a transient process has a form of a line connecting the initial point with the point of stable equilibrium, which corresponds to a zero value of the phase disorder. In the optimal on response time PLLSD MAST, a duration of the transient process will be $\Theta = 1$ s [17].

According to the simulation results by using Simulink / MATLAB software package, it becomes obvious that in the optimal on response time PLLSD the duration of the transient process is shorter than in the traditional scheme. The transient process has a form of a straight line, while its duration is minimal, PLLSD is optimal on response time [17] and, as a result, the efficiency of the frequency auto-tuning of RVCO MAST increases.

Thus, the use of ADPD in MAST provides an ability to control the form of discriminant characteristics of the PD, that is, adaptation of the subsystem structure to the time interval of the transient process depending on a sign of the phases' difference. In its turn, the ADPD control channel provides the fastest transition of the reflecting point to the position of stable equilibrium and, as a result, the optimal on response time control is realized in the subsystem PLLSD MAST.

Further experimental studies of the developed optimal on response time subsystem of the frequency auto-tuning of the reference oscillator MAST will be performed in the Simulink package MATLAB software environment by using a method of comparative analysis and simulation.

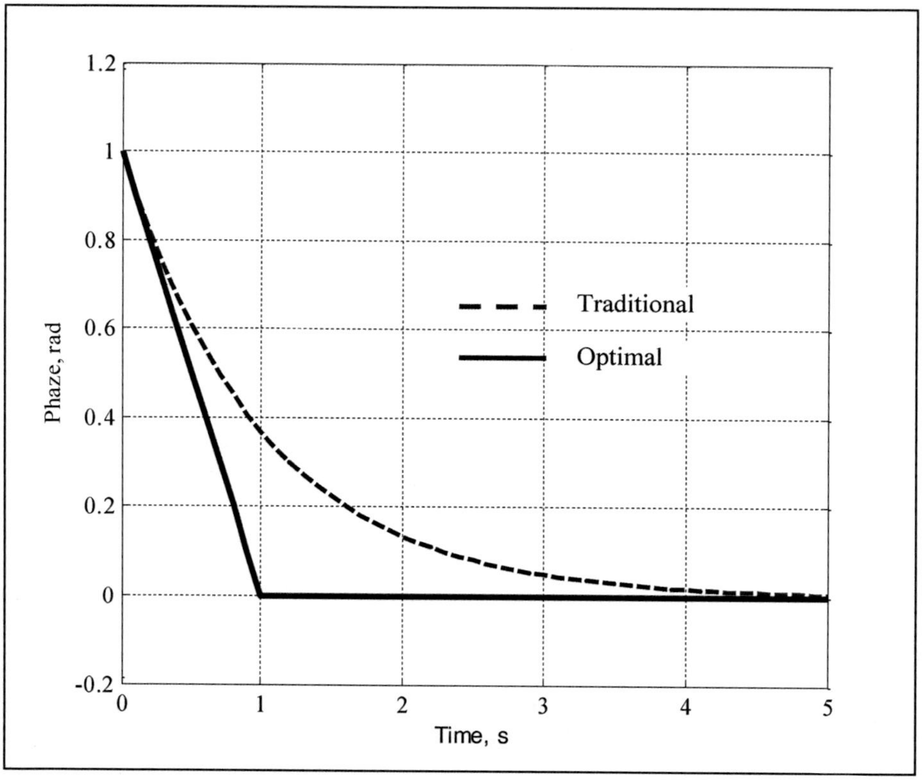

Fig. 7. Oscillograms of the signals at the outputs of RVCO of PLLSD MAST (initial value of phase error $\varphi_0 = 1$). (Reproduced from Kalian. 2018)

For comparison, we will investigate two simulation models presented in Fig. 8. The upper diagram shows a developed scheme of the frequency auto-tuning of the reference oscillator MAST, which has two hierarchical control levels and ADPD. The lower diagram (Fig. 8) shows a classical (non-optimized) scheme in which static characteristic of the PD has a sinusoidal shape.

Figure 8 shows the phase discriminator PD presented in the form of three blocks: comparison, amplifier "Gain1" and block of trigonometric functions "Trigonometric Function", which is configured to implement sinusoidal dependence of its output value on the input. RVCO "DCG" of the optimized scheme of PLLSD with ADPD in Fig. 8 is implemented using the "Integrator1" unit, and in the scheme PLLSD with PD, which has a static characteristic of a sinusoidal shape, a similar oscillator is implemented using the "Integrator" unit.

To study the PLLS subsystem in the space of states on the phase plane, the differentiation units "Derivative" and graph builders "XY Graph" are introduced into the scheme. The differentiation units "Derivative" and "Derivative1" provide formation of a signal corresponding to the derivative of the phase $\varphi_{RVCO}(t)$ of the oscillator signal which is controlled by the voltage "DCG". On a graph, displayed in a window of the

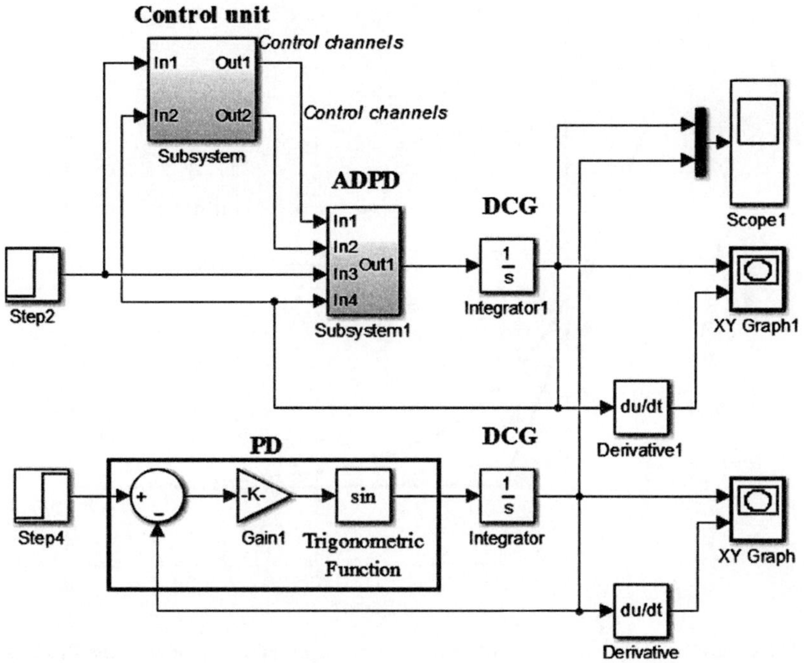

Fig. 8. Simulation models of the optimal on response time PLLSD of the reference oscillator MAST and PLLS system with PD, which has a static characteristic of sinusoidal shape. (Reproduced from Kalian. 2020)

graph builder "XY Graph", the X-axis displays the values of the phase error of the autotuning subsystem RVCO - $\varphi_{RVCO}(t)$, and the Y-axis displays the values of its derivative, that is, $\omega_{RVCO}(t)$. The time analysis of the signals is carried out by using the oscilloscope "Scope1", in the window of which the oscillograms of the signals from oscillators controlled by voltage "DCG" are displayed in one time frame.

The results of experimental studies are shown in Fig. 9 and Fig. 10 for the case when the initial value of the phase error is $\varphi_0 = 3$, and in Fig. 11 and Fig. 12 for the case when the initial value of the phase error is $\varphi_0 = -3$.

In order to summarize the results of experimental studies of PLLSD of the reference oscillator of MAST, let us perform an analysis and determine the efficiency of the optimization on response time process.

As initial data, we will use the indicator of transient process duration in PLLSD. The transient process duration in the non-optimized PLLSD, which will be taken as a base, is denoted by Θ, and Θ_{opt}, - in the subsystem, which is optimal on response time.

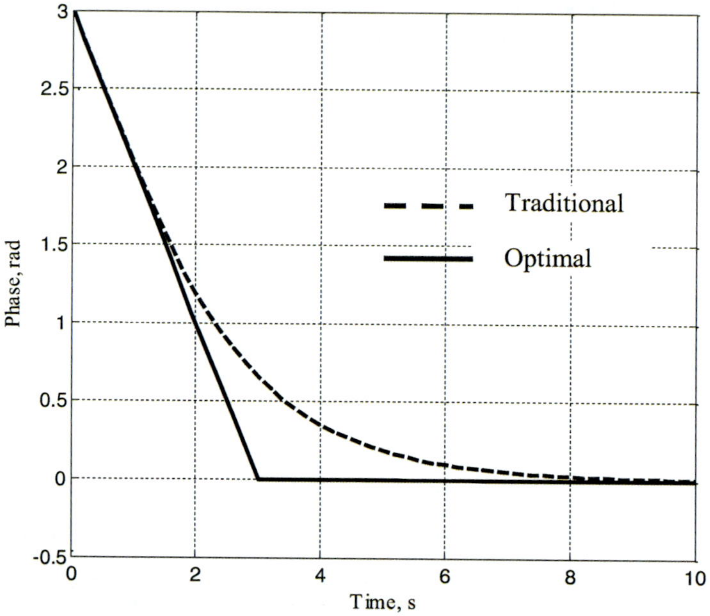

Fig. 9. Oscillograms of the signals at the outputs of the optimal on response time PLLSD of the reference oscillator MAST and non-optimal PLLS system (the initial value of phase error $\varphi_0 = 3$). (Reproduced from Kalian. 2020)

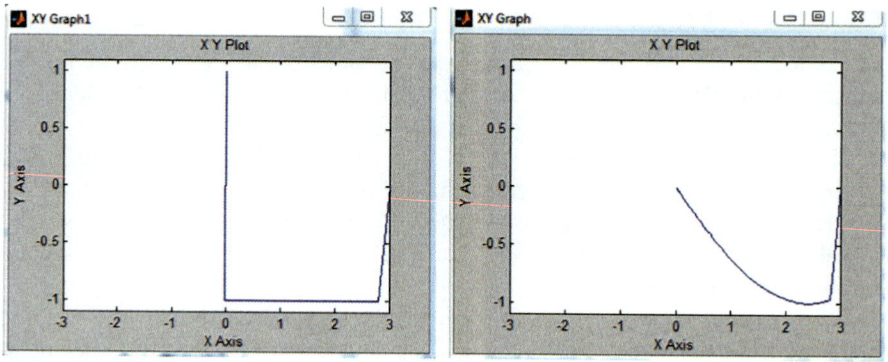

Fig. 10. Phase portraits of the optimal on response time PLLSD of the reference oscillator of MAST and the non-optimal PLLS system by the initial value of the phase error $\varphi_0 = 3$ (graph on the left is optimal on response time PLLSD, and on the right - not optimized). (Reproduced from Kalian. 2020)

Let us apply the method of graphical-analytical analysis to the results of schemes' simulation of the studied PLLS, determined from the graphs of oscillograms, at the initial values of the phase error $\varphi_0 = -3$ and $\varphi_0 = 3$, shown respectively in Fig. 9 and Fig. 11, as well as at the initial values set within the phase error $-3 < \varphi_0 < 3$.

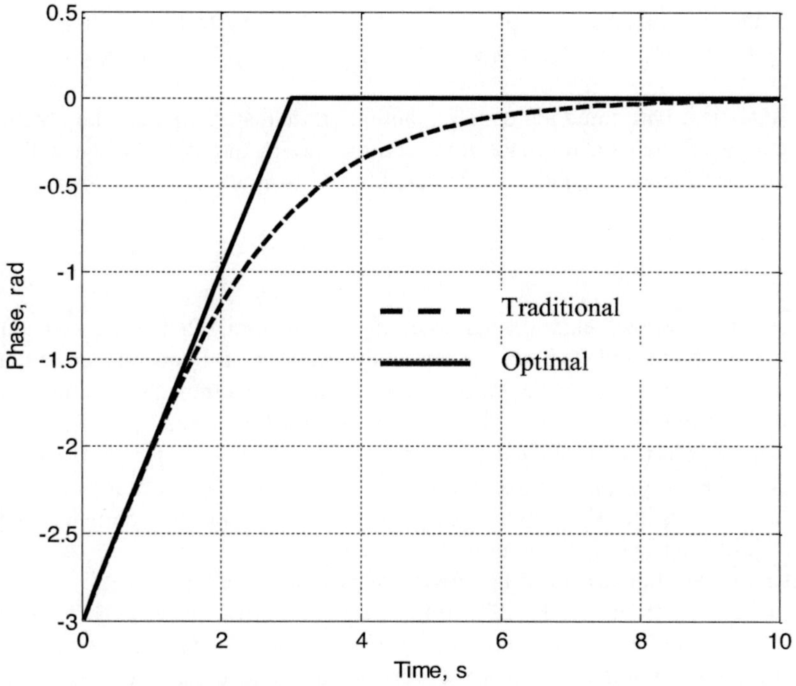

Fig. 11. Oscillograms of the signals at the outputs of the optimal on response time PLLSD of the reference oscillator of MAST and non-optimal PLLS system (the initial value of phase error is $\varphi_0 = -3$). (Reproduced from Kalian. 2020)

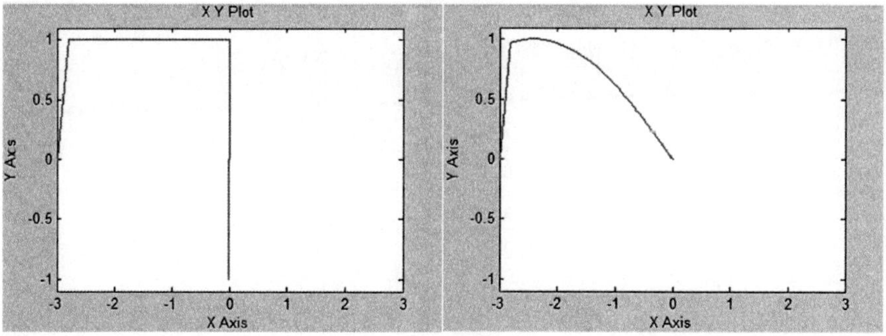

Fig. 12. Phase portraits of the optimal on response time PLLSD of the reference oscillator of MAST and the non-optimal PLLS system by the initial value of the phase error $\varphi_0 = -3$ (graph on the left - optimal on response time PLLSD, graph on the right - not optimized). (Reproduced from Kalian. 2020)

Based on the graphical analysis results, we will compile a table (Table 1), in which we will enter the determined values of the transient process duration Θ_T and Θ_{opt} taken from the graphs of oscillogram signals at the RVCO outputs. Despite the asymptotic nature of the transient process in the non-optimized PLLS system, we will determine its duration when the transient curve reaches 5% of the constant value of the phase error (φ = 0).

For analytical determination of the quantitative indicators that can be obtained by implementing optimal on response time control law in the PLLSD, we will use an optimization efficiency coefficient, calculated by the formula:

$$K_{opt} = \frac{\Theta_T}{\Theta_{opt}}. \tag{15}$$

In the Table 1 we will enter the values of the optimization efficiency coefficient K_{opt} determined by the formula (15) and construct the graphs of the dependence of transient process duration (Fig. 13) and the graph of the optimization efficiency coefficient on the values of the initial phase error φ_0 (Fig. 14) by means of the Excel spreadsheet.

From the graph analysis of the dependence of transient process duration on the values of the initial phase error φ_0 (Fig. 13), it can be concluded that the duration of the transient process in the investigated optimal on response time PLLSD of the reference oscillator of MAST is shorter than in the not optimized scheme.

A quantitative comparison of the subsystem parameters of the frequency auto-tuning of the reference oscillator of MAST, in the case of optimal on response time control law implementation, based on the optimization efficiency coefficient of K_{opt} (Fig. 14), leads to the conclusion that the transient process duration decreases from 2.7 to 4 times depending on the value of initial conditions.

Table 1. Dependences of the transient process duration of RVCO of PLLS of MAST and the optimization efficiency coefficient on different values of the initial conditions

φ_0, rad	Θ_T, s	Θ_{opt}, s	K_{opt}
−3,0	8,0	3,0	2,7
−2,0	7,0	2,0	3,5
−1,0	4,0	1,0	4,0
1,0	4,0	1,0	4,0
2,0	7,0	2,0	3,5
3,0	8,0	3,0	2,7

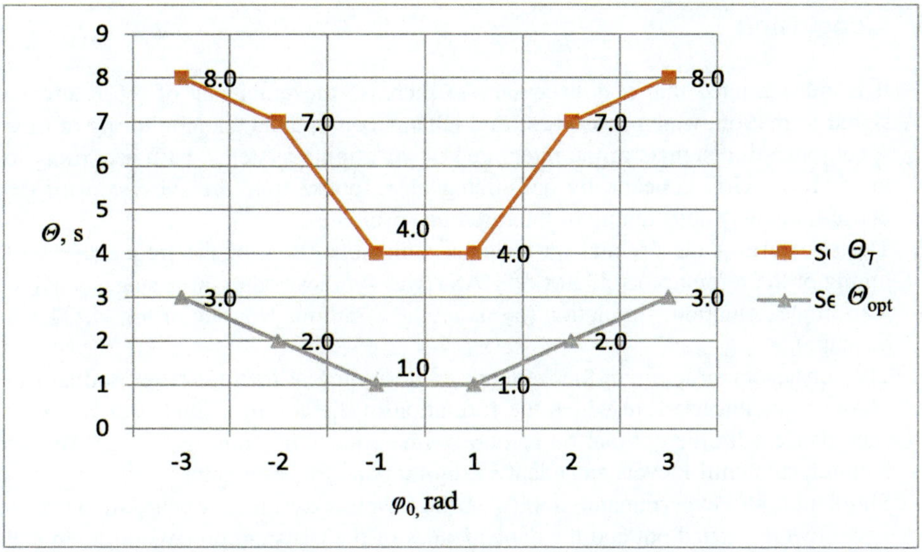

Fig. 13. Graphs of the dependence of the transient process duration of RVCO of PLLS of MAST on the values of the initial phase error $\varphi_{0.}$ (Reproduced from Kalian. 2020)

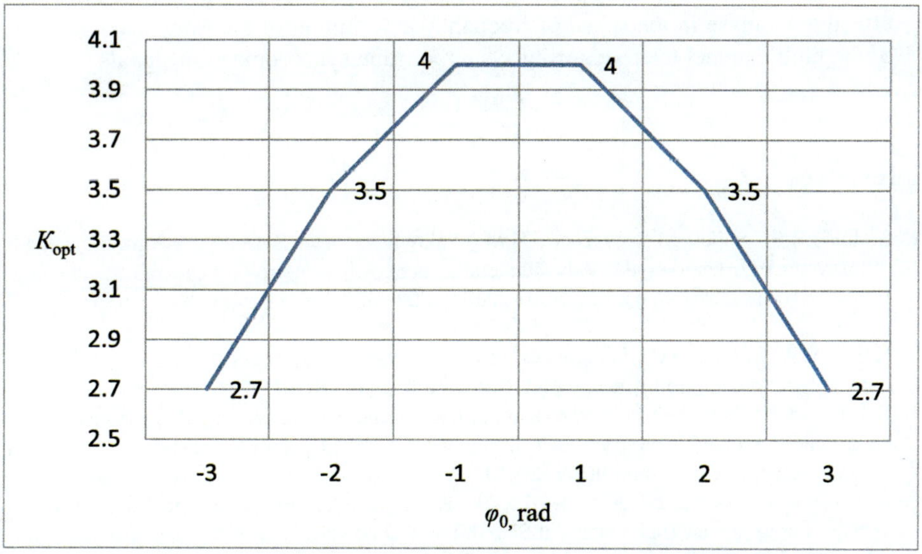

Fig. 14. Dependence of the implementation efficiency in the PLLS MAST subsystem of the optimal on response time control law on the values of the initial conditions. (Reproduced from Kalian. 2020)

5 Conclusion

1. It is substantiated that it is necessary to increase the efficiency of the reference signal formation, which is used for multichannel automated telemonitoring of time synchronization signals of the integrated power supply systems, built according to the SMART-Grid concept, by optimizing on response time the subsystem of the phase frequency auto-tuning of the reference oscillator.
2. The structure of the optimal on response time subsystem of the frequency auto-tuning of the reference oscillator of MAST was synthesized in three stages: analytical; implementation; simulation (by using the Simulink program of the MATLAB package).
3. The subsystem structure of the frequency auto-tuning of the reference oscillator of MAST is synthesized, in which the formation of the control signal is provided in accordance with the optimal on response time control law through the use of two hierarchical control levels and adaptive digital phase discriminator.
4. Simulation and experimental studies of the developed scheme of the MAST subsystem were carried out and the dependences of the transient process duration and the optimization efficiency coefficient on the values of the initial phase error were obtained.
5. The simulation results clearly confirm the obtained analytical results on the possibility of implementing the optimal on response time control law, and demonstrate effective increase in the speed of frequency auto-tuning of the reference oscillator of the multichannel telemonitoring system of time synchronization signals.

References

1. Stohnii, B.S., Sopel, M.F., Varskyi, H.M., Yakovlieva, I.V.: Systemy synkhronizovanykh vymiriv v elektroenerhetytsi. Pidvyshchennia tochnosti ta metrolohichne zabezpechennia. Pratsi Instytutu elektrodynamiky Natsionalnoi akademii nauk Ukrainy, Vyp. 35, pp. 37–47 (2013)
2. Koval, V.V., Kalian, D.O., Samkov, O.V.: Avtomatyzovana systema peredachi synkhrosyhnaliv z vykorystanniam IP-merezh. NUBiP Ukrain, Kyiv (2016)
3. DSTU: IEC/IEEE 61850-9-3:2018: Komunikatsiini merezhi ta systemy dlia avtomatyzatsii elektroenerhetychnykh pidpryiemstv. Chastyna 9-3. Profil protokolu tochnoho chasu dlia avtomatyzatsii enerhosystem. Kyiv (2018)
4. Kalian, D., Kazakova, N., Kravchenko, B., Koval, V.: Automated system for monitoring synchronizing precise time signals at SMART-GRID power plants. In: Projectowanie, badania, i eksploatacja. Tom 1, pp. 155–160. Wydawnictwo naukowe Akademii techniczno-humanistycznej w Bielsku-Bialey. Bielsku-Bialey (2019)
5. Koval, V., Lysenko, V., Osinskiy, O., Samkov, O., Khudyntsev, M.: Infocommunication technologies and networks for multichannel monitoring of synchronization signals of SMART Grid and microgrid electrical systems. In. International Scientific-Practical Conference «Problems of Infocommunications. Science and Technology PICS&T-2019, pp. 153–156. Borys Grinchenko Kyiv University, Kyiv (2019)

6. Koval, V., Kazakova, N., Samkov, O., Osinskyi, O., Kalian, D., Samoilenko, V.: Multi-channel digital discriminator of automated monitoring system of smart-grid power networks time synchronization tools. In. X International Conference of Students, PhD Students and Yjung Scientists "Engineer of XXI Century". pp. 135–144. Bielsku-Bialey (2020)
7. Koval, V., Samoilenko, V., Osinskyi, O., Samkov, A.: Bahatokanalna avtomatyzovana systema telemonitorynhu syhnaliv synkhronizatsii chasu intehrovanykh system elektropostachannia. In: Materialy desiatoi mizhnarodnoi nak.-prakt. konf, «Infokomunikatsii – suchasnist ta maibutnie», pp. 208–210. ONAZ, Odesa (2020)
8. Koval, V.V., Kalian, D.O., Kilchytskyi, Y.V., et al.: Avtomatychni prystroi ta systemy taktovoi synkhronizatsii infokomunikatsiinykh merezh. NUBiP Ukrain, Kyiv (2015)
9. Koval, V.V., Samkov, O.V., Khudyntsev, M.M., Kalian, D.O.: Avtomatyzovana systema synkhronizatsii tsyfrovykh syhnaliv. TOV TsP «Komprynt», Kyiv (2018)
10. Borshch, V.I., Haidar, V.P., Koval, V.V., Lesovoi Y.P.: Taktovaia synkhronyzatsyia v yntehralnыkh tsyfrovыkh setiakh эlektrosviazy. "Naukova dumka", Kyiv (1998)
11. Borshch, V.I., Donets. V.A., Koval, V.V., Mykhailov, N.K., Tumanov Yu.H.: Yerarkhycheskye formyrovately setky chastot setei telekommunykatsyi. "Naukova dumka", Kyiv (2001)
12. Shahgil'dyan, V.V., Lyahovkin, A.A.: Sistemy fazovoj sinhronizacii s elementami diskretizacii. Radio i svyaz', Moscow (1989)
13. Koval, V.V.: Vedeni prystroi synkhronizatsii z periodychnym avtopidstroiuvanniam telekomunikatsiinykh merezh: monohrafiia. TsP "KOMPRYNT", Kyiv (2017)
14. Steklov, V.K., Rudenko, A.A., Yudyn A.K.: Kombinirovannye sistemy fazovoj avtopodstrojki. "Tekhnika", Kyiv (2004)
15. Kalian, D.O.: Automation of the process of monitoring quality parameters of digital signal synchronization devices of flexible manufacturing system. Dissertation, National University of Life and Environmental Sciences of Ukraine (2020)
16. Kalian, D., Koval, V., Maksymenko, Y., Samkov, O., Dubovych-Kostetskyi, V.: Optymizatsiia za shvydkodiieiu avtopidstroiuvannia chastoty kvartsovoho heneratora systemy avtomatyzovanoho kontroliu synkhrosyhnaliv. Elektronne vydannia «Enerhetyka i avtomatyka» (2018). https://journals.nubip.edu.ua/index.php/Energiya/article/view/10897/9540 (2018). Accessed 25 Jan 2019
17. Kalian, D.: Performance-optimal phase-locked system for device of synchronization and formation of the exact time signal. Information and Telecommunication Sciences. National Technical University of Ukraine "Igor Sikorsky Kyiv Politechnik Institute". Numer 1, pp. 11–18 (2018)
18. Yanitskyi, I.: Optymizatsiia formuvacha periodychnykh vidlikiv chasu z kiltsem fazovoho avtopidstroiuvannia vyokremlenoho prystroiu synkhronizatsii telekomunikatsiinykh merezh. Dissertation, Derzhavnyi universytet informatsiino-komunikatsiinykh tekhnolohii (2007).
19. Pontryagin, L.S.: Princip maksimuma v optimal'nom upravlenii. Editoria-lURSS, Moscow (2004)
20. Pontryagin, L.S., Boltyanskij, V.G., Gamkrelidze, R.V., Mishchenko, E.F.: Matematicheskaya teoriya optimal'nyh processov. Glavnaya redakciya fiziko-matematicheskoj literatury, Nauka, Moscow (1983)
21. Ivanov, V.A., Faldin N.V.: Teoriya optimal'nyh sistem avtomaticheskogo upravleniya. Glavnaya redakciya fiziko-matematicheskoj literatury, Nauka, Moscow (1981)
22. Fel'dbaum, A.A.: Osnovy teorii optimal'nyh avtomaticheskih sistem. Nauka, Moscow (1966)
23. Fel'dbaum, A.A., Butkovskij, A.G.: Metody teorii avtomaticheskogo upravleniya. Nauka, Moscow (1971)
24. Koval, V., Kalian, D., Koval, V.: Adaptyvnyi tsyfrovyi fazovyi dyskryminator. Patent Ukraina 25 January 2017

Methods of Signal Detection and Recognition in Cognitive Radio Networks

Valeriy Bezruk(✉), Stanislav Ivanenko(✉), and Oleksii Fedorov(✉)

National University of Radio Electronics, Kharkiv, Ukraine
{valerii.bezruk,stanislav.ivanenko,oleksii.fedorov}@nure.ua

Abstract. The paper employs techniques of signal detection and recognition to perform spectrum sensing in cognitive radio networks. The main focus is on extensive use of signal recognition techniques, which is not the case yet in the context of cognitive radio networks. In particular, the use of unconventional methods of signal recognition allows us to not only divide received signals into classes of those which belong to primary or secondary users but, additionally, register appearance of new, uncategorized, sources of radiation. As a consequence, all considered in the paper procedures may be employed by the local frequency control authorities to monitor radio frequency resource utilization. Performance analysis of presented in the paper algorithms for signal selection and recognition is performed by the method of Monte-Carlo.

Keywords: Cognitive radio networks · Spectrum sensing · Signal · Detection · Recognition

1 Introduction

Nowadays, wireless radio technologies are widely implemented. At the same time, the demand increases rapidly and the frequency resource gets to be quite overfilled. Nevertheless a thorough analysis indicates poor efficiency of frequency band utilization [17]. For example, a large frequency band has been allocated for TV and radio broadcasting services. There are rarely used subbands which may be provided for organizing communication with the aim of collecting data from various sensors of public utilities and security systems, personnel warning systems in hospitals, etc.

The IEEE 802.22 standard has become one of the possible solutions to the problem. It specifies radio systems operation in the 54–698 MHz frequency range. The standard was developed with the concept of cognitive radio (CR) [4,8,11,16,20] in mind. In this case, the system searches for "spectral holes", that is, subbands of the frequency range that are not used at the moment by primary (licensed) users, and provides secondary users with them. In other words, the main task of the CR system [10] is to monitor the dynamics of changes in the broadcast environment of a given frequency range. There are two ways of acquiring information on frequency resource utilization [16], namely

P. Vorobiyenko et al. (Eds.): IPF 2020, LNNS 212, pp. 80–92, 2021.
https://doi.org/10.1007/978-3-030-76343-5_5

(i) from the database of the cognitive radio system;
(ii) by direct sensing a given frequency range.

Main methods of processing collected sensing information include:

(i) detection of radio emissions of the primary user transmitter (non-cooperative method);
(ii) detection of radio emissions of the heterodyne of the receiving device of the primary user (cooperative method);
(iii) interference detection method.

To identify unoccupied frequency channels, an analysis of the signal/noise environment in the frequency channels should be carried out first. As a result, a decision is made on the presence in the analyzed subband of either a signal and noise mixture, or noise only. That is, we actually deal with the problem of detecting a signal mixed with noise. It should be noted that signals and noise in frequency channels are normally random. Thus, we usually face the problem of detecting random signals mixed with random noise. And as a consequence there emerges a priori uncertainty with regard to probabilistic characteristics of signals and noise. A literature survey [4,14,16,20,21] yields the following list of the most frequently used methods to detect primary user radio emissions:

(i) energy method;
(ii) method of matched filtering;
(iii) search for cyclostationarity.

The second and third detection methods are based on employing certain information about the signal, while the energy method does not. The energy detector is the most used, due to its low computational complexity and ease of implementation. The main peculiarity of such the method is low detection accuracy at low SNRs, as well as the difficulty of determining the required operation threshold, taking into account the constantly changing signal and noise environment in frequency channels.

The advantage of the matched filtering method is the short observation time to obtain the required values of false alarm or missed detection probabilities. The disadvantages include the need to use a certain type of receiving device to retrieve signals of each class of primary users. This makes the method impractical for widespread use in CR networks.

The method based on discovering cyclostationarity features, makes use of the initial redundancy inherent in the signal. The advantage of the method is that it is capable of discriminating the signal from noise quite well. This is possible even when the wanted signal is below the noise level. The disadvantages of the method include high computational complexity and significant observation time.

Though, there is another option we have in order to perform spectrum sensing with the aim to find free subbands in CR. Namely, we may employ methods of statistical theory of signal detection [6,9,12]. However they require proper learning samples of signals to be provided, which is hardly implementable as we are unable of acquiring learning samples for every radio emission type. There are lots of different simultaneously active emission sources, possessing different types of modulation with peculiar parameters [18]. Thus there arise a need in such the detection algorithms which would be robust to both errors in estimating unknown parameters of signals and their abrupt changes in a wide frequency range.

On the other hand, an insufficient attention is paid to the problem of recognizing detected signals in CR networks, although a successful solution may be of great value. This is going to make it possible to determine not only a user's type (primary or secondary), but also a signal's characteristics. If such an information acquired, this should allow prioritizing access to the frequency resource for secondary users, sensitive to possible transmission delays.

Thus, the main CR tasks can be formulated as follows:

(i) frequency resource detection;
(ii) prioritizing access to the frequency resource;
(iii) optimization of the frequency resource utilization.

Summarizing, we conclude that listed above methods have some drawbacks and using them for the sake of spectrum sensing is not that beneficial. Therefore, the problem of designing algorithms for detecting and recognizing radio emissions, suitable for operating efficiently in conditions of a priori uncertainty about existing in the frequency band of interest signals, and meeting necessity of making possible frequency resource sharing is actual. The rest of the paper is devoted to investigating the performance characteristics of non-conventional methods for detection and recognition of signals by simulation with real-world samples of signals and noise, obtained during spectrum sensing in CR networks.

2 Spectrum Sensing

2.1 Decision Rules of Detecting Unknown Signals

In our formalization of the problem of detecting unknown signals in the presence of noise we put forward two hypothesis. Namely, H^0, the hypothesis, which is valid if there is no signal in the frequency band to analyze and the alternative hypothesis H^1, which is valid if H^0 fails, i.e. the hypothesis about the presence of a signal mixed with noise.

Let $\boldsymbol{X} \sim W(\boldsymbol{X}|\boldsymbol{\alpha})$ be an observation vector of length L and $W(\boldsymbol{X}|\boldsymbol{\alpha})$ is a multivariate PDF specified by a vector of parameters $\boldsymbol{\alpha}$. For the problem of detecting known signals in the presence of noise, a decision rule based on comparing the likelihood ratio with a certain threshold value can be used [9].

Ref. [2] suggests methods for detecting unknown signals in the presence of specified in a probabilistic sense noise. Namely, the following decision rule is applicable to detect unknown signals in the frequency channel to analyze:

$$W(\boldsymbol{X}|\boldsymbol{\alpha}^0) \underset{H^1}{\overset{H^0}{\gtrless}} \Delta^0, \tag{1}$$

where $\boldsymbol{\alpha}^0$ is the distribution density parameter corresponding to the null hypothesis; Δ^0 is a certain threshold value selected to provide the specified probability of false alarm.

If $\boldsymbol{X}$ comes from a multivariate Gaussian distribution, decision rule for detecting unknown signals in the presence of noise is [2,7,19]

$$(\boldsymbol{X} - \boldsymbol{\mu}^0)^T (\mathbf{R}^0)^{-1} (\boldsymbol{X} - \boldsymbol{\mu}^0) \underset{H^1}{\overset{H^0}{\lessgtr}} \Delta^0, \tag{2}$$

where $\boldsymbol{\mu}^0$ and $\mathbf{R}^0$ are respectively mean vector and covariance matrix of the noise. Estimates for $\boldsymbol{\mu}^0$, $\mathbf{R}^0$ are determined from a learning sample of noise $\boldsymbol{X}^0$.

If components of $\boldsymbol{X}^0$ are uncorrelated, covariance matrix $\mathbf{R}^0$ becomes diagonal and the decision rule (2) reduces to comparing to the threshold Δ^0 Euclidean distances between the noise reference and signals.

In turn, when for decision-making a set of observation vectors $\{\boldsymbol{X}_r\}_{r=1}^{v}$ is used, the decision rule (2) takes the form:

$$\sum_{r=1}^{v} (\boldsymbol{X}_r - \boldsymbol{\mu}^0)^T (\mathbf{R}^0)^{-1} (\boldsymbol{X}_r - \boldsymbol{\mu}^0) \underset{H^1}{\overset{H^0}{\lessgtr}} \Delta_v^0. \tag{3}$$

Another variation of the rule (3) is

$$\mathrm{Tr}\,(\mathbf{R} - \mathbf{R}^0)(\mathbf{R} - \mathbf{R}^0)^T \underset{H^1}{\overset{H^0}{\lessgtr}} \Delta_{\mathbf{R}}^0, \tag{4}$$

where $\mathbf{R}^0 = \frac{1}{n_0}\sum_{r=1}^{n_0}(\boldsymbol{X}_r^0 - \boldsymbol{\mu}_r^0)(\boldsymbol{X}_r^0 - \boldsymbol{\mu}_r^0)^T$, $\mathbf{R} = \frac{1}{v}\sum_{r=1}^{v}(\boldsymbol{X}_r - \boldsymbol{\mu}_r^0)(\boldsymbol{X}_r - \boldsymbol{\mu}_r^0)^T$, n_0 is for the size of a training sample and $v \ll n_0$ is the size of a control sample; $Tr(\cdot)$ designates matrix trace operator.

For the sake of comparison, a conventional energy detector

$$\boldsymbol{X}^T \boldsymbol{X} \underset{H^1}{\overset{H^0}{\lessgtr}} \Delta^0 \tag{5}$$

was used.

The above decision rules (2)–(5) define some possible algorithms for detecting unknown signals in presence of noise, and can be used for acquisition of unoccupied frequency channels in CR networks.

2.2 Spectrum Sensing System

In the study, we used samples of noise and signals peculiar to the IEEE 802.22 standard frequency band, that is, from the band where CR networks operate. Digital recordings of signal and noise realizations were obtained using an SDR equipment, which allows scanning and analyzing a given frequency range, as well as making digital recordings of observations emerging from selected frequency channels. A DVB-T TV USB tuner with an open source SDR# program was employed as an SDR receiver, Fig. 1.

Fig. 1. An SDR receiver.

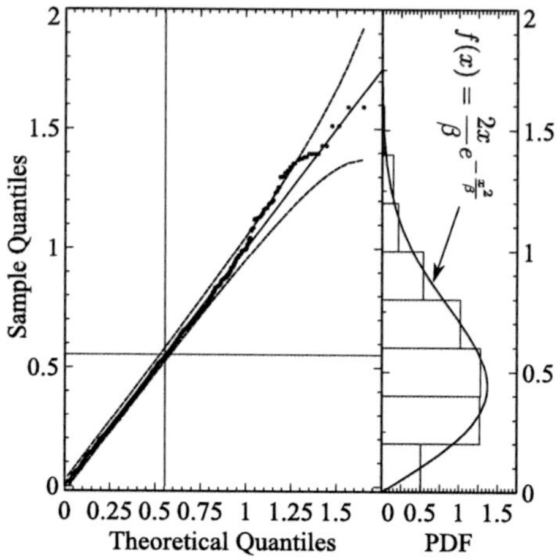

Fig. 2. Square root of exponential distribution Q-Q plot and histogram of magnitude spectrum of noise.

Such a tuner is capable of receiving radio emissions in the frequency range from 24 MHz to 1710 MHz and analyzing the spectra of signals with various types of modulations: AM, FM, WFM, NFM, CW, SSB, which are peculiar to the specified frequency range.

To obtain estimates of the signal detection rules operating characteristics by the Monte-Carlo method, control samples of signal and noise realizations of length 1000 and composed of $L \leq 512$ time bins were used.

2.3 Statistical Analysis of Noise

Since the decision rules for detecting unknown signals are based on assumptions about probabilistic properties of noise, let us briefly discuss the matter. Our previous research [1] has shown that readouts of noise can be treated as independent normally distributed random variables. This substantiates the use of decision rules (2)–(4) as they are based on the assumption that distribution of noise has a good agreement with the Gaussian model.

On the other hand, when performing spectrum sensing, we analyze frequency bands, thus let us consider statistical properties of the noise in the case of representing observations in the DFT basis (which we are going to use in what follows). We stick to the magnitude spectrum rather then to the power one. For the normally distributed noise, power spectrum appears to have χ_2^2 distribution [3], i.e. bins of the power spectrum are exponentially distributed and thus bins of the magnitude spectrum follow square root of exponential distribution model:

$$f(x) = \frac{2x}{\beta} e^{-\frac{x^2}{\beta}}, \quad x > 0,\ \beta > 0. \tag{6}$$

Figure 2 depicts Q-Q plot of the magnitude spectrum of a sample of noise. The Q-Q plot corresponds to model (6). Confidence bands (dashed lines) are drawn to meet significance level $\alpha = 0.01$. Also, Fig. 2 contains the histogram of the magnitude spectrum. Considering Fig. 2 we conclude that model (6) provides quite an accurate description of the magnitude spectrum of noise, however decision rules (2) and (3), being applied to the magnitude spectrum, loose their optimality as probability distribution (6) remarkably deviates from the normal model. Note that model (6) is a special case of the Rician distribution, which mimics a broader class of probability distributions.

2.4 Procedure to Determine Threshold Value Δ^0

In the paper, the following procedure of selecting threshold values Δ^0 for rules (2) and (3) was used.

1. Based on the learning sample $\boldsymbol{X}^0$, estimates for $\boldsymbol{\mu}^0$ and $\mathbf{R}^0$ are found. We are going to treat these values as the reference ones.
2. For every testing sample of noise, values of decision statistics of rules (2) and (3) are computed (i.e. left-hand sides of (2) and (3)).
3. Obtained on the previous step values of decision statistics are arranged in order of magnitude, which yields us order statistics $\Delta^0_{(1)}, \dots, \Delta^0_{(N)}$. Here N is the number of testing samples.
4. For the specified value of the probability of false alarm $\mathsf{P}(1|0)$, threshold Δ^0 is chosen to be equal to order statistics $\Delta^0_{(n)}$, where n is the smallest sequence number for which $1 - \frac{n}{N} \leq \mathsf{P}(1|0)$.

Figure 3 plots values of the rule (2) decision statistic, obtained on the second step of the procedure. As the testing samples, magnitude spectrum bins of noise were used. The length of the DFT block was 512 bins. Solid line in Fig. 3 corresponds to the threshold value $\Delta^0 \approx 306$ determined for $N = 1000$, $\mathsf{P}(1|0) = 0.04$.

Similar results were obtained for the decision statistics corresponding to other proposed detection rules.

2.5 Probability of Unknown Signal Correct Detection

This subsection contains results of performance evaluation of decision rules (2) through (5) for the case of detecting unknown signal in the presence of noise. Results of research are provided for different representations of signal and noise: in the time and frequency domain as well as for narrow and broadband signals.

As a narrowband unknown signal in our simulations we used AFS FSK 130 Bd signal [13]. Magnitude spectrum of this signal is shown in Fig. 4. We performed simulations as follows. At the learning stage, parameters of the decision rules (2) and (3) were determined with respect to accumulated samples of noise. Threshold values were selected to provide the requested value of the false alarm probability, $\mathsf{P}(1|0) = 0.04$.

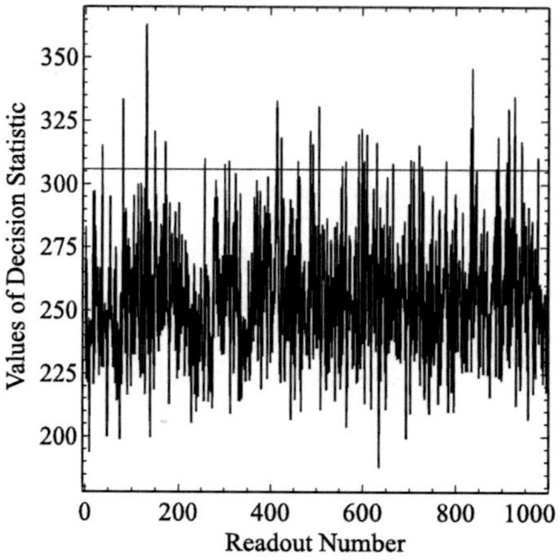

Fig. 3. Values of the rule (2) decision statistic for magnitude spectra of noise and the threshold value Δ^0 (solid line).

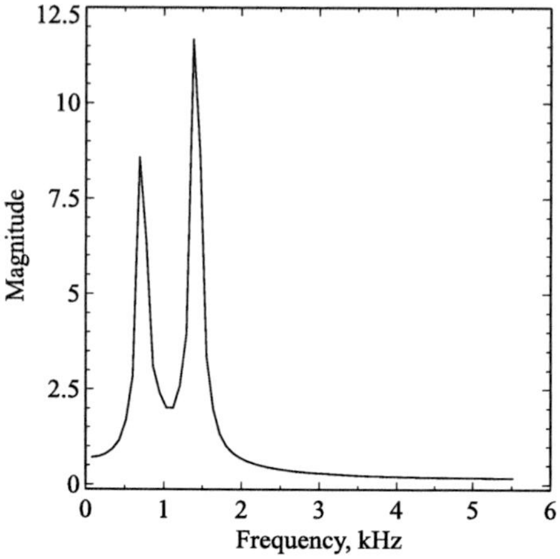

Fig. 4. Magnitude spectrum of AFS FSK 130 Bd signal.

Figures 5a and 5b plot dependencies of estimates for probabilities of correct detections, $\hat{\mathsf{P}}(1|1)$, versus signal to noise ratio, SNR, for the decision rule (3), which was implemented in the time and frequency domain. In our simulations we used additive mixtures of signal and noise, acting in the frequency band of interest. Probabilities $\mathsf{P}(1|1)$ were estimated as proportions, i.e. ratios of the number of successful trials to the total number of trials, which was 1000. As successful ones we treated those trials which yielded correct decisions on the presence of unknown signal. The estimates were obtained for $v = 1, 2, 3$, where v is the number of realizations used to make a decision.

For the sake of comparison Fig. 5a contains results related to a conventional energy detector, ED (implemented in the frequency domain) (5). It can be seen easily that rule (3) significantly outperforms the conventional energy detector (5). Also, Fig. 5b contains data to compare performance of different implementations of rule (3): in the time and frequency domain. Accuracy of detection in the frequency domain is substantially higher compared to the one achieved in the time domain. The difference becomes drastic for low SNRs. This can be explained by the fact that switching to the frequency domain allows us to localize the signal and thereby increase the difference between the mixture of signal+noise and noise itself. In addition, the normal model is typical not only for noise, but also for lots of everyday signals, if we process them in the time domain. Note, error bars in Figs. 5a and 5b match confidence bands for proportions [15], confidence level 0.05.

Considering the above, in what follows to characterize signals we are going to employ magnitude spectrum. Let us focus now on broadband signals from the IEEE 802.22 frequency range.

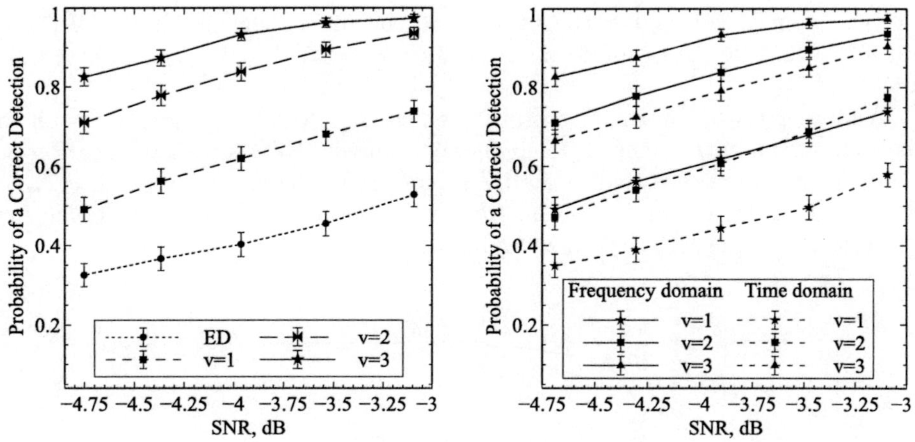

(a) $\hat{\mathsf{P}}(1|1)$ vs SNR, frequency domain.

(b) $\hat{\mathsf{P}}(1|1)$ vs SNR; frequency vs time domain.

Fig. 5. Detection of unknown signal in the presence of noise.

Figure 6 depicts a WFM signal magnitude spectrum. We treated this signal as unknown, which might appear in the presence of noise. To compare operating characteristics of rules (3) and (4), ROC curves were used (Fig. 7). These curves are drawn under the following assumptions: SNR ≈ -1.6 dB and the DFT block size $L = 128$ bins. Also, Fig. 7 illustrates the influence of the parameter v selection (size of a control sample) on the value of probability of a correct detection

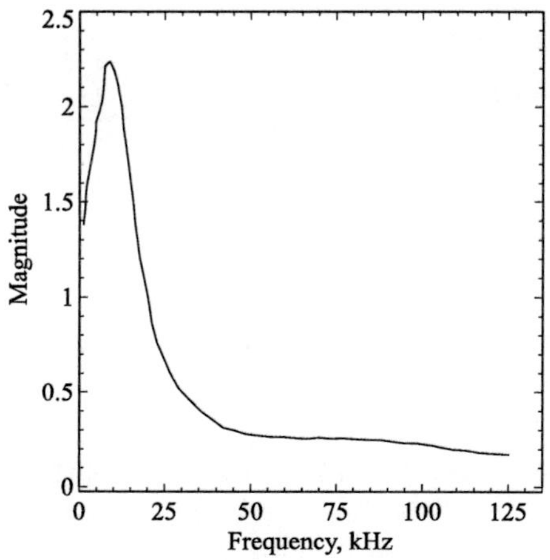

Fig. 6. Magnitude spectrum of a wide-band FM (WFM) signal.

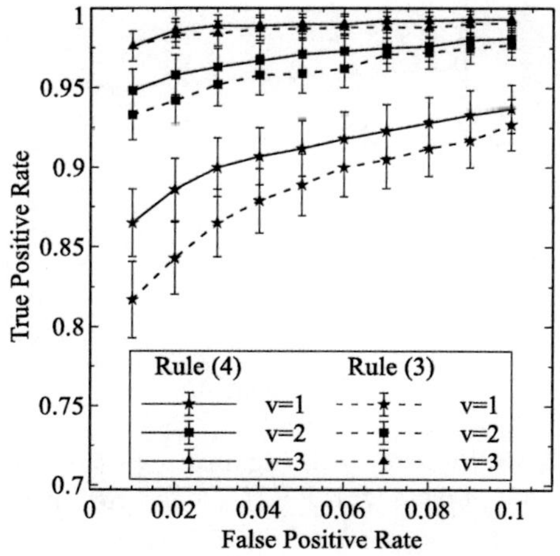

Fig. 7. ROC curves drawn for the DFT block size $L = 128$ bins and SNR ≈ -1.6 dB.

$P(1|1)$. Namely, the higher the value of v, the higher the probability $P(1|1)$. For $\nu = 1, 2$, rule (4) outperforms rule (3); for $v = 3$ the difference is not that drastic, if any.

A more detailed information on the behaviour of rules (3) and (4) as well as the comparison with a conventional energy detector (ED) is presented in Fig. 8. All of the rules are sensitive to the DFT block size selection ($L = 64$, $L = 128$). Both of the rules (3) and (4) are superior to the energy detector and rule (4) outperforms (3), especially at low SNRs.

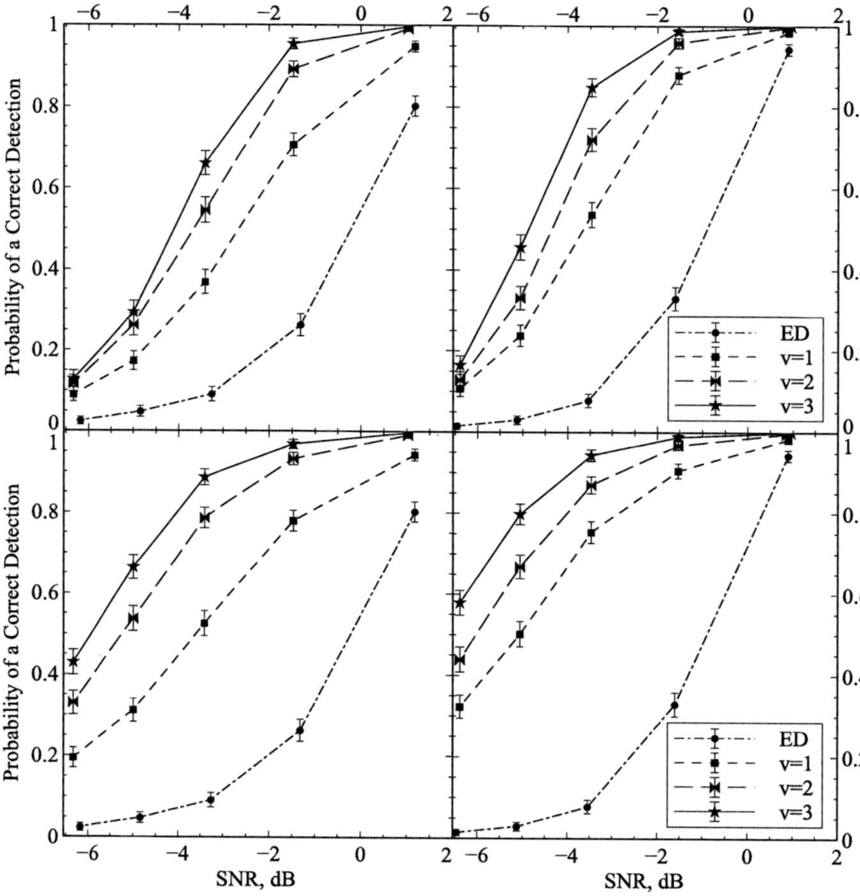

Fig. 8. Plots of $\hat{P}(1|1)$ against SNR for a range of values of v and $L = 64$ (left), $L = 128$ (right); first row: rule (3), second row: rule (4).

The paper [5] plots shown in Fig. 9 dependencies of the probability of a correct detection on SNR for the method of matched filtering as well as for the method employing a cyclostationary test. Such the dependencies may serve the goal of comparing methods based on the suggested in subsection 2.1 rules to these, state-of-the-art ones. In particular, one may conclude that at low SNRs the matched filter detector and cyclostationary detector have superior detection abilities with respect to the considered in the current paper methods. On the other hand, the matched filter method relies on a certain amount of prior information, and the cyclostationary detector requires a substantial amount of time to compute autocovariance sequences.

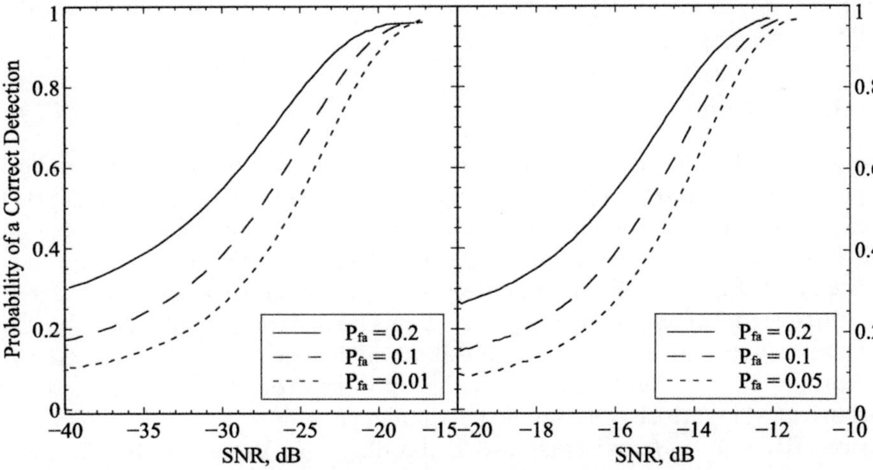

Fig. 9. Plots of $P(\hat{1}|1)$ against SNR for a range of probability of a false alarm values, $L = 1024$; the plot to the left is for the matching filter detector whereas the right one corresponds to the cyclostationary detector.

3 Selection and Recognition of Specified Signals in the Presence of Unknown Signals

3.1 Problem Formalization

To solve the problem of selection and recognition of specified signals in the presence of unknown signals we used the following decision rule, based on comparing the Mahalanobis distance to the threshold [2]:

$$H^i : \begin{cases} D^i < \Delta^i, \\ D^i \leq D^l, \ i, l = \overline{1, M}, \ i \neq l; \end{cases} \tag{7a}$$

$$H^{(M+1)} : \quad D^i \geq \Delta^i, \ i = \overline{1, M}, \tag{7b}$$

where H^i is the hypothesis about the presence of the i-th known signal, in turn $H^{(M+1)}$ is the hypothesis, which is valid if H^i fails, i.e. in the frequency band to analyze an unknown signal is present; $D^i = (\boldsymbol{X} - \boldsymbol{\mu}^i)^T(\mathbf{R}^i)^{-1}(\boldsymbol{X} - \boldsymbol{\mu}^i)$, $\boldsymbol{\mu}^i$

and $\mathbf{R}^i$ are the mean vector and covariance matrix of the i-th known signal represented by its training sample and Δ^i is some threshold value.

Decision rule (7) allows us to formalize the problem of selection and recognition of M specified signals and referencing unknown signals to the $(M+1)$-th class. This rule introduces a two stage procedure for making a decision in favour of the i-th specified signal, namely, it requires both of the inequalities (7a) to be true, otherwise the input signal is treated as an unknown one from the $(M+1)$-th class. Note, rule (7) does not require any information about probability distribution of the $(M+1)$-th class signals for whose we are unable to acquire learning samples.

3.2 Evaluation

Research into peculiarities of selection and recognition of specified signals in the presence of unknown signals was performed by simulation in MATLAB. In our experiment we used samples of 6 kHz signals [13]. Accumulated learning and testing samples included 100 realizations each. Every realization was composed of 128 time bins. Figure 10 plots normalized averaged magnitude spectra of signals. First three signals composed the group of specified signals, while another three – represented the $(M+1)$-th class of unknown signals.

Learning samples of specified signals were used to estimate parameters of decision rule (7), which was applied to magnitude spectrum bins; covariance matrices $\mathbf{R}^i$, $i = \overline{1, M}$ were assumed to be diagonal. Values of the thresholds Δ^i, $i = \overline{1, 3}$ were chosen to keep the probability of false recognition being equal to 0.04. Results of recognition are summarized in Table 1. There are definitely some signals for which probabilities of correct detection are on the high side as well.

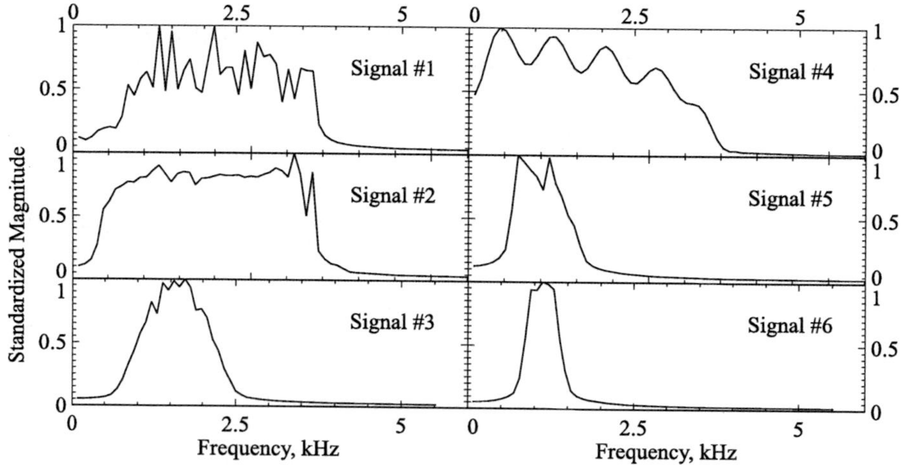

Fig. 10. Standardized averaged magnitude spectra of radio signals.

Table 1. Results of signal selection and recognition

j	Probability of referencing signal j to the given class			
	P(1|j)	P(2|j)	P(3|j)	P($M+1$|j)
1	0.86	0.09	0	0.05
2	0.1	0.89	0	0.01
3	0	0	1	0
4	0	0.01	0	0.99
5	0	0	0	1
6	0	0	0	1

4 Conclusion

A nonconventional approach to solving for the problem of signal detection and recognition in cognitive radio networks for the case when both known and unknown signals are submitted for processing is considered in the paper. The main peculiarity of the proposed approach is that it allows to reference unknown signals to a special class of signals. Research was performed with respect to samples of real signals, typical for IEEE 802.22 standard frequency band.

References

1. Bezruk, V.M., Ivanenko, S.A., Fedorov, O.V., Němec, Z., Pidanič, J.: Detection and recognition of signals in HF radio monitoring. In: 2019 29th International Conference Radioelektronika (RADIOELEKTRONIKA), pp. 1–5. IEEE, April 2019. https://doi.org/10.1109/RADIOELEK.2019.8733445
2. Bezruk, V.M., Pevtsov, G.V.: Theoretical Foundations of Signal Recognition Systems Design for Automated Radio Monitoring. Collegium, Kharkov (2007). (in Russian)
3. Brillinger, D.R.: Time Series: Data Analysis and Theory, vol. 36. SIAM (1981)
4. Cabric, D., Mishra, S.M., Brodersen, R.W.: Implementation issues in spectrum sensing for cognitive radios. In: Conference Record of the Thirty-Eighth Asilomar Conference on Signals, Systems and Computers, 2004, vol. 1, pp. 772–776 (2004). https://doi.org/10.1109/ACSSC.2004.1399240
5. Diaz, L.M.G., Marrero, L.M., Gómez, J.T.: Performance comparison of spectrum sensing techniques for cognitive radio networks. In: 2016 16th International Convention and Fair Informática, pp. 1–7. Havana, Cuba, March 2016
6. Duda, R.O., Hart, P.E., Stork, D.G.: Pattern Classification. Wiley, Hoboken (2000)
7. Gallego, G., Cuevas, C., Mohedano, R., García, N.: On the Mahalanobis distance classification criterion for multidimensional normal distributions. IEEE Trans. Sig. Process. **61**(17), 4387–4396 (2013). https://doi.org/10.1109/TSP.2013.2269047
8. Haykin, S.: Cognitive radio: brain-empowered wireless communications. IEEE J. Sel. Areas Commun. **23**(2), 201–220 (2005). https://doi.org/10.1109/JSAC.2004.839380
9. Helstrom, C.W.: Statistical Theory of Signal Detection. International Series of Monographs in Electronics and Instrumentation, Pergamon (1968)

10. Khattab, A., Perkins, D., Bayoumi, M.: Cognitive Radio Networks. Springer, New York (2013). https://doi.org/10.1007/978-1-4614-4033-8
11. Mishra, S.M., Sahai, A., Brodersen, R.W.: Cooperative sensing among cognitive radios. In: 2006 IEEE International Conference on Communications. vol. 4, pp. 1658–1663, June 2006. https://doi.org/10.1109/ICC.2006.254957
12. Poor, H.V.: An Introduction to Signal Detection and Estimation. Springer, New York (1994)
13. Proesch, R., Daskalaki-Proesch, A.: Technical handbook for radio monitoring VHF/UHF. Books on Demand (2017)
14. Rembovsky, A.M., Ashikhmin, A.V., Kozmin, V.A., Smolskiy, S.M.: Radio Monitoring. Signals and Communication Technology. Springer (2018). https://doi.org/10.1007/978-3-319-74277-9
15. Stine, R., Foster, D.: Statistics for Business: Decision Making and Analysis. Pearson Education, London (2014)
16. Tabaković, Ž.: A survey of cognitive radio systems, January 2008. https://www.researchgate.net/publication/228450166_A_Survey_of_Cognitive_Radio_Systems
17. UCRF: Register of radio frequency assignments (centralized assignments) (2021)
18. Weber, C., Peter, M., Felhauer, T.: Automatic modulation classification technique for radio monitoring. Electron. Lett. **51**(10), 794–796 (2015). https://doi.org/10.1049/el.2015.0610
19. Wickens, T.D.: Elementary Signal Detection Theory. Oxford University Press, Oxford (2002)
20. Wild, B., Ramchandran, K.: Detecting primary receivers for cognitive radio applications. In: First IEEE International Symposium on New Frontiers in Dynamic Spectrum Access Networks, DySPAN 2005, pp. 124–130 (2005). https://doi.org/10.1109/DYSPAN.2005.1542626
21. Yücek, T., Arslan, H.: A survey of spectrum sensing algorithms for cognitive radio applications. IEEE Commun. Surv. Tutor. **11**(1), 116–130 (2009). https://doi.org/10.1109/SURV.2009.090109

Research of Telecommunication Traffic and Its Performance Indicators

Research of the QoE Fast ReRouting Processes with Differentiated R-Factor Maximization for VoIP-Flows Using the Tensor Model of the Corporate Telecommunication Network

Oleksandr Lemeshko(✉), Oleksandra Yeremenko(✉), Maryna Yevdokymenko(✉), and Tamara Radivilova(✉)

Kharkiv National University of Radio Electronics, 14 Nauky Avenue, Kharkiv, Ukraine
{oleksandr.lemeshko.ua,oleksandra.yeremenko.ua, maryna.yevdokymenko}@ieee.org, tamara.radivilova@nure.ua

Abstract. In the work, the flow-based model of the QoE Fast ReRouting is proposed. The model is based on the implementation of the single path or multipath routing, the conditions of flow conservation, which are introduced for routing variables that regulate the construction of both primary and backup paths. In addition, restrictions have been introduced to prevent the overloading of communication links with packet flows, the implementation of which actually provided bandwidth protection. The model was supplemented by the conditions of structural network elements protection (node, link, and route). The peculiarity of these conditions is the consideration of possible packet losses due to congestion of router interfaces. In order to obtain analytical expressions for the calculation of the R-factor for each of the VoIP-flows, a tensor generalization of the mathematical model of routing has been performed. Based on the tensor description of the network, it was possible to obtain expressions for calculating the average multipath end-to-end delay and packet loss probability, which allowed to formulate in analytical form the R-factor calculation expressions for each of the VoIP-Flows. The novelty of the proposed model is the formulation of the problem of QoE Fast ReRouting in the optimization form when the optimality criterion was the maximum of the additive form, represented by the sum of weighted according to the IP-priority values of R-factor for each of the VoIP-Flows. The results of the study of the proposed model confirmed its efficiency and adequacy, which was especially evident in the conditions of complex network topologies, high network congestion, and flow differentiation relative to the IP-priority values of packets.

Keywords: Quality of experience · Fast ReRouting · R-Factor · VoIP · Tensor model · Optimization · Corporate network

1 Introduction

Building reliable, fault-tolerant telecommunication networks (TCN) is an important scientific and applied problem, the solution of which is associated with the need to

P. Vorobiyenko et al. (Eds.): IPF 2020, LNNS 212, pp. 95–110, 2021.
https://doi.org/10.1007/978-3-030-76343-5_6

review existing mathematical models, methods, and protocols for traffic management and allocation of network (link and buffer) resources [1–7]. The main reasons that cause denials of service in TCN, traditionally include breaches of the level of confidentiality (security) in the network, the overload of network elements, software and hardware failures of network devices (servers, switches, routers), etc.

Routing equipment occupies an important place among the solutions for ensuring resilience, network security, and Quality of Service (QoS) in TCN. In turn, fault-tolerant routing protocols based on the implementation of a method (model) of route calculation provide protection of structural (node, link, path) and functional parameters (e.g., bandwidth) [8, 9].

The analysis showed that the routing protocols used in IP networks (Internet Protocol) and MPLS (MultiProtocol Label Switching) implement very limited functions to protect network elements in case of failure, and the protection of a given Quality of Service level for them is the task of the nearest perspectives. Scientists have proposed a number of solutions to protect such an important QoS-indicator as bandwidth [8–11]. A number of scientific papers are aimed at ensuring the protection of a given level of quality on a variety of indicators, for example, bandwidth, average end-to-end delay, probability of packet loss [12–15]. However, with the growing share of multimedia flows in the structure of modern network traffic, the tasks of protection during routing of the set values of multimedia quality indicators, which is perceived at the user level (Quality of Experience, QoE), come to the fore. When transmitting VoIP packet flows, it is the value of the R-factor [16–26].

The work proposes a flow-based model of QoE Fast ReRouting. The model is based on the conditions for the implementation of single- or multipath routing, the conditions of flow conservation, which are introduced for routing variables that regulate the construction of both the primary and backup routes. Besides, conditions have been introduced to prevent the overload of communication links with packet flows, the implementation of which actually protected such an important QoS indicator as bandwidth. The model was supplemented by the protection conditions of the structural network elements – node, link, and path (route). The peculiarity of these conditions is the consideration of possible packet losses that occur due to overloading the interfaces of TCN routers.

2 QoE Fast ReRouting Flow-Based Model with Differentiated R-Factor

In order to obtain analytical expressions for the calculation of the R-factor for each of the VoIP-Flows, a tensor generalization of the mathematical model of routing is conducted. The structure of the TCN determined the geometric space, where the number of communication links determined the size of the introduced discrete space. The metric of the space directly depended on the number of VoIP-Flows, their characteristics (bit rate, packet size), as well as interface parameters (service discipline, bandwidth, maximum queue buffer capacity). Each VoIP-Flow of packets circulating in the TCN was modeled with a bivalent mixed tensor, which took into account the average flow intensities and average packet delays in the coordinate paths of the network. On the structure of TCN, i.e. within the introduced geometric space, two coordinate systems (CS) are introduced:

the basis of network edges and the basis of interpolar paths and internal node pairs. Based on the tensor description of TCN, it was possible to obtain expressions for calculating the average multipath end-to-end delay and packet loss probability, which ultimately allowed formulating in analytical form the R-factor calculation expressions for each of the VoIP-Flows.

2.1 Basic Multipath Fast ReRouting Model

In this work, the tensor approach for mathematical modeling of rerouting under QoS/QoE has been chosen as an effective tool for a complex network description [27–30]. Therefore, the flow-based fast rerouting model has been selected for further tensor generalization in order to obtain the conditions for QoE assurance under R-factor with control of the end-to-end average packet delay and the probability of packet loss.

The following parameters are introduced into the basic QoE-centric multipath routing model:

$S = (U, V)$	one-dimensional network with a set of zero-dimensional simplexes $U = \{u_i, i = \overline{1, m}\}$ modeling network nodes and a set of one-dimensional simplexes modeling links $V = \{v_z = (i, j);\ z = \overline{1, n};\ i, j = \overline{1, m};\ i \neq j\}$;
m	number of nodes in the network S;
n	number of links in the network S;
$v_z = (i, j)$	z th network link between i th and j th routers;
$\varphi_{i,j}$, φ_z	link $v_z = (i, j) \in V$ capacity (packets per second, 1/s);
K	set of flows in the network S ($k \in K$);
$x_{i,j}^k$	control variables that determine fraction of intensity of the k th flow in the link (i, j) of the primary path;
$\overline{x}_{i,j}^k$	control variables that determine fraction of intensity of the k th flow in the link (i, j) of the backup path;
s_k, d_k	pair of source and destination nodes;
b^k	the fraction of the k th flow served by the network using the primary path/multipath, i.e. packets which are successfully delivered to the destination node;
$\overline{b}^k$	the fraction of the k th flow served by the network using the backup path/multipath, i.e. packets which are successfully delivered to the destination node;
$p_{i,j}^k$	the probability of packet loss of the k th flow on the j th interface of the i th router of the primary path;
$\overline{p}_{i,j}^k$	the probability of packet loss of the k th flow on the j th interface of the i th router of the backup path;
$\lambda_k^{\langle req \rangle}$	average intensity of the k th flow (flow rate) arriving to the network – bandwidth requirements for the k th flow (packets per second, 1/s);
ρ	utilization factor;
$r_{i,j}^k$	intensity of lost packets of the k th flow on the interface (i, j), which belongs to the primary path;
$\overline{r}_{i,j}^k$	intensity of lost packets of the k th flow on the interface (i, j), which belongs to the backup path;

$N = \Theta_{buf} + 1$	maximum number of packets that can be on the interface, including the buffer (Θ_{buf}) and the link itself;
R	value of R-factor;
$R_{\langle req \rangle}$	requirements for R-factor (limit value);
$I_{dd}(T_a)$	coefficient of quality degradation caused by delay as a function of network delay;
$I_{e-eff}(P_{pl})$	coefficient of quality degradation caused by the loss of voice packets;
I_e	coefficient of quality degradation caused by the use of low-speed codecs;
T_a	average end-to-end packet delay;
P_{pl}	percent of lost packets in the network;
B_{pl}	coefficient that takes into account the robustness to losses of codec;
$BurstR$	coefficient of «burst» of losses.

In the case of single path and multipath routing, the following conditions are imposed on routing variables:

$$x_{i,j}^k \in \{0; 1\} \, and \, \overline{x}_{i,j}^k \in \{0; 1\}, \tag{1}$$

$$0 \le x_{i,j}^k \le 1 \, and \, 0 \le \overline{x}_{i,j}^k \le 1. \tag{2}$$

Conditions of the flow conservation on network nodes (routers), taking into account the possible packet loss caused by overloading the queue buffer, take the form [18, 19]:

$$\begin{cases} \sum\limits_{j:(i,j)\in V} x_{i,j}^k = 1, & \text{if } k \in K, \quad i = s_k; \\ \sum\limits_{j:(i,j)\in V} x_{i,j}^k - \sum\limits_{j:(j,i)\in V} x_{j,i}^k\left(1 - p_{j,i}^k\right) = 0, & \text{if } k \in K, \quad i \ne s_k, d_k; \\ \sum\limits_{j:(j,i)\in V} x_{j,i}^k\left(1 - p_{i,j}^k\right) = b^k, & \text{if } k \in K, \quad i = d_k; \end{cases} \tag{3}$$

$$\begin{cases} \sum\limits_{j:(i,j)\in V} \overline{x}_{i,j}^k = 1, & \text{if } k \in K, \quad i = s_k; \\ \sum\limits_{j:(i,j)\in V} \overline{x}_{i,j}^k - \sum\limits_{j:(j,i)\in V} \overline{x}_{j,i}^k\left(1 - \overline{p}_{j,i}^k\right) = 0, & \text{if } k \in K, \quad i \ne s_k, d_k; \\ \sum\limits_{j:(j,i)\in V} \overline{x}_{j,i}^k\left(1 - \overline{p}_{i,j}^k\right) = \overline{b}^k, & \text{if } k \in K, \quad i = d_k. \end{cases} \tag{4}$$

For example, if the operation of the j th interface of the i th router of the primary or backup routes is modeled by a queuing system M/M/1/N with failures, then the probability of packet loss of the k th flow can be calculated as follows:

$$p_{i,j}^k = \frac{(1 - \rho_{i,j})(\rho_{i,j})^N}{1 - (\rho_{i,j})^{N+1}} \, and \, \overline{p}_{i,j}^k = \frac{\left(1 - \overline{\rho}_{i,j}^k\right)\left(\overline{\rho}_{i,j}^k\right)^N}{1 - \left(\overline{\rho}_{i,j}^k\right)^{N+1}}, \tag{5}$$

where $\rho_{i,j} = \frac{\sum_{k\in K} \lambda_k^{\langle req \rangle} x_{i,j}^k}{\varphi_{i,j}}$ and $\overline{\rho}_{i,j} = \frac{\sum_{k\in K} \lambda_k^{\langle req \rangle} \overline{x}_{i,j}^k}{\varphi_{i,j}}$ are the utilization factors of the j th interface on the i th router of the primary or backup routes.

To ensure controllability of the process of preventing links and queues overload, the next restrictions are used [12, 13]:

$$\sum_{k\in K}\lambda_k^{\langle req\rangle}\max[x_{i,j}^k,\overline{x}_{i,j}^k]\leq\varphi_{i,j}\quad under\ (i,j)\in V. \tag{6}$$

Consequently, the intensity of successfully transmitted packets of the k th flow (i.e. without losses) in the network link (i,j) of the primary or backup routes is defined in a form:

$$\lambda_{i,j}^k=\lambda_k^{\langle req\rangle}x_{i,j}^k\left(1-p_{i,j}^k\right)\ and\ \overline{\lambda}_{i,j}^k=\lambda_k^{\langle req\rangle}\overline{x}_{i,j}^k\left(1-\overline{p}_{i,j}^k\right). \tag{7}$$

In this case, the intensity of lost packets of the k th flow on the interface (i,j), which belongs to the primary or backup path, is defined as

$$r_{i,j}^k=\lambda_k^{\langle req\rangle}x_{i,j}^k p_{i,j}^k\ and\ \overline{r}_{i,j}^k=\lambda_k^{\langle req\rangle}\overline{x}_{i,j}^k\overline{p}_{i,j}^k.$$

With the aim of ensuring a differentiated increase in the level of QoE VoIP-Flows, there is a need to formalize the expression for calculating the R-factor [19, 31]:

$$R=R_0-I_{dd}(T_a)-I_{e-eff}\left(P_{pl}\right), \tag{8}$$

where corresponding quality degradation factors are determined as follows:

$$I_{dd}(T_a)=\begin{cases}0, & T_a\leq 100\ \text{ms};\\ 25\left[\left(1+X^6\right)^{\frac{1}{6}}-3\left(1+\left[\frac{X}{3}^6\right]^{\frac{1}{6}}\right)+2\right], & T_a>100\ \text{ms};\end{cases} \tag{9}$$

$$I_{e-eff}\left(P_{pl}\right)=I_e+(95-I_e)\frac{P_{pl}}{\frac{P_{pl}}{BurstR}+B_{pl}}, \tag{10}$$

where $X=\log\frac{\left(\frac{T_a}{100}\right)}{\log 2}$.

Ensuring that conditions (8) are met, taking into account (9) and (10), it is necessary to have analytical expressions for the relation of routing variables (1), traffic characteristics, network parameters, end-to-end delay, and packet loss probability. Taking into account the results [18, 19], it is proposed to apply an improved tensor approach to modeling routing processes in TCN.

2.2 Conditions for Ensuring the Quality of Experience

With reference to the tensor network modeling approach [19, 28–30], the network S poles are nodes (routers), through which a particular packet flow incomes to or outgoes from the network (SDN data plane). The following network S parameters and characteristics will also be used to further formalize the model:

$\kappa(S)$	number of basic interpolar paths in the network S, $\kappa(S)=n-m+2$;
$\vartheta(S)$	number of basic internal node pairs in the network S, $\vartheta(S)=m-2$;

$\{\nu_z, z = \overline{1, n}\}$	coordinate system of edges;
$\{\gamma_i, i = \overline{1, \kappa}\}$	set of linearly independent interpolar paths;
$\{\varepsilon_j, j = \overline{1, \vartheta}\}$	set of internal node pairs;
$\otimes$	operator of direct tensor product;
$[\cdot]^{-1}$	matrix inversion operation;
T	univalent covariant tensor of the average packet delays;
Λ	univalent contravariant tensor of the average flow intensities in the coordinate paths of the network;
$\Lambda_{\gamma\varepsilon}$	projection of tensor Λ in the CS of interpolar paths and internal node pairs;
$T_{\gamma\varepsilon}$	projection of tensor T in the CS of interpolar paths and internal node pairs;
$E_{\gamma\varepsilon}$	projection of twice covariant metric tensor E in the CS of interpolar paths and internal node pairs ($n \times n$ diagonal matrix);
$G_{\gamma\varepsilon}$	projection of twice contravariant metric tensor G in the CS of interpolar paths and internal node pairs ($n \times n$ diagonal matrix);
λ_i	total intensity of all packet flows transmitted in the i th network link;
λ_v^i	intensity of the packet flow that is considered in terms of building a tensor model in the i th network link;
x_p	fraction of the packet flow that was successfully delivered to the destination router via the p th path;
τ_p	average delay of packets transmitted along the p th path in the network;
$\|P\|$	number of paths available for routing;
λ_p	intensity of the packet flow that was successfully delivered to the destination router via the p th path;
λ^*	intensity of the packet flow that was successfully delivered to the destination router using all available paths from the set P (in the absence of packet loss in the network $\lambda^* = \lambda^{\langle req \rangle}$).

The network structure introduces a discrete n-dimensional geometric space, which is determined by the number of network links. Appropriate coordinate systems (CS) can be introduced in the space, in which the coordinate axes can act as [28–30]: edges, circuits, node pairs, cuts, etc. In this paper, orthogonal CSs will be adopted, in which the tensors projections of the main functional network parameters are considered interrelated:

- CS of edges of the network S (tensors projections will be denoted by the index ν);
- CS of interpolar paths and internal node pairs of the network S (tensors projections will be denoted by the index $\gamma\varepsilon$) where $n = \vartheta(S) + \kappa(S)$.

Then in the n-dimensional space the mixed divalent tensor can be introduced [24]:

$$Q = T \otimes \Lambda. \tag{11}$$

In the CS of interpolar paths and internal node pairs of the network, tensor equations of the interconnection the components of the mixed divalent tensor (9) by means of the corresponding metric tensors will have the following form [14]:

$$T_{\gamma\varepsilon} = E_{\gamma\varepsilon}\Lambda_{\gamma\varepsilon} \; and \; \Lambda_{\gamma\varepsilon} = G_{\gamma\varepsilon}T_{\gamma\varepsilon}. \tag{12}$$

Then the coordinates of the projection of a twice contravariant metric tensor G in the CS of edges can be represented by the values of the diagonal elements of the matrix G_v [18, 19]:

$$g_v^{ii} = \frac{\lambda_i\left(1-\rho_i^{N+1}\right)(1-\rho_i)\lambda_v^i}{\rho_i - \rho_i^{N+2} - (N+1)\rho_i^{N+1}(1-\rho_i)}, \tag{13}$$

where ρ_i is the utilization factor of the i th link.
In addition, the following rule applies:

$$E_v = [G_v]^{-1}. \tag{14}$$

Next, suppose that average end-to-end delay of packets transmitted between network poles using a set of routes P is obtained in the following way:

$$\mathrm{T}_a = \sum_{p=1}^{|P|} x_p \tau_p. \tag{15}$$

Generally, x_p can be defined as follows:

$$x_p - \frac{\lambda_p}{\lambda^*}, \tag{16}$$

The law of transformation of projections of twice covariant tensor E at a change of the coordinate system from the CS of edges to the CS of interpolar paths and internal nodal pairs is written as follows [29, 30]:

$$E_{\gamma\varepsilon} = \left(C_{\gamma\varepsilon}^{v}\right)^t E_v C_{\gamma\varepsilon}^{v}, \tag{17}$$

where $C_{\gamma\varepsilon}^{v}$ is the matrix of contravariant coordinate transformation when changing the CS.

Then decomposition representation of the metric tensor projection in the CS of interpolar paths and internal node pairs takes the form [18]:

$$\left\| \begin{array}{c|c} E_{\gamma\varepsilon}^{\langle 1\rangle} & E_{\gamma\varepsilon}^{\langle 2\rangle} \\ \hline E_{\gamma\varepsilon}^{\langle 3\rangle} & E_{\gamma\varepsilon}^{\langle 4\rangle} \end{array} \right\| = E_{\gamma\varepsilon}, \tag{18}$$

where $E_{\gamma\varepsilon}^{\langle 1\rangle}$ is the $\kappa \times \kappa$ square submatrix; $E_{\gamma\varepsilon}^{\langle 4\rangle}$ is the $\vartheta \times \vartheta$ square submatrix; $E_{\gamma\varepsilon}^{\langle 2\rangle}$ is the $\kappa \times \vartheta$ submatrix; $E_{\gamma\varepsilon}^{\langle 3\rangle}$ is the $\vartheta \times \kappa$ submatrix.

Taking into account (14)–(16), the expression for calculating the average end-to-end packet delay in the network (13) (in milliseconds), which has the physical meaning of T_a in (7), has been obtained:

$$\mathrm{T}_a = \frac{1000}{\lambda_k^{\langle req\rangle} b^k}\left(\Lambda_\gamma^t E_{\gamma\varepsilon}^{\langle 1\rangle}\Lambda_\gamma + \Lambda_\gamma^t E_{\gamma\varepsilon}^{\langle 2\rangle}\Lambda_\varepsilon\right). \tag{19}$$

Within the presented model (2), the expression for calculating the percent of lost packets of the k th flow (P_{pl}) in the network can be defined as follows:

$$P_{pl} = (1 - b^k) \cdot 100\%. \tag{20}$$

In turn, projections of the metric tensors E and G depend on the values of the routing variables in a way:

$$\lambda_z = \sum_{k\in K} \lambda_k^{\langle req\rangle} x_{i,j}^k \; and \; \lambda_v^z = \lambda_k^{\langle req\rangle} x_{i,j}^k \left(1 - p_{i,j}^k\right). \tag{21}$$

Expressions (19) determine the intensities of the aggregate flow and distinct k th flow in the same network link $v_z = (i, j)$.

2.3 Tensor Model of Fast ReRouting with Quality Assurance According to the R-Factor Indicator

The flow-based model (1)–(21) is used as a basis for protection of the QoE level by the R-factor indicator in the conditions of fast rerouting in the telecommunication network. That is, to implement the QoE FRR scheme in the network, it is also necessary to provide the calculation of two types of routing variables $x_{i,j}^k$ and $\overline{x}_{i,j}^k$, each of which characterizes the fraction of intensity of the k th flow sent from node u_i to node u_j through the link (i, j) of the primary or backup path, respectively. Within this tensor model, represented by expressions (1)–(21), restrictions on the use of the primary and backup paths were:

- conditions for implementation of single path and multipath routing strategy (1), (2);
- flow conservation conditions (3), (4);
- conditions of the link, node, and path (route) protection;
- conditions for the implementation of the network bandwidth protection scheme (6).

The following constraints are used for the implementation of the link protection [12, 14]:

- single path routing

$$\overline{x}_{i,j}^k \in \left\{0;\ \delta_{i,j}^k\right\}, \tag{22}$$

• multipath routing

$$0 \leq \overline{x}_{i,j}^{k} \leq \delta_{i,j}^{k}, \tag{23}$$

where

$$\delta_{i,j}^{k} = \begin{cases} 0, & under\ (i,j) \in V\ link\ protection; \\ 1, & otherwise. \end{cases} \tag{24}$$

In the case of using the conditions (5)–(7), the backup path will not include the protected link $(i,j) \in V$. The use of such a formulation of the link protection conditions allows one to reduce the computational complexity of the implementation of the presented model.

The next conditions are responsible for the *node* $u_i \in U$ *protection* and its incident links [12, 14]:

• single path routing

$$\overline{x}_{i,j}^{k} \in \left\{0;\ \delta_{i,j}^{k}\right\} at\ u_j \in u_i^*,\ j = \overline{1,m}, \tag{25}$$

• multipath routing

$$0 \leq \overline{x}_{i,j}^{k} \leq \delta_{i,j}^{k}\, at\ u_j \in u_i^*,\ j = \overline{1,m}, \tag{26}$$

where $u_i^* = \left\{u_j : \exists\ (i,j) \in V;\ i \neq j\right\}$ is the subset of routers that are adjacent to the router u_i, while $\delta_{i,j}^{k}$ is determined using (24).

Here the node $u_i \in U$ protection is guaranteed by the prevention of using the links outgoing from the given node in the backup route. Due to the fact that only transit nodes are under protection, if the outgoing links are not included in the backup route, then the incoming links will also not be used to form the backup path. This is the essence of the implementation of the node u_i protection scheme as a whole. The linearity of protection schemes for link and node is the advantage of the presented model underlying the redundancy cyber resiliency technique. Moreover, it will not affect computation complexity while obtaining the resulting routing order for both types of variables taking part in the formation of corresponding primary and backup routes.

If the *path protection* scheme is implemented, it is necessary to ensure that there are no common nodes and links in both the primary and backup routes. Then, by analogy with the results obtained in [14], for multipath routing, the multipath protection conditions have a nonlinear form:

$$\sum_{u_j \in u_i^*} \sum_{u_p \in u_i^*} x_{j,i}^{k} \overline{x}_{p,i}^{k} = 0,\ \forall u_i \in U \backslash \{s_k, d_k\}. \tag{27}$$

In analogy to (8), for the ensuring a differentiated increase in the level of QoE VoIP-Flows the expression for calculating the R-factor for backup path is following:

$$\overline{R} = R_0 - I_{dd}(\overline{T}_a) - I_{e-eff}(\overline{P}_{pl}), \tag{28}$$

where $I_{dd}(\overline{T}_a)$ is the coefficient of quality degradation caused by average end-to-end delay $\overline{T}_a$ along the backup path; $I_{e-eff}(\overline{P}_{pl})$ is the coefficient of quality degradation caused by the loss of voice packets on the nodes when using the backup path; $\overline{P}_{pl}$ is the percent of lost packets in the network when using the backup path.

Then according to the recommendations [31] similarly as to the primary path (9), (10), the coefficients $I_{dd}(\overline{T}_a)$ and $I_{e-eff}(\overline{P}_{pl})$ for the backup path are determined by following expressions:

$$I_{dd}(\overline{T}_a) = \begin{cases} 0, & \overline{T}_a \le 100\,ms; \\ 25\left[\left(1+\overline{X}^6\right)^{\frac{1}{6}} - 3\left(1+\left[\frac{\overline{X}}{3}\right]^6\right)^{\frac{1}{6}} + 2\right], & \overline{T}_a > 100\,ms; \end{cases} \tag{29}$$

$$I_{e-eff}(\overline{P}_{pl}) = I_e + (95 - I_e)\frac{\overline{P}_{pl}}{\frac{\overline{P}_{pl}}{BurstR} + B_{pl}}, \tag{30}$$

where $\overline{X} = \lg\frac{\left(\frac{\overline{T}_a}{100}\right)}{\lg 2}$.

Percent of lost packets $\overline{P}_{pl}$ in analogy to (20) will be calculated under the next expression:

$$\overline{P}_{pl} = (1 - \overline{b}^k) \cdot 100\%. \tag{31}$$

When using the tensor model of TCN, presented in the basis of interpolar paths and internal node pairs, to calculate the average end-to-end delay and the probability of packet loss to determine the R-factor the implementation of the strategy of QoE Fast ReRouting can be used that is presented in detail in Subsects. 2.1 and 2.2. Thus, when using the backup path, the calculation of the average end-to-end delay will be performed using an expression

$$\overline{T}_a = \frac{1000}{\lambda_k^{\langle req\rangle}\overline{b}^k}\left(\overline{\Lambda}_\gamma^t \overline{E}_{\gamma\varepsilon}^{\langle 1\rangle}\overline{\Lambda}_\gamma + \overline{\Lambda}_\gamma^t \overline{E}_{\gamma\varepsilon}^{\langle 2\rangle}\overline{\Lambda}_\varepsilon\right). \tag{32}$$

The percent of lost packets $\overline{P}_{pl}$ along the backup path will still be calculated with expression (31). Using the tensor model of TCN, presented in the basis of interpolar paths and internal node pairs, in the process of solving the problem of QoE Fast ReRouting, the protection schemes of not only link (22)–(24) and node (25), (26), but also the path (27) can be implemented. This is due to the fact that at the stage of geometrization of the TCN structure, the network elements that are subject to protection are not excluded.

When solving the problem of QoE Fast ReRouting in corporate geographically distributed TCN, it is advisable to use the criterion of type

$$J = \sum_{k \in K} \left(f(pr^k)R^k + f(\overline{pr}^k)\overline{R}^k \right) \to \max, \tag{33}$$

where $f(\overline{pr}^k)$ is the weight linearly or nonlinearly increasing function of the values of the IP priority (class) of the k th VoIP packet flow when it is routed by backup. The use of criterion (33) aims at a differentiated increase in the level of QoE in terms of R-factor VoIP-Flows by redistributing the network resource (without saving it) and load balancing along both the primary and backup paths.

Therefore, the problem of QoE Fast ReRouting was formulated in the optimization form, where condition (33) was used as a criterion of optimality. As a restrictions for the two types of tensor models, the general constraints imposed on the routing variables were the conditions for the implementation of single path or multipath routing (1), (2), the flow conservation conditions (3), (4), the links overload prevention conditions (6), the protection schemes of the link (22)–(24), node (25), (26), and path (27) when using the primary and backup routes, respectively. The nonlinearity of most constraints, for example (3), (4), (6), (27), allows us to refer the formulated optimization problem to the class of nonlinear programming problems. In addition, whith implementing single path routing, when routing variables took only Boolean values (1), the optimization task belonged to the class of mixed integer nonlinear programming (MINLP).

3 Numerical Research QoE FRR Flow-Based Model with Differentiated R-Factor in Coordinate System of Interpolar Paths and Internal Node Pairs

To evaluate the efficiency of the tensor model of QoE Fast ReRouting, represented by expressions (1)–(32) with the criterion of optimality (33), a study on a fragment of the telecommunication network is presented, which is shown in Fig. 1. In this case, the network consisted of nine routers and twelve communication links. The operation of each of the router interfaces was modeled by the M/M/1/N queuing system, and the buffer capacity was 30 packets ($N = 30$).

The intensity of the VoIP packet flow that arrive to the network on the router R_1 and was destined to the router R_9 was $\lambda^{\langle req \rangle} = 150$ 1/s. Suppose that in the course of fast rerouting it was necessary to implement path protection schemes and provide a differential level of QoE. Then Figs. 1 and 2 show the order of QoE Fast ReRouting of one VoIP-Flow using the primary (Fig. 1) and backup (Fig. 2) routes. For the considered numerical research, a single path routing strategy is implemented. In this figure, the breaks in the communication links indicated (from top to bottom): the intensity of the packet flow in this communication link (1/s), bandwidth (1/s) and the average packet delay in the link (ms).

The characteristics of the calculated paths (primary and backup) are given in more detail in Tables 1 and 2. It is also noted that the value of the R-factor when using the

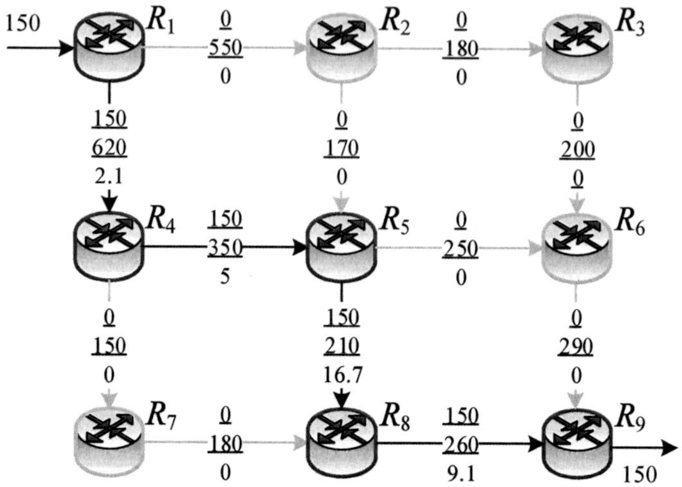

Fig. 1. The order of QoE Fast ReRouting of one VoIP-Flow that is transmitted over the primary path.

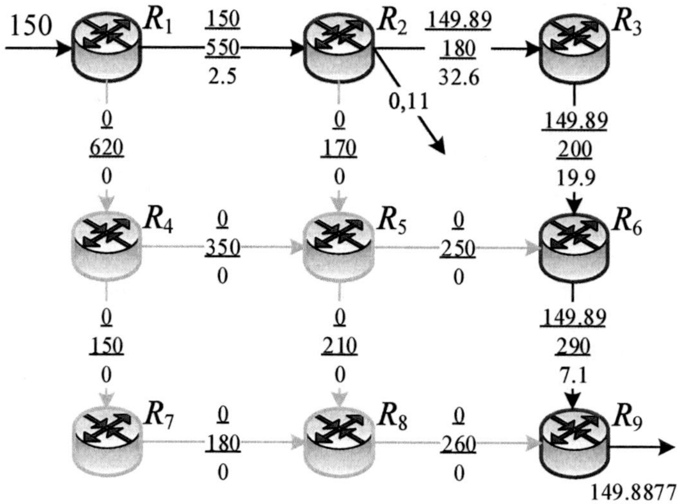

Fig. 2. The order of QoE Fast ReRouting of one VoIP-Flow that is transmitted over the backup path.

primary route was $R = 93.1739$, and when using the backup path $\overline{R} = 91.5749$ that met the requirements ($R_{\langle req \rangle} = 90$).

It should be noted that the primary path did not intersect with the backup, i.e. all the necessary protection schemes were fully implemented.

According to the content of Table 1 and Table 2, users of the VoIP-Flow, both when using the primary and backup paths, are very satisfied.

Table 1. Results of solving the problem of QoE Fast ReRouting of VoIP-Flow in TCN.

Link#	Primary route				Backup route				$\varphi_{i,j}$ (1/s)
	$\lambda_{i,j}$ (1/s)	$r_{i,j}$ (1/s)	$\tau_{i,j}$ (ms)	$p_{i,j}$	$\overline{\lambda}_{i,j}$ (1/s)	$\overline{r}_{i,j}$ (1/s)	$\overline{\tau}_{i,j}$ (ms)	$\overline{p}_{i,j}$	
(1, 2)	0	0	0	0	150	0	2.5	0	550
(2, 3)	0	0	0	0	149.89	0.11	32.6	0.0007	180
(1, 4)	150	0	2.1	0	0	0	0	0	620
(2, 5)	0	0	0	0	0	0	0	0	170
(3, 6)	0	0	0	0	149.89	0	19.9	0	200
(4, 5)	150	0	5	0	0	0	0	0	350
(5, 6)	0	0	0	0	0	0	0	0	250
(4, 7)	0	0	0	0	0	0	0	0	150
(5, 8)	150	0	16.7	0	0	0	0	0	210
(6, 9)	0	0	0	0	149.89	0	7.1	0	290
(7, 8)	0	0	0	0	0	0	0	0	180
(8, 9)	150	0	9.1	0	0	0	0	0	260

Table 2. Characteristics of the paths obtained in the process of solving the problem of QoE Fast ReRouting of VoIP-flow in TCN.

Path Type	$\lambda^{\langle req \rangle}$ (1/s)	λ^* (1/s)	Characteristics of the calculated paths			T_a (ms)	P_{pl} (%)	R
			Path	λ_p (1/s)	τ_p (ms)			
Primary	150	150	$R_1 \to R_4 \to R_5 \to$ $\to R_8 \to R_9$	150	32.9	32.9	0	93.1739
Backup	150	149.8877	$R_1 \to R_2 \to R_3 \to$ $\to R_6 \to R_9$	149.8877	62.1	62.1	0.07	91.5749

4 Conclusion

The novelty of the proposed model is the formulation of the problem of QoE Fast ReRouting in the optimization form when the criterion of optimality was the maximum additive form, represented by the sum of weighted values of R-factors concerning the IP-priority for each of the VoIP-Flows transmitted over the primary or backup routes. The backup routes did not include elements that were protected, for example, due to either compromisation or overload. The formulated optimization problem belonged to the class of nonlinear programming problems since both the objective function that was maximized and the constraints were represented by nonlinear functions from routing variables. The proposed approach to the formulation and solution of the problem of fast rerouting focuses on the most complete and efficient use of available network resources, which identifies as the preferred field of its use is the corporate networks. The results of the study of the proposed tensor model of QoE Fast ReRouting of VoIP-Flows confirmed its efficiency and adequacy, which was especially evident in complex network topologies, high TCN load and flow differentiation relative to IP-priority packet values.

References

1. Dodd, A.Z.: The Essential Guide to Telecommunications (Essential Guide Series). Prentice Hall, Hoboken (2019)
2. Mas-Machuca, C., Musumeci, F., Vizarreta, P., Pezaros, D., Jouët, S., Tornatore, M., Hmaity, A., Liyanage, M., Gurtov, A., Braeken, A.: Reliable control and data planes for softwarized networks. In: Guide to Disaster-Resilient Communication Networks, pp. 243–270. Springer, Cham (2020). https://doi.org/10.1007/978-3-030-44685-7_10
3. Strelkovskaya, I., Solovskaya, I.: Using spline-extrapolation in the research of self-similar traffic characteristics. J. Electr. Eng. **70**(4), 310–316 (2019). https://doi.org/10.2478/jee-2019-0061
4. Strelkovskaya, I., Solovskaya, I., Makoganiuk, A.: Spline-extrapolation method in traffic forecasting in 5G networks. J. Telecommun. Inf. Technol. **3**, 8–16 (2019). https://doi.org/10.26636/jtit.2019.134719
5. Strelkovskaya, I., Solovskaya, I., Makoganiuk, A.: Different extrapolation methods in problems of forecasting. In: Ilchenko, M., Uryvsky, L., Globa, L. (eds.) Advances in Information and Communication Technology and Systems. MCT 2019. Lecture Notes in Networks and Systems, vol. 152, pp. 217–228. Springer, Cham (2021). https://doi.org/10.1007/978-3-030-58359-0_12
6. Globa, L., Skulysh, M., Romanov, O., Nesterenko, M.: Quality control for mobile communication management services in hybrid environment. In: Ilchenko, M., Uryvsky, L., Globa, L. (eds.) Advances in Information and Communication Technologies. UKRMICO 2018. Lecture Notes in Electrical Engineering, vol. 560, pp. 76–100. Springer, Cham (2019). https://doi.org/10.1007/978-3-030-16770-7_4
7. Semenov, A.A., Semenova, O.O., Voznyak, O.M., Vasilevskyi, O.M., Yakovlev, M.Y.: Routing in telecommunication networks using fuzzy logic. In: 2016 17th International Conference of Young Specialists on Micro/Nanotechnologies and Electron Devices (EDM) Proceedings, pp. 173–177. IEEE (2016). https://doi.org/10.1109/EDM.2016.7538719
8. Chiesa, M., Kamisiński, A., Rak, J., Rétvári, G., Schmid, S.: Fast recovery mechanisms in the data plane. TechRxiv. Preprint, pp. 1–46 (2020). https://doi.org/10.36227/techrxiv.12367508.v2
9. Papán, J., Segeč, P., Palúch, P., Mikuš, Ľ., Moravčík, M.: The survey of current IPFRR mechanisms. In: Federated Conference on Software Development and Object Technologies, pp. 229–240. Springer, Cham (2015). https://doi.org/10.1007/978-3-319-46535-7_18
10. Monge, A.S., Szarkowicz, K.G.: MPLS in the SDN Era: Interoperable Scenarios to Make Networks Scale to New Services. O'Reilly Media Inc., Sebastopol (2015)
11. Cruz, P., Gomes, T., Medhi, D.: A heuristic for widest edge-disjoint path pair lexicographic optimization. In: 2014 6th International Workshop on Reliable Networks Design and Modeling (RNDM) Proceedings, pp. 9–15. IEEE (2014). https://doi.org/10.1109/RNDM.2014.7014925
12. Lemeshko, O., Yeremenko, O.: Enhanced method of fast re-routing with load balancing in software-defined networks. J. Electr. Eng. **68**(6), 444–454 (2017). https://doi.org/10.1515/jee-2017-0079

13. Lemeshko, O., Yevdokymenko, M., Yeremenko, O., Hailan, A.M., Segeč, P., Papán, J.: Design of the fast ReRoute QoS protection scheme for bandwidth and probability of packet loss in software-defined WAN. In: 2019 IEEE 15th International Conference on the Experience of Designing and Application of CAD Systems (CADSM) Proceedings, pp. 1–5. IEEE (2019). https://doi.org/10.1109/CADSM.2019.8779321
14. Lemeshko, O., Yeremenko, O., Tariki, N.: Solution for the default gateway protection within fault-tolerant routing in an IP network. Int. J. Electr. Comput. Eng. Syst. **8**(1), 19–26 (2017). https://doi.org/10.32985/ijeces.8.1.3
15. Yeremenko, O., Lemeshko, O., Persikov, A.: Secure routing in reliable networks: proactive and reactive approach. In: Shakhovska, N., Stepashko, V. (eds.) Advances in Intelligent Systems and Computing II. CSIT 2017. Advances in Intelligent Systems and Computing, vol. 689, pp. 631–655. Springer, Cham (2018). https://doi.org/10.1007/978-3-319-70581-1_44
16. Lemeshko, O., Yevdokymenko, M., Yeremenko, O., Nevzorova, O., Snihurov, A., Kovalenko, T.: Fast ReRoute model with VoIP quality of experience protection. In: 2019 3rd International Conference on Advanced Information and Communications Technologies (AICT) Proceedings, pp. 16–21 (2019). https://doi.org/10.1109/aiact.2019.8847918
17. Lemeshko, O., Yevdokymenko, M., Yeremenko, O.: Fast ReRoute model with multimedia quality protection. In: 2020 IEEE 15th International Conference on Advanced Trends in Radioelectronics, Telecommunications and Computer Engineering (TCSET) Proceedings, pp. 287–292. IEEE (2020). https://doi.org/10.1109/TCSET49122.2020.235441
18. Yevdokymenko, M.: Enhancement of the adaptive routing tensor model in the infocommunication network with providing quality of experience by the R-factor. Technol. Audit Prod. Reserves **3**, (2(53)), 15–22 (2020). https://doi.org/10.15587/2706-5448.2020.206031
19. Lemeshko, O., Yevdokymenko, M., Yeremenko, O., Al-Dulaimi, A.M.K., Segeč, P., Papán, J.: QoE-centric adaptive multipath routing under R-Factor with load balancing in SDN. In: 2020 IEEE International Conference on Problems of Infocommunications. Science and Technology (PIC S&T) Proceedings, pp. 1–6. IEEE (2020)
20. Barakabitze, A.A., Barman, N., Ahmad, A., Zadtootaghaj, S., Sun, L., Martini, M.G., Atzori, L.: QoE management of multimedia streaming services in future networks: A tutorial and survey. IEEE Commun. Surv. Tutor. **22**(1), 526–565 (2019). https://doi.org/10.1109/COMST.2019.2958784
21. Barakabitze, A.A., Sun, L., Mkwawa, I.H., Ifeachor, E.: A novel QoE-centric SDN-based multipath routing approach for multimedia services over 5G networks. In: 2018 IEEE International Conference on Communications (ICC) Proceedings, pp. 1–7. IEEE (2018). https://doi.org/10.1109/ICC.2018.8422617
22. Oliveira, T., Sargento, S.: QoE-based load balancing of OTT video content in SDN networks. In: 2019 IEEE Symposium on Computers and Communications (ISCC) Proceedings, pp. 1–6. IEEE (2019). https://doi.org/10.1109/ISCC47284.2019.8969720
23. Robinson, M., Milosavljevic, M., Kourtessis, P., Fisher, S., Stafford, G.P., Treiber, J., Burrell, M.J., Senior, J.M.: QoE based holistic traffic engineering in SDN enabled heterogeneous transport networks. In: 2017 19th International Conference on Transparent Optical Networks (ICTON) Proceedings, pp. 1–4. IEEE (2017). https://doi.org/10.1109/ICTON.2017.8024878
24. Liotou, E., Samdanis, K., Pateromichelakis, E., Passas, N., Merakos, L.: QoE-SDN APP: a rate-guided QoE-aware SDN-APP for HTTP adaptive video streaming. IEEE J. Sel. Areas Commun. **36**(3), 598–615 (2018). https://doi.org/10.1109/JSAC.2018.2815421
25. Abar, T., Letaifa, A.B., Elasmi, S.: Enhancing QoE based on machine learning and DASH in SDN networks. In: 2018 32nd International Conference on Advanced Information Networking and Applications Workshops (WAINA) Proceedings, pp. 258–263. IEEE (2018). https://doi.org/10.1109/WAINA.2018.00095

26. Ramakrishnan, S., Zhu, X., Chan, F., Kambhatla, K.: SDN based QoE optimization for HTTP-based adaptive video streaming. In: 2015 IEEE International Symposium on Multimedia (ISM) Proceedings, pp. 120–123. IEEE (2015). https://doi.org/10.1109/ISM.2015.53
27. Strelkovskaya, I.V., Solovskaya, I.N.: Tensor model of multiservice network with different classes of traffic service. Radioelectron. Commun. Syst. **56**(6), 296–303 (2013). https://doi.org/10.3103/S0735272713060058
28. Lemeshko, O.V., Yeremenko, O.S., Hailan, A.M.: QoS solution of traffic management based on the dynamic tensor model in the coordinate system of interpolar paths and internal node pairs. In: 2016 International Conference Radio Electronics & Info Communications (UkrMiCo) Proceedings, pp. 1–6. IEEE (2016). https://doi.org/10.1109/UkrMiCo.2016.7739625
29. Lemeshko, O., Yeremenko, O.: Routing tensor model presented in the basis of interpolar paths and internal node pairs. In: 2016 Third International Scientific-Practical Conference Problems of Infocommunications Science and Technology (PIC S&T) Proceedings, pp. 201–204. IEEE (2016). https://doi.org/10.1109/INFOCOMMST.2016.7905381
30. Lemeshko, O.V., Yevseyeva, O.Y.O., Garkusha, S.V.: A tensor model of multipath routing based on multiple QoS metrics. In: 2013 International Siberian Conference on Control and Communications (SIBCON) Proceedings, pp. 1–4. IEEE (2013). https://doi.org/10.1109/SIBCON.2013.6693645
31. ITU-T G.107: The E-model: a computational model for use in transmission planning (2015)

Kolmogorov–Wiener Filter Weight Function for Stationary Traffic Forecasting: Polynomial and Trigonometric Solutions

V. Gorev(✉), A. Gusev(✉), V. Korniienko(✉), and M. Aleksieiev(✉)

Dnipro University of Technology, av. Dmytra Yavornytskoho, 19, Dnipro 49005, Ukraine

Abstract. The paper is devoted to obtaining the Kolmogorov–Wiener filter weight function for stationary telecommunication traffic forecasting. The problem of telecommunication traffic forecasting is an urgent problem for telecommunications. There are a variety of traffic models and approaches to traffic forecasting, and in rather simple models traffic is considered to be a stationary random process. The Kolmogorov–Wiener filter may be used for the forecasting of stationary random processes, so it may be used for stationary traffic forecasting. However, the corresponding approach has not been sufficiently developed in the literature. Two models of stationary traffic are considered in the paper: a model where traffic is considered to be a process with a power-law structure function and a model where traffic is considered to be fractional Gaussian noise. In the case of a large amount of data, traffic may be described as a continuous random process, and the Kolmogorov–Wiener filter weight function is a solution to the corresponding Wiener–Hopf integral equation. An approximate solution multiply to the corresponding equation may be sought in the form of a truncated expansion in an orthogonal functional series. The paper contains a description of the authors' previous results obtained on the basis of a truncated polynomial expansion and new results obtained on the basis of an expansion in trigonometric functions.

The aim of the work is to obtain an approximate solution to the Kolmogorov–Wiener filter weight function for stationary traffic forecasting in the framework of the model of a power-law structure function and the model of fractional Gaussian noise and to investigate the validity of the obtained results.

The object of the work is the Kolmogorov–Wiener filter used for the stationary telecommunication traffic forecasting.

The subject of the work is the weight function of the filter under consideration.

The scientific novelty of the work is the fact that for the first time the corresponding weight function is sought as a truncated trigonometric Fourier series. Trigonometric and polynomial solutions are compared. It is shown that trigonometric solutions are better than the polynomial ones for the model where the telecommunication traffic is considered as a process with a power-law structure function.

Keywords: Kolmogorov–Wiener filter weight function · Truncated orthogonal polynomial series · Truncated trigonometric Fourier series · Power-law structure function · Fractional Gaussian noise · Telecommunication traffic

P. Vorobiyenko et al. (Eds.): IPF 2020, LNNS 212, pp. 111–129, 2021.
https://doi.org/10.1007/978-3-030-76343-5_7

1 Introduction

The problem of telecommunication traffic forecasting is an important problem for telecommunications. For example, it allows one to optimize the network resources and to detect security attacks by comparing the actual traffic with the predicted one [1].

In the telecommunication systems with data burst transfer, traffic is considered to be a self-similar random process. There are a variety of corresponding telecommunication traffic models [2], and it should be stressed that in rather simple models traffic is considered as a stationary random process. In the case of large amount of data, traffic may be expressed as a continuous random process [3], so in rather simple models telecommunication traffic may be treated as a continuous stationary random process.

In this work we consider two models of stationary traffic. The first one is a model where the traffic is treated as a stationary random process with a power-law structure function [3]. In the second model, the traffic is treated as fractional Gaussian noise. In the case where the Hurst exponent $H > 0.5$, the traffic correlation function in the framework of this model is a positively-defined function [4]. For simplicity, in the framework of the fractional Gaussian noise model we restrict ourselves only to the case where $H > 0.5$.

The Kolmogorov–Wiener filter may be used for the forecasting of stationary processes [5]. This filter is rather simple, it is linear and stationary. So it is natural enough to apply the corresponding filter to the stationary traffic forecasting. There are a variety of different approaches to telecommunication traffic forecasting [1, 6], but the approach based on the Kolmogorov–Wiener filter is not sufficiently developed in the literature. So the construction of this filter for stationary traffic forecasting contains scientific novelty.

The weight function of the filter under consideration is the solution of the corresponding Wiener–Hopf integral equation [4]. This equation is a Fredholm integral equation of the first kind, and a search for its exact analytical solution meets difficulties. That is why we search for an approximate solution to this equation. The corresponding approximate solution may be found with the help of the Galerkin method [7], in the framework of which the unknown function is expanded into a truncated orthogonal functional series.

Our previous works [8–11] are devoted to the investigation of polynomial expansions. It should be stressed that the truncated polynomial expansion method of solving the Fredholm integral equation is popular in the literature; see, for example, its use in the framework of statistical physics [12, 13]. Papers [8–10] are devoted to the power-law structure function model. In paper [8] we used orthonormal polynomials without weight, in paper [9] we used the Chebyshev polynomials of the second kind orthogonal on the required time interval, and in paper [10] we used the Chebyshev polynomials of the first kind orthogonal on the required time interval. In paper [11] we used the truncated polynomial expansion method based on the Chebyshev polynomials of the second kind in the framework of the fractional Gaussian noise model. The results are as follows. The truncated polynomial expansion method is convergent for the fractional Gaussian noise model, but it is not necessary convergent in the framework of the power-law structure function model. Moreover, in the framework of the power-law structure model some approximations are reliable, but some approximations completely fail. So, the following question occurs: May some non-polynomial expansion give better results than a polynomial one?

The trigonometric Fourier series is a well-known example of non-polynomial orthogonal functional series. So in this work we search for the unknown weight function under consideration on the basis of a truncated trigonometric Fourier series expansion. The aim of the work is to obtain the Kolmogorov–Wiener filter weight function by the corresponding expansion and to compare the results of the polynomial and trigonometric solutions for the above-mentioned stationary continuous traffic models.

The paper is organized as follows. In Sect. 1, the introduction is given. In Sect. 2, the Wiener–Hopf integral equation and traffic correlation functions in the framework of the models under consideration are described. In Sect. 3, our previous results in the framework of the truncated polynomial expansion method are described. In Sect. 4, our new results in the framework of the truncated trigonometric Fourier series expansion method are given, and in Sect. 5 conclusions are made.

2 Wiener–Hopf Integral Equation and Traffic Correlation Function

The Kolmogorov–Wiener filter weight function obeys the following Wiener–Hopf integral equation [5]:

$$\int_0^T d\tau h(\tau) R(t-\tau) = R(t+z) \tag{1}$$

where $h(\tau)$ is the unknown filter weight function, $R(\tau)$ is the correlation function of the stationary process for which the filter is used, the input data is given for the time interval $t \in [0, T]$, and z is the time interval for which the forecast is made. It should be stressed that the non-noisy case is investigated, i.e., there is no noise added to the stationary process under consideration.

In this work we consider two models of the stationary traffic. The first one is a model where the traffic is treated as stationary random process with a power law structure function [3]

$$C(\tau) = \left\langle (x(t) - x(t-\tau))^2 \right\rangle = \alpha|\tau|^{2H} \tag{2}$$

where is the proportionality constant, and H is the Hurst exponent; the corresponding traffic correlation function in the framework of this model is as follows

$$R(\tau) = \sigma^2 - \frac{1}{2}\alpha|\tau|^{2H} \tag{3}$$

where σ^2 is the process variance. In the second model, the traffic is treated as fractional Gaussian noise. In the case where the Hurst exponent $H > 0.5$, the continuous traffic correlation function in the framework of this model is as follows [4]

$$R(\tau) = 2H(2H-1)\sigma^2|\tau|^{2H-2}, \tag{4}$$

the correlation function (4) is a positively-defined function. For simplicity, in the framework of the fractional Gaussian noise model we restrict ourselves only to the case where $H > 0.5$.

A search for an exact solution for the integral Eq. (1) meets difficulties, so we investigate an approximate solution for Eq. (1). In the framework of the Galerkin method [7], such a solution is sought in the form of a truncated series in an orthogonal function system. The following sections are devoted to the investigation of the corresponding solutions based on orthogonal polynomial systems and trigonometric Fourier series.

3 Polynomial Solutions for the Weight Function

This section is devoted to the description of the corresponding polynomial solutions. The unknown weight function is sought as a truncated polynomial series:

$$h(\tau) = \sum_{s=0}^{l-1} g_s S_s(\tau) \tag{5}$$

where $S_s(\tau)$ are polynomials orthogonal on $t \in [0, T]$, and g_s are unknown coefficients multiplying the polynomials. The solution (5) is also called the solution in the l-polynomial approximation.

On substitution of (5) into (1), one can obtain

$$\sum_{s=0}^{l-1} g_s \int_0^T d\tau S_s(\tau) R(t-\tau) = R(t+z) \tag{6}$$

which after multiplying by $S_k(t)$ and integrating over t leads to the following set of linear equations in the unknown coefficients g_s:

$$\sum_{s=0}^{l-1} g_s G_{ks} = B_k,\ k = \overline{0, l-1} \tag{7}$$

where

$$G_{ks} = \int_0^T \int_0^T dt d\tau S_s(\tau) S_k(t) R(t-\tau),\ B_k = \int_0^T dt S_k(t) R(t+z), \tag{8}$$

the quantities G_{ks} are called the integral brackets. The system (7) may be solved with the help of the matrix method:

$$\begin{pmatrix} g_0 \\ g_1 \\ \vdots \\ g_{l-1} \end{pmatrix} = \begin{pmatrix} G_{00} & G_{01} & \cdots & G_{0,l-1} \\ G_{10} & G_{11} & \cdots & G_{1,l-1} \\ \vdots & \vdots & \ddots & \vdots \\ G_{l-1,0} & G_{l-1,1} & \cdots & G_{l-1,l-1} \end{pmatrix}^{-1} \cdot \begin{pmatrix} B_0 \\ B_1 \\ \vdots \\ B_{l-1} \end{pmatrix}. \tag{9}$$

In our previous works, the following orthogonal polynomial systems orthogonal on $t \in [0, T]$ were used. In paper [8], we used the polynomials $S_s^{(1)}(\tau)$ orthonormal without weigh:

$$S_s^{(1)}(\tau) = \frac{S_s'(\tau)}{\sqrt{\int_0^T dt\left(S_s'(t)\right)^2}}, \quad S_s'(\tau) = \begin{vmatrix} \mu_0 & \mu_1 & \mu_2 & \cdots & \mu_s \\ \mu_1 & \mu_2 & \mu_3 & \cdots & \mu_{s+1} \\ \vdots & \vdots & \vdots & \ddots & \vdots \\ \mu_{s-1} & \mu_s & \mu_{s+1} & \cdots & \mu_{2s-1} \\ 1 & \tau & \tau^2 & \cdots & \tau^s \end{vmatrix} \tag{10}$$

where

$$\mu_n = \int_0^T x^n dx = \frac{T^{n+1}}{n+1}. \tag{11}$$

In papers [9, 11], the polynomials $S_s^{(2)}$ were used:

$$S_s^{(2)}(\tau) = U_s\left(\frac{2\tau}{T} - 1\right) \tag{12}$$

where $U_s(x)$ are the Chebyshev polynomials of the second kind. In paper [10], the polynomials $S_s^{(2)}$ were used:

$$S_s^{(2)}(\tau) = T_s\left(\frac{2\tau}{T} - 1\right) \tag{13}$$

where $T_s(x)$ are the Chebyshev polynomials of the first kind.

In order to check the validity of the obtained results for the weight function, we numerically compare the left-hand side and the right-hand side of Eq. (1) for the obtained weight functions $h(\tau)$.

The correlation function (3) is not a positively-defined one. So, in order to check the validity of the obtained solutions in the framework of the power-law structure function model, we calculate the mean average error (MAE) as follows:

$$\text{MAE} = \frac{1}{T}\int_0^T dt\left|\text{Left}(t) - \text{Right}(t)\right| \tag{14}$$

where Left(t) and Right(t) are the left-hand side and the right-hand side of the integral Eq. (1) with the correlation function (3):

$$\begin{gathered} \text{Left}(t) = \sum_{s=0}^{l-1} g_s \int_0^T d\tau S_s(\tau)\left(\sigma^2 - \tfrac{\alpha}{2}|t-\tau|^{2H}\right) \\ = \sum_{s=0}^{l-1} g_s \int_0^t d\tau S_s(\tau)\left(\sigma^2 - \tfrac{\alpha}{2}(t-\tau)^{2H}\right) + \sum_{s=0}^{l-1} g_s \int_t^T d\tau S_s(\tau)\left(\sigma^2 - \tfrac{\alpha}{2}(\tau-t)^{2H}\right), \\ \text{Right}(t) = R(t+z) = \sigma^2 - \tfrac{\alpha}{2}(t+z)^{2H}. \end{gathered} \tag{15}$$

The following parameters were investigated in [8–10]:

$$T = 100,\ z = 3,\ H = 0.8,\ \sigma = 1.2,\ \alpha = 3 \cdot 10^{-3}. \quad (16)$$

These parameters are chosen in order to meet the inequalities $z \ll T$, $|R(t)| \le R(0)$. All the calculations are made in the Wolfram Mathematica package. The following MAE errors are obtained, see Table 1, Table 2 and Table 3.

Table 1. The results for the MAE for approximations of different numbers of polynomials $S_s^{(1)}(t)$ for parameters (16) in the framework of the power-law structure model

l	MAE	l	MAE	l	MAE	l	MAE	l	MAE
1	0.64	5	$8.8 \cdot 10^{-3}$	9	$1.6 \cdot 10^{2}$	13	$2.1 \cdot 10^{5}$	17	$2.5 \cdot 10^{-4}$
2	$2.8 \cdot 10^{-2}$	6	$8.1 \cdot 10^{-3}$	10	$2.3 \cdot 10^{2}$	14	$6.3 \cdot 10^{6}$	18	$1.0 \cdot 10^{-3}$
3	$7.9 \cdot 10^{-2}$	7	$2.8 \cdot 10^{-3}$	11	$6.4 \cdot 10^{3}$	15	$3.3 \cdot 10^{7}$		
4	$7.7 \cdot 10^{-2}$	8	$2.5 \cdot 10^{-3}$	12	$4.3 \cdot 10^{4}$	16	$2.9 \cdot 10^{-4}$		

Table 2. The results for the MAE for approximations of different numbers of polynomials $S_s^{(2)}(t)$ for parameters (16) in the framework of the power-law structure model

l	MAE	l	MAE	l	MAE	l	MAE	l	MAE
1	0.64	5	$8.8 \cdot 10^{-3}$	9	$1.7 \cdot 10^{2}$	13	$2.1 \cdot 10^{5}$	17	$2.7 \cdot 10^{-4}$
2	$2.8 \cdot 10^{-2}$	6	$8.1 \cdot 10^{-3}$	10	$2.3 \cdot 10^{2}$	14	$1.1 \cdot 10^{6}$	18	$1.1 \cdot 10^{-3}$
3	$7.9 \cdot 10^{-2}$	7	$2.8 \cdot 10^{-3}$	11	$6.4 \cdot 10^{3}$	15	$5.9 \cdot 10^{6}$		
4	$7.7 \cdot 10^{-2}$	8	$2.4 \cdot 10^{-3}$	12	$4.3 \cdot 10^{4}$	16	$2.8 \cdot 10^{-4}$		

Table 3. The results for the MAE for approximations of different numbers of polynomials $S_s^{(3)}(t)$ for parameters (16) in the framework of the power-law structure model

l	MAE	l	MAE	l	MAE	l	MAE	l	MAE
1	0.64	5	$8.8 \cdot 10^{-3}$	9	$1.7 \cdot 10^{2}$	13	$5.3 \cdot 10^{5}$	17	$2.5 \cdot 10^{-4}$
2	$2.8 \cdot 10^{-2}$	6	$8.1 \cdot 10^{-3}$	10	$2.3 \cdot 10^{2}$	14	$6.3 \cdot 10^{6}$	18	$1.1 \cdot 10^{-3}$
3	$7.9 \cdot 10^{-2}$	7	$2.8 \cdot 10^{-3}$	11	$6.3 \cdot 10^{3}$	15	$3.3 \cdot 10^{7}$		
4	$7.7 \cdot 10^{-2}$	8	$2.5 \cdot 10^{-3}$	12	$4.3 \cdot 10^{4}$	16	$2.9 \cdot 10^{-4}$		

As can be seen, the results for all the three polynomial systems are almost identical. It should be stressed that here and in what follows the numbers in the tables are rounded off to 2 significant digits.

The accuracy of the approximations is not necessarily increasing with the number of polynomials, and some approximations completely fail [approximations from 9 to 15 polynomials for parameters (16)]. However, some approximations give a good agreement between the left-hand side and the right-hand side of the integral equation under consideration. For example, the corresponding graphs for the 18-polynomial approximation for the polynomials $S_s^{(2)}(t)$ are given on Fig. 1.

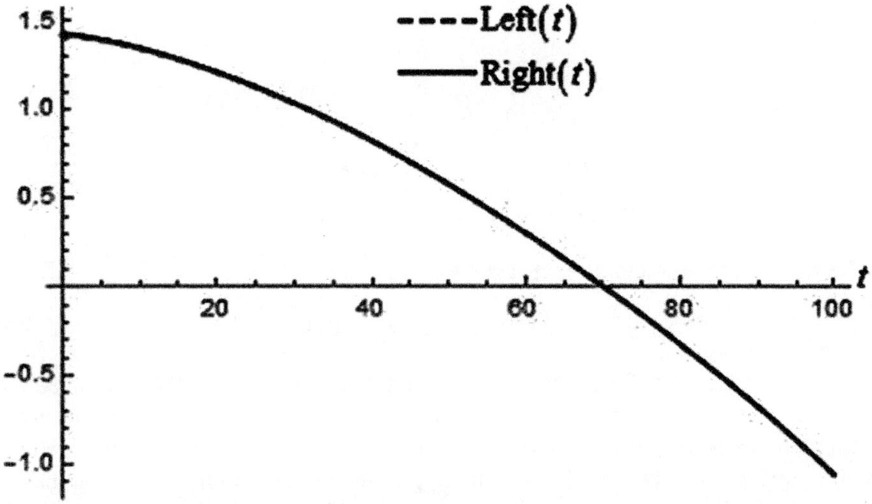

Fig. 1. Graphs of Left(t) and Right(t) for the 18-polynomial approximation.

As can be seen from Fig. 1, the corresponding coincidence of the left-hand and the right-hand sides of the integral equation under consideration is almost ideal.

The following set of parameters

$$T = 1000,\ z = 3,\ H = 0.8,\ \sigma = 1.2,\ \alpha = 8 \cdot 10^{-5} \tag{17}$$

was also investigated in [8, 9]. It is obtained that for these parameters, in fact, only the two-polynomial approximation is rather accurate, while the approximations starting from the five-polynomial one completely fail.

Our paper [11] is devoted to obtaining an approximate solution for the Kolmogorov–Wiener filter weight function in the framework of the fractional Gaussian noise model. The polynomials $S_s^{(2)}(\tau)$ are used in [11]. The substitution of the corresponding correlation function (4) into (1) leads to the following integral equation in $h(\tau)$:

$$\int_0^T d\tau h(\tau)|t-\tau|^{2H-2} = (t+z)^{2H-2} \tag{18}$$

So, if we redefine the function $R(t)$ as follows:

$$R(\tau) = |\tau|^{2H-2} \tag{19}$$

we will obtain the solution of the integral Eq. (18) by the above-mentioned algorithm. The function (19) is positively-defined, so the mean average percentage error (MAPE) is calculated:

$$\mathrm{MAPE} = \frac{1}{T}\int_0^T dt \left| \frac{\mathrm{Left}(t) - \mathrm{Right}(t)}{\mathrm{Right}(t)} \right| \cdot 100\% \tag{20}$$

where

$$\begin{gathered} \mathrm{Left}(t) = \sum_{s=0}^{l-1} g_s \int_0^T d\tau S_s(\tau)|t-\tau|^{2H-2} \\ = \sum_{s=0}^{l-1} g_s \int_0^t d\tau S_s(\tau)(t-\tau)^{2H-2} + \sum_{s=0}^{l-1} g_s \int_t^T d\tau S_s(\tau)(\tau-t)^{2H-2}, \\ \mathrm{Right}(t) = R(t+z) = |t+z|^{2H-2}. \end{gathered} \tag{21}$$

The following parameters are investigated:

$$T = 100,\ z = 3,\ H = 0.8. \tag{22}$$

The following results for the MAPE errors are obtained, see Table 4. As can be seen from Table 4, in the framework of the fractional Gaussian noise model the truncated polynomial expansion method is convergent, the accuracy of approximations increases with the number of polynomials, and the approximations or rather large numbers of polynomials are rather accurate, the corresponding MAPE error is less than 1%. The corresponding graphs for the 19-polynomial approximation are given in Fig. 2; as can be seen from the figure, the curves nearly coincide.

Table 4. The results for the MAPE for approximations of different numbers of polynomials $S_s^{(2)}(t)$ for parameters (22) in the framework of the fractional Gaussian noise model

l	MAPE, %	l	MAPE, %	l	MAPE, %	l	MAPE, %	l	MAPE, %
1	28	5	6.7	9	2.5	13	1.2	17	0.71
2	19	6	5.3	10	2.1	14	1.1	18	0.64
3	13	7	4.0	11	1.7	15	0.91	19	0.57
4	9.3	8	3.2	12	1.5	16	0.81		

Now let us summarize the results of the truncated polynomial expansion method. For the power-law structure function model: the method may not be convergent, some approximations are not valid, but some approximations give reliable results. So the method can be applied, but the obtained results should be checked numerically. For the fractional Gaussian noise model: the accuracy of the method increases with the number of polynomials, and the method is convergent. In our opinion, the reason is as follows. As is known [14], the convergence of a method is guaranteed if the kernel of

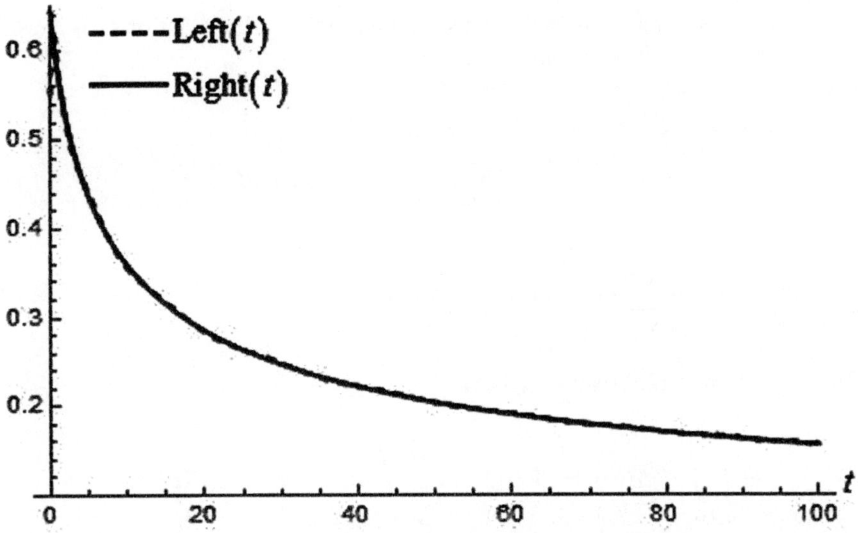

Fig. 2. Graphs of Left(t) and Right(t) for the 19-polynomial approximation.

the corresponding integral equation is a positively-defined function. The kernel of the Wiener-Hopf integral equation is the traffic correlation function. The correlation function in the framework of the fractional Gaussian noise model is positively defined in contrast to that of the power-law structure function model. So the method is convergent for the fractional Gaussian noise model and it may not be convergent in the framework of the power-law structure function model.

As can be seen from Tables 1, 2 and 3, for parameters (16) some polynomial approximations lead to huge errors. In addition, as indicated above, most of the polynomial approximations are not valid for parameters (17). Maybe, the reason is that the Wolfram Mathematica does not calculate the integrals adequately because of the product of very small and very large numbers. Anyway, the question arises: May some non-polynomial orthogonal function system give better results than a polynomial one in the framework of the power-law structure function model?

So the following section is devoted to the investigation of the unknown weight function based on the truncated trigonometric Fourier series.

4 Trigonometric Solutions for the Weight Function

In this section we search for the unknown weight function in the form of the following truncated trigonometric Fourier series:

$$h(\tau) = a_0 + \sum_{s=1}^{l} (a_s \cos(s\omega\tau) + b_s \sin(s\omega\tau)), \tag{23}$$

where

$$\omega = \frac{2\pi}{T} \tag{24}$$

and a_0, a_s, b_s are unknown coefficients. In what follows, the solution (23) is called the solution in the l–harmonic approximation.

The substitution of (23) into (1) leads to

$$a_0 \int_0^T d\tau R(t-\tau) + \sum_{s=1}^{l} a_s \int_0^T d\tau \cos(s\omega\tau) R(t-\tau) + \sum_{s=1}^{l} b_s \int_0^T d\tau \sin(s\omega\tau) R(t-\tau) = R(t+z). \quad (25)$$

The integration of (25) over t leads to

$$a_0 \int_0^T \int_0^T d\tau dt R(t-\tau) + \sum_{s=1}^{l} a_s \int_0^T \int_0^T dt d\tau \cos(s\omega\tau) R(t-\tau) + \sum_{s=1}^{l} b_s \int_0^T \int_0^T dt d\tau \sin(s\omega\tau) R(t-\tau) = \int_0^T dt R(t+z). \quad (26)$$

After multiplying (25) by $\cos(k\omega t)$, $k=\overline{1,l}$, one can obtain

$$a_0 \int_0^T \int_0^T d\tau dt \cos(k\omega t) R(t-\tau) + \sum_{s=1}^{l} a_s \int_0^T \int_0^T dt d\tau \cos(s\omega\tau)\cos(k\omega t) R(t-\tau) + \sum_{s=1}^{l} b_s \int_0^T \int_0^T dt d\tau \sin(s\omega\tau)\cos(k\omega t) R(t-\tau) = \int_0^T dt \cos(k\omega t) R(t+z). \quad (27)$$

After multiplying (25) by $\sin(k\omega t)$, $k=\overline{1,l}$, one can obtain

$$a_0 \int_0^T \int_0^T d\tau dt \sin(k\omega t) R(t-\tau) + \sum_{s=1}^{l} a_s \int_0^T \int_0^T dt d\tau \cos(s\omega\tau)\sin(k\omega t) R(t-\tau) + \sum_{s=1}^{l} b_s \int_0^T \int_0^T dt d\tau \sin(s\omega\tau)\sin(k\omega t) R(t-\tau) = \int_0^T dt \sin(k\omega t) R(t+z). \quad (28)$$

Equations (26)–(28) nay be summarized in matrix form:

$$Gg = B \quad (29)$$

where g is the column vector of the unknown coefficients, B is the column vector of the free terms:

$$g = \begin{pmatrix} a_0 \\ a_1 \\ \vdots \\ a_l \\ b_1 \\ \vdots \\ b_l \end{pmatrix}, \; B = \begin{pmatrix} B_1 \\ B_2 \\ \vdots \\ B_l \\ B_{l+1} \\ \vdots \\ B_{2l+1} \end{pmatrix} \tag{30}$$

and G is the matrix of the integral brackets:

$$G = \begin{pmatrix} G_{11} & G_{12} & \cdots & G_{1,l+1} & G_{1,l+2} & \cdots & G_{1,2l+1} \\ G_{21} & G_{22} & \cdots & G_{2,l+1} & G_{2,l+2} & \cdots & G_{2,2l+1} \\ \vdots & \vdots & \ddots & \vdots & \vdots & \ddots & \vdots \\ G_{l+1,1} & G_{l+1,2} & \cdots & G_{l+1,l+1} & G_{l+1,l+2} & \cdots & G_{l+1,2l+1} \\ G_{l+2,1} & G_{l+2,2} & \cdots & G_{l+2,l+1} & G_{l+2,l+2} & \cdots & G_{l+2,2l+1} \\ \vdots & \vdots & \ddots & \vdots & \vdots & \ddots & \vdots \\ G_{2l+1,1} & G_{2l+1,2} & \cdots & G_{2l+1,l+1} & G_{2l+1,l+2} & \cdots & G_{2l+1,2l+1} \end{pmatrix}. \tag{31}$$

The solution for the unknown coefficients is as follows:

$$g = G^{-1}B. \tag{32}$$

As can be seen from (26)–(30), the explicit expressions for the components of the column B are as follows

$$\begin{gathered} B_1 = \int_0^T dt R(t+z), \; B_{k+1} = \int_0^T dt \cos(k\omega t) R(t+z), \\ B_{k+l+1} = \int_0^T dt \sin(k\omega t) R(t+z), \; k = \overline{1,l} \end{gathered} \tag{33}$$

and the components of the integral brackets matrix G are as follows:

$$\begin{gathered} G_{11} = \int_0^T\int_0^T dt d\tau R(t-\tau), \; G_{1,s+1} = \int_0^T\int_0^T dt d\tau \cos(s\omega\tau) R(t-\tau), \\ G_{k+1,1} = \int_0^T\int_0^T dt d\tau \cos(k\omega t) R(t-\tau), \; G_{1,s+l+1} = \int_0^T\int_0^T dt d\tau \sin(s\omega\tau) R(t-\tau), \\ G_{k+l+1,1} = \int_0^T\int_0^T dt d\tau \sin(k\omega t) R(t-\tau), \; G_{k+1,s+1} = \int_0^T\int_0^T dt d\tau \cos(k\omega t)\cos(s\omega\tau) R(t-\tau), \\ G_{k+l+1,s+l+1} = \int_0^T\int_0^T dt d\tau \sin(k\omega t)\sin(s\omega\tau) R(t-\tau), \\ G_{k+1,s+l+1} = \int_0^T\int_0^T dt d\tau \cos(k\omega t)\sin(s\omega\tau) R(t-\tau), \\ G_{k+l+1,s+1} = \int_0^T\int_0^T dt d\tau \sin(k\omega t)\cos(s\omega\tau) R(t-\tau), \; k = \overline{1,l}, \; s = \overline{1,l}. \end{gathered} \tag{34}$$

The integral brackets (34) obey some properties, which follow from the fact that the correlation function of a stationary random process is an even one. First of all, let us show that the integral brackets (34) are symmetrical ones:

$$G_{ij} = G_{ji},\ i,j = \overline{1, 2l+1}. \tag{35}$$

In order to prove the property (35), we should prove the following properties on the basis of (34):

$$\begin{gathered} G_{1,s+1} = G_{s+1,1},\ G_{1,s+l+1} = G_{s+l+1,1}\ G_{k+1,s+1} = G_{s+1,k+1}, \\ G_{k+1,s+l+1} = G_{s+l+1,k+1},\ G_{k+l+1,s+l+1} = G_{s+l+1,k+l+1},\ k = \overline{1,l},\ s = \overline{1,l}. \end{gathered} \tag{36}$$

Let us prove that $G_{1,s+1} = G_{s+1,1}$.

$$\begin{gathered} G_{1,s+1} = \int_0^T \int_0^T dt d\tau \cos(s\omega\tau) R(t-\tau)^{2H-2} = \{\tau \leftrightarrow t\} \\ = \int_0^T \int_0^T dt d\tau \cos(s\omega\tau) R(\tau - t) = \int_0^T \int_0^T dt d\tau \cos(s\omega\tau) R(t-\tau) = G_{s+1,1}. \end{gathered} \tag{37}$$

see (34); the evenness of the correlation function is used in (37). The proof of the identity $G_{1,s+l+1} = G_{s+l+1,1}$ is similar to (37). Let us prove that $G_{k+1,s+1} = G_{s+1,k+1}$:

$$\begin{gathered} G_{k+1,s+1} = \int_0^T \int_0^T dt d\tau \cos(k\omega\tau) \cos(s\omega\tau) R(t-\tau) = \{t \leftrightarrow \tau\} \\ = \int_0^T \int_0^T dt d\tau \cos(k\omega\tau) \cos(s\omega\tau) R(\tau - t) = \int_0^T \int_0^T dt d\tau \cos(k\omega\tau) \cos(s\omega\tau) R(t-\tau). \end{gathered} \tag{38}$$

The proof of the other identities in (36) is similar to (38). So the property (35) is proved.

The second property of the integral brackets is as follows

$$G_{1,s+l+1} = 0,\ G_{k+1,s+l+1} = 0,\ k = \overline{1,l},\ s = \overline{1,l}. \tag{39}$$

This property also comes from the fact that the correlation function is an even one. Let us prove that $G_{1,s+l+1} = 0$:

$$\begin{gathered} G_{1,s+l+1} = \int_0^T \int_0^T dt d\tau \sin(s\omega\tau) R(t-\tau) = \left\{x = t - \tfrac{T}{2}, y = \tau - \tfrac{T}{2}\right\} \\ = \int_{-\frac{T}{2}}^{\frac{T}{2}} \int_{-\frac{T}{2}}^{\frac{T}{2}} dx dy \sin\left(s\omega y + s\omega \tfrac{T}{2}\right) R(x-y). \end{gathered} \tag{40}$$

On the basis of (24) we can rewrite (40):

$$G_{1,s+l+1} = \int_{-\frac{T}{2}}^{\frac{T}{2}} \int_{-\frac{T}{2}}^{\frac{T}{2}} dx dy \sin(s\omega y + s\pi) R(x-y). \tag{41}$$

As is known,

$$\sin\left(s\omega y+s\pi\right)=\begin{cases}-\sin\left(s\omega y\right), s\not\vdots\, 2\\ \sin\left(s\omega y\right), s\vdots\, 2\end{cases}=\left(-1\right)^{s}\sin\left(s\omega y\right)\ , \tag{42}$$

which leads to

$$\begin{gathered}G_{1,s+l+1}=(-1)^{s}\int\limits_{-\frac{T}{2}}^{\frac{T}{2}}\int\limits_{-\frac{T}{2}}^{\frac{T}{2}}dxdy\sin(s\omega y)R(x-y)=\left\{x'=-x,y'=-y\right\}\\ =-\int\limits_{-\frac{T}{2}}^{\frac{T}{2}}\int\limits_{-\frac{T}{2}}^{\frac{T}{2}}dx'dy'\sin\left(-s\omega y'\right)R\left(y'-x'\right)=-(-1)^{s}\int\limits_{-\frac{T}{2}}^{\frac{T}{2}}\int\limits_{-\frac{T}{2}}^{\frac{T}{2}}dx'dy'\sin\left(s\omega y'\right)R\left(x'-y'\right)\\ =-G_{1,s+l+1}\Rightarrow G_{1,s+l+1}=-G_{1,s+l+1}\Rightarrow G_{1,s+l+1}=0.\end{gathered} \tag{43}$$

Now let us prove that $G_{k+1,s+l+1}=0$:

$$\begin{aligned}G_{k+1,s+l+1}&=\int\limits_{0}^{T}\int\limits_{0}^{T}dtd\tau\cos(k\omega t)\sin(s\omega\tau)R(t-\tau)=\left\{x=t-\tfrac{T}{2},y=\tau-\tfrac{T}{2}\right\}\\ &=\int\limits_{-\frac{T}{2}}^{\frac{T}{2}}\int\limits_{-\frac{T}{2}}^{\frac{T}{2}}dxdy\cos\left(k\omega x+k\omega\tfrac{T}{2}\right)\sin\left(s\omega y+s\omega\tfrac{T}{2}\right)R(x-y),\end{aligned} \tag{44}$$

which with account for (24) can be rewritten as

$$G_{k+1,s+l+1}=\int\limits_{-\frac{T}{2}}^{\frac{T}{2}}\int\limits_{-\frac{T}{2}}^{\frac{T}{2}}dxdy\ \cos(k\omega x+\pi k)\sin(s\omega y+\pi s)R(x-y) \tag{45}$$

As is known,

$$\cos\left(k\omega x+\pi k\right)=\begin{cases}-\cos\left(k\omega t\right), k\not\vdots\, 2\\ \cos\left(k\omega t\right), k\vdots\, 2\end{cases}=\left(-1\right)^{k}\cos\left(k\omega x\right)\ , \tag{46}$$

which with account for (45) and (42) leads to

$$\begin{aligned}G_{k+1,s+l+1}&=(-1)^{k+s}\int\limits_{-\frac{T}{2}}^{\frac{T}{2}}\int\limits_{-\frac{T}{2}}^{\frac{T}{2}}dxdy\cos(k\omega x)\sin(s\omega y)R(x-y)=\left\{x'=-x,y'=-y\right\}\\ &=-(-1)^{k+s}\int\limits_{-\frac{T}{2}}^{\frac{T}{2}}\int\limits_{-\frac{T}{2}}^{\frac{T}{2}}dx'dy'\cos\left(k\omega x'\right)\sin\left(s\omega y'\right)R\left(y'-x'\right)\\ &=-(-1)^{k+s}\int\limits_{-\frac{T}{2}}^{\frac{T}{2}}\int\limits_{-\frac{T}{2}}^{\frac{T}{2}}dx'dy'\cos\left(k\omega x'\right)\sin\left(s\omega y'\right)R\left(x'-y'\right)=-G_{k+1,s+l+1}\Rightarrow G_{k+1,s+l+1}=0\end{aligned} \tag{47}$$

So the property (39) is proved.

On the basis of the properties (35) and (39), one can conclude that the integral bracket matrix G may be rewritten as follows:

$$G = \begin{pmatrix} G_{11} & G_{12} & \cdots & G_{1,l+1} & 0 & \cdots & 0 \\ G_{12} & G_{22} & \cdots & G_{2,l+1} & 0 & \cdots & 0 \\ \vdots & \vdots & \ddots & \vdots & \vdots & \ddots & \vdots \\ G_{1,l+1} & G_{2,l+1} & \cdots & G_{l+1,l+1} & 0 & \cdots & 0 \\ 0 & 0 & \cdots & 0 & G_{l+2,l+2} & \cdots & G_{l+2,2l+1} \\ \vdots & \vdots & \ddots & \vdots & \vdots & \ddots & \vdots \\ 0 & 0 & \cdots & 0 & G_{l+2,2l+1} & \cdots & G_{2l+1,2l+1} \end{pmatrix}, \quad (48)$$

so a straightforward calculation is needed not for $(2l+1)^2$ brackets, but only for the number of brackets equal to

$$\begin{aligned} ((l+1)+l+(l-1)+\ldots+1)+(l+(l-1)+\ldots+1) &= \tfrac{(l+1)(l+2)}{2}+\tfrac{l(l+1)}{2} \\ &= (l+1)(l+2+l)/2=(l+1)^2, \end{aligned} \quad (49)$$

which significantly reduces the computation time.

In the framework of the power-law structure model, the corresponding MAE error is given by formula (14) where

$$\begin{aligned} \mathrm{Left}(t) &= \int_0^T \left(a_0+\sum_{s=1}^{l} a_s\cos(s\omega\tau)+\sum_{s=1}^{l} b_s\sin(s\omega\tau)\right)R(t-\tau)d\tau, \\ \mathrm{Right}(t) &= R(t+z)=\sigma^2-\tfrac{\alpha}{2}(t+z)^{2H}. \end{aligned} \quad (50)$$

The function $\mathrm{Left}(t)$ in (50) after a rather cumbersome calculation may be expressed in terms of the hypergeometric function ${}_1F_1$:

$$\begin{aligned} \mathrm{Left}(t) &= \sigma^2 T a_0 - a_0\tfrac{\alpha}{2}\tfrac{(T-t)^{2H+1}+t^{2H+1}}{2H+1} \\ &-\tfrac{\alpha}{4(2H+1)}\sum_{s=1}^{l}\left[\{a_s\cos(s\omega t)+b_s\sin(s\omega t)\}(F_+(s,t)+F_+(s,T-t))\right] \\ &+\tfrac{\alpha}{4(2H+1)}\sum_{s=1}^{l}\left[\{a_s\sin(s\omega t)-b_s\cos(s\omega t)\}(F_-(s,t)-F_-(s,T-t))\right] \end{aligned} \quad (51)$$

where

$$F_+(s,t)=t^{2H+1}\{{}_1F_1(2H+1,2H+2,is\omega t)+{}_1F_1(2H+1,2H+2,-is\omega t)\}, \quad (52)$$

and

$$F_-(s,t)=it^{2H+1}\{{}_1F_1(2H+1,2H+2,is\omega t)-{}_1F_1(2H+1,2H+2,-is\omega t)\}, \quad (53)$$

it should be stressed that the functions (52) and (53) are real-valued. The expression (51) may be obtained from (50) with the help of the following tabulated integrals [15]:

$$\begin{aligned}&\int_0^1 x^{\mu-1}\sin(ax)dx = -\frac{i}{2\mu}({}_1F_1(\mu,\mu+1,ia) - {}_1F_1(\mu,\mu+1,-ia)),\\&\int_0^1 x^{\mu-1}\cos(ax)dx = \frac{1}{2\mu}({}_1F_1(\mu,\mu+1,ia) + {}_1F_1(\mu,\mu+1,-ia)).\end{aligned} \tag{54}$$

The corresponding trigonometric solutions for the parameters (16) are investigated, see the MAE in Table 5:

Table 5. The results for the MAE for approximations of different numbers of harmonics for parameters (16) in the framework of the power-law structure model

l	MAE	l	MAE	l	MAE	l	MAE	l	MAE
1	$4.7\cdot10^{-2}$	5	$2.5\cdot10^{-3}$	9	$7.1\cdot10^{-4}$	13	$5.7\cdot10^{-4}$	17	$2.1\cdot10^{-3}$
2	$3.3\cdot10^{-2}$	6	$1.7\cdot10^{-3}$	10	$6.4\cdot10^{-4}$	14	$6.6\cdot10^{-4}$	18	$9.7\cdot10^{-3}$
3	$8.1\cdot10^{-3}$	7	$1.2\cdot10^{-3}$	11	$5.8\cdot10^{-4}$	15	$7.9\cdot10^{-4}$		
4	$4.1\cdot10^{-3}$	8	$9.1\cdot10^{-4}$	12	$5.5\cdot10^{-4}$	16	$1.1\cdot10^{-3}$		

As can be seen, the accuracy may not increase with the number of harmonics, but most of the harmonic approximations give a very good agreement between the left-hand side and the right-hand side of the integral equation under consideration. So, as can be seen from Tables 1, 2, 3 and Table 5, the truncated trigonometric Fourier series expansion method is much better than the truncated polynomial expansion method for parameters (16) in the framework of the power-law structure function model.

The following results are obtained for the parameters (17):

Table 6. The results for the MAE for approximations of different numbers of harmonics for parameters (17) in the framework of the power-law structure model

l	MAE	l	MAE	l	MAE	l	MAE	l	MAE
1	$7.1\cdot10^{-2}$	5	$2.0\cdot10^{-3}$	9	$5.4\cdot10^{-4}$	13	$3.6\cdot10^{-4}$	17	$1.4\cdot10^{-3}$
2	$1.5\cdot10^{-2}$	6	$1.3\cdot10^{-3}$	10	$4.6\cdot10^{-4}$	14	$4.0\cdot10^{-4}$	18	$1.2\cdot10^{-2}$
3	$6.1\cdot10^{-3}$	7	$9.4\cdot10^{-4}$	11	$4.0\cdot10^{-4}$	15	$5.1\cdot10^{-4}$		
4	$3.2\cdot10^{-3}$	8	$7.1\cdot10^{-4}$	12	$3.7\cdot10^{-4}$	16	$7.3\cdot10^{-4}$		

As can be seen from Table 6, the use of the truncated Fourier series for the parameters (17) is valid in contrast to the truncated polynomial expansion. The accuracy of the harmonic approximations for parameters (17) is rather high.

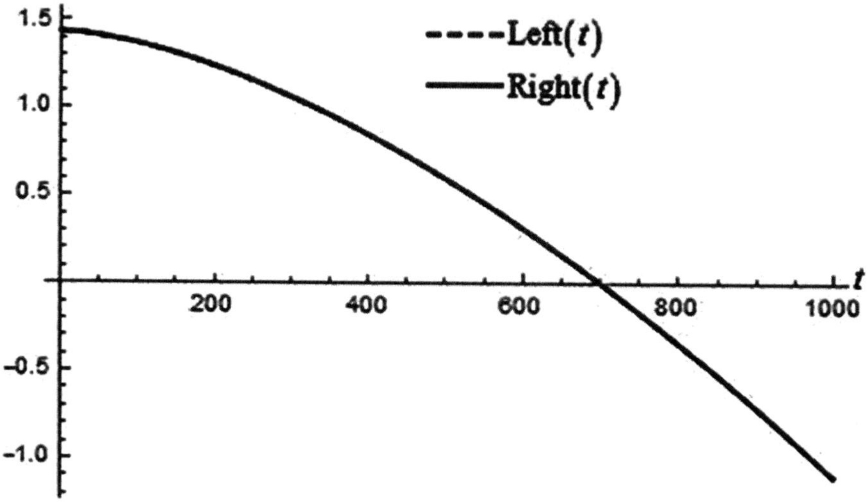

Fig. 3. Graphs of Left(t) and Right(t) for the 13-harmonic approximation for parameters (17)

For example, the corresponding graphs for the 13-harmonic approximation are shown on Fig. 3; as can be seen, the curves are nearly identical.

The trigonometric solutions are also investigated in the framework of the fractional Gaussian noise model. The parameters (22) are investigated, the MAPE error (20) is calculated where

$$\begin{gathered}
\mathrm{Left}(t) = \int_0^T \left(a_0 + \sum_{s=1}^{l} a_s \cos(s\omega\tau) + \sum_{s=1}^{l} b_s \sin(s\omega\tau) \right) |t-\tau|^{2H-2} d\tau \\
= a_0 \frac{(T-t)^{2H-1} + t^{2H-1}}{2H-1} \\
+ \frac{1}{2(2H-1)} \sum_{s=1}^{l} \left[\{a_s \cos(s\omega t) + b_s \sin(s\omega t)\}(Q_+(s,t) + Q_+(s,T-t)) \right] \\
- \frac{1}{2(2H-1)} \sum_{s=1}^{l} \left[\{a_s \sin(s\omega t) - b_s \cos(s\omega t)\}(Q_-(s,t) - Q_-(s,T-t)) \right]; \\
\mathrm{Right}(t) = R(t+z) = |t+z|^{2H-2}
\end{gathered} \tag{55}$$

where $Q_+(s,t)$ and $Q_-(s,t)$ are the following real-valued functions:

$$\begin{gathered}
Q_+(s,t) = t^{2H-1}\{{}_1F_1(2H-1, 2H, is\omega t) + {}_1F_1(2H-1, 2H, -is\omega t)\}, \\
Q_-(s,t) = it^{2H-1}\{{}_1F_1(2H-1, 2H, is\omega t) - {}_1F_1(2H-1, 2H, -is\omega t)\}.
\end{gathered} \tag{56}$$

The corresponding MAPE errors are given in Table 7.

As can be seen, the method is convergent, the accuracy of the approximations increases with the number of harmonics. But, as can be seen from Table 7 and Table 4, the accuracy of the polynomial solutions is higher than the accuracy of the trigonometric solutions in the case of the fractional Gaussian noise model. Moreover, let us consider the graphs for the 15-harmonic approximation.

Table 7. The results for the MAPE for approximations of different numbers of harmonics for parameters (22) in the framework of the fractional Gaussian noise model

l	MAPE, %	l	MAPE, %	l	MAPE, %	l	MAPE, %	l	MAPE, %
1	18	4	8.7	7	5.7	10	4.2	13	3.3
2	13	5	7.4	8	5.1	11	3.8	14	3.0
3	11	6	6.4	9	4.6	12	3.5	15	2.8

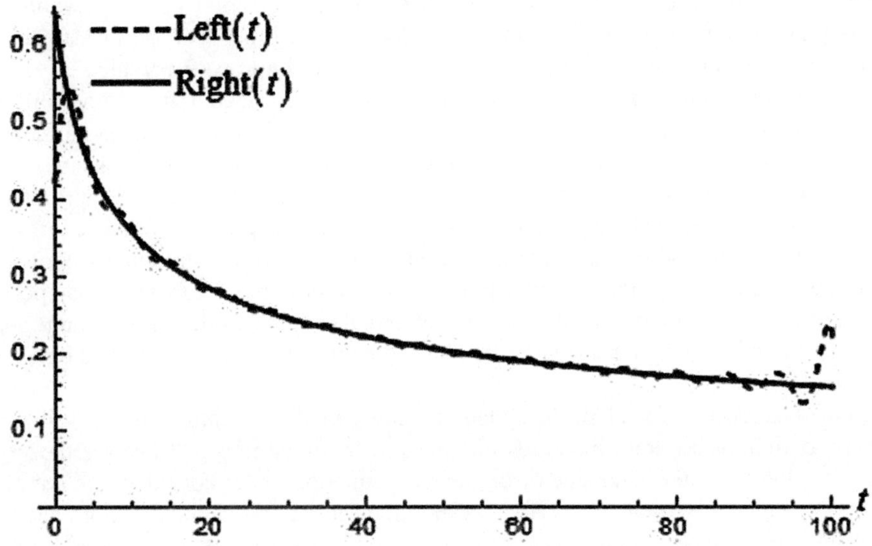

Fig. 4. Graphs of Left(t) and Right(t) for the 15-harmonic approximation for parameters (22)

As can be seen from Fig. 4, the curves are in good agreement in the middle part of the interval $t \in (0, T)$, but the agreement is not good near the interval boundaries. So one can conclude that the polynomial solutions are better than the trigonometric ones in the framework of the fractional Gaussian noise model.

Let is briefly summarize the results of this section. The trigonometric solutions are much better than the polynomial ones in the framework of the power-law structure function model, but in the framework of the fractional Gaussian noise model the polynomial solutions are better than the trigonometric ones.

5 Conclusion

The paper is devoted to the problem of stationary telecommunication traffic forecasting, which is an urgent problem for telecommunications. The Kolmogorov–Wiener filter is used. The theoretical fundamentals of the filter weight function derivation are developed.

Two models of stationary traffic are investigated. The first one is a model where traffic is treated as a random process with a power-law structure function. In the second model, traffic is treated as fractional Gaussian noise. In the case of a large amount of data, traffic may be described as a continuous process, so in the framework of both models we consider traffic as a continuous stationary random process.

The integral equation for the filter weight function is a Wiener-Hopf integral equation. An exact solution for this equation meets difficulties, so we search for an approximate solution on the basis of the Galerkin method, in the framework of which the solution is sought as a truncated orthogonal functional series.

Our previous results based on the polynomial expansions are described. The polynomial expansion method works well in the framework of the fractional Gaussian noise model, but in the framework of the power-law structure function model polynomial results may be not valid for some approximations. So in this paper we investigate a truncated expansion into a trigonometric Fourier series to improve the results.

Our new results for the trigonometric expansions are described. In the framework of the power-law structure function model, the situation is as follows. The one-harmonic approximation is not sufficiently accurate, the accuracy of the approximations may not increase with the number of harmonics. However, from the two-harmonic approximation and on, the accuracy of approximations is rather high. So one can conclude that the truncated trigonometric Fourier expansion method works well in the framework of the power-law structure function model, trigonometric solutions are much better than polynomial ones. The reason may be that in the polynomial case the product of very large and very small numbers may occur, which can hardly be treated by computer mathematics packages.

In the framework of the fractional Gaussian noise model, the situation is as follows. The approximation accuracy increases with the number of harmonics. However, the left-hand side and the right-hand side of the integral equation under consideration are not in good agreement near the boundaries of the time interval on which the input data is given. Moreover, the accuracy of polynomial approximations is higher than the accuracy of trigonometric ones. So one can conclude that the polynomial solutions are better than the trigonometric ones in the framework of the fractional Gaussian noise model.

Our future plans are as follows. First of all, a search for an orthogonal function system that works well in the framework of both models is of interest. Another plan is to use the proposed theoretical approach in practical traffic forecasting.

References

1. Katris, C., Daskalaki, S.: Comparing forecasting approaches for internet traffic. Expert Syst. Appl. **42**(21), 8172–8183 (2015). https://doi.org/10.1016/j.eswa.2015.06.029
2. Al-Azzeh, J.S., Al Hadidi, M., Odarchenko, R., Gnatyuk, S., Shevchuk, Z., Hu, Z.: Analysis of self-similar traffic models in computer networks. Int. Rev. Model. Simul. **10**(5), 328–336 (2017). https://doi.org/10.15866/iremos.v10i5.12009
3. Bagmanov, V.Kh., Komissarov, A.M., Sultanov, A.Kh.: Teletraffic forecast on the basis of fractal fliters. Bull. Ufa State Aviat. Tech. Univ. **9**(6(24)), 217–222 (2007). (in Russian)
4. Quian, H.: Fractional Brownian motion and fractional Gaussian noise. In: Rangarajan, G., Ding, M. (eds.) Processes with Long-Range Correlations. Theory and Applications. LNP, vol. 621, pp. 22–33. Springer, Heidelberg (2003). https://doi.org/10.1007/3-540-44832-2

5. Miller, S., Childers, D.: Probability and Random Processes with Applications to Signal Processing and Communications, 2nd edn. Elseiver/Academic Press, Amsterdam (2012). https://doi.org/10.1016/C2010-0-67611-5
6. Xu, Y., Li, Q., Meng, S.: Self-similarity analysis and application of network traffic. In: Yin, Y., Li, Y., Gao, H., Zhang, J. (eds.) Proceedings of the MobiCASE 2019. LNICST,vol. 290, pp. 112–125. Springer, Cham (2019). https://doi.org/10.1007/978-3-030-28468-8_9
7. Polyanin, A.D., Manzhirov, A.V.: Handbook of Integral Equations, 2nd edn. Chapman & Hall/CRC Press, Taylor & Francis Group, Boca Raton (2008)
8. Gorev, V.N., Gusev, A.Y., Korniienko, V.I.: Polynomial solutions for the Kolmogorov-Wiener filter weight function for fractal processes. Radio Electron. Comput. Sci. Control **2**, 44–52 (2019). https://doi.org/10.15588/1607-3274-2019-2-5
9. Gorev, V., Gusev, A., Korniienko, V.: Investigation of the Kolmogorov–Wiener filter for treatment of fractal processes on the basis of the Chebyshev polynomials of the second kind. In: CEUR Workshop Proceedings, vol. 2353, pp. 596–606 (2019)
10. Gorev, V., Gusev, A., Korniienko, V.: Investigation of the Kolmogorov-Wiener filter for continous fractal processes on the basis of the Chebyshev polynomials of the first kind. IAPGOS **10**(1), 58–61 (2020). https://doi.org/10.35784/iapgos.912
11. Gorev, V., Gusev, A., Korniienko, V.: On the telecommunication traffic forecasting in a fractional Gaussian noise model. In: CEUR Workshop Proceedings, vol. 2623, pp. 164–173 (2020)
12. Gonzalez, R.G., Khalil, N., Garzo, V.: Enskog kinetic theory for multicomponent granular suspensions. Phys. Rev. E **101**, 012904 (2020). https://doi.org/10.1103/PhysRevE.101.012904
13. Sokolovsky, A.I., Sokolovsky, S.A., Hrinishyn, O.A.: On relaxation processes in a completely ionized plasma. East. Eur. J. Phys. **3**, 19–30 (2020). https://doi.org/10.26565/2312-4334-2020-3-03
14. Ziman, J.M.: Electrons and Phonons. The Theory of Transport Phenomena in Solids. Oxford University Press, Oxford (2001). https://doi.org/10.1093/acprof:oso/9780198507796.001.0001
15. Gradshteyn, I.S., Ryzhik, I.M., Geronimus, Yu.V., Tseytlin, M.Yu., Alan, J.: Table of Integrals, Series, and Products, 8th edn. Elsevier/Academic Press, Amsterdam (2014). Ed. by D. Zwillinger and V. Moll. https://doi.org/10.1016/C2010-0-64839-5

QOS of Data Networks Analyzing Based on the Fuzzy Knowledge Base

L. Globa[1(✉)], Z. Savchuk[1(✉)], O. Vasylenko[2(✉)], and E. Siemens[2(✉)]

[1] Institute of Telecommunication Systems, National Technical University of Ukraine "Igor Sikorsky Kyiv Polytechnic Institute", Kyiv, Ukraine
lgloba@its.kpi.ua
[2] Anhalt University of Applied Sciences, Köthen, Germany
eduard.siemens@hs-anhalt.de

Abstract. With the rapid growth of data traffic from various information sources and the complexity of services increasing there is a current new trend in the ICT field that has been called the Big Data processing trend. There are a lot of up-to-day intelligent technics and systems used for overcoming this trend but the computational and data processing complexity under real-time requirements continues to be one of the important disadvantages for many engineering fields. The paper deals with an approach of Big Data stream structuring into fuzzy logic rules for fuzzy knowledge base development that has no large data processing complexity. To guarantee the correctness of the fuzzy knowledge base the metagraph theory apparatus is used based on control conflicting and duplicate rules under consideration of their logical inter-connections. Usage of the fuzzy knowledge base during Big Data stream processing helps to decrease data processing time for decision-making systems in real engineering fields.

Keywords: Data Mining · Big Data analysis · The quality of service delivery by telecom operator · Fuzzy logic · Fuzzy knowledge base · Metagraph theory apparatus

1 Introduction

According to research conducted by a number of leading companies in the world [1] the telecom operators are faced with the need to comprehensively consider the influence of various services characteristics (technical, economic, experience of services using by the end user) to clearly understand and manage the processes occurring between the telecom operator and his subscribers. This is especially important due to the rapidly expanding range of provided services and the transition to digital space.

The whole range of this characteristics is too large and complex to collect, process and analyze using the current computing infrastructure. They are characterized by:

- significant volume (from terabytes to petabytes);
- the need for high-speed processing in real time taking into account that their storage volumes should be reduced;

P. Vorobiyenko et al. (Eds.): IPF 2020, LNNS 212, pp. 130–149, 2021.
https://doi.org/10.1007/978-3-030-76343-5_8

- heterogeneity (data may be structured, unstructured, partially structured);
- the need to meet the requirements of authenticity (this requirement can be broken due to the variety of data sources and methods of their processing, breach of security requirements);
- value (the usage of forecasting and analysis methods to predict the direction of business development).

Telecom companies are investing great money in the development of analytical tools and services. At the same time, the data analysis to solve the urgent tasks is performed often based on the data obtained as a result of telecom operator economic activity, or based on the sociological surveys, or based on the technical parameters of the operator's infrastructure functioning. Very often deciding whether to invest in a particular part of the system or infrastructure does not take into account the impact and analysis of all possible factors and consequences. As an example it's possible to consider the task concerning the degree of customer satisfaction with the services provision by the operator, it is quite obvious that technical failures affect the degree of satisfaction. So, operator's pricing policy and quality of services provision together influence to the final evaluation of the quality of services by the subscriber.

Previously, applied statistics was widely used in the field of processing knowledge in telecom industry. The statistics evaluated, tested hypotheses, but gave rough and average results. Thanks to technical progress, people began to store huge amounts of information, which was heterogeneous, and naturally, had to be processed. The methods of statistics did not make it possible to predict and control processes and had a highly computational complexity from telecom operator point of view. The telecom operator could no longer solve the tasks by applying statistics. It is necessary to structure large amounts of data and to extract useful information from them in order to:

- analyze this data successfully in the future;
- visualize these data for easy human perception;
- move from data analysis to knowledge application for effective business management, especially in the telecom industry.

Thus, the telecom industry faces the urgency to process, analyze and predict significant amounts of information, characterized by significant unstructured, non-simultaneous receiving data from various sources. Such large amounts of data are called Big Data.

The purpose of this paper is to identify the features of existing mathematical methods named Data Mining that can be used by telecom operators for solving problems arising from new information technology such as IoT, M2M and others. In more detail the method of Big Data processing based on designing the fuzzy knowledge base used fuzzy logic will be considered.

The paper is structured as follows: Sect. 2 contains state of the art of Big data analysis features and the specifics of this process for telecom operators. Section 3 introduces the proposed approach to Big Data analysis on the example of the telecom operator data. Section 4 presents the results of fuzzy knowledge base development for Big data of telecom operator analysis. Section 5 includes the summary and outlook on future work.

2 State of the Art and Background

Nowadays, the Data Mining solutions are applied by telecom operators such as network operators and Internet providers with the aim to improve processing of different kind unstructured large data volumes (Big Data) for getting decisions that are more intelligent. Analytics enables communication and Internet providers to increase economic effectiveness and efficiency in fields of service provisions significantly.

Big Data, defined as data too large and complex to capture, process, and analyze using current computing infrastructure, is now popularly characterized by at least five V's (initially it was described as having three, but two have since been added to emphasize the need for data authenticity and business value):

- *Volume* - data measurement is in terabytes (240) or even petabytes (250), and is rapidly heading toward exabytes (260);
- *Velocity* - data production occurs at very high rates, and, because of this sheer volume, some applications require real-time data processing to determine whether to store a piece of data;
- *Variety* - data is heterogeneous and can be highly structured, semi-structured, or totally unstructured;
- *Veracity* - due to intermediary processing, diversity among data sources and in data evolution raises concerns about security, privacy, trust, and accountability, creating a need to verify secure data provenance;
- *Value* - through predictive models that answer what-if queries, analysis of this data can yield counterintuitive insights and actionable intelligence.

Big Data enables new directions for scientific research once limited by the volume of available data. To better address this problem, statistical machine-learning to require training data for designing and evaluating models are often used [2–5].

Thanks Big Data the paradigm for solving complicated problems have been shifted; now the accurate selection of a mathematical model loses its importance because there are big enough data to compensate the model accuracy [3, 6].

Modern Big Data analytics require a transformation from unstructured to structured data that in fact forms the process of "compressing data to their sense" and develops Big Data processing strategy as "data - information - knowledge – prediction". Under such understanding there are the following elements of data processing:

- data - flows of raw facts such as business operations;
- information - clusters of facts that are relevant for people such as decision making;
- knowledge - data/information that are organized to express understanding, experience and learning using previous experience.

Data mining is the process of detecting in raw data previously unknown non-trivial but practically useful and accessible interpretation of knowledge necessary for decision-making in various spheres of human activity. Modern technology helps to identify hidden links in databases of very big sizes. Since Data Mining has developed based on applied statistics, methods of artificial intelligence, database theory and so there are a lot of application for data processing. However, all this methods and algorithms are not very effective for Big Data processing [1].

Data Mining uses the concepts of averaging over a sample, the concept of templates and so on. However, there is no universal way to determine the meaning and connection of data among themselves, and additional studies are required for each subject area in order to identify these links.

The purpose of Data Mining is to find such models that can not to be found by the usual methods. There are two types of models: predictive and descriptive. Predictive models: positioned on a data set with known results.

These are classification models (describe the rules by which the description of an object can be attributed to one of the classes) and the sequence model (they describe the functions by which the change of continuous numerical parameters can be predicted).

Descriptive models: they pay special attention to the nature of the dependencies in the data set, the mutual influence of various factors, the construction of empirical models. They are easy for human perception.

The types of patterns that used by Data Mining technology are presented in the paper [2]:

- *Association* - applies when several events are related. For example, it happens that the subscriber stops using the operator's telecom services because his family uses the services of another telecom company. It's understandable that the subscriber starts to use the services of telecom company which his family prefers;
- *Classification* - the identification of features that will characterize the group to which the object belongs according training based on the already classified objects;
- *Clustering* - differs from the classification in that groups are not known in advance and Data Mining tools independently identify different homogeneous groups of data;
- *Sequence* - applies when there is a chain of events related in time;
- *Forecasting* - creating or finding templates that will show truly the trend of the required indicators behavior in terms of time series. With the help of them one can predict the behavior of the system in the future.

Because Data Mining has developed at the junction of many areas, it is possible to reuse classes and methods of this technology: neural networks, decision trees, genetic algorithms, limited search algorithms, associative rules, cluster analysis and much more [4]. Moreover, methods of data mining allow to solve the problems of structural engineering design for innovative technical systems effectively in the telecom industry. These methods have much in common with methods for solving problems of classification, diagnosis and pattern recognition. Nevertheless, one of their main distinctive features is the function of the regularities interpreting that form the basis rules for the objects inclusion in equivalence classes. Therefore, logical methods are becoming more common today. There is another important reason to determine the priority of logical methods. It lies in the complex systemic organization of the areas that constitute the application subject of modern information technologies.

The development of intelligent data processing methods for telecom companies is necessary for:

- reducing the computational complexity of tools for Big Data processing in the services provision by operator to a subscriber with a given quality of service;
- operator should to predict the risks that may arise from the operation of the telecom system;
- the operator should be able to identify faults in the system and find out the cause of their occurrence.

Some of the existing algorithms can be adapted to compute large distributed information arrays. At the same time serious difficulties can arise with a visual representation of the results - because of the huge amount of information entering the input the number of dissimilar reports at the output sharply increases. For their convenient presentation new mathematical methods are needed which are fundamentally different from report generators used for traditional Big Data processing and storage technologies.

3 Approach to Big Data Analysis on the Example of the Telecom Operator Data

Taking into account the peculiarities of Big Data processing for the telecom industry, an approach based on fuzzy logic methods is proposed for jointly accounting for the influence of both clear and fuzzy parameters, as well as using the transition from unstructured data to a fuzzy knowledge base (FKB) with clearly structured rules.

3.1 Calculation of the Integral Quality Index of Service Provision by the Technical Infrastructure of the Network

To solve this problem, it is proposed to use decision-making methods based on fuzzy logic, for which fuzzy expert rules are formed, which are the basis of the expert system. Fuzzy rules in such a system can be periodically adjusted to the current state of the technical infrastructure by re-forming (clarifying) them during the accumulation of monitoring system data during the provision of services by the telecom operator.

Since the telecom operator faces quite diverse analysis tasks, it is necessary to use an approach that applies a group of different mathematical methods and allows to solve problems using a certain set of software tools based on them.

In the proposed approach, the assessment of the current state of the process of providing services and services by the telecom operator is performed by determining the fuzzy value ("good", "bad", etc.) of the integral quality functioning index of the technical network infrastructure and consist the following stages:

1. Getting the integral quality index (Y).
   It is difficult for the operator to assess the complex impact of individual indicators of the service quality provision due to the fact that it is necessary to know the degree of influence of each parameter on the overall state of the system of their provision as a whole, so obtaining an integral quality index is a non-trivial task.

2. Formation of FKB using the current values of the integral quality index.
   Based on the integral quality index Y, it is necessary to get an assessment of the system state as a whole and understand the algorithms of its behavior, as well as compress information into knowledge or trends (based on logical or functional dependencies) that can be controlled, as well as make decisions. To compress large amounts of data, it is proposed to form a FKB in the form of fuzzy logic rules, which uses fuzzy logic methods to obtain conclusions.
3. Use of the formed FKB for continuous assessment of the quality of services provided by the telecom operator using the integral quality index of the network technical infrastructure during operation.

The main steps of this approach are:

Step 1. Pre-clean statistics from randomly occurring errors, using a combination of existing effective approaches into a single algorithm.
Step 2. Forming the data flow (extracting functional and content dependencies, forming meta-descriptions, determining the rate of influence of parameters).
Step 3. Creating a fuzzy knowledge base using clustering algorithms.
Step 4. Reorganize the FKB, taking into account duplicate and conflicting fuzzy logical rules.
Step 5. Representation of FKB using metagraph theory.
Step 6. Fuzzy logical conclusion.

Output Data. From the point of view of the task of assessing the dependence of the degree of satisfaction with the quality of services by end users and the possible outflow of subscribers, the parameters that affect this process will be divided into the following groups:

- parameters that characterize the technical condition of the system;
- cost of using voice and internet services;
- quality of services provided.
  To collect data, it is proposed to use the following data as sources of initial statistical data to solve this problem:
- technical – functioning operator monitoring system;
- economic – use the calculations of economic indicators available in companies;
- sociological – use the subscriber survey data available in companies.

This data is combined into one or more tables according to certain groups of parameters to form a fuzzy rule base. Structuring and classifying this data allows to get a Table 1.

The input data that enters the system and is processed is influenced by a number of different factors, such as collecting and forming a set from various sources, transmitting it through the World Wide Web, conditions for collecting information from sensors (temperature, mechanical influences), software failure, and many others. Each of these factors potentially reduces the quality of the data to be processed. As a result, invalid, false, or missing values can occur in the final data sets, which makes it almost impossible

Table 1. Structuring and classifying data flows to form fuzzy knowledge base rules.

Subscriber satisfaction (Y_1)	Technical parameters (Y_2)			Economic parameters (Y_3)		
	Connection success rate	Connection block rate	…	Churn rate	Appetency	…
0,953	0,993	0,999	…	0,563	0,859	…
0,986	0,456	0,745	…	0,947	0,495	…
…	…	…	…	…	…	…

to obtain high-quality results of big data calculations. In this case, the first step is used to increase the reliability of the result when processing large calculated data by clearing the data from accidentally occurring errors [7].

All the most common preprocessing methods are essentially aimed at correcting only one aspect of data contamination. The main idea of the modified dataset clearing algorithm is to add an additional step to the base process, which aims to combine different methods into a single processing algorithm (see Fig. 1).

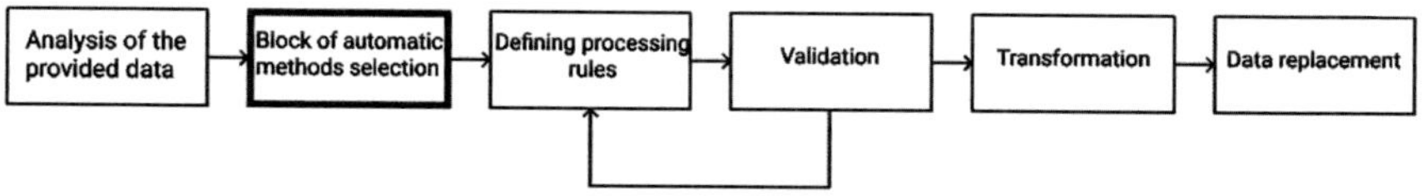

Fig. 1. Flowchart of the modified data cleaning process.

The proposed modified algorithm aims to analyze the input data and adaptively create a unique cleaning procedure. To select algorithms for processing, clear selection criteria must be formed. The purpose of applying the modified algorithm is to formulate conditions for using a particular method of data preprocessing. Thus, an algorithm is proposed that is dynamically able to create the order and components of processing, and thus not only perform effective data cleaning, but also reduce the load on the system and the amount of time and software resources required (Fig. 2).

Based on the obtained statistical data, rules are formed that represent certain knowledge (for example, a forecast of changes in the degree of subscriber satisfaction for future periods), such as:

$$\text{"IF } Y_2 \text{ AND } Y_3 \ldots \text{THEN } Y_1\text{"}.$$

Using these rules, it is possible to forecast the technical and economic condition of the telecom operator for the next period in accordance with the methodology given in [6].

Consider an approach to solving the problem of assessing the dependence of the satisfaction quality degree of services by subscribers. First of all, it is necessary to determine an integral indicator that can characterize the quality of services provided and how it is possible to get it from the data of the network monitoring system. This indicator

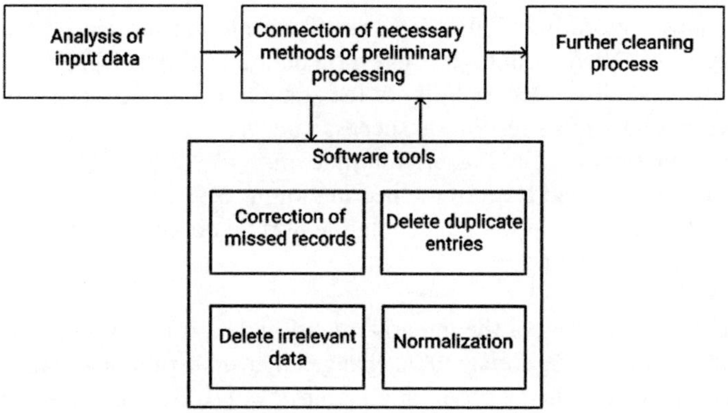

Fig. 2. Block diagram of data cleaning tools using a modified algorithm.

is considered to be the integral quality index of network functioning. This integrated indicator allows to determine further actions to improve the quality of service, based on its current value. The integrated network infrastructure quality index can be used, for example, to estimate the required amount of cloud resources that need to be connected if the problem of insufficient resources of the existing infrastructure is identified.

One of the most difficult steps of the proposed approach is the second step, where it is necessary to form, define, structure and classify data flows in order to obtain the possibility of applying algorithms for forming rules of the fuzzy knowledge base, which will allow its further use for decision-making and prediction.

To do this, you need to define parameters for creating a knowledge base, as well as ways to get the values of these parameters.

Output data: measurement tables provided by the monitoring system at the telecom operator's network nodes. The measurement table is a set of parameters that denote X_1 … X_n.

Note that $X_1 \ldots X_n$ is a set of parameters that the operator uses to evaluate the overall state of the system, but none of them characterizes the quality of the system as a whole, i.e. the output parameter (Y).

1. Generalized algorithm for determining the unknown value of Y:

Getting Y_d using the desirability function based on M_1 data. The following parameters of the monitoring system were used to determine the integral quality index (Y):

- Connection Success Rate – success of 2G/3G data connections, %;
- Connection Block Rate – percentage of blockages due to 2G/3G overload, %;
- Connection Drop Rate – percentage of 2G/3G data connection failures, %;
- PS Attach Success Rate – percentage of successfully completed Attach 2G/3G procedures, %;
- PDP Context Activation Success Rate – percentage of successfully completed PDP Context 2G/3G activation procedures, %;

- Speed DL – average daily HSDPA data transfer rate to the subscriber, Kbit/s;
- IUB Congestion – share with high congestion on the IUB interface, %;
- Backhaul Accessibility – accessibility of the zonal transport network, %;
- DNS Success Rate – DNS resolution success rate, %;
- DNS response Latency – DNS resolution time, ms;
- Peering Utilization – load level of the peering joint, %;
- Backbone Abnormal Latency – exceeding the normal delay between nodal points of the main transport network, %;

Based on the parameters of the integral quality index, it is possible to form a FKB for evaluating and predicting trends in its changes in a short time interval.

The input information is the observation table $T = \{t_{M1}\}$, where the i-th element $t_i = (x_i, y_i)$, $x_i = [x_{i1}, x_{i2}, \ldots, x_{ik}]$ – vector of input values, y_i – output value, M_1 – number of observations, k—number of input variables, $x_i \in DX$, $y_i \in DY$.

The data values in the table do not have the same content and different dimensions, so they need to be normalized from 0 to 1, that is, converted in such a way that the system can assess the level of quality of service delivery in general. To calculate the integral quality index of service delivery, it is proposed to use the desirability function of E. K. Harrington [8].

2. In order to build a FKB, you need to divide the measurement table into 2 samples:

- *training* sample with M_1 sets of data, where $M_1 = \{1, k\}$, which is necessary to form fuzzy logical rules of the knowledge base. In order to form the FKB rules, it is necessary to determine the value of the integral quality index obtained using the desirability function, denote it as Y_D;
- *test* sample with M_2 data sets, where $M_2 = \{k + 1, n\}$, which is necessary to check the quality of fuzzy logical rules of the knowledge base.

In this case, the higher the accuracy and completeness of the fuzzy rule database, the better. But in real life, these parameters are both unattainable and have to find a balance. Therefore, use the F-measure metric, which is the harmonic mean between the accuracy and completeness of our algorithm. It approaches zero if the accuracy or completeness tends to zero [9]:

$$F = 2 * \frac{Pr * Rec}{Pr + Rec} \tag{1}$$

3. Forming the FKB using M_1 and getting the Y_{FKB} value.

FKB is formed by a set of rules like: "IF $X_1, X_2, X_3 \ldots$ THEN Y ".
To do this, follow these steps:

3.1. Clustering.
3.2. Selecting the membership function.
3.3. Obtaining a Y_{FKB} based on the selected membership function.

3.4. If $|Y_D - Y_{FKB}| > \varepsilon$, then go to point 3.2.
3.5. If $|Y_D - Y_{FKB}| \leq \varepsilon$, then the membership function selected correctly.

Let's take a closer look at the main steps of the proposed approach.

Forming a Fuzzy Knowledge Base. In accordance with the features of big data processing, which are widely discussed in a number of papers [10, 11], an approach based on fuzzy logic methods is proposed to jointly take into account the influence of both clear and fuzzy parameters, as well as using the transition from unstructured data to FKB with clearly structured rules, which can significantly reduce the amount of calculations during the transmission and processing of traffic data in the network.

Output Data. A fuzzy model is defined as a system with input variables $X = \{X_{k+1}, X_{k+2},\ldots, X_n\}$, defined on the input domain of reasoning $DX = DX_{k+1} \cdot DX_{k+2} \cdot \ldots \cdot DX_n$, and one resulting variable Y, which is defined on the resulting reasoning domain DY. The clear value that the original variable X_i, takes is denoted as x_i and as y for the resulting variable Y [12, 13].

Clustering. There are a number of clustering methods. In this paper, consider the fuzzy C-mean algorithm, which is based on minimizing the distance between the observed data, x and cluster centers. For this purpose, the Euclidean distance is calculated [14].

Let's take a closer look at the fuzzy C-mean algorithm [14–16]:

Given: table of observations $\boldsymbol{T}$, number of clusters c, number of observations m, stop parameter ε.

Get: fuzzy partitioning matrix $\boldsymbol{F}$ and cluster center matrix $\boldsymbol{V}$.

Step 1. Randomly initialize the fuzzy partitioning matrix F, which satisfies conditions (2).

$$\forall_{k=\overline{1,M}}\left(\sum\nolimits_{i=1}^{c} \mu_{ki} = 1\right), \forall_{k=\overline{1,c}}\left(0 < \sum\nolimits_{i=1}^{c} \mu_{ki} < M\right) \tag{2}$$

Step 2. Calculate the center vector of clusters $\boldsymbol{v}_i$.
Step 3. Calculate the d_{ki} distance between the k -th observation from the matrix $\boldsymbol{T}$ and the i-th center of the cluster $d_{ki} = \sqrt{\boldsymbol{t}_k - \boldsymbol{v}_i}$.
Step 4. Calculate the next approximation of the matrix F;
Step 5. If $\|\boldsymbol{F}—\boldsymbol{F}^*\|^2$, then exit the algorithm, otherwise go to step 2.

F* is the fuzzy partitioning matrix obtained from the previous iteration of the algorithm.

As a result of the algorithm, the matrix of cluster centers is calculated $\boldsymbol{V} = \{\boldsymbol{v}_1, \boldsymbol{v}_2,\ldots, \boldsymbol{v}_c\}$, where each $\boldsymbol{v}_k = \{v_{k1}, v_{k2}, \ldots, v_{kn}, v_{kn+1}\}$, v_{ki} is the coordinate value of the i-th variable in the k-th cluster, v_{kn+1}– coordinate value of the resulting variable in the k-th cluster.

Each selected cluster corresponds to one linguistic rule of the fuzzy system rule base.

Selecting the Membership Function. Each linguistic term located in the antecedent of fuzzy rules is defined by its own membership function.

In the course of the study, the Gaussian membership function was determined to be the best option for evaluating the generalized quality indicator of service provision by a telecom operator.

Based on the results of cluster analysis, it is possible to determine the parameters of the membership function using formulas (3) and (4).

$$a = v_{ij} \tag{3}$$

$$\sigma_{ij}^2 = \frac{\sum_{k=1}^{M} (\mu_{ki})^m (x_{kj} - y_{ij})^2}{\sum_{k=1}^{M} (\mu_{ki})^m} \tag{4}$$

Obtaining YFKB Based on the Selected Membership Function. After determining the necessary parameters and constructing the membership functions, FKB rules are formed.

The fuzzy concept of $LX_{i,k}$ corresponds to the fuzzy definition of "X_i is approximately equal to v_{ij}". The general view of the rule is constructed as a model of Mamdani rules type:

$$R_j : X_1 \approx v_{j1} \text{ AND} \ldots \text{AND } X_n \approx v_{jn} \rightarrow y \approx v_{jn+1},$$

where v_{ji} is the coordinate value of the i-th input variable in the j-th cluster, v_{jn+1} - coordinate value of the resulting variable in the j-th cluster, $1 \le j \le c$.

Check the correctness of the obtained rules "IF … AND … THEN" based on the set of data $\boldsymbol{M_2}$.

If the value of the F-measure indicator is reliable, then proceed to step 4. Otherwise, reconfigure $\boldsymbol{M_1}$and $\boldsymbol{M_2}$.

3.2 Reorganization FKB, Taking into Account Duplicate and Conflicting Fuzzy Logic Rules

It is worth noting that with this approach of rule formation, anomalies can be formed in the form of contradictory or redundant rules. Fuzzy models that contain two or more identical rules (i.e. rules that have the same conditions and output) can be caused by:

- an error was made when designing the rule base (with a large number of rules);
- in the case of a self-organized fuzzy model, generate additional rules that are identical to the existing ones in order to strengthen their conclusions.

It is proposed to use the rule clearing approach in order to conduct an additional analysis of the established rules. The rule clearing mechanism can be displayed as follows (Fig. 3):

Forming a unique array of fuzzy knowledge base rules begins with combining duplicate rules – this is the task of finding and combining, taking into account the importance

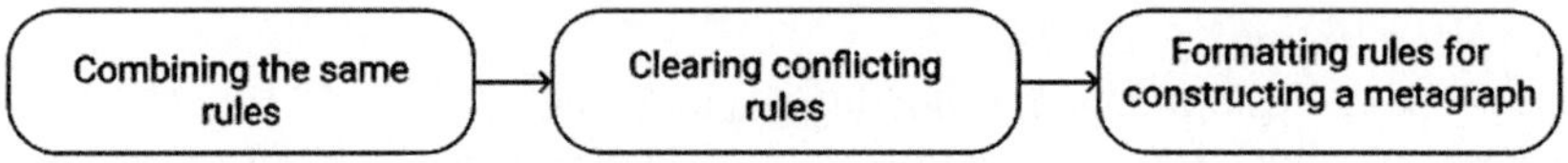

Fig. 3. Rule clearing mechanism.

of the rule's influence on the system, in an array of group objects with the same set of IF … AND … THEN [17].

When passing a loop through the entire array of rules, each rule is added to an intermediate sample with an initial weight of 1.

$$Wght_r = 1, \tag{5}$$

where $Wght_r$ is the weight of the rule.

Each duplicate rule that will occur in the sample increases the weight of the duplicate by the growth factor, which is usually k = 1 and, accordingly, is not included in the resulting array [18].

$$Wght_{r_{i+1}} = Wght_{r+k}, \tag{6}$$

where $Wght_{r_i}$ is the weight of the rule at the i-th step before weight gain, k is the growth coefficient of the rule weight.

Each conflicting rule in this set is entered into an intermediate sample as a separate rule and processed in the same way.

Keep in mind that there are conflicting rules in the resulting intermediate set of rules. To eliminate them, the second stage is carried out, at which the set of rules is cleared of conflict.

From a set of conflicting rules (there can be more than two of them), the rule with the highest weight is selected and this rule will remain in the result set, but its weight becomes equal to the arithmetic mean of all the weights of the conflicting rules.

$$Wght_{r_{conf}} = \frac{\sum_{k=1}^{n} Wght_{r_k}}{n}, \tag{7}$$

where n is the number of conflicting rules, $Wght_{r_k}$ is the weight of the k-th conflicting rule, $Wght_{r_{conf}}$ is the resulting weight of the rule that replaces all conflicting rules.

Using FKB to Assess the Quality of Service Delivery. As a result of processing the rule set, a new set is obtained that does not contain duplicate or conflicting rules. In this paper, it is proposed to use the metagraph theory apparatus to visualize the fuzzy knowledge base, since a visual representation of the results of Big Data analysis is of fundamental importance for their interpretation. It is no secret that human perception is limited, and scientists continue to conduct research in the field of improving modern methods of presenting data in the form of images, diagrams or animations.

The most common methods of data mining are: artificial neural networks, decision trees, genetic algorithms, the nearest neighbor method, and rule induction [19, 20].

Big Data can be processed using Knowledge Discovery in databases systems, which also includes some preliminary steps:

- purchase, collection;
- preprocessing, clustering, and evaluating the template;
- compression using fuzzy logic and fuzzy knowledge bases, as well as real analysis (data output) with processing and prediction [21].

The proposed combined method for constructing a database of fuzzy logical rules and checking them for correctness using metagraph theory has the following advantages:

- simplified procedure for forming rules;
- reduced computational complexity during rule formation;
- high-quality verification allows to remove conflicts, which increases the reliability of the received rules.

In the proposed approach, metagraphs are used to check the correctness and reliability of generated meta-descriptions of fuzzy knowledge bases and clear them of abnormal data. Assessment of the quality of service provision by a telecom operator is unclear and requires the use of a special mathematical apparatus to make decisions close to the engineer's thinking. In this case, it is necessary to have expert rules in the form of statements IF … AND … THEN. Metagraphs are used to visualize such fuzzy rules.

Formally, the metagraph can be represented as follows [22]:

$V_g = \left\{v\left(t_{j1}^{k1}\right), v\left(t_{j2}^{k2}\right), \ldots, v\left(t_{jn_j}^{kn_j}\right), v\left(t_l^z\right)\right\}$ – vertices corresponding to the terms on the left side of the rules and the resulting term;
$M_g = \{m_g\}$ – set of metavertices;
$m_g = \left\{v\left(t_{j1}^{k1}\right), v\left(t_{j2}^{k2}\right), \ldots, v\left(t_{jn_j}^{kn_j}\right)\right\}$– metavertices that include vertices corresponding to the terms on the left side of the rule;
$E_g = \{e_g\}$ – set of edges;
$e_g = \left\{m_g, v\left(t_l^z\right)\right\}$– an edge connects the metavertices to the vertices corresponding to the resulting terms.

Using a metagraph, you can describe a fuzzy knowledge base. The rules of IF … AND … THEN format are described from the metagraph side (Fig. 4):

$$\left(P_l^z\right)_j = If\left(X_{j1} = t_{j1}^{k1}\right)And\left(X_{j2} = t_{j2}^{k2}\right)And \ldots \left(X_{jn_j} = t_{jn_j}^{kn_j}\right)Then\left(X_l = t_l^z\right)$$

The presented metagraph visualization method allows to use methods of graphical representation, structuring, and data analysis in problems that use metagraphs. This device can detect anomalies such as redundancy, inconsistency, and incompleteness in the subject area model that are difficult to detect in a textual or formal representation. Frequent changes in information in the fuzzy knowledge base lead to the need to update the metagraph image automatically, since this is difficult to implement manually [23]. As a result, the relevance of automatic construction for metagraph representation plays an important role in decision support systems.

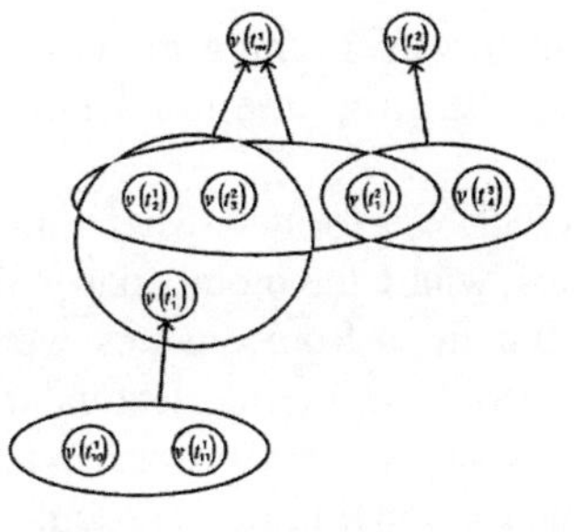

Fig. 4. Metagraph example.

4 The Results of Fuzzy Knowledge Base Development for Big Data of Telecom Operator Analysis Based on Example of Determining the Quality of Service Delivery

Input variables are fed to the input of the fuzzy logical output algorithm, which carry information obtained by measuring a certain physical quantity (telecom operator monitoring system).

At the output of the algorithm, the final variables of the fuzzy logical conclusion are formed, for example, the level of satisfaction with the quality of service.

The fuzzy logical inference algorithm converts the values of input variables of the monitoring process into output variables based on certain fuzzy product rules.

To test the proposed approach, we used data obtained from one of the Ukrainian telecom operators. Natural values were normalized in the range from 0 to 1. Fragment of the data is presented in Table 2.

Table 2. Example of output data from Ukrainian telecom operator.

X_1	X_2	X_3	...	X_{11}	X_{12}
0,01	0,99	0,98	...	0,008	0,99
0,001	0,995	0,999	...	0,003	0,996
0,001	0,994	0,999	...	0,003	0,996
0,0006	0,998	0,999	...	0,003	0,998
0,088	0,9989	0,9947	...	0,007	0,997
0,083	0,9966	0,9979	...	0,0073	0,996
0,0014	0,9949	0,9924	...	0,0072	0,9973
0,0054	0,9978	0,9934	...	0,0044	0,999
0,0008	0,9962	0,9841	...	0,0037	0,9986
0,0014	0,9923	0,9984	...	0,0055	0,9984
0,004	0,9944	0,9943	...	0,0074	0,997
...	...	...	...	...	...

According to the algorithm for determining the unknown Y, the data was divided into two samples: training and test. From the first sample, Y was obtained using the desirability function.

From the second sample, a FKB was formed, where numerical values were converted into terms of linguistic variables, where the membership functions obtained at the stage of constructing membership functions from clusters were used. For the term of the corresponding linguistic variable for each numerical value from the statistical data set, the procedure for "mapping" these data to the corresponding terms were performed. Values refer to terms due to their belonging to a particular cluster. The initial data will look like this (see Table 3).

Table 3. Fuzzy knowledge base for Ukrainian telecom operator.

X_1	X_2	X_3	X_4	...	X_{12}	Y
Middle	Not low	Not low	Not low	...	Not low	Not low
Middle	Middle	Not low	Not low	...	Not low	Not low
Very low	High	High	High	...	High	High
Very low	High	High	High	...	Very low	High
Low	High	High	High	...	High	High
Middle	High	High	High	...	Low	High
Very low	High	High	Very high	...	Low	High
Low	High	High	Very high	...	Very low	Very high
Low	High	High	Very high	...	Very low	Middle
Very low	Middle	Middle	Very high	...	Not low	Middle
Very low	Middle	Middle	Very high	...	Low	Very high
...	...	...	...	...	...	...

There are examples of product rules that are included in the developed FKB based on the data of the Ukrainian telecom operator:

IF X_1 = middle AND X_2 = not low AND X_3 = not low AND X_4 = not low AND X_5 = not low AND X_6 = not low AND X_7 = not low AND X_8 = not low AND X_9 = not low AND X_{10} = not low AND X_{11} = not low AND X_{12} = not low **THEN *Y*= not low**
IF X_1 = middle AND X_2 = middle AND X_3 = not low AND X_4 = not low AND X_5 = not low AND X_6 = not low AND X_7 = not low AND X_8 = not low AND X_9 = not low AND X_{10} = not low AND X_{11} = not low AND X_{12} = not low **THEN *Y*= not low**

IF X_1 = very low AND X_2 = high AND X_3 = high AND X_4 = high AND X_5 = high AND X_6 = high AND X_7 = high AND X_8 = high AND X_9 = very low AND X_{10} = high AND X_{11} = very low AND X_{12} = high **THEN *Y*= high**
IF X_1 = very low AND X_2 = high AND X_3 = high AND X_4 = high AND X_5 = high AND X_6 = high AND X_7 = high AND X_8 = high AND X_9 = very low AND X_{10} = high AND X_{11} = high AND X_{12} = high **THEN *Y*= high**

However, you should pay attention to the fact that at this step of converting data into terms of linguistic variables, conflicting and duplicate rules were formed, due to the different nature of the data, or the data during clustering was not correctly distributed between clusters at the boundaries of the cluster distribution.

At the FKB reorganization step, fuzzy logic rules are presented as a collection of pairs of a two-dimensional array of objects in the "name: value" format, for example, "Iub Congestion (X_7): low" At the first stage of the rule clearing mechanism, duplicates are removed and a new *weight* property is added, which reflects the frequency of finding the rule in the intermediate set. After that, the rule collection is cleared of conflicting rules, where the left side of the rule (IF…) is checked for a duplicate. If there are two or more identical parts of the rule where the result values contradict each other, then the object with the weight parameter has a higher value is selected and it is replaced with the arithmetic mean of all the weights of the conflicting rules. The end result of the rule clearing mechanism will be a unique array of objects collection without anomalies with the specified weight of the rule's impact on the decision support system.

Before starting to implement the metagraph construction implementation, perform an additional automatic rule formatting procedure for a file with cleared rules based on the *nodes* and *links* principle (see Table 4).

Parameter links must define "source" and "target", which indicate the direction of dependency of rule objects, as well as the "value" parameter, which determines the weight of the rule among all the rules of the fuzzy knowledge base. At the output of the system, we get a ready-made metagraph, with which can visually edit logical structures, define and analyze its properties that would be difficult to detect using a text representation (Fig. 5).

As can be seen from Fig. 5, a metagraph was constructed based on the rules of the fuzzy knowledge base products built for Ukrainian telecom operator, where three groups of vertices are distinguished: a set of vertices corresponding to the terms on the left side of the rules (green); a set of metavertices including vertices corresponding to the terms on the left side of the rule (blue); a set of vertices corresponding to the resulting terms (orange). Figure 6 shows a separate rule, where the "value" parameter is represented as the arc width connecting the resulting term and the metavertex – the larger the arc width, the greater influence of the rule on the system.

The metagraph can be considered as a solution to the problem of predicting medium- and large-size relationships, in which the goal is to accurately train models that quickly adapt to new data. The presented visualization method in comparison with other algorithms for structuring information can be considered a new productive method of graphical analysis of complex logical structures, since it allows improving the quality and reliability of the obtained fuzzy logic rules that are in the fuzzy knowledge base. In addition, the presentation of business processes and workflows as metagraphs can be used to

Table 4. Format of rules based on the nodes and links principle.

{"nodes": [{name: Connection Success Rate_low, group: 1}, {name: Connection Block Rate_middle, group: 1}, {name: Connection Drop Rate_high, group: 1}, … {name: Rule_0, group: 2}, {name: Rule_1, group: 2}, … {name: Rule_N, group: 2}, … {name: QoS_low, group: 3}, {name: QoS_middle, group: 3}, {name: QoS_high, group: 3}]}
{"links": [{source: *Connection Success Rate_low*, target: *Rule_0, value: 5*}, {source: *Connection Block Rate_middle*, target: "*Rule_0, value: 5*}, {source: *Connection Drop Rate_high*, target: "*Rule_0, value: 5*}, … {source: *Speed DL_middle*, target: *Rule_10, value: 5*}, {source: *Iub Congestion_high*, target: *Rule_10, value: 5*}, {source: *Backhaul Accessibility_low*, target: *Rule_10, value: 5*}, … {source: *DNS Success Rate_middle*, target: *Rule_N, value: 5*}, {source: *DNS Response Latency_high*, target: *Rule_N, value: 5*}, {source: *Peering Utilization_high*, target: *Rule_N, value: 5*}, … {source: *Rule_0*, target: QoS_low, *value: 15*}, {source: *Rule_10*, target: QoS_middle, *value: 24*}, {source: *Rule_N*, target: QoS_high, *value: 8*},]}

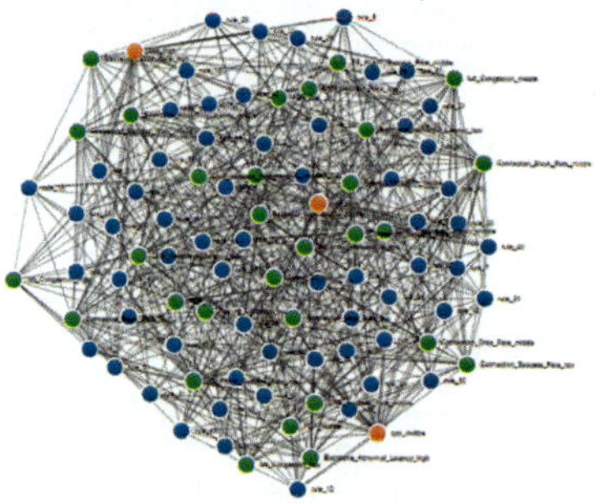

Fig. 5. Ready-made metagraph.

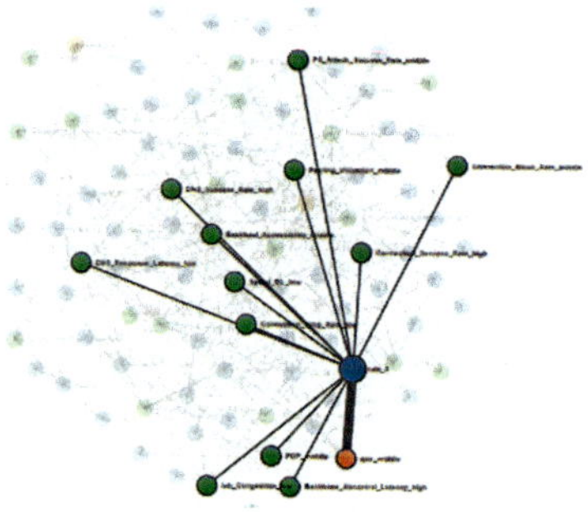

Fig. 6. Separate rule of metagraph.

share knowledge about process structure and planning across organizational boundaries in a way that supports not only useful visualization and abstraction of processes, but also structural analysis of these processes.

The proposed approach to analyzing data from a telecom operator requires time to set up and generate a FKB, but this is compensated by a decrease in the computational load on the network during its operation. This approach makes it possible to reduce computational complexity in the process of classifying large amounts of data, to solve the problem of obtaining an integral index of the quality of the technical infrastructure of a telecom operator for a number of parameters provided by its monitoring system and which are logically unrelated. In addition, the proposed method of visualization of the metagraph makes it possible to work with FKB in graphical form, which simplifies the formation of new FKB, and check existing ones for compliance with properties by using a graphical representation on which it is possible to clearly identify dependencies and anomalies.

5 Conclusion

In this overview the Big Data processing methods relevant for telecom operator were analyzed. There are lot of problems for data processing methods choosing:

- the problem of big data - need mathematical methods for structuring data and knowledge (in future work it's proposed to use ontology) and very important to find methods that do not have high computational complexity;
- the problem of prediction – proposed to use the knowledge base patterns and fuzzy logic.

All of the above problems can be eliminated by developing new methods for Big Data processing that will: successfully analyze these data, a fuzzy knowledge base designing, and visualize these data for easy human perception.

As a result of the conducted research, an approach is proposed to assess the quality of services provided by the telecom operator, by determining the integral quality index of service, a comprehensive assessment of structured data, the advantage of which is to reduce the amount of information processed in the process of Big Data Analysis, by reducing this amount of information to knowledge.

To carry out the proposed analysis, it takes time to set up and form the FKB, but this is compensated by a decrease in the computational load on the system during its operation.

For the database of fuzzy logical rules, it is proposed to use visualization of the metagraph, which allows the expert to qualitatively check the obtained rules and their reliability.

The considered approach makes it possible to reduce computational complexity in the process of classifying large amounts of data, to solve the problem of obtaining an integral quality index of the technical telecom operator infrastructure for a number of parameters provided by its monitoring system and which are logically unrelated.

The further work consists in studying methods of data structuring and fuzzy knowledge base designing for the telecom operator. As well as in the further revision of the proposed method, it is planned to work on a clear display of metavertices, a new model for determining multiple cross-sections of metavertices, and adapt the method to training.

References

1. Ulema, M.: Big Data and Telecommunications [Electronic resource]. In: 4th International Black Sea Conference on Communications and Networking (2016). (Professor of Computer Information Systems Manhattan College, Riverdale New York, USA)
2. Wei Fan, A.B.: Mining big data: current status, and forecast to the future. ACM SIGKDD Explor. News **14**(2), 1–5 (2012). https://doi.org/10.1145/2481244.2481246
3. Jagadish, H.V., Gehrke, J., Labrinidis, A., Papakonstantinou, Y., Patel, J.M., Ramakrishnan, R., Shahabi, C.: Big data and its technical challenges. Assoc. Comput. Mach. Commun. ACM **57**(7), 86 (2014)
4. Chen, M., Mao, S., Liu, Y.: Big data: a survey. Mob. Netw. Appl. **19**(2), 171–209 (2014). https://doi.org/10.1007/s11036-013-0489-0
5. Pedrycz, W.: Granular computing for data analytics: a manifesto of human-centric computing. IEEE/CAA J. Autom. Sinica **5**(6), 1025–1034 (2018)
6. Globa, L., Svetsynska, I., Luntovskyy, A.: Case studies on big data. J. Theor. Appl. Comput. Sci. **10**(2), 41–52 (2018)
7. Grebinichenko, M.V.: Methods of pre-processing of large data/M.V. Grebinichenko. Kyiv, 54 p. (2020) [in Ukrainian]
8. Pichkalev, A.V.: Generalized desirability function of Harrington for comparative analysis of technical means. Res. Sci. City №. 1 (1)/A.V. Pichkalev, January–March 2012
9. Yutaka Sasaki The truth of the F-measure/Y. Sasaki, 26 October 2007
10. What is big data? A consensual definition and a review of key research topics. In: De Mauro, A., Greco, M., Grimaldi, M. (eds.) 4th International Conference on Integrated Information, AIP Proceedings (2014)
11. Using SMART Big Data, Analytics and Metrics To Make Better Decisions and Improve Performance. John Wiley, Sons Ltd. (2015)
12. Sugeno, M., Yasukawa, T.: A fuzzy-logic-based approach to qualitative modeling. IEEE Trans. Fuzzy Syst. **1**, 7–31 (1993)
13. Takagi, T., Sugeno, M.: Fuzzy identification of systems and its application to modeling and control. IEEE Trans. Syst. Man Cybern. **15**, 116–132 (1985)
14. Espinosa, J., Vandewalle, J., Wertz, V.: Fuzzy Logic, Identification and Predictive Control, 263 p. Springer-Verlag, London (2005)
15. Piegat, A.: Fuzzy modeling and control. In: Piegat, A. (ed.) 2nd (edn.) 798 p. BINOM. Knowledge Laboratory, Moscow (2018). [in Russian]
16. Lyashenko, A.V.: Method of construction of fuzzy logical rules for big data. In: Lyashenko, A.V. Kyiv, 67 p. (2020). [in Ukrainian]
17. Savchuk, Z.R.: Application of fuzzy logical rules for analysis and structuring of big data. In: Savchuk, Z.R. Kyiv, 2020, 80 p. (2020). [in Ukrainian]
18. Zakharchuk, A.G.: Methods of fuzzy logic for data processing in the Internet of Things. In: Zakharchuk, A.G. (ed.) – Kyiv, 2019. – 97 p. (2019). [in Ukrainian]
19. Keberle, N.: Modeling of dynamic domains under use of the ontologies. Bull. Kharkiv Air Force Univ. **3**, 121–127 (2009)
20. Konys, A., Rogoza, W.: Big data and ontologies. In: Talk at ACS International Conference 2016 in Międzyzdroje, October 2016, 3 p. (2016)
21. Kuiler, E.: From big data to knowledge: an ontological approach to big data analytics. Rev. Pol. Res. **31**(4), 311–318 (2014). https://onlinelibrary.wiley.com/doi/full/10.1111/ropr.12077#reference

22. Shtogrina, O.S.: Information Technology of creating and using fuzzy knowledge bases with the use of metagraphs : dis. candidate. technical sciences / O.S. Shtogrina. Kiev, 2016. – 157 p (2016). [in Ukrainian]
23. Savchuk, Z.R.: Means of visualization of complex logical structures. In: Savchuk, Z.R. (ed.) Proceedings of the International Scientific and Technical Conference "Prospects for Telecommunications". 15–19 April 2019, pp. 261–263 (2019). [in Ukrainian]

Modeling of Telecommunication Components of Automated Control Systems in Low-Bandwidth Radio Networks

I. V. Strelkovskaya(✉), R. V. Zolotukhin(✉), and A. O. Makoganiuk(✉)

O. S. Popov Odesa National Academy of Telecommunications, kuznechnaya Street, 1, Odessa 65029, Ukraine
{strelkovskaya,a.makoganyuk}@onat.edu.ua

Abstract. The main problem, when designing and planning a communication system for automated control systems with high requirements for security, survivability and reliability, is the reasonable choice of the necessary telecommunications equipment to build communication networks with the required functional characteristics to provide access to services. The paper presents simulations of the telecommunication system at the lower level of management aimed at maintaining communication during the active movement of users. It also includes the further studies of traffic parameters based on analytical, numerical methods and methods of statistical modeling. For public networks, the characteristics of voice, interactive, streaming, and background traffic are well known and described in ITU Recommendations and reports. However, ultra and very high frequency (UHF/VHF) radio stations of low-bandwidth communication networks demonstrate low speed and high data latency, high probability of data loss in the channel, which complicates the use of standard data transmission protocols. The work proposes the analysis of STANAG-4677 and ADatP-36 standards. As well, it offers the concept of their use to design telecommunication components of the automated control system for UHF/VHF radio communication networks. The software implementations of the JDSS, NFFI, FFI protocols has been developed using.NET framework and C# programming language. The data has been obtained on the average bitrate of traffic for the standards, possible packet lengths and average length, the time interval between incoming packets. The distribution law of time intervals between requests for service and the density of this distribution has been determined.

Keywords: STANAG 4677 · NFFI · FFI · ADatP-36 · ACS · Traffic characteristics · UHF/VHF radio station · Low-bandwidth network · Average package length · Distribution law

1 Introduction

Rapid development of technologies at the end of XX and the beginning of XXI centuries [1–3] has forced the creation of digital automated control systems (ACS). Modern ACS's are widely used to manage enterprises, production, public organizations and government. This article focuses on governmental ACS of the low echelon management

P. Vorobiyenko et al. (Eds.): IPF 2020, LNNS 212, pp. 150–170, 2021.
https://doi.org/10.1007/978-3-030-76343-5_9

level with the constant movements of users. Such type of systems is intended to solve specifically important problems and have high requirements to reliability, efficiency, security, survivability and scalability. Such systems are mainly based on Ultra and Very High Frequency (UHF/VHF) radio stations. The obtained QoS characteris-tics [4] show that the usage of standard data transmission protocols in UHF/VHF radio networks is impossible due to low throughput and significant transmission delay, high probability of data loss. There was created a set of protocols specific for such communication channels. For example, JDSS compliant to the STANAG 4677standard [5] is the one of them, developed specifically to operate in low throughput communication networks of tactical echelon of army management under conditions of NATO nations cooperation. This standard gives the ability to transmit text messages, visual map objects, geometric drawings (point, line, polygon), targeting, special messages of nuclear, biological and chemical situations, medical evacuation needs, etc. Beside this, the majority of modern UHF/VHF radio stations that support AEP-76, VOL. III [6], also are compliant with AdatP-36 [7] standard, which is used to transmit information about own forces locations. This article gives the concept of the JDSS and AdatP-36 protocols being used to build the government ACS of low echelon with fast users' movements.

The important problem of necessity to predict traffic services characteristics arises at the stage of planning the low throughput networks based on UHF/VHF radio stations. This problem leads to complication of the adequate appliance choice and its' functional characteristics along with prediction of services availability. The article [8] uses special software based on statistical modeling and mathematical simulations of different traffic types. There was emulated and calculated network services availability. Every service type has its own requests intervals distribution law, for example, voice traffic demonstrates exponential law, video conference – Pareto distribution law, high velocity data traffic – lognormal one. Duration of service for modeling was chosen among the values given in the ITU-R M.1768 [9] Recommendation for the Uniform law. However, this data is not enough to calculate the services availability and predict traffic parameters in order to construct the government ACS for the low echelon with fast movements of users. The article considers simulations of telecommunication components of low management echelon to provide connectivity during rapid movements users together with further research of the parameters by means of analytical, mathematical and static modeling methods.

2 Algorithms and Protocols for Data Transmission Model of ACS

2.1 Analysis of the STANAG 4677 Standard

It is necessary to define the following parameters of ACS telecommunication components in low management echelon in order to perform simulations:

- algorithms and protocols of data transmission system;
- the modeled network structure;
- the transmitted information flows;
- the set of variable parameters of communication system.

First of all, it is necessary to define algorithms and protocols used in low throughput radio networks when performing modeling of telecommunication components of the ACS. The goal of the STANAG 4677 standard is to enable cooperation between different NATO members armies in order to increase the situational awareness about the battlefield by means of tactical information exchange between different army groups in low throughput communication channels based on UHF/VHF radio. The standard is mostly focused on the dismounted soldier system in tactical management level. The concept of this standard is shown in the Fig. 1.

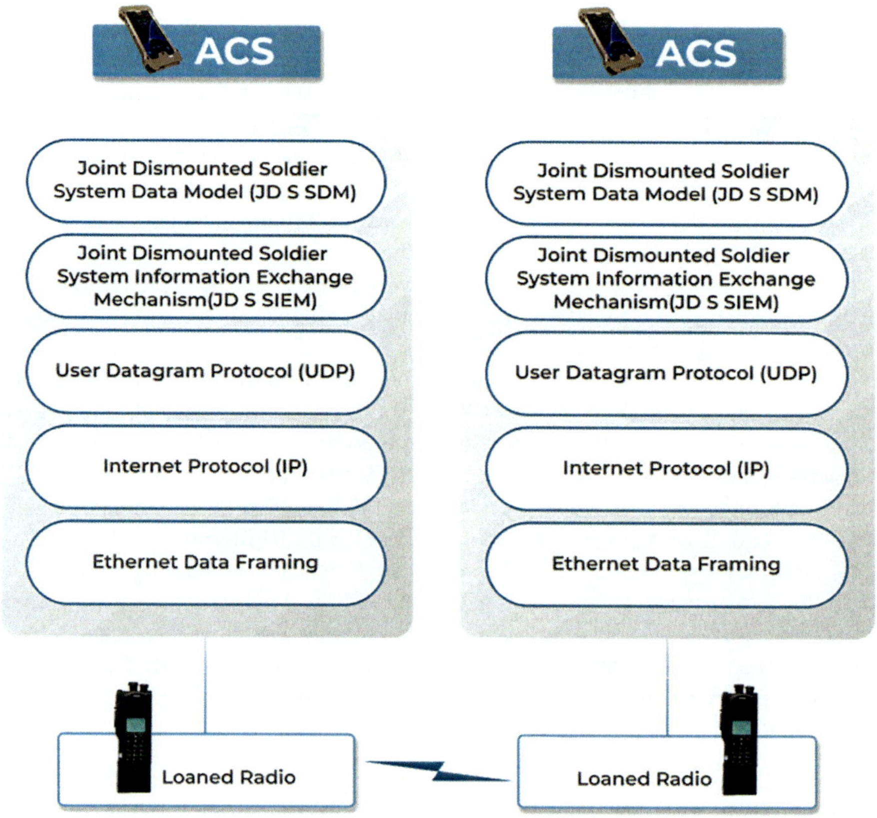

Fig. 1. Dismounted soldier system C4 interoperability solution

The standard states that the main source and recipient of data is the national ACS. The information, that comes from national ACS, is then transformed into the form of Joint Dismounted Soldier System Data Model (JDSSDM). JDSSDM describes a set of data models for different message types, namely:

1. Tactical messages, those containing current valid data (identification and text messages, graphical marks of objects on the map, objects geographical position on

the map, geometrical figures, targeting, special messages of nuclear, biological and chemical situations, medical evacuation needs.
2. Communicational messages, that are responsible for information exchange and its status.
3. Extension messages that are used to expand system functionality that goes beyond the standard specification.

JDSSDM data models are described using XML language, the exact tags and structure of messages is strictly defined in the STANAG 4677 standard. JDSSDM itself is incapsulated into another data structure Joint Dismounted Soldier Information Exchange Mechanism (JDSSIEM). This mechanism is dedicated for use in broadcast and multicast communication systems. This means that the packets, sent into IP-network, use multicast IP-addresses. The standard implies the usage of Loaned Radio from allied country. Loaned Radio has to support this standard or use the special software and hardware to interface JDSS gateway between national ACSs of NATO members. Loaned Radio has to:

- provide a communication link for 1500 m in line of sight conditions and 500 m in urban area;
- ensure data retransmission and connection to several nodes;
- support errors correction;
- support barrier detection;
- avoid hardware malfunctioning;
- encrypt all traffic.

The STANAG 4677 protocols stack is shown in the Fig. 2. Data transmission is per-formed using the ordinary UDP protocol in IP multicast communication network. JDSSDM is used as XML messages payload.

NATO Friendly Force Information (NFFI) is a 16-bytes binary header that is used for data marking for segmentation and addressing between different ACS. It also shows the compression of data, processing algorithm and packets priorities. NFFI header is shown in the Fig. 3 [5]. JDSSIEM just like JDSSDM, are the XML messages, but are responsible for messages synchronization and data compression.

The mechanism of guaranteed data transmission according to the STANAG 4677 has one special property: unlike standard TCP/IP stack protocols, it uses special mechanism of missed messages synchronization. Several service message types are used to build up this mechanism:

- “HeartBeat”;
- “MessageSyncRequest”;
- “MessageSyncReply”;
- “FullSyncRequest”;
- “FullSyncReply”.

“HeartBeat” messages are sent by every gateway with the same time interval (normally 60 s). It lets all nodes on the network be aware of the sent messages quantity

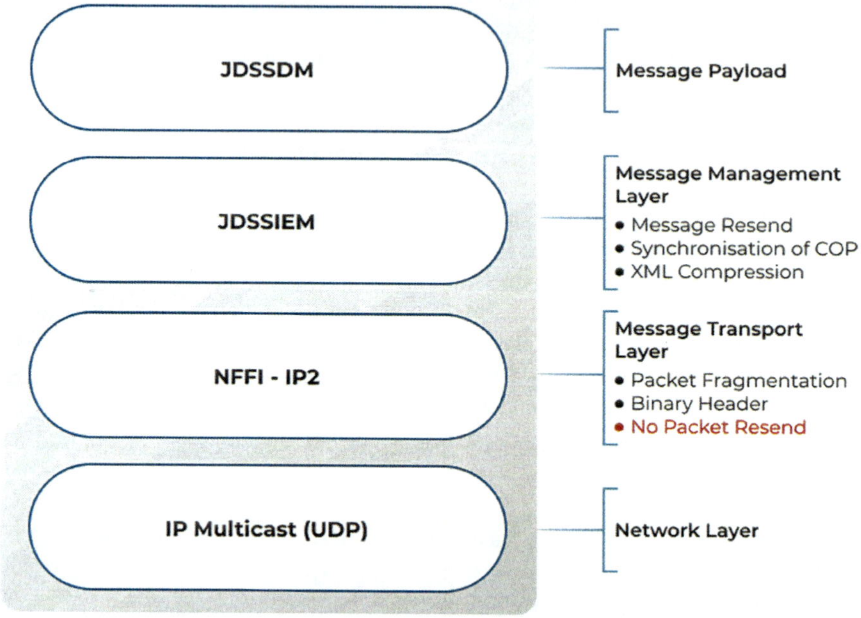

Fig. 2. STANAG 4677 transport and network layers

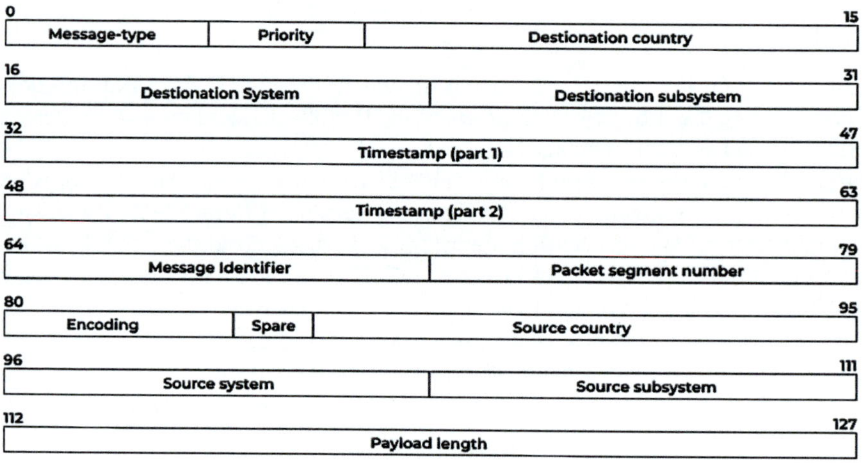

Fig. 3. NFFI Header definition

according to message category. "Heartbeat" messages invoke in all nodes the verification of received messages from the source of the "Heartbeat" message. Depending on the number of missed messages, "MessageSyncRequest" is sent to request single missed message or "FullSyncRequest" to request all current data. Moreover, if the request is sent to the address of a particular gateway, the response is sent to all nodes in a multicast packet to prevent repeated requests and possible loss of other nodes. In addition, it should be noted that not all messages are re-sent or requested because some messages become obsolete faster than one "HeartBeat" interval. These messages are:

- "PresenceMsg";
- "HeartBeat";
- "MessageSyncRequest";
- "MessageSyncReply";
- "FullSyncRequest";
- "FullSyncReply".

In order to reduce the load on the communication network and obtain only relevant information, the standard provides the maximum number of messages for each type of message that can be re-requested.

Based on the analysis, we can identify the main aspects of the STANAG 4677 standard:

1. The JDSS protocol provides data model for various messages – JDSSDM and the mechanism for transmitting these messages – JDSSIEM.
2. The NFFI protocol is used as a binary header. It is responsible for fragmentation, addressing between systems and countries, data compression and uses only 16 bytes of header.
3. The STANAG 4677 standard is designed to organize the interaction of existing digital ACS of NATO members for mission cooperation in radio networks based on "loaned radio".
4. Each message is sent to a multicast group, which allows receiving data by multiple subscribers simultaneously.
5. The standard is designed specifically to receive and send data on low-bandwidth radio networks.
6. The JDSS information exchange mechanism uses message synchronization unlike the standard queries on request, which reduces the load on the radio network.

Since the concept of the STANAG 4677 standard provides the JDSS protocol as a digital gateway between the existing NATO members ACS, it is necessary to consider the concept in which the JDSS and NFFI protocols appear to be the basic algorithms and protocols for the governmental ACS of the low echelon management level dealing with the constant movements of users of the low-bandwidth communication networks. In this case, the JDSS and NFFI protocols will not be used on gateways between different ACS. Instead, they will be used as a system-forming mechanism for information transfer for an ACS with high requirements for efficiency, reliability, survivability and scalability. The role of the Loaned Radio can be played by any UHF/VHF radio station that has the ability to transmit data using IP, UDP, IGMP [10, 11] and supports multicast addressing. All users use their ACS terminals to communicate via UHV/VHF radio stations to the low-bandwidth radio network. The number of users is limited only by the bandwidth of this network.

The concept of ACS construction based on JDSS and NFFI protocols is presented in the Fig. 4.

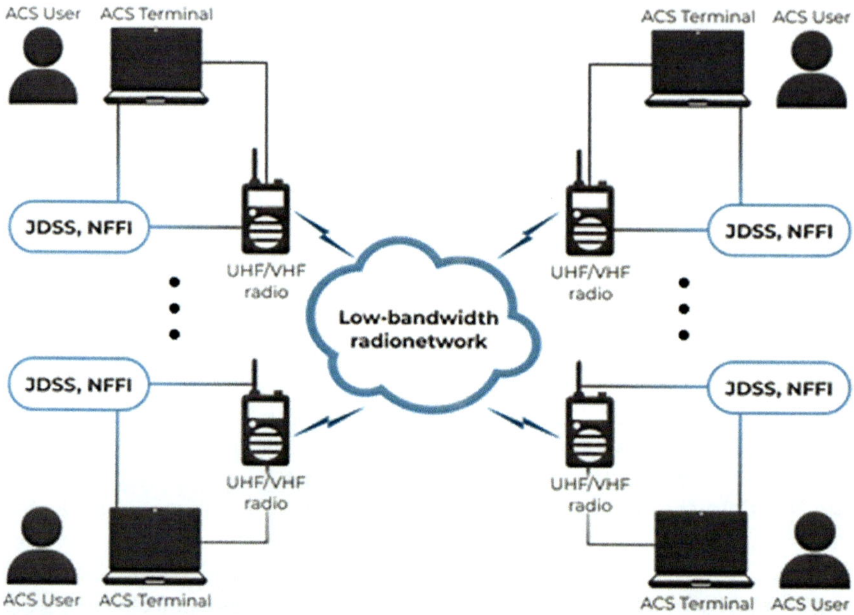

Fig. 4. The concept of using JDSS and NFFI protocols for governmental ACS of the low echelon management level

2.2 AdatP-36 Standard Analysis

In Subsect. 2.1 was presented the concept of using JDSS and NFFI protocols to build ACS. The important function for the governmental ACS of the low echelon management level with fast users' relocations is geolocation of the users. JDSSDM data model has a special "Presence Message" to transmit the position information of their users. Another NATO standard, AdatP-36 [7], needs to be considered. AdatP-36 was created specifically to transmit the positions of objects on the map and the status of these objects. This standard duplicates the function of JDSS geolocation message, but unlike the STANAG 4677 standard, a number of specialized UHF/VHF radio stations with data transmission supports the AdatP-36. That is why it is necessary to analyze this standard for the purpose of its possible use in the ACS of the lower echelon management level.

The stack of protocols of the AdatP-36 standard is shown in the Fig. 5. AdatP-36 is a data model described by XML. This model describes the object, its status and position on the map. The Friendly Force Information (FFI) binary header is added to this model in the same way as for JDSS. The data is sent to the multicast address using the UDP transport protocol. Therefore, this standard does not provide a mechanism for data synchronization or guaranteed delivery, but according to the rules of STANAG 4677, geolocation notifications for JDSSDM objects are also not repeated by a request. This is because object position data is sent at regular intervals and becomes obsolete faster than it is re-invited and resent. FFI defines also a 16-byte header that is identical to the NFFI header and performs the same functions, but does not use the "Destination country" and "Source Country" fields (Fig. 3).

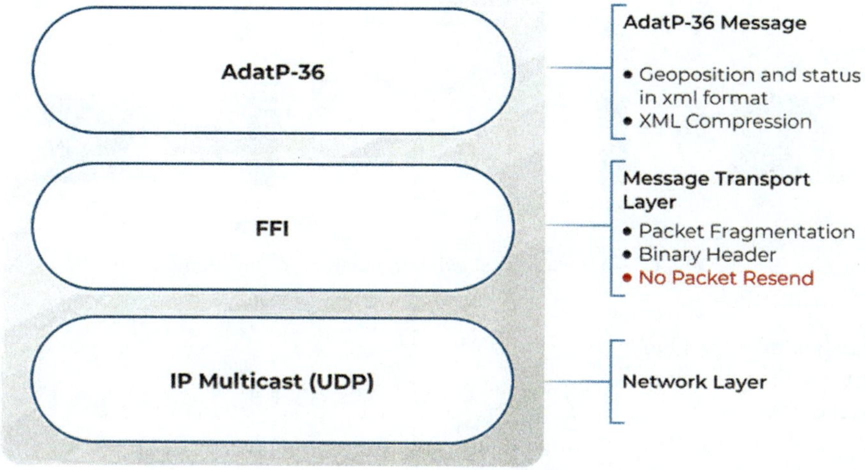

Fig. 5. Protocol stack for the AdatP-36

Based on the above, AdatP-36 performs the same functions as the "Presense Message" JDSSDM. Thus, if the ACS terminal uses supports the AdatP-36 standard, then it is preferable to avoid the use of "Presence Message" in order to reduce the load on the network with unnecessary duplicated information.

3 Initial Data for Modeling of Telecommunication Components of the ACS

The modelling of the telecommunication components of the governmental ACS of the low echelon management level with active relocation of users requires to determine the structure of the network and information flows that will circulate in this system. Based on the proposed concept, shown in the Fig. 4, the network structure shown in the Fig. 6 is used in simulations. This scheme is the basis for construction of governmental ACS of the low echelon management level.

According to this scheme, personal computers act as terminals of ACS of the low echelon management level. Each personal computer is connected to the Harris RF-7850M-HH radio station via an Ethernet cable. The UHF/VHF radio station is set to M-TNW mode. Radio stations are located at a distance of 100 m from each other. The power of radio stations is set to 1 W. According to [4], the following parameters were obtained before the simulation was performed, and the QoS of the low-speed radio network was measured:

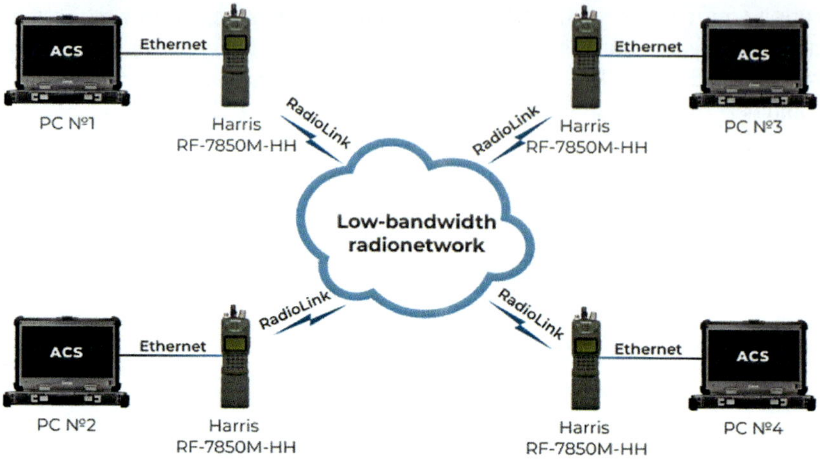

Fig. 6. Communication network diagram for modeling

- bandwidth – 338 kbit/s;
- jitter – 42 ms;
- packet loss 1%;
- minimal ping - 270 ms, maximal ping – 360 ms, average ping – 315 ms.

Harris RF-7850M-HH radio stations have a built-in GPS receiver and with special settings and firmware can transmit the coordinates of their geolocation to a multicast network using IP, UDP, FFI protocols and XML message that meet the requirements of the AdatP-36 standard. The fact that the AdatP-36 and STANAG 4677 standards allow the use of multicast addresses in range of 239.192.0.0/14 and any UDP port, gives the possibility to perform modelling of ACS telecommunication component using 239.192.1.1 multicast group and 49001 port.

The specialized software was created by us in order to perform modelling of information flows. It is based on.NET framework and C# programming language. This software performs sending and reception of JDSS protocol messages according to STANAG-4677. It also is able to receive AdatP-36 messages from Harris RF-7850M-HH UHF/VHF radio station. The structure of information flows is shown in the Fig. 7.

The software implements the following JDSSDM messages:

- “IdentificationMsg”, which uniquely identifies the users of ACS;
- “CasevacreqMsg”, which is used to request the medical evacuation procedure;
- “ReceiptMsg”, which serves to confirm the reception or reading by users;
- “GenInfoMsg”, which is used to send text messages;
- “ScetchMsg”, which gives the possibility to transfer arbitrary sketches;
- “NBCMsg”, which is used to inform the users of ACS about nuclear,
- biological or chemical danger;
- “ContactSightingMsg”, which is used to transfer objects marks on the map;
- “CoordinationMsg”, which is used to implement targeting.

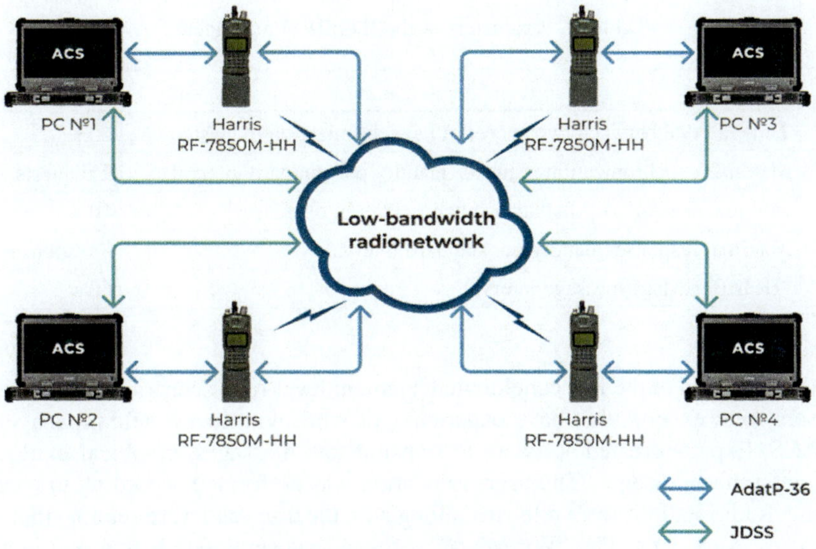

Fig. 7. Structure of information flows

According to the STANAG 4677 standard, each message type has its own maximum number of recent messages that can be repeatedly requested. In compliance with the recommendations of [5] the parameters presented in the Table 1 were selected.

Table 1. Parameters of the JDSSDM messages

№	JDSSDM Message Types	Syncable	Repair window
1	IdentificatonMsg	Yes	1
2	CasevacreqMsg	Yes	10
3	ReceiptMsg	Yes	10
4	GenInfoMsg	Yes	10
5	SketchMsg	Yes	50
6	NBCMsg	Yes	50
7	ContactSightingMsg	Yes	50
8	CoordinationMsg	Yes	50

According to the STANAG 4677 standard, timers and parameters for sending special JDSSIEM messages were set. These are shown in the Table 2. All data transmitted via JDSS and NFFI protocols are compressed using GZIP.

Harris RF-7850M-HH UHF/VHF radio station is configured to send objects geolocation messages every 10 s. The "Wireshark" software is installed on personal computers, which is used to analyze computer network traffic.

Table 2. Parameters of the JDSSIEM messages

№	Parameter	Value
1	Time interval between two successful synchronization requests	60 s
2	Maximal synchronization requests quantity per standard interval	2 requests
3	Minimal time interval between synchronization responses	10 s
4	Maximal response quantity per standard interval	3 responses
5	“HeartBeatMsg” message interval	60 s

The simulation of the low echelon management level ACS scenario was performed. The users were experts, that have experience of work with low echelon management level ACS. Experts created software to transmit text messages, graphical marks, targeting and other messages. The users relocation was performed according to received targeting. Radio stations were relocated along with the users and were sending their own coordinates every 10 s. The “Wireshark” software has captured all sent and received messages. The dump of received packets shown in the Fig. 8 was captured in this way.

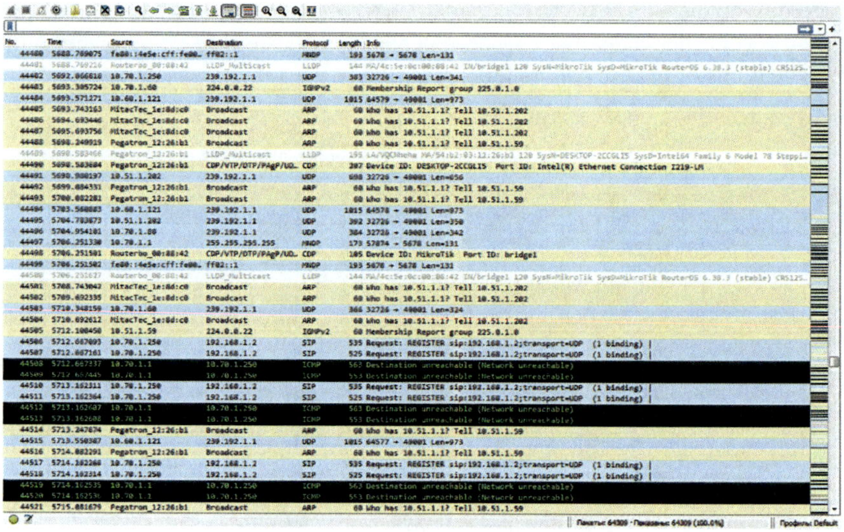

Fig. 8. Simulation packets dump

4 Initial Data and Results of Modelling for Telecommunication Components of ACS in Low Bandwidth Communication Networks

4.1 Analysis of Traffic Indicators According to the STANAG 4677 and AdatP-36 Standards

Only packets according to the STANAG 4677 and AdatP-36 standards were chosen among the packet dump for analysis with "Wireshark" software filters. The result is the packets (Fig. 9) from low bandwidth UHF/VHF radio network.

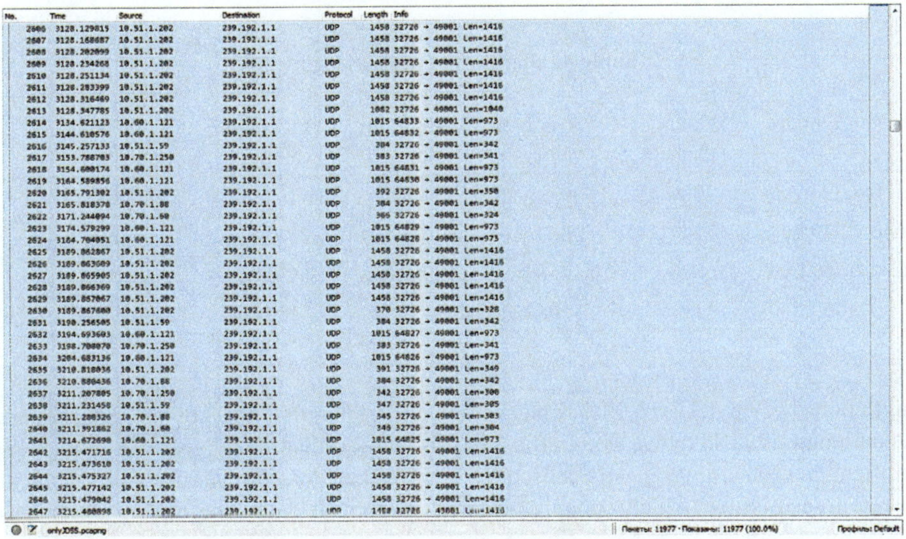

Fig. 9. Dump of packets from UHF/VHF radionetwork

Statistics of the traffic from communication network is shown in the Table 3. In this way, 11977 packets were sent over low bandwidth communication network. The total of 12,6 Mbytes were sent. The average packet length was 1108 bytes, the intensity of packets for service is 1,07 and the average bandwidth consumed by the traffic was 5,34 Kbps.

According to [4], the data rate in the UHF/VHF radio network is 69,4–471 Kbps. The obtained average traffic bandwidth of STANAG 4677 and AdatP-36 standards is much lower. This shows that the use of the proposed concept for the construction of governmental ACS of the low echelon management level with active relocation of users allows to implement the necessary functionality in low-bandwidth radio networks. The lengths of packets transmitted over the network has been analyzed. Statistics of packet lengths received for maintenance in the UHF/VHF radio network have been obtained (Table 4) using the "Packet Length" tool of the "Wireshark" software.

The received data demonstrates that the most significant part of packets had length 1290–1459 bytes – 56,99%, packets of length 0–639 bytes – 24,27%, and packets of

Table 3. Statistics of the traffic compliant with STANAG 4677 and AdatP-36 standards

№	Traffic parameter	Value
1	Count of packets	11977
2	Total packets size	12,6 Mbytes
3	The average packet length	1108 bytes
4	The average time between packets	0,937 s
5	Intensity of packets for service	1,07 packet/sec
6	Average bandwidth	5,34 Kbps

Table 4. Statistics of packet lengths

Packets length	Count of packets	Average value	Minimum value	Maximum value	Percentage of total
0–639 byte	2907	397 byte	321 byte	638 byte	24,27%
640–1279 byte	2244	966 byte	640 byte	1279 byte	18,74%
1290–1459 byte	6826	1458 byte	1290 byte	1459 byte	56,99%
0–1459	11977	1108 byte	321 byte	1459 byte	100%

length 640–1279 bytes – 18,74%. The average packet length is 1108 bytes. The average packet length is 1108 bytes. The traffic analysis showed that the biggest part of packets with length 1290–1459 bytes is caused by high packets loss probability and graphical objects transmission, which occupied a large amount of data. The high packets loss probability led to the increase of synchronization requests quantity, which in turn led to the increase of synchronization responses that contained several information messages that were fragmented to fit the Ethernet network MTU [12].

4.2 The Research of Statistical Characteristics of Random Value – Quantity of Packets Coming for Service into Communication Network

Let us research the data that was obtained experimentally. The X is assumed to be a random value which denotes the packets quantity that are coming for service into communication network. The hit quantity p_i^* of observed X values every interval $[x_i, x_{i+1})$ is calculated by the formula [13, 14]:

$$p_i^* = P(x_i \le X \le x_{i+1}) = \frac{m_i}{N}, \tag{1}$$

where X – random value that characterizes the serviced packets quantity,

$P(x_i \le X \le x_{i+1})$ – the probability of random value X coming from interval $[x_i, x_{i+1})$i,

m_i – the quantity of observed values that hit the i-th partial interval,

N – the total packets quantity.

Alongside, the sum of partial intervals relative frequencies is 1 [13, 14]:

$$\sum_{i=1}^{n} p_i^* = 1, \tag{2}$$

where n – quantity of time intervals.

The results of initial data calculation are shown in the Table 5. It is now possible to build the histogram of random value X by the following formula:

$$h_i = \frac{p_i^*}{x_{i+1} - x_i}, \tag{3}$$

where h_i – histogram height of i-th time interval.

The calculation results are shown in the Table 5.

Table 5. Parameters of JDSSIEM messages

Number of i-th interval	Interval limits $[x_i, x_{i+1})$	Packets quantity in interval m_i	Relative frequencies p_i^*	Histogram heights h_i	$F^*(x_i)$
1	[0;0,25)	7220	0,6028	2,4112	0
2	[0,25;0,5)	1305	0,109	0,436	0,6028
3	[0,5;0,75)	753	0,0628	0,2512	0,7118
4	[0,75;1)	501	0,0418	0,1672	0,7746
5	[1;1,25)	318	0,0266	0,1064	0,8164
6	[1,25;1,5)	257	0,0215	0,086	0,843
7	[1,5;1,75)	202	0,0169	0,0676	0,8645
8	[1,75;2)	157	0,0131	0,0524	0,8814
9	[2;2,25)	158	0,0132	0,0528	0,8945
10	[2,25;2,5)	136	0,0114	0,0456	0,9077
11	[2,5;2,75)	117	0,0098	0,0392	0,9191
12	[2,75;3)	102	0,0085	0,034	0,9289
13	[3;3,25)	97	0,0081	0,0324	0,9374
14	[3,25;3,5)	85	0,0071	0,0284	0,9455
15	[3,5;3,75)	76	0,0063	0,0252	0,9526
16	[3,75;4)	70	0,0058	0,0232	0,9589
17	[4;4,25)	57	0,0048	0,0192	0,9647
18	[4,25;4,5)	51	0,0043	0,0172	0,9695
19	[4,5;4,75)	44	0,0037	0,0148	0,9738
20	[4,75;5)	42	0,0035	0,014	0,9775
21	[5;5,25)	28	0,0023	0,0092	0,981
22	[5,25;5,5)	25	0,0021	0,0084	0,9833
23	[5,5;5,75)	22	0,0018	0,0072	0,9854

(continued)

Table 5. (*continued*)

Number of i-th interval	Interval limits $[x_i, x_{i+1})$	Packets quantity in interval m_i	Relative frequencies p_i^*	Histogram heights h_i	$F^*(x_i)$
24	[5,75;6)	20	0,0017	0,0068	0,9872
25	[6;6,25)	27	0,0023	0,0092	0,9889
26	[6,25;6,5)	23	0,0019	0,0076	0,9912
27	[6,5;6,75)	16	0,0013	0,0052	0,9931
28	[6,75;7)	17	0,0014	0,0056	0,9944
29	[7;7,25)	16	0,0013	0,0052	0,9958
30	[7,25;7,5)	14	0,0012	0,0048	0,9971
31	[7,5;7,75)	11	0,0009	0,0036	0,9983
32	[7,75;8]	10	0,0008	0,0032	0,9992
					1

The Fig. 10 shows the histogram of random value *X*. The histogram approaches some curve that represents the graph of random value *X* distribution density. This way the replacement of step line with continuous curve gives the assumption about possible distribution law of random value *X* (Fig. 10).

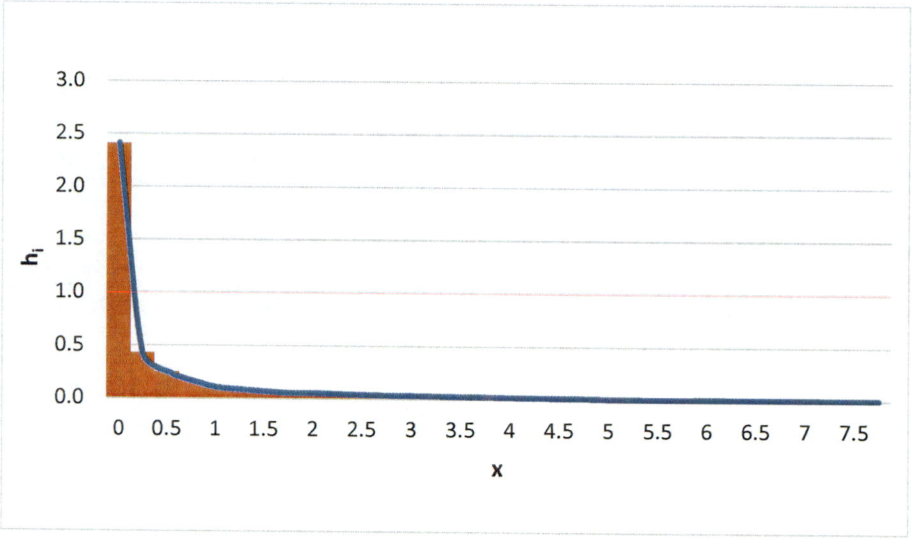

Fig. 10. Histogram of random value X

The data about samples distribution allows to approximately build the empirical function of random value X distribution. Then it is obvious that,

$$F^*(x) = \begin{cases} 0, & x \in [x_1, x_2), \\ p_1^*, & x \in [x_2, x_3), \\ p_1^* + p_2^*, & x \in [x_3, x_4), \\ & \vdots \\ \sum_{k=1}^{n} p_k^*, & x \in [x_{n-1}, x_n]. \end{cases} \quad (4)$$

Calculation results are shown in the Table 5. The connection of resulting points into a line gives us the approximate graph of empirical function that describes the random value X distribution. It is shown in Fig. 11.

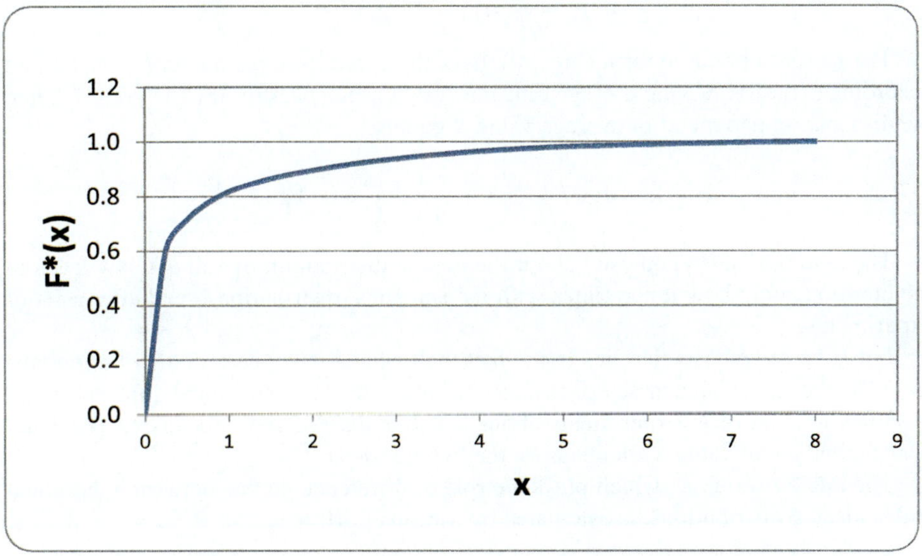

Fig. 11. Graph of empirical function that describes the random value X distribution

The shape of the curve allows to make assumption (hypothesis) that the random value X is distributed by the Weibull law [15, 16]. Its' density is described by formula [13]:

$$f(x) = \frac{k}{\lambda}\left(\frac{x}{\lambda}\right)^{k-1} e^{-\left(\frac{x}{\lambda}\right)^k}, x \geq 0 \quad (5)$$

where k– curve form parameter,
λ – scale parameter,
x – time interval between service requests.
Weibull distribution law is described by formula [13, 14]:

$$F(x) = 1 - e^{-\left(\frac{x}{\lambda}\right)^k}, x \geq 0. \quad (6)$$

k parameter of distribution curve form is defined by formula [13, 15, 16]:

$$k = 2 - 2H, \tag{7}$$

where H – Hurst exponent.

Distribution parameter λ is described by formula [13, 15, 16]:

$$\lambda = \frac{M(X)}{\Gamma\left(1 + \frac{1}{k}\right)}, \tag{8}$$

where $M(X)$ – mathematical expectation of X value,

$\Gamma(k) = \int_0^\infty t^{k-1} e^{-t} dt$ - Euler gamma function.

The probability of packets receipts hit into time intervals $\left[x_i; x_{i+1}\right)$ is defined by formula [13]:

$$p_i = F(x_{i+1}) - F(x_i). \tag{9}$$

The research of curve form (Fig. 10) gives the curve form parameter $k \approx 0{,}43$, and according to the Table 3 the average time interval between packets $M(X) \approx 0{,}937$. Then the distribution parameter of average value X equals:

$$\lambda \approx 1,07 * \Gamma\left(1 + \frac{1}{0,43}\right) \approx 0,34.$$

The assumption (hypothesis) about theoretical distribution of value X law leads to the question about how it correlates with its' statistical distribution found after a set of experiments.

Let H be hypothesis that lies in the fact that value X obey certain distribution law, and U be the value that characterizes degree of theoretical and statistical difference, that is shown in a set of N experiments about X value observe and depends on predicted distribution law of value X (depends on the hypothesis H).

The random value U, which plays the role of difference degree between theoretical and statistical distributions, is calculated by formula [14]:

$$U = N \sum_{i=1}^{n} \frac{\left(p_i^* - p_i\right)^2}{p_i}, \tag{10}$$

where N – samples volume,

n – quantity of intervals that form the range of observed values X,

p_i^* – relative frequency of observed values X hit into i-th partial interval $[x_i, x_{i+1})$,

p_i– the probability of X value hit into this interval, if the true distribution law coincides the predicted one.

Formula (10) is Pearson's «criteria χ^2», which is called agreement criteria. N can be introduced under sum sign in formula (10) and it can be taken into account that $p_i^* = m_i/N$, where p_i^* – number of observed values X in i-th interval, then this formula is transformed into next form:

$$\chi^2 = \sum_{i=0}^{n} \frac{(m_i - Np_i)^2}{Np_i}, \tag{11}$$

where n – intervals quantity.

The theoretical probabilities p_i of X value hit and probable packets quantity x_i. The results are shown in Table 6.

Table 6. Parameters of Weibull distribution

Interval number i	x_i	Frequency, p_i	$N * p_i$
1	0	0,58712	7032
2	0,25	0,10643	1275
3	0,5	0,06027	722
4	0,75	0,04046	485
5	1	0,02953	354
6	1,25	0,02258	272
7	1,5	0,01804	216
8	1,75	0,01472	176
9	2	0,01225	147
10	2,25	0,01035	124
11	2,5	0,00886	106
12	2,75	0,00766	92
13	3	0,00669	80
14	3,25	0,00588	70
15	3,5	0,00521	62
16	3,75	0,00464	55
17	4	0,00415	50
18	4,25	0,00374	45
19	4,5	0,00338	40
20	4,75	0,00306	37
21	5	0,00279	33
22	5,25	0,00254	30
23	5,5	0,00233	28
24	5,75	0,00214	26
25	6	0,00197	24
26	6,25	0,00182	22
27	6,5	0,00168	20
28	6,75	0,00156	19
29	7	0,00145	17
30	7,25	0,00135	16
31	7,5	0,00125	15
32	7,75	0,00117	14

Formula (11) and calculations in the Table 5 and Table 6 conclude that $\chi^2 \approx 42{,}09$.

In order to confirm the assumption about predicted distribution law it is necessary to define point u, such that [14]:

$$P(\chi^2 \geq u) = \alpha, \tag{12}$$

where α – level of significance.

In order to define u it is necessary to find the degree of freedom number r [13, 14]:

$$r = n - s - 1, \tag{13}$$

where s – quantity of distribution parameters.

The Weibull distribution has parameters quantity $s = 2$, while the intervals quantity is $n = 32$. The theoretical distribution has the degree of freedom number $r = 29$. Assuming that, $\alpha = 0,05$ while $r = 29$ and according to previously calculated values [13], critical point is $u \approx 42,56$. This way, it can be stated that $\chi^2 < u$ while the significance level is $\alpha = 0,05$. Then such difference degree between theoretical and statistical distributions of the X value can be defined as insignificant and it can be stated that the predicted distribution law of random value X at value $\alpha = 0,05$ (Weibull law) corresponds to researched data. This way the density of time intervals between packets service distribution for STANAG 4677 and AdatP-36 standards equals:

$$f(x) \approx 0,68x^{-0,57}e^{-1,59x^{0,43}}, x \geq 0.$$

The Fig. 12 shows the packets reception during simulation and packets quantity according to Weibull distribution.

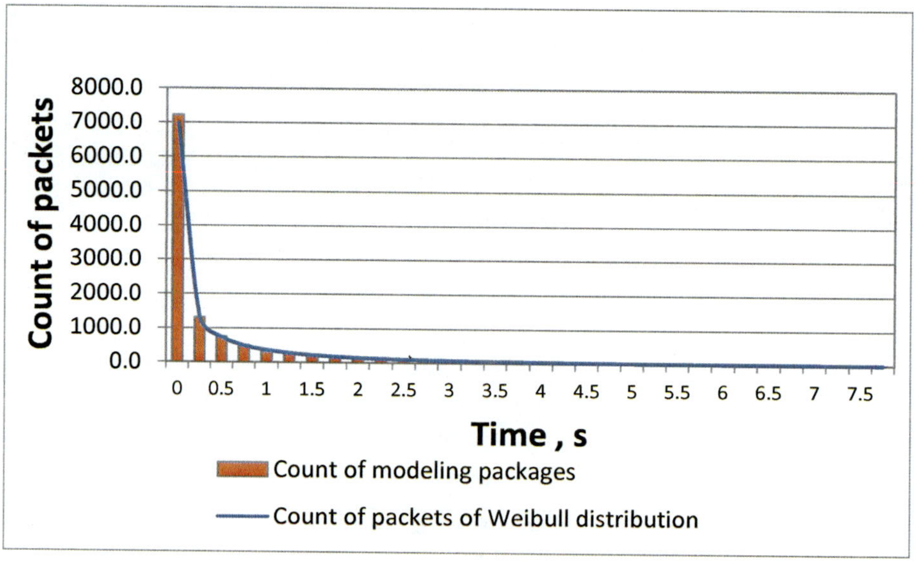

Fig. 12. Count of modeling and Weibull distribution packets.

5 Conclusions

1. The analysis of protocols JDSS, NFFI, FFI, AdatP-36 and standards
2. STANAG 4677 and AdatP-36 was performed, and the concept of these protocols usage was offered as the backbone mechanism of data transmission by the national low management echelon ACS with fast users relocation.
3. The software for simulations of telecommunication components of national low management echelon ACS with fast users relocation by means of C# programming language and.NET framework was developed.
4. The traffic parameters of STANAG 4677 and AdatP-36 standards were obtained: average packet length – 1108 byte, average time interval between service requests – 0,937 s, average bandwidth used by traffic – 5,34 Kbps.
5. It was found that the distribution density of time intervals between packets service corresponds to the Weibull distribution law.
6. The received traffic characteristics of STANAG 4677 and AdatP-36 standards allow the definition of service availability in low bandwidth communication networks of low management echelon ACS with fast users relocation, which gives the ability to increase the quality of calculations about load on communication channels and decrease the service refusals probability on the communication system planning stage.

References

1. Ilchenko, M., Uryvsky, L., Osypchuk, S.: The main directions of improving infocommunications in the global tendencies. In: Advances in Information and Communication Technologies. Lecture Notes in Networks and Systems, pp. 3–22. Springer, Cham (2020). https://doi.org/10.1007/978-3-030-58359-0
2. Vorobiyenko, P.P.: Industry 4.0 and information communication technologies. In: International Conference on Information and Telecommunication Technologies and Radio Electronics (UkrMiCo 2017), 11–15 September 2017, Odessa, Ukraine, pp. 15–18 (2017)
3. Skulysh, M., Romanov, O., Globa, L., et al.: Managing the Process of Servicing Hybrid Telecommunications Services. Quality Control and Interaction Procedure of Service Subsystems, pp. 244–256. Springer, Heidelberg (2019). https://www.mendeley.com/catalogue/47719483-0379-34e3-b439-7815799f7e11.
4. Strelkovskaya, I., Zolotukhin, R.: Research of low-bandwidth radionetworks QoS parameters. Inf. Telecommun. Sci. Int. Res. J. **11**(1(20)), (2020). https://doi.org/10.20535/2411-2976.12020.77-81
5. STANAG 4677: 2014 Dismounted soldier systems standards and protocols for command, control, communications and computers(C4) interoperability/NATO 2014
6. AEP-76, VOL.III Specifications Defining the Joint Dismounted Soldier System Interoperability Network (JDSSIN) – LOANED RADIO (STANAG 4677) Edition A/NATO 2014
7. NATO – ADATP-36: Friendly force tracking systems (FFTS) interoperability/NATO (2017)
8. Solovskaya, I.M., Biryukov, S.V., Gonsiorovsky, D.M., Gutsova, K.S., Zolotukhin, R.V., Fomichov, P.A.: Research of LTE/MVNO network traffic when operated jointly by several mobile operators. Scientific works of O. S. Popov ONAT 2012'1 - O. S. Popov ONAT, Odessa, pp. 167–176 (2012)

9. ITU-R MCE-R M.1768. Methodology for calculation of spectrum requirements for the future development of the terrestrial component of IMT-2000 (2003)
10. Internet Group Management Protocol, Version 2. https://tools.ietf.org/html/rfc2236. Accessed 21 Nov 2020
11. Internet Group Management Protocol, Version 3. https://tools.ietf.org/html/rfc3376. Accessed 21 Nov 2020
12. A Standard for the Transmission of IP Datagrams over IEEE 802 Networks. https://tools.ietf.org/html/rfc1042. Accessed 21 Nov 2020
13. Strelkovskaya, I.V.: Probability theory and random processes: textbook. [for specialists in the IT industry]. In: Strelkovskaya, I.V., Paskalenko, V.M. (eds.) Odessa: O. S. Popov ONAT, 384 p. (2018)
14. Strelkovskaya, I.V.: Textbook. Mathematical statistics. In: Strelkovskaya, I.V., Paskalenko, V.M. (eds.) Odessa: O. S. Popov ONAT, 384 p. (2018)
15. Strelkovskaya, I.V.: Self-similar traffic in G/M/1 queue defined by the Weibull distribution. In: Strelkovskaya, I.V., Grygoryeva, T.I., Solovskaya, I.N. (eds.) Radioelectronics and Communications Systems, vol. 61, no. 3, pp. 173–180 (2018). https://doi.org/10.20535/S0021347018030056
16. Strelkovskaya, I.: Probabilistic and time characteristics of the G/M/1 QS with the Weibull distribution of arrivals. In: Strelkovskaya, I., Solovskaya, I. (eds.) International Conference on Information and Telecommunication Technologies and Radio Electronics UkrMiCo'2017: proceedings of the Second International Conference, Odessa, Ukraine, 11–15 September, 2017: Odessa National Academy of Telecommunication named after O.S. Popov, pp. 452–455. https://doi.org/10.1109/UkrMiCo.2017.8095416

Modeling Unconditional Forwarding Decision Within Switching Lattice

Tatiana R. Shmeleva[1](✉) and Inna V. Stetsenko[2](✉)

[1] A.S. Popov, Odessa National Academy of Telecommunications, Kuznechna 1, Odessa 65029, Ukraine
t.shmeleva@onat.edu.ua

[2] National Technical University of Ukraine "Igor Sikorsky Kyiv Polytechnic Institute", Peremoga Ave. 37, Kyiv 03056, Ukraine

Abstract. A detailed model of a realistic up-to-date switching (routing) device has been built that combines cut-through and store-and-forward switching techniques. The model is based on a non-deterministic unconditional forwarding decision made with regard to the input packet. After the forwarding decision, two alternative procedures are launched: either cut-through direct transmission to the destination port or buffering the packet into the corresponding section of the internal buffer. The packet is buffered only in the case the destination port is busy. A rectangular (square) switching lattice model is composed in the form of an Infinite Petri net using dedicated transitions to connect devices. A parametric representation of p-invariants is obtained as a solution of an infinite Diophantine linear system of equations. P-invariance has been proven for a lattice an any size, that allows the conclusion about boundedness and conservativeness which are properties of an ideal networking protocol model. Some comparisons with the alternative technique of Object Petri nets and the corresponding simulation systems have been obtained which acknowledge mutual complementary of the two approachcs.

Keywords: Computing lattice · Packet switching · Cut-through · Store-and-forward · Infinite Petri net · Object Petri net

1 Introduction

Communication systems of supercomputers and clusters represent multidimensional torus [1, 2]. Communication lattices [3] with other kinds of edge conditions are widely applied for numerical solving of partial differential equations in manifold application areas. Plain (2D) lattices play a key role as networks of service providers [4]; for instance, a hexagonal lattice models cellular communications and triangular lattice is applied for radio broadcasting. An open lattice represents a construct for further applying various edge conditions [5].

Infinite Petri nets [2, 5, 6], successfully applied for studying communication grids, were using a simplified model of switching device. At first, devices with compulsory buffering of packets have been studied [7], then modern cut-through devices have been

P. Vorobiyenko et al. (Eds.): IPF 2020, LNNS 212, pp. 171–186, 2021.
https://doi.org/10.1007/978-3-030-76343-5_10

modeled [5]. Though while modeling cut-through devices, packet forwarding decision depended on the availability of the destination port. This procedure is correct for some kinds of lattices, for instance a torus. In the general case, a switching device should combine store-and-forward and cut-through procedures providing nondeterministic packet forwarding decisions [4].

The present paper refines the early studied models of switching devices. Open plane square lattices are composed and studied using Infinite Petri nets. We combine verification of lattice protocols and its performance evaluation using Object Petri nets technique [8, 9]. The obtained results illustrate practical benefits of using both Infinite and Object Petri nets.

The second section contains specification and description of a device model, in which the unconditional decision of packets forwarding is realized. In the third section, a switching square lattice of an arbitrary size was composed; a direct parametric expression of Infinite Petri nets with a regular structure has been used for the specification of infinite systems of linear equations and computing p-invariants. The fourth section contains a brief introduction of Petri-object simulation, a description of the rectangular communication grid Petri-object model construction, and presentation of simulation results.

2 Combined Cut-Through and Store-and-Forward Switching Device

2.1 Specification of a Switching Device Model

The Petri net model of the four-ports switching device with combined cut-through and store-and-forward switching modes is shown in Fig. 1, ports are enumerated clockwise starting from the upper port. There are four ports in the model, each port contains input and output channels; internal buffer, which is described by four buffer sections, one for each port, respectively, and the buffer size limitation. The example of device model, shown in Fig. 1, contains five packets (tokens) in each port buffer section, total 20 packets in the switching node, and the available buffer size is five. The switching mode of packet transmitting [4] is selected by simple rules:

- a switching node receives packet;
- the next hop destination of switching is a random choice between three output ports (excepts current input port);
- if the chosen output port is free, then the packet is switched to the output port without buffering by the cut-through switching procedure;
- if the chosen output port is busy, then the packet is stored in a node buffer (internal buffer section of the chosen output port), store-and-forward switching procedure is implemented.

For the description of device model in Petri net form, we construct a direct parametric expression [7]. The parametric expression (1) of the device model (see Fig. 1) contains four rows, the first's three rows describe transitions of all input port channels (index v

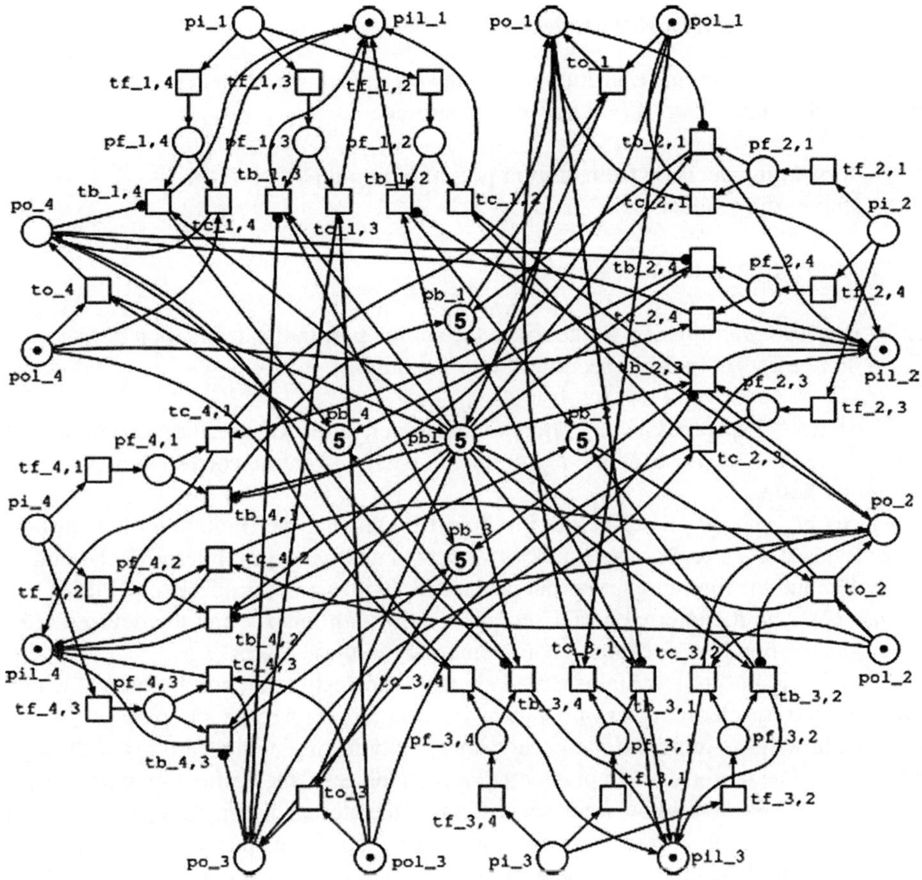

Fig. 1. A model of a switching device with combined cut-through and store-and-forward switching procedures.

is a number of the next output port for packet transmitting), the fourth row describes transitions of all output port channels (index u is a number of the current port). In the notations of device model, we use standard specifications [5]: the letter "t" specifies a transition; the letter "p" specifies a place; the letter "b" specifies a buffer.

$$\left(\begin{pmatrix}\begin{pmatrix} tf_{u,v}: pi_u \rightarrow pf_{u,v} \\ tc_{u,v}: pf_{u,v}, pol_v \rightarrow po_v, pil_u \\ tb_{u,v}: pf_{u,v}, pbl, po_v \rightarrow po_v, pb_v, pil_u \end{pmatrix}, 1 \le v \le 4, v \ne u \\ to_u: pb_u, pol_u \rightarrow po_u, pbl \end{pmatrix}, 1 \le u \le 4\right) \quad (1)$$

The direct parametric expression (1) of the device model (see Fig. 1) is used for construction of communication square lattice [5, 7] with an arbitrary size k in Infinite Petri nets form and with regular structure.

2.2 Description of a Device Model

To the names of places and transitions of the device model, we add letters, which specify rules of switching. Letter "*f*" for packet forwarding:

$tf_{u,v}$ – forwarding decision from current port u to output port v;
$pf_{u,v}$ – forwarding indicator.

Letter "*c*" for cut-through switching:

$tc_{u,v}$ – direct forwarding of a packet from the source port to the destination port;
$tb_{u,v}$ – store a packet within a section of internal buffer.

The letter "i" specifies an input tract of the model; letter "o" specifies an output tract of the device model, for example to_u outputs a packet from the corresponding section of internal buffer.

Places pi_u and pil_u describe the input port buffer and the limitation of input port buffer (with size 1), respectively; places po_u (po_v) and pol_u (pol_v) describe the output port buffer and the limitation of output port buffer (with size 1), respectively. Places pb_v and pbl describe the internal buffer section of port with index v and the device buffer limitation, the number of tokens is complementary in these places.

Specification of $tb_{u,v}$ contains a read arc from the destination port buffer po_v which is represented by a loop (a pair of counter direction arcs with respect to place po_v). The read arc allows us to represent the lower priority of buffering with regard to cut-through forwarding. The buffering is implemented only in the case, the destination port is busy. When there are packets within a certain section of the internal buffer and a packet arrived from the same port, the choice between transmitting the arrived packet or a packet from the buffer is implemented in nondeterministic way. It is the best possible solution that does not lead to cumbersome models and it is justified by real-life procedures when we do not model priority classes of packets.

Number of places, transitions, and arcs in the switching device model is represented in Table 1.

Table 1. Number of places, transitions, and arcs in the switching device model.

Number of places	Number of transitions	Number of arcs
$(5 + 2)\cdot 4 + 5$	$(3 + 6 + 1)\cdot 4$	$(30 + 6 + 4)\cdot 4$
33	40	160

Calculation results from Table 1 are used in model checking, which is methodical compliment to the theoretical investigation of lattice models.

3 Open Computing Lattice

3.1 Specification of an Open Lattice

A visual representation of a 3 × 3 open lattice model with a combined cut-through and store-and-forward switching device is shown in Fig. 2. The open lattice model is a model without terminal devices on lattice boards.

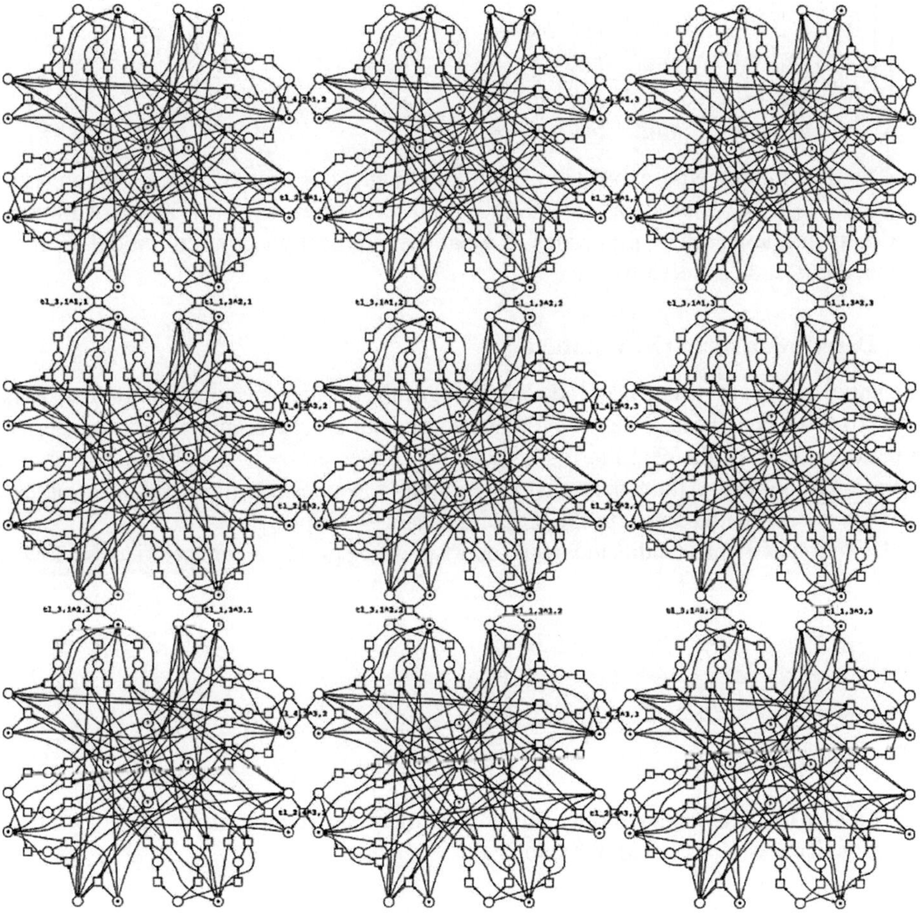

Fig. 2. A model of 3 × 3 open square lattice.

The direct parametric expression (1) of the device model (see Fig. 1) is used for composition of open communication square lattice with size k [5, 7]. According to composition rules [5] the specification of an open lattice model (2) is the following:

$$\left(\left(\begin{array}{l}\left(\begin{array}{l}\left(\begin{array}{l}tf_{u,v}^{i,j}: pi_u^{i,j} \to pf_{u,v}^{i,j}\\ tc_{u,v}^{i,j}: pf_{u,v}^{i,j}, pol_v^{i,j} \to po_v^{i,j}, pil_u^{i,j}\\ tb_{u,v}^{i,j}: pf_{u,v}^{i,j}, pbl^{i,j}, po_v^{i,j} \to po_v^{i,j}, pb_v^{i,j}, pil_u^{i,j}\end{array}\right), \begin{array}{c}1 \le v \le 4,\\ v \ne u\end{array}\\ to_u^{i,j}: pb_u^{i,j}, pol_u^{i,j} \to po_u^{i,j}, pbl^{i,j}\end{array}\right), 1 \le u \le 4\\ \left(tl_1^{i,j}: po_1^{i,j}, pil_3^{i-1,j} \to pi_3^{i-1,j}, pol_1^{i,j}\right), i \ne 1\\ \left(tl_4^{i,j}: po_4^{i,j}, pil_2^{i,j-1} \to pi_2^{i,j-1}, pol_4^{i,j}\right), j \ne 1\\ \left(tl_3^{i,j}: po_3^{i,j}, pil_1^{i+1,j} \to pi_1^{i+1,j}, pol_3^{i,j}\right), i \ne k\\ \left(tl_2^{i,j}: po_2^{i,j}, pil_4^{i,j+1} \to pi_4^{i,j+1}, pol_2^{i,j}\right), j \ne k\end{array}\right), \begin{array}{l}1 \le i \le k,\\ 1 \le j \le k\end{array}\right) \tag{2}$$

The direct parametric expression (2) of communication square lattice with an arbitrary size k [5, 10] is used for calculation of place invariants (p-invariants) of Infinite Petri nets and analyses of properties of ideal communication protocol model such as boundedness and conservativeness.

3.2 Description of an Open Lattice

The lattice represents a matrix of indexed copies of the device; the upper pair of indexes is used. To connect devices, a matrix of dedicated transitions, representing links, is inserted in between devices; the neighboring devices are connected by a pair of transitions: one transition for input and the other transition for output. Names of unidirectional (simplex) links are chosen with respect to the output device.

Direction of a link is indicated by the corresponding port number specified as a lower index:

$tl_1^{i,j}$ – up;
$tl_4^{i,j}$ – left;
$tl_3^{i,j}$ – down;
$tl_2^{i,j}$ – right.

Number of places, transitions, and arcs in the open square lattice model with an arbitrary size k is represented in Table 2.

Table 2. Number of places, transitions, and arcs in the open lattice.

Number of places	Number of transitions	Number of arcs
	$40 \cdot k^2 + 2 \cdot k \cdot (k-1) + 2 \cdot k \cdot (k-1)$	$160 \cdot k^2 + 16 \cdot k \cdot (k-1)$
$33 \cdot k^2$	$44 \cdot k^2 - 4 \cdot k$	$176 \cdot k^2 - 16 \cdot k$

In Fig. 1 and Fig. 2 the models obtained by the modeling system Tina [11, 12] are shown. For the correspondence of net element names in Petri net model and in modeling system Tina, we use the following notation: name of net element, the character "_", lower indexes, the character "^", upper indexes. For example, the transition $tf^{i,j}_{u,v}$, which describes forwarding decision from current port u to output port v in the node with (i, j) indexes, looks like {*tf_u,v^i,j*}. Tina [12] uses curly brackets to separate names of nodes. The notation is compatible with LaTeX rules for specification of formulae.

3.3 Parametric Solution for p-invariants

The direct parametric expression of lattice model and rules [10] are used for calculations of p-invariants, where a transition from parametric expression (2) corresponds to equation. An equation contains equal sums of input and output arcs, and a common notation of equation is the following:

$$-apin_{j_k} \cdot xpin_{j_k} - \ldots + apout_{j_l} \cdot xpout_{j_l} + \ldots = 0;\ indices\_range,$$

where $xpin_{j_k}, xpout_{j_l}$ are unknown variables, which correspond to places of a Petri net; coefficients $apin_{j_k}, apout_{j_l}$ describe multiplicity of corresponding arcs. An infinite system of linear algebraic Eqs. (3) for calculations of p-invariants of open lattice (2) is represented in the following parametric form:

$$\left(\left(\left(\begin{array}{l} \left(\begin{array}{l} tf^{i,j}_{u,v}: pi^{i,j}_u - pf^{i,j}_{u,v} = 0 \\ tc^{i,j}_{u,v}: pf^{i,j}_{u,v} + pol^{i,j}_v - po^{i,j}_v - pil^{i,j}_u = 0 \\ tb^{i,j}_{u,v}: pf^{i,j}_{u,v} + pbl^{i,j} + po^{i,j}_v - po^{i,j}_v - pb^{i,j}_v - pil^{i,j}_u = 0 \end{array}\right), \begin{array}{c} 1 \le v \le 4, \\ v \ne u \end{array} \\ to^{i,j}_u: pb^{i,j}_u + pol^{i,j}_u - po^{i,j}_u - pbl^{i,j} = 0 \end{array}\right), 1 \le u \le 4 \\ \left(tl^{i,j}_1: po^{i,j}_1 + pil^{i-1,j}_3 - pi^{i-1,j}_3 - pol^{i,j}_1 = 0\right), i \ne 1 \\ \left(tl^{i,j}_4: po^{i,j}_4 + pil^{i,j-1}_2 - pi^{i,j-1}_2 - pol^{i,j}_4 = 0\right), j \ne 1 \\ \left(tl^{i,j}_3: po^{i,j}_3 + pil^{i+1,j}_1 - pi^{i+1,j}_1 - pol^{i,j}_3 = 0\right), i \ne k \\ \left(tl^{i,j}_2: po^{i,j}_2 + pil^{i,j+1}_4 - pi^{i,j+1}_4 - pol^{i,j}_2 = 0\right), j \ne k \end{array}\right), \begin{array}{l} 1 \le i \le k, \\ 1 \le j \le k \end{array}\right) \tag{3}$$

For open communication square lattice (2) with size k, the parametric solution (4) of system (3) was obtained:

$$\left(\begin{array}{c} \left(po^{i,j}_u, pol^{i,j}_u\right), 1 \le u \le 4, 1 \le i \le k, 1 \le j \le k \\ \left(\left(pb^{i,j}_u, 1 \le u \le 4\right), pbl^{i,j}\right), 1 \le i \le k, 1 \le j \le k \\ \left(\left(pf^{i,j}_{u,v}, 1 \le v \le 4, v \ne u\right), pi^{i,j}_u, pil^{i,j}_u\right), 1 \le u \le 4, 1 \le i \le k, 1 \le j \le k \\ \left(\left(pbl^{i,j}, \left(\left(pil^{i,j}_u, pol^{i,j}_u\right), 1 \le u \le 4\right), 1 \le i \le k, 1 \le j \le k\right)\right. \\ \left(\left(pb^{i,j}_u, \left(pf^{i,j}_{u,v}, 1 \le v \le 4, v \ne u\right), pi^{i,j}_u, po^{i,j}_u\right), 1 \le u \le 4, 1 \le i \le k, 1 \le j \le k\right) \end{array}\right) \tag{4}$$

According to the following Lemma 1 and Theorem 1, each row of (4) is a solution of system (3) [5], therefore Petri net model is p-invariant, and consequently the model of lattice possesses boundedness and conservativeness properties [10] for an arbitrary size k. Calculated number of basis invariants of various types as well as their components are represented in a Table 3 specified with formula.

Total number of p-invariants in a lattice with size k is

$$\mathrm{N} = 9 \cdot k + 2 \tag{5}$$

Table 3. Number of basis invariants.

Invariant	Lines	Components
Invariant 1	$4{\cdot}k^2$	2
Invariant 2	k^2	5
Invariant 3	$4{\cdot}k^2$	5
Invariant 4	1	$9{\cdot}k^2$
Invariant 5	1	$24{\cdot}k^2$

Sum of 1, 2, and 3 lines (4) contains all the places once, $2{\cdot}4{\cdot}k^2 + 5{\cdot}k^2 + 5{\cdot}4{\cdot}k^2 = 33{\cdot}k^2$; and sum of 4 and 5 lines contains all the places once, $9{\cdot}k^2 + 24{\cdot}k^2 = 33{\cdot}k^2$ (see Table 2 and Table 3).

Lemma 1. Each line of the sparse matrix (4) is a solution of equations system (3).

Proof. Let us substitute each parametric line of parametric solution (4) into each parametric equality of equations system (3). When substituting, different values of indices are selected, as a result, this gives us the correct statement. For instance, let us substitute the first line of parametric solution (4).

$\left(po_u^{i,j}, pol_u^{i,j}\right)$, $1 \le u \le 4,\ 1 \le i \le k,\ 1 \le j \le k,$
into the fourth equality of system (3).
$to_u^{i,j} : pb_u^{i,j} + pol_u^{i,j} - po_u^{i,j} - pbl^{i,j} = 0,\ 1 \le u \le 4,\ 1 \le i \le k,\ 1 \le j \le k$
and we obtain:
when $i' \ne i$ or $j' \ne j$: $0 + 0 - 0 - 0 = 0$, then $0 = 0$;
when $i' = i$ or $j' = j$: $0 + 1 - 1 - 0 = 0$, then $0 = 0$.
In the same way, all other combinations are checked.

Theorem 1. The lattice model (4) is a *p*-invariant Petri net for an arbitrary natural number *k*.

Proof. The sum of 1st, 2nd, and 3d lines of parametric solution (4) contains all the places once, $2{\cdot}4{\cdot}k^2 + 5{\cdot}k^2 + 5{\cdot}4{\cdot}k^2 = 33{\cdot}k^2$; or sum of 4th and 5th lines contains all the places once, $9{\cdot}k^2 + 24{\cdot}k^2 = 33{\cdot}k^2$ (see Table 2 and Table 3). According to *Lemma 1*, these lines are solutions of system (3).

An example of invariants for 3 × 3 lattice is represented in Table 4 (invariants 1, 2, 3) and Table 5 (invariants 4, 5); invariants are calculated by modeling system Tina [11, 12]. Total number of invariants is 83 (Tina analysis) and according to formula (5) is equal to $9{\cdot}3^2 + 2 = 83$ solutions, results are the same (coincide).

Table 4. Example of invariants for 3 × 3 lattice, invariants 1, 2, 3.

Invariant	Lines	List of solutions
Invariant 1	36	{po_2^1,1} {pol_2^1,1}, {po_2^1,2} {pol_2^1,2},{po_2^1,3} {pol_2^1,3}, {po_2^2,1} {pol_2^2,1}, {po_2^2,2} {pol_2^2,2}, {po_2^2,3} {pol_2^2,3}, {po_2^3,1} {pol_2^3,1}, {po_2^3,2} {pol_2^3,2},{po_2^3,3} {pol_2^3,3}, {po_3^1,1} {pol_3^1,1},{po_3^1,2} {pol_3^1,2}, {po_3^1,3} {pol_3^1,3}, {po_3^2,1} {pol_3^2,1},{po_3^2,2} {pol_3^2,2},{po_3^2,3} {pol_3^2,3}, {po_3^3,1} {pol_3^3,1}, {po_3^3,2} {pol_3^3,2},{po_3^3,3} {pol_3^3,3}, {po_4^1,1} {pol_4^1,1},{po_4^1,2} {pol_4^1,2}, {po_4^1,3} {pol_4^1,3}, {po_4^2,1} {pol_4^2,1},{po_4^2,2} {pol_4^2,2},{po_4^2,3} {pol_4^2,3}, {po_4^3,1} {pol_4^3,1},{po_4^3,2} {pol_4^3,2},{po_4^3,3} {pol_4^3,3}, {po_1^1,1} {pol_1^1,1}, {po_1^1,2} {pol_1^1,2},{po_1^1,3} {pol_1^1,3}, {po_1^2,1} {pol_1^2,1},{po_1^2,2} {pol_1^2,2}, {po_1^2,3} {pol_1^2,3}, {po_1^3,1} {pol_1^3,1},{po_1^3,2} {pol_1^3,2},{po_1^3,3} {pol_1^3,3}
Invariant 2	9	{pb_1^1,1} {pb_2^1,1} {pb_3^1,1} {pb_4^1,1} {pbl^1,1}, {pb_1^1,2} {pb_2^1,2} {pb_3^1,2} {pb_4^1,2} {pbl^1,2}, {pb_1^1,3} {pb_2^1,3} {pb_3^1,3} {pb_4^1,3} {pbl^1,3}, {pb_1^2,1} {pb_2^2,1} {pb_3^2,1} {pb_4^2,1} {pbl^2,1},{pb_1^2,2} {pb_2^2,2} {pb_3^2,2} {pb_4^2,2} {pbl^2,2}, {pb_1^2,3} {pb_2^2,3} {pb_3^2,3} {pb_4^2,3} {pbl^2,3}, {pb_1^3,1} {pb_2^3,1} {pb_3^3,1} {pb_4^3,1} {pbl^3,1}, {pb_1^3,2} {pb_2^3,2} {pb_3^3,2} {pb_4^3,2} {pbl^3,2}, {pb_1^3,3} {pb_2^3,3} {pb_3^3,3} {pb_4^3,3} {pbl^3,3}
Invariant 3	36	{pf_2,1^1,2} {pf_2,3^1,2} {pf_2,4^1,2} {pi_2^1,2} {pil_2^1,2},{pf_2,1^1,3} {pf_2,3^1,3} {pf_2,4^1,3} {pi_2^1,3} {pil_2^1,3},{pf_2,1^2,1} {pf_2,3^2,1} {pf_2,4^2,1} {pi_2^2,1} {pil_2^2,1},{pf_2,1^2,2} {pf_2,3^2,2} {pf_2,4^2,2} {pi_2^2,2} {pil_2^2,2},{pf_2,1^2,3} {pf_2,3^2,3} {pf_2,4^2,3} {pi_2^2,3} {pil_2^2,3},{pf_2,1^3,1} {pf_2,3^3,1} {pf_2,4^3,1} {pi_2^3,1} {pil_2^3,1},{pf_2,1^3,2} {pf_2,3^3,2} {pf_2,4^3,2} {pi_2^3,2} {pil_2^3,2},{pf_2,1^3,3} {pf_2,3^3,3} {pf_2,4^3,3} {pi_2^3,3} {pil_2^3,3}, {pf_3,1^1,1} {pf_3,2^1,1} {pf_3,4^1,1} {pi_3^1,1} {pil_3^1,1},{pf_3,1^1,2} {pf_3,2^1,2} {pf_3,4^1,2} {pi_3^1,2} {pil_3^1,2},{pf_3,1^1,3} {pf_3,2^1,3} {pf_3,4^1,3} {pi_3^1,3} {pil_3^1,3}, {pf_3,1^2,1} {pf_3,2^2,1} {pf_3,4^2,1} {pi_3^2,1} {pil_3^2,1},{pf_3,1^2,2} {pf_3,2^2,2} {pf_3,4^2,2} {pi_3^2,2} {pil_3^2,2},{pf_3,1^2,3} {pf_3,2^2,3} {pf_3,4^2,3} {pi_3^2,3} {pil_3^2,3},{pf_3,1^3,1} {pf_3,2^3,1} {pf_3,4^3,1} {pi_3^3,1} {pil_3^3,1},{pf_3,1^3,2} {pf_3,2^3,2} {pf_3,4^3,2} {pi_3^3,2} {pil_3^3,2}, {pf_3,1^3,3} {pf_3,2^3,3} {pf_3,4^3,3} {pi_3^3,3} {pil_3^3,3},{pf_4,1^1,1} {pf_4,2^1,1} {pf_4,3^1,1} {pi_4^1,1} {pil_4^1,1},{pf_4,1^1,2} {pf_4,2^1,2} {pf_4,3^1,2} {pi_4^1,2} {pil_4^1,2},{pf_4,1^1,3} {pf_4,2^1,3} {pf_4,3^1,3} {pi_4^1,3} {pil_4^1,3},{pf_4,1^2,1} {pf_4,2^2,1} {pf_4,3^2,1} {pi_4^2,1} {pil_4^2,1},{pf_4,1^2,2} {pf_4,2^2,2} {pf_4,3^2,2} {pi_4^2,2} {pil_4^2,2},{pf_4,1^2,3} {pf_4,2^2,3} {pf_4,3^2,3} {pi_4^2,3} {pil_4^2,3},{pf_4,1^3,1} {pf_4,2^3,1} {pf_4,3^3,1} {pi_4^3,1} {pil_4^3,1}, {pf_4,1^3,2} {pf_4,2^3,2} {pf_4,3^3,2} {pi_4^3,2} {pil_4^3,2},{pf_4,1^3,3} {pf_4,2^3,3} {pf_4,3^3,3} {pi_4^3,3} {pil_4^3,3},{pf_1,2^1,1} {pf_1,3^1,1} {pf_1,4^1,1} {pi_1^1,1} {pil_1^1,1},{pf_1,2^1,2} {pf_1,3^1,2} {pf_1,4^1,2} {pi_1^1,2} {pil_1^1,2},{pf_1,2^1,3} {pf_1,3^1,3} {pf_1,4^1,3} {pi_1^1,3} {pil_1^1,3},{pf_1,2^2,1} {pf_1,3^2,1} {pf_1,4^2,1} {pi_1^2,1} {pil_1^2,1},{pf_1,2^2,2} {pf_1,3^2,2} {pf_1,4^2,2} {pi_1^2,2} {pil_1^2,2},{pf_1,2^2,3} {pf_1,3^2,3} {pf_1,4^2,3} {pi_1^2,3} {pil_1^2,3}, {pf_1,2^3,1} {pf_1,3^3,1} {pf_1,4^3,1} {pi_1^3,1} {pil_1^3,1},{pf_1,2^3,2} {pf_1,3^3,2} {pf_1,4^3,2} {pi_1^3,2} {pil_1^3,2},{pf_1,2^3,3} {pf_1,3^3,3} {pf_1,4^3,3} {pi_1^3,3} {pil_1^3,3},{pf_2,1^1,1}
		{pf_2,3^1,1} {pf_2,4^1,1} {pi_2^1,1} {pil_2^1,1}

There are 36 lines (solutions) and 2 components in 1th invariant for output port places, 9 lines and 5 components in 2nd invariant for buffers places, 36 solutions and 5 components in 3[d] invariant (see Table 4), 1 line and 81 components in 4th invariant, 1 solution and 216 components in 5th invariant (see Table 5).

We can check obtained results by following way: if we put each solution from Table 4 (or Table 5) into each parametric Eq. (3), then as result we obtain correct equalities. For example, we select the first invariant ({po_2^1,1} {pol_2^1,1}) from Table 4, Invariant 1, where ($i = 1, j = 1, v = 2$ or $u = 2$), and substitute it into linear algebraic Eqs. (3) to obtain:

$0 - 0 = 0.$
$0 + 1 - 1 - 0 = 0.$
$0 + 0 + 1 - 1 - 0 - 0 = 0.$
$0 + 1 - 0 - 1 = 0.$
$0 + 0 - 0 - 0 = 0.$
$0 + 0 - 0 - 0 = 0.$

Table 5. Example of invariants for 3 × 3 lattice, 4th and 5th invariants

Invariant	Lines	List of solutions
Invariant 4	1	{pbl^1,1} {pbl^1,2} {pbl^1,3} {pbl^2,1} {pbl^2,2} {pbl^2,3} {pbl^3,1} {pbl^3,2} {pbl^3,3} {pil_1^1,1} {pil_1^1,2} {pil_1^1,3} {pil_1^2,1} {pil_1^2,2} {pil_1^2,3} {pil_1^3,1} {pil_1^3,2} {pil_1^3,3} {pil_2^1,1} {pil_2^1,2} {pil_2^1,3} {pil_2^2,1} {pil_2^2,2} {pil_2^2,3} {pil_2^3,1} {pil_2^3,2} {pil_2^3,3} {pil_3^1,1} {pil_3^1,2} {pil_3^1,3} {pil_3^2,1} {pil_3^2,2} {pil_3^2,3} {pil_3^3,1} {pil_3^3,2} {pil_3^3,3} {pil_4^1,1} {pil_4^1,2} {pil_4^1,3} {pil_4^2,1} {pil_4^2,2} {pil_4^2,3} {pil_4^3,1} {pil_4^3,2} {pil_4^3,3} {pol_1^1,1} {pol_1^1,2} {pol_1^1,3} {pol_1^2,1} {pol_1^2,2} {pol_1^2,3} {pol_1^3,1} {pol_1^3,2} {pol_1^3,3} {pol_2^1,1} {pol_2^1,2} {pol_2^1,3} {pol_2^2,1} {pol_2^2,2} {pol_2^2,3} {pol_2^3,1} {pol_2^3,2} {pol_2^3,3} {pol_3^1,1} {pol_3^1,2} {pol_3^1,3} {pol_3^2,1} {pol_3^2,2} {pol_3^2,3} {pol_3^3,1} {pol_3^3,2} {pol_3^3,3} {pol_4^1,1} {pol_4^1,2} {pol_4^1,3} {pol_4^2,1} {pol_4^2,2} {pol_4^2,3} {pol_4^3,1} {pol_4^3,2} {pol_4^3,3}
Invariant 5	1	{pb_1^1,1} {pb_1^1,2} {pb_1^1,3} {pb_1^2,1} {pb_1^2,2} {pb_1^2,3} {pb_1^3,1} {pb_1^3,2} {pb_1^3,3} {pb_2^1,1} {pb_2^1,2} {pb_2^1,3} {pb_2^2,1} {pb_2^2,2} {pb_2^2,3} {pb_2^3,1} {pb_2^3,2} {pb_2^3,3} {pb_3^1,1} {pb_3^1,2} {pb_3^1,3} {pb_3^2,1} {pb_3^2,2} {pb_3^2,3} {pb_3^3,1} {pb_3^3,2} {pb_3^3,3} {pb_4^1,1} {pb_4^1,2} {pb_4^1,3} {pb_4^2,1} {pb_4^2,2} {pb_4^2,3} {pb_4^3,1} {pb_4^3,2} {pb_4^3,3} {pf_1,2^1,1} {pf_1,2^1,2} {pf_1,2^1,3} {pf_1,2^2,1} {pf_1,2^2,2} {pf_1,2^2,3} {pf_1,2^3,1} {pf_1,2^3,2} {pf_1,2^3,3} {pf_1,3^1,1} {pf_1,3^1,2} {pf_1,3^1,3} {pf_1,3^2,1} {pf_1,3^2,2} {pf_1,3^2,3} {pf_1,3^3,1} {pf_1,3^3,2} {pf_1,3^3,3} {pf_1,4^1,1} {pf_1,4^1,2} {pf_1,4^1,3} {pf_1,4^2,1} {pf_1,4^2,2} {pf_1,4^2,3} {pf_1,4^3,1} {pf_1,4^3,2} {pf_1,4^3,3} {pf_2,1^1,1} {pf_2,1^1,2} {pf_2,1^1,3} {pf_2,1^2,1} {pf_2,1^2,2} {pf_2,1^2,3} {pf_2,1^3,1} {pf_2,1^3,2} {pf_2,1^3,3} {pf_2,3^1,1} {pf_2,3^1,2} {pf_2,3^1,3} {pf_2,3^2,1} {pf_2,3^2,2} {pf_2,3^2,3} {pf_2,3^3,1} {pf_2,3^3,2} {pf_2,3^3,3} {pf_2,4^1,1} {pf_2,4^1,2} {pf_2,4^1,3} {pf_2,4^2,1} {pf_2,4^2,2} {pf_2,4^2,3} {pf_2,4^3,1} {pf_2,4^3,2} {pf_2,4^3,3} {pf_3,1^1,1} {pf_3,1^1,2} {pf_3,1^1,3} {pf_3,1^2,1} {pf_3,1^2,2} {pf_3,1^2,3} {pf_3,1^3,1} {pf_3,1^3,2} {pf_3,1^3,3} {pf_3,2^1,1} {pf_3,2^1,2} {pf_3,2^1,3} {pf_3,2^2,1} {pf_3,2^2,2} {pf_3,2^2,3} {pf_3,2^3,1} {pf_3,2^3,2} {pf_3,2^3,3} {pf_3,4^1,1} {pf_3,4^1,2} {pf_3,4^1,3} {pf_3,4^2,1} {pf_3,4^2,2} {pf_3,4^2,3} {pf_3,4^3,1} {pf_3,4^3,2} {pf_3,4^3,3} {pf_4,1^1,1} {pf_4,1^1,2} {pf_4,1^1,3} {pf_4,1^2,1} {pf_4,1^2,2} {pf_4,1^2,3} {pf_4,1^3,1} {pf_4,1^3,2} {pf_4,1^3,3} {pf_4,2^1,1} {pf_4,2^1,2} {pf_4,2^1,3} {pf_4,2^2,1} {pf_4,2^2,2} {pf_4,2^2,3} {pf_4,2^3,1} {pf_4,2^3,2} {pf_4,2^3,3} {pf_4,3^1,1} {pf_4,3^1,2} {pf_4,3^1,3} {pf_4,3^2,1} {pf_4,3^2,2} {pf_4,3^2,3} {pf_4,3^3,1} {pf_4,3^3,2} {pf_4,3^3,3} {pi_1^1,1} {pi_1^1,2} {pi_1^1,3} {pi_1^2,1} {pi_1^2,2} {pi_1^2,3} {pi_1^3,1} {pi_1^3,2} {pi_1^3,3} {pi_2^1,1} {pi_2^1,2} {pi_2^1,3} {pi_2^2,1} {pi_2^2,2} {pi_2^2,3} {pi_2^3,1} {pi_2^3,2} {pi_2^3,3} {pi_3^1,1} {pi_3^1,2} {pi_3^1,3} {pi_3^2,1} {pi_3^2,2} {pi_3^2,3} {pi_3^3,1} {pi_3^3,2} {pi_3^3,3} {pi_4^1,1} {pi_4^1,2} {pi_4^1,3} {pi_4^2,1} {pi_4^2,2} {pi_4^2,3} {pi_4^3,1} {pi_4^3,2} {pi_4^3,3} {po_1^1,1} {po_1^1,2} {po_1^1,3} {po_1^2,1} {po_1^2,2} {po_1^2,3} {po_1^3,1} {po_1^3,2} {po_1^3,3} {po_2^1,1} {po_2^1,2} {po_2^1,3} {po_2^2,1} {po_2^2,2} {po_2^2,3} {po_2^3,1} {po_2^3,2} {po_2^3,3} {po_3^1,1} {po_3^1,2} {po_3^1,3} {po_3^2,1} {po_3^2,2} {po_3^2,3} {po_3^3,1} {po_3^3,2} {po_3^3,3} {po_4^1,1} {po_4^1,2} {po_4^1,3} {po_4^2,1} {po_4^2,2} {po_4^2,3} {po_4^3,1} {po_4^3,2} {po_4^3,3}

$0 + 0 - 0 - 0 = 0.$
$1 + 0 - 0 - 1 = 0.$

The next step of analysis of square communication lattice models with switching device, which realizes unconditional forwarding decision of packets, is studying possible deadlocks [13] of the models that represent a future research direction.

4 The Petri-Object Simulation of Cut-Through Switching Communication Grid

4.1 The Basics of Petri-Object Simulation

Petri-object simulation technique is described in work [9]. It has advantages when the model is built of many typical elements as it provides replication of objects with similar behavior. The model structure is based on object-oriented technology. Its key idea is that the *PetriObjModel* class, used to compose Petri-object model, aggregates classes inheriting specific *PetriSim* class, which is able to reproduce discrete event behavior of stochastic Petri net. By definition, Petri-object is an object of *PetriSim* class (or its subclasses). The use of specified links between Petri-objects guarantees that the behavior of the model will be described by the stochastic Petri net, which is the union of Petri nets of all its Petri-objects.

Suppose a researcher has to simulate a discrete event system, the elements of which can be distinguished on *s* types. If he uses stochastic Petri net, he should create each transition, place, arc and set the parameters of each. Imagine that the model consists of hundreds of elements. Even if the researcher uses the Petri net graph editor simplifying creation of places and transitions and connections between them, it would be a time-consuming process. If he uses Petri-object model technique, he should create significantly smaller stochastic Petri net for each type of the model elements and use once-developed stochastic Petri net to create a set of elements. As soon as the Petri net is saved as a method, it can be created with given parameters. Then, the Petri net, being set to the constructor of *PetriSim* class, is used to simulate the behavior of the model element. To be able to interact with each other, the links between nets of Petri-objects should be established.

Petri-object simulation software helps to create net, using Petri net editor, and save it as a Java method with determined parameters. Calling the method with given parameters, the constructor of *PetriSim* class (or its descenders) creates Petri-objects. Thus, the researcher can concentrate his attention only on parameters of the created elements and the connections between them. When the list of linked Petri-objects is created, the model can be created by the constructor of *PetriObjModel* class. Simulation can be launched by calling *go()* method for the created model. It is proved that the Petri-object simulation algorithm has much less complexity compared to stochastic Petri net [14]. Therefore, the use of this algorithm for the model with a big number of elements provides not only fast construction of the model but also fast performance.

4.2 The Communication Grid Petri-Object Model

To construct rectangular communication grid Petri-object model the following types of Petri-objects are determined: *Port*, *Device*, *Connector*, *Generator*, *UserBuffer*. The Petri-object *Port* is the main constructive element of the model as each communication device is composed of four ports and an interaction between different devices is performed by means of ports connections. All events *input*, *tc*, *th*, *output*, representing the functioning of one port, are included to the net of Petri-object *Port* (Fig. 3). The places *deviceBuffer*, *portBufferFirst*, *portBufferSecond*, *portBufferThird*, *portBufferForth*, representing the current state of device buffers and port buffers, creates the net of Petri-object *Device* (Fig. 3). The group of one *Device* object and four *Port* object simulates one communication device. Each *Port* object interacts with other *Port* objects of this device as its first, second, or third ports. For example, the 'east' port of the device has 'north', 'west', 'south' ports as its first, second, third ports. The 'north' port has 'west', 'south', 'east' ports as its first, second, third ports. The *Port* object uses appropriate places of *Device* object as the buffers of its first, second and third ports. It also uses appropriate places *po, pol* of the other *Port* objects of the same group.

The *Connector* object includes events of a packet sending to the nearby device. It has share places with two *Port* objects of the appropriate communication devices (Fig. 4). The Petri-object *Generator* simulates the arrival of new packet into the communication grid, and the Petri-object *UserBuffer* produces its leaving (Fig. 4).

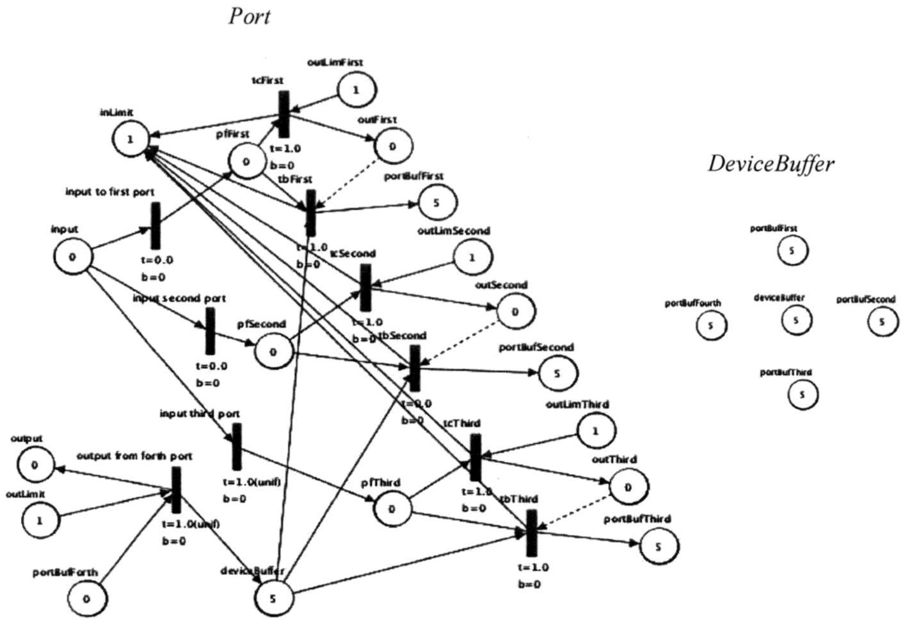

Fig. 3. Models of port and buffer device.

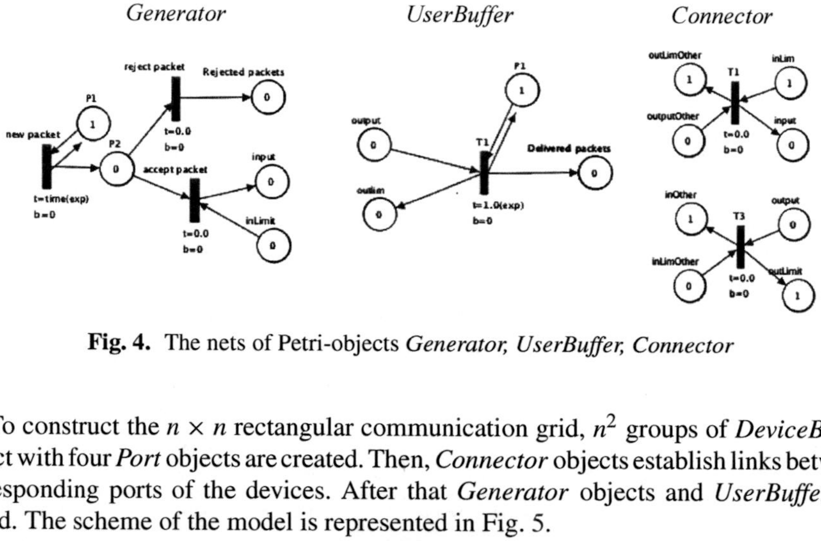

Fig. 4. The nets of Petri-objects *Generator, UserBuffer, Connector*

To construct the $n \times n$ rectangular communication grid, n^2 groups of *DeviceBuffer* object with four *Port* objects are created. Then, *Connector* objects establish links between corresponding ports of the devices. After that *Generator* objects and *UserBuffer* are added. The scheme of the model is represented in Fig. 5.

Thus, the process of the $n \times n$ grid model development includes the following actions:

- creation of n^2 *Device* objects,
- creation of four *Port* objects for each *Device* object linking them with each other and with *Device* object,

- connecting of $(n-1)^2$ 'east' *Port* objects with 'west' *Port* objects of nearby devices using *Connector* objects,
- connecting of $(n-1)^2$ 'north' *Port* objects with 'south' *Port* objects of nearby devices using *Connector* objects,
- connecting of $4n$ *Generator* and $4n$ *UserBuffer* objects with 'east' *Port,* 'west' *Port,* 'north' *Port,* 'south' *Port* objects,
- adding all created Petri-objects to the list of Petri-objects,
- creation of the Petri-object model with given list of Petri-objects.

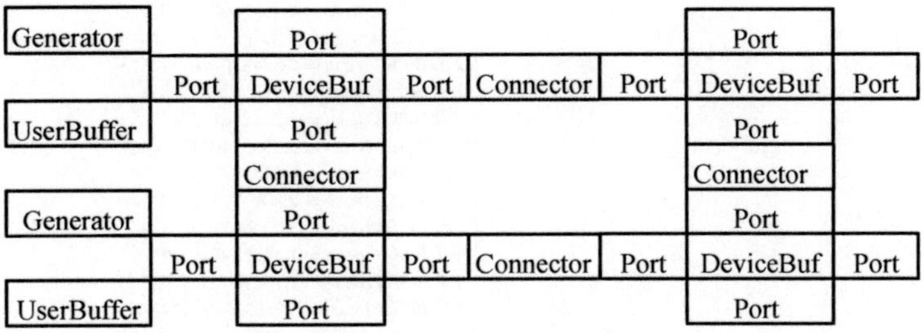

Fig. 5. The scheme of links between Petri-objects.

The model simulation is performed for the given simulation time, during which the algorithm reproduces a time-ordered sequence of events. The simulation results are represented by the number of delivered packets, rejected packets and the mean value of device buffer state.

4.3 The Simulation Results

Comparing with classical Petri net, simulation by means of stochastic Petri net provides the process observation in time. The impact of time delay, being set for every elementary operation, can be investigated. In addition, the statistics about observed values is provided.

The results, obtained for the grid of size 3×3 and 6×6, are represented in Table 6. The time delay of sending packet to the other port of device has been set to 10 ns, the time delay of sending packet to the other device has been set to 10^4 ns, and the time delay of generating new packet has been set to $1.5 \cdot 10^4$ ns. The buffer size has been set to 5 packets. The probability of rejection and the mean value of buffers are the characteristics of the model.

The impact on rejection probability has been investigated. The result in case of 3×3 grid is represented in Fig. 6. Decreasing the delay of new packet causes a valuable decrease in the value of rejection probability.

Table 6. Simulation results

3 × 3 grid	6 × 6 grid
Rejection probability = .143192488262911	Rejection probability = 0.539613716988569
0_0 Buffer mean = 4.594350976140457	0_0 Buffer mean = 3.4563000966841884
0_1 Buffer mean = 4.477316812124258	0_1 Buffer mean = 3.5220056231354815
0_2 Buffer mean = 4.545383333068567	0_2 Buffer mean = 3.558251743666928
1_0 Buffer mean = 4.191165538568087	0_3 Buffer mean = 3.463055906109721
1_1 Buffer mean = 3.9821954741769607	0_4 Buffer mean = 3.3533412754819047
1_2 Buffer mean = 4.537030809755917	0_5 Buffer mean = 3.6865858536310894
2_0 Buffer mean = 4.382081927132834	1_0 Buffer mean = 3.084179365513917
2_1 Buffer mean = 4.271692241026342	1_1 Buffer mean = 2.990154906444918
2_2 Buffer mean = 4.5636484832971504	1_2 Buffer mean = 2.64397153192779
	1_3 Buffer mean = 2.7139728314910956
	1_4 Buffer mean = 2.7593509090277966
	1_5 Buffer mean = 3.3990415376946044
	2_0 Buffer mean = 2.8848803207114675
	2_1 Buffer mean = 2.4435624788585417
	2_2 Buffer mean = 2.231256162775022
	2_3 Buffer mean = 2.3553940479304467
	2_4 Buffer mean = 2.2555677532764347
	2_5 Buffer mean = 2.8093773126103248
	3_0 Buffer mean = 3.004615369683636
	3_1 Buffer mean = 1.6522297480095867
	3_2 Buffer mean = 1.768194251379255
	3_3 Buffer mean = 1.7850328711292243
	3_4 Buffer mean = 1.7158953924970628
	3_5 Buffer mean = 2.4841077485343144
	4_0 Buffer mean = 2.9933880458735427
	4_1 Buffer mean = 2.9613309545760025
	4_2 Buffer mean = 2.412275698617538
	4_3 Buffer mean = 2.36285856341043
	4_4 Buffer mean = 1.9398417284909473
	4_5 Buffer mean = 3.0249965062668283
	5_0 Buffer mean = 3.2084769938050957
	5_1 Buffer mean = 3.652674963102253
	5_2 Buffer mean = 3.495215516733797
	5_3 Buffer mean = 2.9398978743056157
	5_4 Buffer mean = 2.610908734988761
	5_5 Buffer mean = 3.1206744427424615

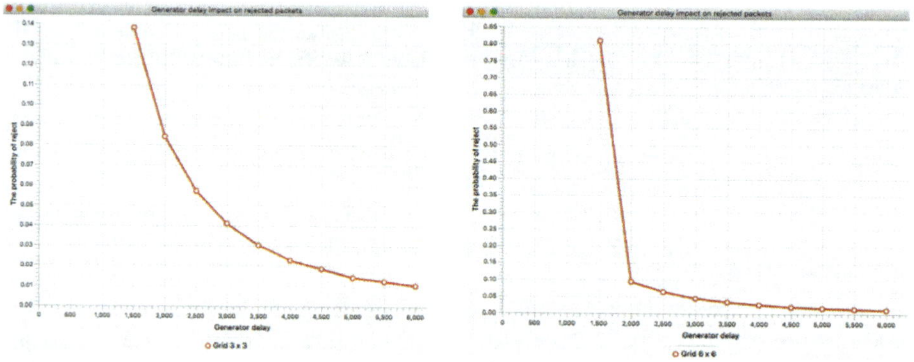

Fig. 6. The dependence of rejection probability on generator time delay (time is measured in 10^{-8}s.

5 Conclusion

In the present paper, we have refined the model of a switching device and applied it for composition of an open square communication lattice. Infinite Petri nets have been applied for the verification of the communication lattice protocols. An infinite system of linear algebraic equations have been composed and solved in parametric form to prove p-invariance of the model.

The refined switching device model implements a nondeterministic packet forwarding decision for both store-and-forward and cut-through switching procedures. Moreover, the lattice has been composed connecting the device models via dedicated transitions which correspond to unidirectional (simplex) cannels.

For the communication lattice performance evaluation, object Petri nets have been applied, such characteristic as the packet rejection probability and the mean buffer size obtained at the result of simulation.

The present study has shown mutual advantages of combined application of Infinite and Object Petri nets.

References

1. Dongarra, J., Tomov, S., Zaitsev, D.: Solving linear diophantine systems on parallel architectures. IEEE Trans. Parallel Distrib. Syst. **30**(5), 1158–1169 (2019)
2. Zaitsev, D.A., Shmeleva, T.R., Groote, J.F.: Verification of hypertorus communication grids by infinite petri nets and process algebra. IEEE/CAA J. Automatica Sinica **6**(3), 733–742 (2019)
3. Preve, N.P. (ed.): Grid Computing: Towards a Global Interconnected Infrastructure. Springer, Heidelberg (2011)
4. Liberzon, D.: Switching in Systems and Control. 230 p. Published by Birkhäuser, Boston (2003)
5. Shmeleva, T.R., Zaitsev, D.A., Zaitsev, I.D.: Infinite petri nets: part 1, modeling square grid structures. Complex Syst. **26**(2), 157–195 (2017)
6. Zaitsev, D.A.: Universality in infinite petri nets. In: Proceedings of 7th International Conference, MCU 2015. Lecture Notes in Computer Science, Volume 9288, September 9–11, pp. 180–197. Famagusta, North Cyprus (2015)
7. Zaitsev, D.A., Zaitsev, I.D., Shmeleva, T.R.: Verification of square communication grid protocols via infinite Petri nets. In: 10th Middle Eastern Simulation Multiconference (MESM) on Proceedings, pp. 53–59, September 27–29, Beirut, Lebanon (2009)
8. Trub, I.I.: Object-oriented simulation on C++. Piter, S.-P.B. (2005)
9. Stetsenko, I.V., Dyfuchyn, A.: Petri-object simulation: technique and software. Inf. Comput. Intell. Syst. **1**, 51–59 (2020)
10. Murata, T.T.: Petri nets: properties, analysis and applications. Proc. IEEE **77**(4), 541–580 (1989)
11. Berthomieu, B., Ribet, O.-P., Vernadat, F.: The tool TINA – construction of abstract state spaces for Petri nets and time Petri nets. Int. J. Prod. Res. **42**(4), 2741–2756 (2004)
12. Tina modeling system. https://www.laas.fr/tina. Accessed 27 Nov 2020
13. Retschitzegger, W., Proll, B., Zaitsev, D.A., Shmeleva, T.R.: Blocking communication grid via ill-intentioned traffic. middle eastern simulation & modelling multiconference. In: Proceedings of 14th Middle Eastern Multiconference (MESM-2014), pp. 63–71. February 3–5, Muscat, Oman (2014)

14. Stetsenko, I.V., Dyfuchyn, A.: Petri-object simulation: software package and complexity. In: Proceedings of the 8th IEEE International Conference on Intelligent Data Acquisition and Advanced Computing Systems: Technology and Applications (IDAACS 2015), pp. 381–385. Warsaw, Poland (2015)

Research of Information Systems Technological Parameters

Wireless Communication Systems of Terahertz Frequency Range

M. Ilchenko(✉), T. Narytnyk(✉), and G. Avdeyenko(✉)

Telecommunication Systems Department, Igor Sikorsky Kyiv Polytechnic Institute, Industrialnyi Lane, 2 (Campus 30), Kyiv 03056, Ukraine
ilch@kpi.ua, director@mitris.com, django2006@ukr.net

Abstract. One of the trends in the development of modern wireless systems is the application of the terahertz frequency range. The terahertz range occupies an intermediate position between the well-studied microwave and optical ranges of the electromagnetic radiation spectrum. The transition to the terahertz range solves both the issue of increasing the data rate transmission by using of wider frequency bands and the issue of electromagnetic compatibility ensuring between wireless communication systems, as well as elimination of intentional interferences. In recent years, the number of basic and applied research on this topic has increased sharply. The aim of this paper is to inform about the results of investigation and development in the terahertz range which completed by the authors in order to determine the possibilities and prerequisites for solving the fundamental problem of wireless communications systems development using world and domestic experience in microwave telecommunications. The paper substantiates the necessity of the terahertz range application in the deployment of wireless communications systems and considers the main factors that lead to the occurrence of signal fading. It is shown that in the terahertz range the main influence on a radio link energy potential exerted by the electromagnetic waves attenuation in hydro meteors and gases. The terahertz frequency range sections that are most suitable for use in radio links are determined. The results of functional design, simulation and experimental research of the receiving and transmitting devices of terahertz wireless communication system with a gigabit data rate in the 130–134 GHz frequency range are presented. A prototype of simplex digital radio relay system of 130–134 GHz band with digital gigabit modem was fabricated and tested. It provides channel data rate up to 1200 Mbps within 1 km distance between corresponding stations. The concept of creating a software-defined radio technology based on Wi-Fi which could be used in order to ensure high data rates, reliability and security of transport distribution networks of the 5G mobile systems in the terahertz range was proposed. The parameter impairments of a DVB-C standard multi-channel signal during its transmission through the terahertz transceiver prototype of 130 GHz band investigated. The paper also presents the simulation results of wireless data transmission and reception in the 130.4–131.5 GHz frequency band by using impulse radio ultra-wideband (IR-UWB) signals. The results of investigation presented in this paper may be useful for scientist and wireless communication equipment manufacturers.

P. Vorobiyenko et al. (Eds.): IPF 2020, LNNS 212, pp. 189–222, 2021.
https://doi.org/10.1007/978-3-030-76343-5_11

Keywords: Terahertz technology · Wireless communication · Radio link · High-speed wireless networks · Channel capability · Transceiver · Terahertz range · Signal-code sequence · DVB-C · Digital signal · IR-UWB

1 Introduction

One of the trends in the development of modern telecommunications is the application of increasingly high-frequency bands. From the beginning of the 1980s to the present, the terahertz range, which approximately occupies the radio frequency band from 0.1 to 3 THz, has attracted wide attention of researchers (Fig. 1) [1, 2].

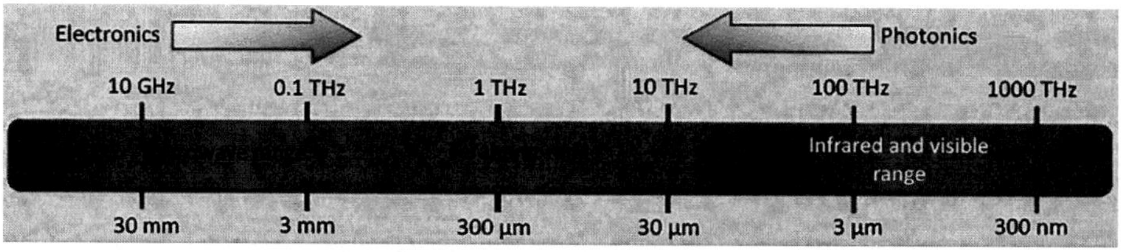

Fig. 1. Terahertz range in the electromagnetic spectrum

The uptaking of the terahertz range was facilitated by the serial development of semiconductor active elements for oscillations generation and its amplification at these frequencies with usage of nanotechnology, as well as new types of vacuum devices to operate in this frequency range.

One of the most promising areas of terahertz technologies application in telecommunications is associated with current trends in wireless transport networks of 5G and 6G, high-precision guidance and control systems and counter-terrorism, which provide noise-immune exchange of large amounts of data.

Modern wireless telecommunication networks must provide high reliability and high speeds of information transmission to the subscriber and between subscribers. This can be technically implemented only if the state centers of radio frequencies provide network operators with a sufficient number of radio frequency resource that have a fairly wide band. Due to the congestion of the licensed radio-frequency spectrum below 30 GHz in most developed states of the world, telecommunication operators need to uptake the higher spectrum bands of non-busy frequencies, namely the unlicensed terahertz band. The transition to the terahertz range solves the issue of increasing the data transmission speed by the usage of a wider band of frequencies, and the issue of the electromagnetic compatibility ensuring of radio electronic devices and systems. It is necessary and expedient to study the propagation of terahertz waves in the atmosphere in a wide range of frequencies for further results application in the design and development of technologies and tools for future high-speed wireless communications systems which could be the basis for 5G and 6G cellular networks deployment [3–5].

According to the Recommendations and Reports of Electronic Communications Committee of CEPT [6–8], the main features of the fixed radio systems operation in the terahertz range of the radio-frequency spectrum can be summarized as follows:

- availability of very wide bandwidth, which allows to ensure low cost of traffic in the area of operation of the multimedia provider and to ensure future market demands for ever higher bandwidth, especially for Internet applications and next generation mobile networks;
- the feasibility and flexibility of deploying wireless communication systems is much simpler compared to alternative solutions related to wired communication systems;
- ability to provide a high level of security due to low possibility of interference effect and signals-of-interest interception;
- main application of fixed services provided in the bands between 92–114.25 GHz, 130–134 GHz, 141–148.5 GHz, 151.5–164 GHz and 167–174.8 GHz and intended for fixed wireless access;
- the expected bandwidth for these systems is from 1 Gbps to more than 10 Gbps (currently 40 Gbps are required), depending on the choice of telecommunications network;
- the usage of terahertz spectrum of aforesaid bands is a viable solution for fixed communication lines to achieve high bandwidth ccapability. Such systems will allow the rapid and efficient deployment of very high-capacity mobile and broadband communications in areas where fiber-optic cables are absent or not cost-effective;
- wireless communication lines of approximately 1 km distance can be deployed much faster and in some cases (for example, when radio links of line-of-sight are deployed between two or several buildings) are more cost-effective than wired networks because aforesaid frequency bands provide sufficient bandwidth capability to compete with or complement fiber-optic access networks;
- it is possible to use terahertz bands indoor where there should be no attenuation of the radio signal due to rain and high reliability of up to 99.999% is possible; therefore, radio communications are considered an alternative to fiber-optic cable, which connects servers in data centers.

So, terahertz frequency range is promising for the high-speed wireless communication networks creation because it has a low level of noise and interference. Despite the difficulties in ensuring the reliable operation of communication networks due to the large attenuation of the signal power during propagation in the atmosphere, it is possible to obtain bandwidths with a width of several tens of GHz, and thus achieve high bandwidth capability.

An example of the successful practical implementation of the terahertz frequency band is the recent launch of the world's first experimental satellite with an terahertz on-board transceiver designed to test 6G mobile networks [9, 14]. Compared to known methods of bandwidth capability increasing, this method is unique because most modern methods (multilevel modulation, antinoise coding, etc.) have contradictions. The contradiction is that when obtaining high values of some indicators (e.g., high data rate), other indicators (e.g., noise immunity) deteriorate.

In Ukraine, based on the results of research conducted by the authors radio relay communication technology in the frequency bands 94.1–100 GHz; 102–105 GHz; 106.5–109.5 GHz; 111.8–113 GHz; 130–134 GHz; 141–148.5 GHz introduced to the Plan for the Use of the Radio Frequency Resource of Ukraine (Sect. 2).

Forecasts of current trends in wireless communication show that by 2024 year the data rate should be at least tens of Gbps. This problem raises the question of the availability of a frequency band capable of meeting such requirements. The main possibility is to increase the available frequency band to several tens of GHz. It is not possible to do this at frequencies below 100 GHz and above 500 GHz.

In recent 10–15 years, research teams around the world have begun researching and prototyping terahertz wireless communication systems [10–14]. For example, in the scientific and practical work [10] of Japanese scientists presented the development of an integrated circuit (IC) of transmitter in the 300 GHz band based on electronic technology 40 nm CMOS. This transmitter uses 32-QAM modulation and allows to provide a data rate of 17.5 Gbps in each of the 6 channels with a frequency band of 5 GHz, while covering the frequency range of 275–305 GHz. The total data rate that can be provided by the transmitter IC is 105 Gbps. A feature of the transmitter IC is the use of high-line subharmonic mixers (cubic mixers), which are based on mis-type devices with n-channel and maximum gain at 280 GHz for the second frequency conversion of radio signals with multilevel modulation (16-QAM, 32-QAM). Due to the small gain of the mis-type devices s in the range of 275–305 GHz, in order to ensure a sufficient level of output power there are active power dividers in the structure of the IC.

In paper [11], a terahertz two-channel wireless communication system in the 375–450 GHz band with a data rate of 5 Gbps and 16-QAM modulation per channel was experimentally demonstrated, which allows to achieve a total bandwidth capability of up to 40 Gbps. Also four-channel wireless communication system with a total bandwidth of up to 80 Gbps was presented.

Scientists of Igor Sikorsky Kyiv Polytechnic Institute also made his own contribution to the research and development of wireless telecommunications systems of the terahertz range. This contribution is represented by the following areas of research:

- estimation of the terahertz waves propagation and attenuation in the atmosphere [15];
- implementation of the concept of software-defined terahertz wireless systems development with usage of Wi-Fi technology [16];
- design of transmitting and receiving devices for radio relay systems of terahertz range [17–20];
- investigation of the DVB-C standard multi-channel signal transmission through the laboratory testbed of the terahertz tranceiver [21];
- simulation of impulse radio ultra-wideband (IR-UWB) signal transmission by terahertz radio line [22, 23].

2 Estimation of Terahertz Wave Propagation and Attenuation in the Atmosphere

For efficient design of wireless terahertz telecommunication systems, knowledge of terahertz wave propagation mechanisms in the atmosphere is extremely important, as they allow to assess the reliability of the radio systems implementation level.

The analysis conducted by the authors showed that in the frequency range of 30–300 GHz of the known types of radio waves fading (refractive fading due to the impact obstruction shielding effect, refractive fading of interference type, interference fading due to reflection from the troposphere layers inhomogeneities, fading due to the shielding effect of inhomogeneities of the atmosphere, fading due to the influence of antenna patterns (for the terahertz range - this is the inaccuracy of antenna adjustment, as well as wind load on the antenna support), fading due to attenuation of the radio signal by hydrometeors (rain, dry and wet snow, hail, fog, clouds) fading due to radio signal absorption in gases, radio fading in sand and dust storms) the most important that should be considered in the design are the following:

- fading due to signal attenuation by hydrometeors;
- fading due to the absorption of the radio signal in the gases (primarily in water vapor and molecular oxygen);
- fading due to the influence of antenna patterns.

The operation of radio links at such high frequencies (especially in the terahertz range) due to the high directivity of the antennas of corresponding stations allows practically ignore the electromagnetic waves interference in the signal propagation area, which occurs especially in the environment of urban buildings. The calculated maximum radius value of the first Fresnel zone at the center of a 5 km radio link is 2.3 m at 140 GHz and not exceeding 1.6 m at 300 GHz, Therefore, this entitles to ignore refraction and interference losses in the calculation of the terahertz radio links energy budget.

Atmospheric attenuation of terahertz waves up to 300 GHz occurs mainly due to the presence of oxygen and water vapours in the atmosphere. Other gases make a negligible contribution to the terahertz wave attenuation. Figure 2 shows the result of the International Telecommunication Union (ITU) study of the frequency dependence of terahertz-wave attenuation, conducted in 1996 for normal atmospheric parameters (temperature 15 °C, pressure 101,3 kPa and water vapour density 7.5 g/m^3) in close proximity to thc Earth. One curve investigated the influence of oxygen, the other - the influence of water vapour.

The results showed that the attenuation peaks due to resonant interaction of the radio waves with the oxygen molecules occurred in the frequency range of 50–70 GHz with a maximum at 60 GHz and at a frequency of 118 GHz. Attenuation peaks due to the interaction of electric moments of water and radio wave were observed at 22.2 GHz and at 183 GHz. Other frequencies exhibited lower attenuation values and the ranges between the peaks were called radio windows. In other words, by selecting frequencies which located in the radio windows, the influence of atmospheric parameters can be significantly reduced for radio link.

Accurate attenuation calculations in oxygen and water molecules were obtained from the same investigations in 1996 (the results were subsequently presented into ITU-R Recommendation P.676). It should be noted that for distances up to 5 km, which are typical for terahertz radio links, a model where the attenuation in gases (dB/km) is taken as the sum of the attenuation in dry air and the attenuation in water vapour is suitable.

Analysing the dependencies in Fig. 1 we can make a logical conclusion that with increasing frequency of terahertz waves the attenuation in the rain also monotonically increases, reaching up to tens of decibels per kilometer.

In accordance with ITU-R Recommendation P.676, the atmospheric gas attenuation is still approximately 2 dB/km below 164 GHz and rises above 3 dB/km only at the upper end of the 174 GHz band. According to ITU-R Recommendation P.838, rain attenuation above approximately 80 GHz is practically constant and is comparable to what is experienced in lower frequency bands, e.g. 71 GHz to 86 GHz, which described by ECC in Recommendation REC (05)07. This makes the 130–174.8 GHz range suitable for short (up to 1 km) radio links with very high bandwidth capabilities.

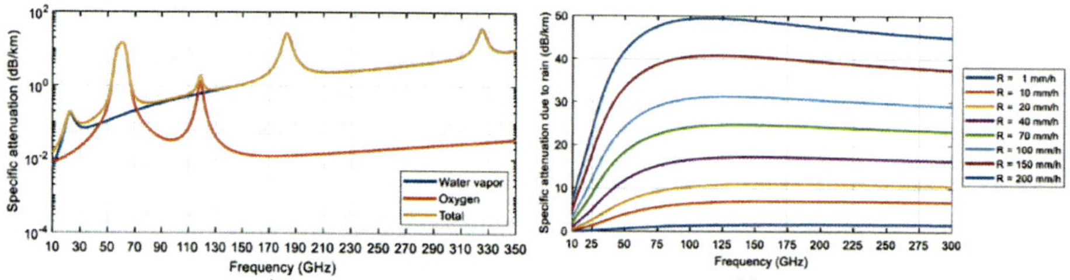

a) b)

Fig. 2. Frequency dependence of attenuation: a) in oxygen and water wapour according to ITU-R P.676; b) due to rain according to ITU-R P.838

The terahertz wave attenuation in dry snow is small. The attenuation in wet snow or rain of the same intensity increases by a factor of ten. As a comparison, investigations showed that for 88 GHz frequency at 1.4 km distance the attenuation in dry snow was only 1 dB, while in wet snow the attenuation was already up to 20 dB. The conclusion is that the snow contribution is insignificant compared to the rain with the same intensity.

In the case of fog, characterised by water droplets of 0.1 mm in diameter or less, concentrated in a confined space and dispersed according to Rayleigh's law, the terahertz wave attenuation to 300 GHz frequencies is negligible.

3 Implementation of the Concept of Software-Defined Terahertz Wireless Systems Development with Usage of Wi-Fi Technology

5G and 6G mobile networks, cognitive networks, big data, optical networks, green communications, terahertz telecommunications systems, distribution transport networks are the most important among the current trends in infocommunications development.

A significant proportion of wireless communication devices within these trends are built on the basis of the software-defined radio (SDR) concept. A well-known examples of the SDR concept today are devices for GSM, UMTS, Wi-Fi, WiMAX, etc.

The goal of any telecommunication system is to achieve, high speed and the required reliability of information transmission within the limits of the available communication channel resources. One well-known means for this objective realization is a combination of multilevel modulation and noiseless coding with specified noise immunity parameters, which is called a signal-code sequence (SCS).

If the communication device selects the appropriate SCS structure automatically using the appropriate program and algorithm, it should be considered to be in line with the SDR concept.

A gigabit modem named G1 was developed for the purpose of SDR concept realization in telecommunication systems of terahertz range on the basis of Wi-Fi technology. It can be used for interconnection of terrestrially distributed networks segments with Ethernet 10/100/1000-BaseTx interfaces. The block diagram of the G1 Gigabit modem is shown in Fig. 3.

The main components of a Gigabit modem are channel routers 1 and 2 as well as a group router. The input stream is automatically divided into all channels and further processed to generate a radio spectrum in a transmission mode.

The G1 gigabit modem (Fig. 4) is housed in a metal case that allows it to be mounted in a rackmount or wall-mounted.

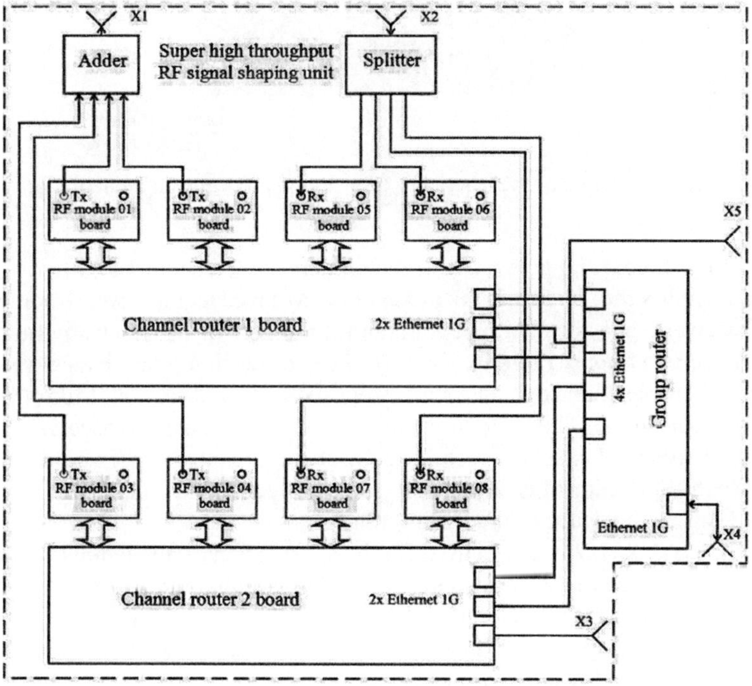

Fig. 3. Block diagram of a G1 gigabit modem

Intermediate frequency (IF) signal of the channel routers is routed to the output via port X1 (Fig. 3). Port X1 (type "N") is connected directly to the radio frequency input of the terahertz range transmitting device. Radio frequency output from the terahertz receiver device (X2 volume) is similarly configured. For Ethernet connectivity, as well as for group router control, port X4 with UTP cables was used. To ensure channel data rste of 1.2 Gbps in terahertz band, eight 802.11n Wi-Fi transceivers of 2.1–2.7 GHz band were installed with 40 MHz bandwidth, each having channel data rate up to 150 Mbps (Fig. 5 and Fig. 6).

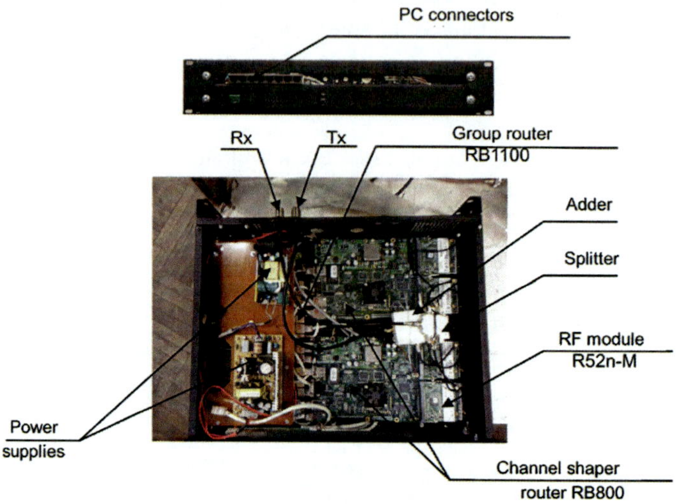

Fig. 4. Photo of the laboratory testbed of the G1 gigabit modem

The main technical characteristics of the modem are shown in Table 1.

The modem has the "dual stream" mode of the Mikrotik equipment, which uses two R52n-M receivers, one for data reception and the other for data transmission, which allow to create one duplex radio channel. In order to reach a total channel rate of 1.2 Gbps, four 150 Mbps duplex channels are proposed in each direction. Mikrotik RB800 routers with four mini-PCI slots with installed Mikrotik R52n-M transceivers are used to create dual stream channels, which is necessary to organize duplex communication and increase the efficiency of each channel. Access to each radio channel is provided by a separate Ethernet interface of the RB800 router.

Mikrotik RB1100Hx2 router is used for the aggregation of all channels, which provides a single interface for external connection. This configuration ensures high performance and features of the modem, while having at the same time relatively low cost of a gigabit modem development. It is also possible to increase data rate of gigabit modem up to 1.2 Gbps in each direction by increasing a number of RB800 routers and Mikrotik R52n-M transceivers.

Table 1. Main technical characteristics of the G1 Gigabit modem

Denomination	Value
Power supply voltage, V	220
Modulation/demodulation mode control interface	Ethernet 10/100 Base-Tx, RJ-45 connector
Router management interface	Ethernet 10/100/1000 Base-T, RJ-45 connector
Data interface	Ethernet 10/100/1000 Base-T, RJ-45 connector
IF input/output interface	Coaxial, 50 Ω, N-type
IF central frequency, MHz	2400
Maximum bandwidth occupied by modulated signal, MHz	not more 40
Type of modulation	QAM-64
Power of the IF signal at the modulator output, dBm	0 … −3
Sensitivity at the IF input of the demodulator, dBm	−70
Maximum permissible level of IF signal at the demodulator input, dBm,	−45
Modem weight, kg, max	4

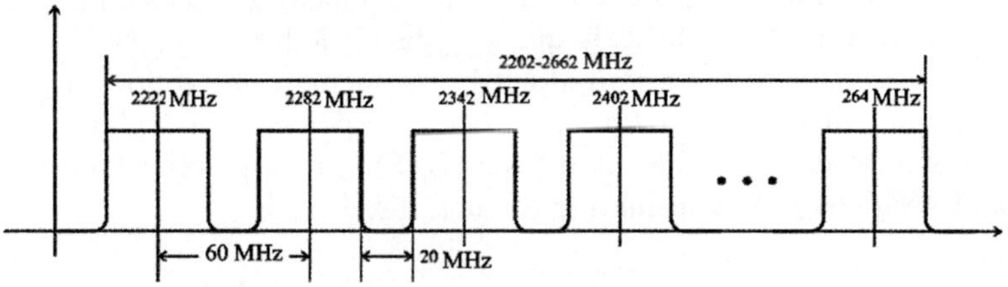

Fig. 5. Gigabit modem G1 frequency plan

To maintain an outdoor aggregate link at 600 Mbps, the RB750 group router must be replaced with a router capable of delivering 1 Gbps or more, e.g. the RB 450G or RB1100Hx2, as provided by the modem design.

On the basis of the G1 gigabit modem with Mikrotik R52n-M transceivers of Wi-Fi technology and the manufactured transmitting and receiving devices [25–33] the laboratory testbed of gigabit wireless communication system for the 130−134 GHz band at a data rate of up to 1.2 Gbps was developed.

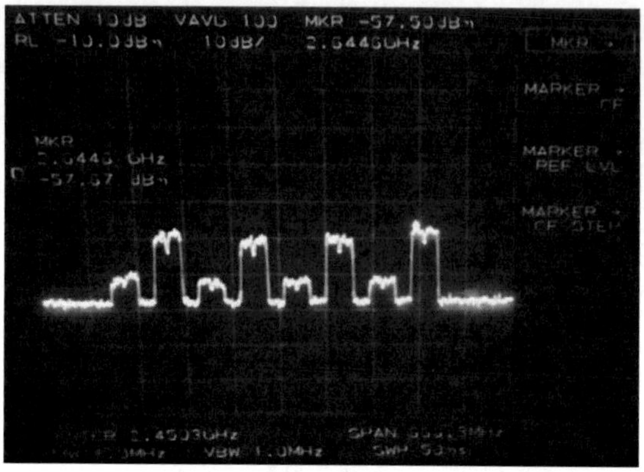

Fig. 6. Amplitude-frequency response of a 4-channel duplex communication

4 Design of Transmitting and Receiving Devices for Radio Relay Systems of Terahertz Range

4.1 Transmission and Reception Devices Block Diagrams

The key elements of a terahertz radio communication system are radio-electronic receiving and transmitting devices, which capable of generating and transmitting modulated signals with data rate of 1 Gbps and accepting signals with an acceptable high sensitivity.

Transmitting (Fig. 7) and receiving (Fig. 8) devices are the analog (linear) part of the radio relay system. These devices based on a heterodyne scheme, and will ensure line-of-sight signals transmission and reception in terahertz frequency range between 130 to 134 GHz. The intermediate frequencies bandwidth is 2…4 GHz. According to Fig. 7 a block diagram the transmitting device includes the following functional units: intermediate frequency amplifier (IFA), frequency converter (FC) (upconverter), oscillator (heterodyne) (Osc.), bandpass filter (BPF), output power amplifier (PA) (if component is available) and transmitting antenna (TA).

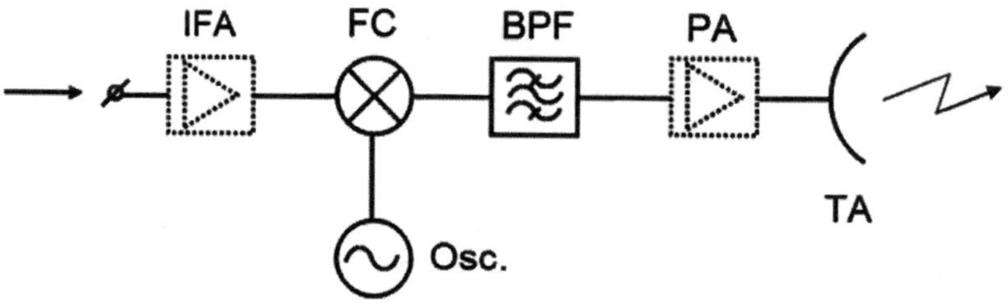

Fig. 7. Block diagram of transmitting device

The modulated signal comes from the modulator or group radio signal generator at the input of the transmitting device. If the signal level is sufficient to obtain the required signal level at the frequency converter output, the circuit does not require the use of an power amplifier.

The block diagram of the receiving device is shown in Fig. 8 and consists of the following components: receiving antenna (RA); input low-noise amplifier (LNA) (if component is available); bandpass filter; frequency converter (downconverter); heterodyne (Osc.); intermediate frequency amplifier.

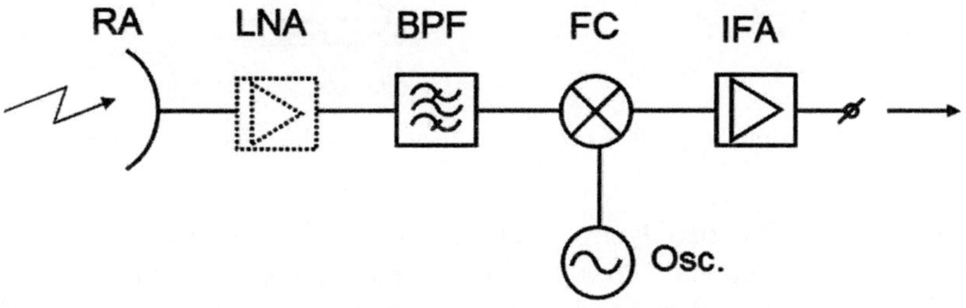

Fig. 8. Block diagram of the receiving device

The implementation of a low-noise amplifier at the input of the receiving device is just as problematic as with a power amplifier at the transmitting device output.

4.2 Functional Units Design of Receiving and Transmitting Devices

4.2.1 Signal Frequency Converters

The frequency converters (upconverter and downconverter) operate in different modes and perform different functions, but have the same unified electric circuit and design. The converters were built on the basis of unpackaged gallium arsenide-Schottky diodes of Ukrainian production (Scientific and Production Enterprise "Saturn", Kyiv,Ukraine). The limiting frequency of these diode is 2.5 THz, which allows them to operate at least in the lower part of the terahertz range. In terms of electrical and design parameters the developed diodes are not inferior to their foreign modern counterparts, e.g. diodes which developed by Hewlett Packard.

For the purpose of subharmonic circuit of the half-frequency pumping converter realization two Schottky diode in parallel to each other are used as a nonlinear element. Such inclusion forms symmetric with respect to the origin of coordinates N-variable resulting volt-ampere characteristic. This condition determines the nonlinear element parameters change with a frequency twice as high as the heterodyne frequency. Functional and structural schemes of the converter are shown in Fig. 9.

The converter design consists of two waveguides connected by a symmetrical stripline with a pair of unpackaged Schottky diode which is mounted on it. The diode pair are selected to be equal by parameters in order to ensure the resulting symmetrical volt-ampere characteristic for a good suppression of the odd harmonics of the local oscillator.

The waveguide channel has a 1.6×0.8 mm^2 cross section and is part of the high-frequency signal circuit. This waveguide channel operating frequencies are out of bounds for local oscillator and IF frequencies.

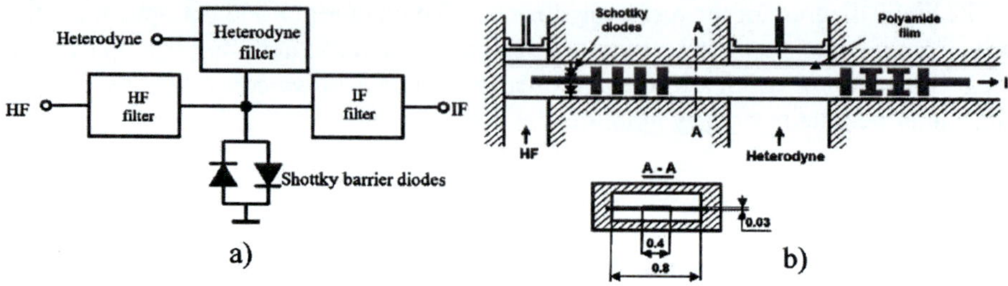

Fig. 9. Functional (a) and structural (b) diagrams of the frequency converter

The local oscillator (heterodyne) signal is fed to the diode pair through a waveguide channel with a cross section of 3.6 × 1.8 mm^2. The oscillator channel is out-of-bounds for the IF signal, and the isolation of the local oscillator from the RF signal provides a low-pass filter (LPF) with a cutoff frequency of 67 GHz. This channel designed as a part of the stripline between the waveguide channels.

The IF signal circuit is fully constructed on a symmetrical stripline with a suspended substrate. As a substrate polyimide film with a thickness of 30 μm was used, which suspended in a rectangular channel with a cross section of 0.8 × 0.4 mm. Such construction prevents the emergence of higher waveguide modes. The output LPF in the IF circuit with a cutoff frequency of 30 GHz blocks the penetration of the local oscillator and RF signals into the intermediate frequency circuit. Short-circuiting pistons in the twin channels allow to carry out adjustment of the corresponding chains of converters.

Figure 10 shows the frequency converter (mixer) photos with connected waveguide sections and without top part and waveguide sections.

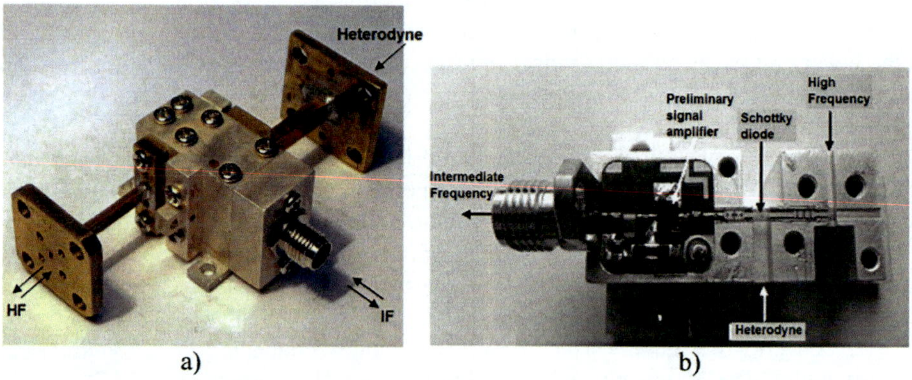

Fig. 10. Photos of frequency converter with connected waveguide sections (a) and without top part and waveguide sections (b)

Diodes are connected to the stripline by means of electrically conductive glue. The converter construction also includes a preamplifier of the intermediate frequency signal. An SMA connector of the instrument type is used at the output/input of the converter IF circuit. The local oscillator power required for frequency converter normal operation did not exceed 15 mW. The measured value of conversion losses is 11 dB, which corresponds to the best achievements of foreign counterparts.

The preamplifier is structurally integrated with the mixer to minimize the loss of a weak signal. The noise temperature of this amplifier is about 50 K and largely determines the sensitivity of the entire receiver. This is where the main signal amplification takes place. The total gain of the IFA is 47 dB.

4.2.2 High-Frequency Amplifiers

The power amplifier (PA) at the transmitter output and the low noise amplifier (LSE) at the receiver input are the functional units which principally determine the energy potential of the entire radio relay system.

Problems of implementation of such amplifiers in terahertz range are caused by high operating frequencies. The success of semiconductor technology in recent years and the creation of an appropriate element base open up prospects for the implementation of high-frequency high-quality amplifiers. The low-noise amplifier created in a 3 mm range is implemented on a monolithic PIn chip in the frequency range of 87… 100 GHz, while ensuring 27 dB signal gain and a noise figure of 5.5 dB. The photo of this amplifier is shown in Fig. 11.

Laboratory designs of PAs and LNAs up to 300 GHz are available. In the near future the appearance of commercially available power amplifiers in the entire millimeter range are expected, which will become a real basis for the construction of power amplifiers in the terahertz range. The use of these amplifiers in a transmitting/receiving device will ensure reliable and high-quality signal transmission/reception over long communication distances.

Fig. 11. Photo of the 3 mm range power amplifier

4.2.3 Heterodyne

The heterodyne appears to be the most challenging device to design digital telecommunications systems in the terahertz frequency range. This is due to both the difficulty of achieving sufficient power level at very high frequencies and the need to ensure high frequency stability of the heterodyne and its low phase noise level.

The subharmonics scheme of the frequency converters allows to reduce the required heterodyne frequency, which to some extent facilitates the oscillator circuit design. However, there are stringent requirements for frequency stability and phase noise level of local oscillator/heterodyne, especially in digital data transmission systems with multilevel modulation types.

A 64.8 GHz heterodyne circuit was designed for a digital radio relay system with a frequency range of 130 to 134 GHz. The heterodyne construction is based on the use of a high-stability crystal oscillator with the following cascades of frequency multiplying and amplification. The power level was set so as to provide the optimal operating mode of the multiplier stages and the required power level at the heterodyne output. This principle of heterodyne construction is much cheaper than a frequency synthesizer, and the signal filtering ignal after each multiplier stage reduces the presence of parasitic harmonic and combinational frequencies.

Block diagram and photo of the heterodyne circuit is shown in Fig. 12. As a reference a 100MHz CCHD-950X-25–100 crystal oscillator manufactured by Crystek Crystals with a phase noise level no higher than −143dB at 1 kHz offset from the central frequency was used. The oscillator parameters mainly determine the frequency stability and phase noise of the heterodyne.

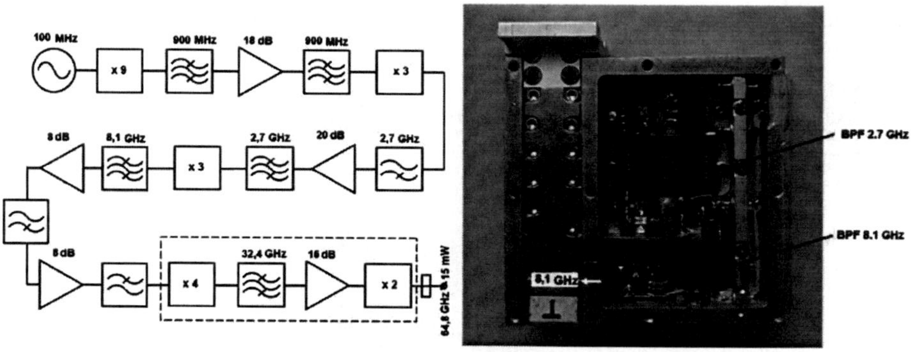

a b

Fig. 12. Block diagram (a) and photo of the heterodyne circuit (b)

A special feature of this circuit is that the transistor-based first multiplier stage (Fig. 12a) allows for the immediate use of the 9th frequency harmonic of the oscillator, i.e. 900 MHz. The signal at this frequency is amplified by 18 dB by the SPF5043Z IC amplifier and is filtered by the FAR-F5KA-897M50 surface acoustic wave filters.

Further increase of the frequency to 8.1 GHz is carried out by two stages of frequency triplers (Fig. 12a). The first stage in relation to the output frequency of 2.7 GHz is built on a transistors, and the second at an output frequency of 8.1 GHz on a monolithic HMC916LP3 chip. At 2.7 GHz, the signal is amplified by 20 dB using the TQP3M9008 chip, and it is filtered using low- and high-pass filters (LPF and HPF) on lumped elements, as well as a bandpass filter at the output of the amplifier. The BPF is mounted on a 0.25 mm thick duroid resonator substrate (RT/Duroid 5880) (Fig. 12b).

At the 8.1 GHz frequency, the signal is also amplified by two NLB-300 stages of amplifier and filtered in LPF and HPF on lumped elements, as well as BPF, which is performed as half-wave elements of stripline on the duroid substrate of 0.125 mm thick.

After each multiplier stage the signal quality was checked for sufficient power level, absence of adjacent parasitic harmonics and low level of phase noise. For example, Fig. 13a shows the measured signal spectrum at the output of the frequency multiplier module in the frequency range of 8.1 GHz.

Figure 13a shows that near the output signal at the frequency of 8.1 GHz there are no noticeable parasitic harmonics of the oscillator, and the 2nd harmonic of the output signal at a frequency of 16.2 GHz is attenuated by more than 40 dB. Frequency multiplier output power is 7 mW and the phase noise level does not exceed −107dB at 100 kHz and −94dB at 10 kHz offset from the center frequency.

The signal with a power of 11 dBm and frequency of 8.1 GHz is fed to the output of higher-frequency stages of the heterodyne circuit, mounted in a separate housing, which, in turn, is built into the common heterodyne housing. The photo of the designed output cascades module of heterodyne is shown in Fig. 13b.

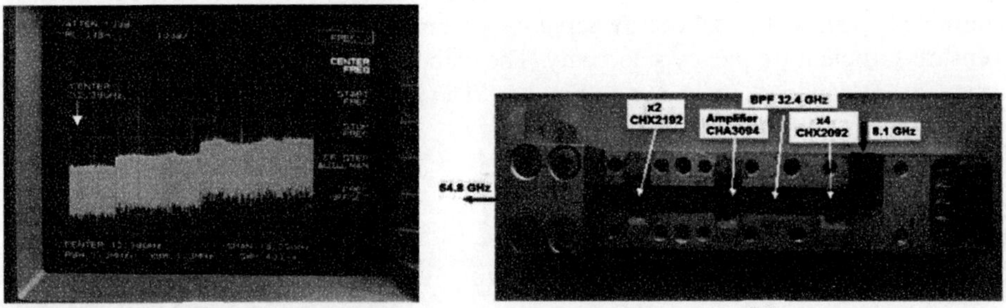

a) b)

Fig. 13. The photos of measured signal spectrum at the frequency multiplier module output (a) and designed output cascades module of heterodyne

The designed module (Fig. 13b) includes a frequency multiplier by four, BPF and amplifier stage at 32.4 GHz, an output frequency doubler and the transition from the microstrip line to the waveguide channel with a cross section of 7.2×3.4 mm^2.

Multipliers and amplifier are built on the basis of unpackaged monolithic chips. The CHX2092 chip was used as a multiplier by four, the amplifier constructed on the basis of the CHA3093 chip and the output frequency doubler was constructed on the CHX2192 chip.

Monolithic chips of harmonic signal multipliers and amplifier are connected to the microstrip lines on a duroid substrate with a thickness of 0.125 mm. Also a waveguide-microstrip transition is performed on the same substrate.

The seven-resonator microstrip BPF is developed on a ceramical substrate of 0.2 mm thick. For better BPF matching with the chips there are fixed 3 dB chip attenuators included on both sides of the BPF.

The measured values of the heterodyne output power exceeded 15 mW, which is fully sufficient for the normal operation of the frequency converters of the receiving and transmitting devices.

Thus, as a result of the research, a heterodyne was designed, manufactured and tested. Heterodyne is not inferior to foreign analogues in terms of its performance, including output power.

4.2.4 Band-Pass Filter

Bandpass filters at the output of the transmitting and the input of the receiving devices must ensure frequency isolation of these devices from each other, as well as sufficient suppression of the mirror channel and the heterodynes harmonics. Low value of bandwidth frequency ($F_{IF} = 2\ldots2.5$ GHz) causes severe requirements for BPF frequency selectivity. So-called septum-filters in terms of low losses and high frequency selectivity characteristics are most acceptable for BPF application in 2 mm range of wavelengths. This is exactly the filter that was developed as part of transmitting and receiving devices.

Structurally, the filter is a thin metal plate inserted in the E plane of the waveguide channel. The plate consist of the resonance windows, the connection between them is determined by the width of the strips separating them. A six-resonator filter was designed to ensure sufficient frequency selectivity. The calculated dimensions of the plate insert with resonant windows are shown in Fig. 14a. The photo of the fabricated septum-filter is shown in Fig. 14b.

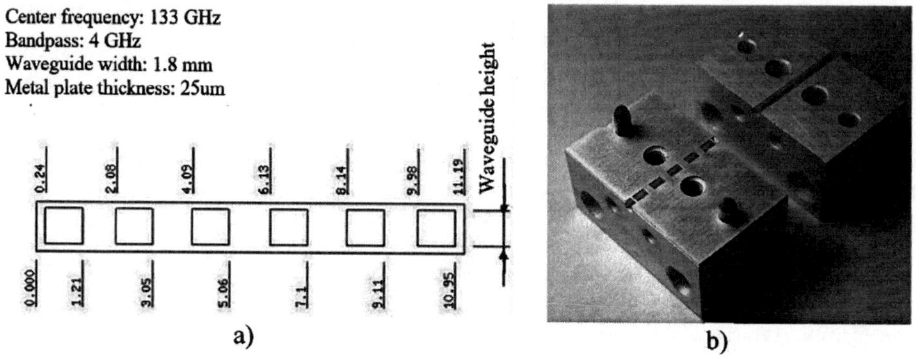

Fig. 14. The calculated dimensions of the six-resonator septum-filter plate (a) and the photo of the fabricated septum-filter (b)

Figure 15 shows the calculated S-parameters as well as the measured frequency response of the fabricated six-resonator BPF (septum-filter).

Experimental results show that the filter satisfies the frequency selectivity requirements. Its provides sufficient (by 20 dB) suppression of the mirror channel and the second harmonic of the heterodyne signal (129.6 GHz).

Filter loss in the terahertz frequency range of the transmitter does not exceed 4 dB, which is an acceptable result for such high frequencies. The irregularity of the frequency response is less than 2 dB.

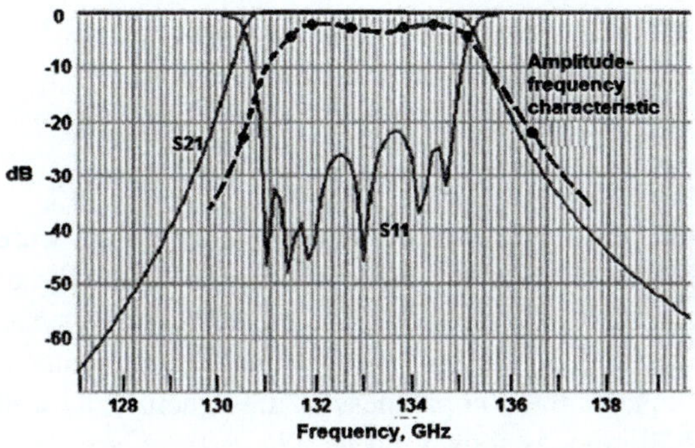

Fig. 15. Calculated (solid line) and measured (dashed line) S-parameters of fabricated BPF

4.2.5 Conical Horn Antenna

In both transmitting and receiving devices a conical horn antenna with a dielectric lens concentrator was used. The antenna construction is shown in Fig. 16.

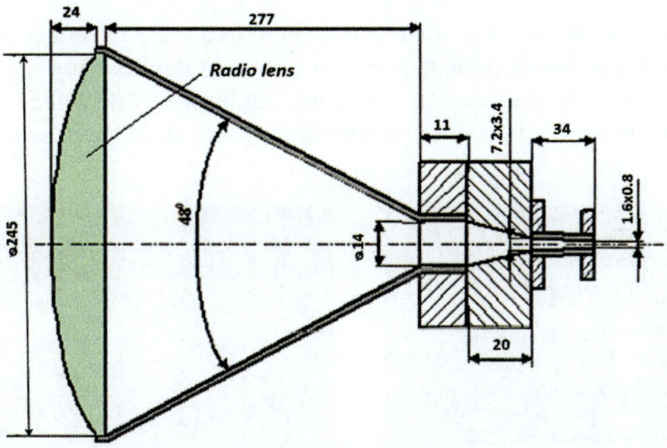

Fig. 16. Construction of the conical horn antenna

In addition to the conical horn, the structure contains a transition from a round waveguide to a rectangular one with a cross section of 7.2×3.4 mm^2. The connection of the antenna to the transmitting or receiving device was carried out by means of a waveguide transition from 7.2×3.4 mm^2 cross section to 1.6×0.8 mm^2 cross section. The antenna diameter is 245 mm. A fluoroplastic dielectric lens was used as a concentrator.

The calculations give the following antenna characteristics:

- operating frequency range from 130 to 134 GHz;
- input waveguide channel with a cross section of 1.6×0.8 mm^2 ($\lambda = 2$ mm);
- the gain factor is at least 47 dB;

– half-power beamwidth no more than 0.60°;
– voltage standing wave ratio approximately equals to 1.15.

4.2.6 Testing of 130–134 GHz Terahertz Radio Relay System

The receiving and transmitting devices of the radio relay system were built on the basis of the developed functional units (frequency converter, heterodyne, BPF etc.) which described above. Receiver and transmitter of the radio relay system have the same structural design. The modular design of individual components with maximum use of monolithic chips ensures the compactness of the structure, as well as the ease of its arrangement and installation, as shown in Fig. 17a.

The devices have coaxial input (output) with SMA sockets at IF frequencies, and rectangular waveguide input (output) of $1.6 \times 0.8\ \text{mm}^2$ cross section at terahertz frequencies. The photo of the receiving (transmitting) device is shown in Fig. 17b.

In order to ensure high quality operation of the receiving and transmitting devices, secondary power supplies have been developed which generate the necessary DC voltages for all units of this devices.

The photo of the transmitting (receiving) device with conical horn antenna is shown in Fig. 18a.

As can be seen from the experimental measured frequency dependence of the transmission coefficient between transmitting and receiving devices (Fig. 18b), radio link between these devices has a transmission gain of at least 18 dB, while the irregularity of the transmission gain in the operating frequency range does not exceed 3 dB.

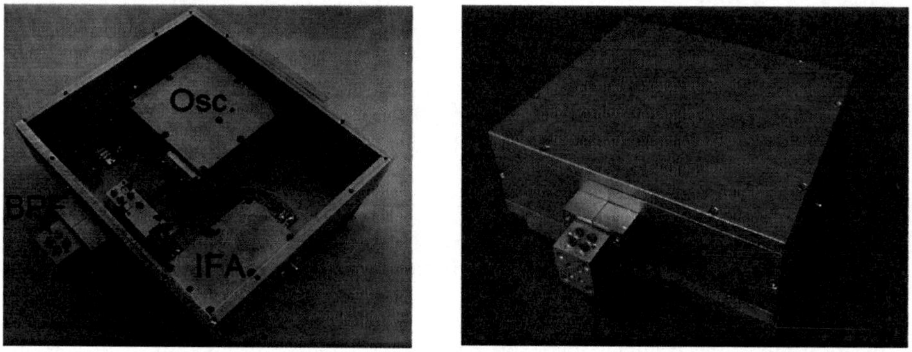

a) b)

Fig. 17. Location of main units (Osc., FC, IFA, BPF) in the receiving (transmitting) device housing (a) and its appearance (b)

In addition, experimental research was carried out for laboratory testbed of digital simplex terahertz radio relay system (Fig. 19) consisting of the 130–134 GHz frequency range receiving and transmitting devices and the G1 gigabit modems.

Experimental investigations of the radio relay system laboratory testbed showed the following values of the main parameters in the 130–134 GHz operating frequency range:

a)

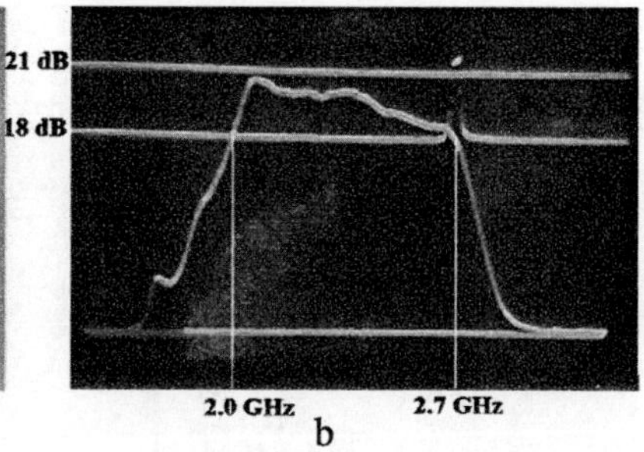

b

Fig. 18. The photo of the transmitting (receiving) device with conical horn antenna (a) and the frequency dependence of the transmission coefficient between transmitting and receiving devices (a)

- noise temperature of the receiving device is 5000 K;
- output power of the transmitting device is 40 μW.

The following results were obtained:

- multichannel data rate up to 1200 Mbit/s;
- BER is not more than 10^{-6};
- communication range in normal conditions is within 1 km of line-of-sight.
- system gain factor of 50 dB.
- type of modulation for testing is QAM-64.

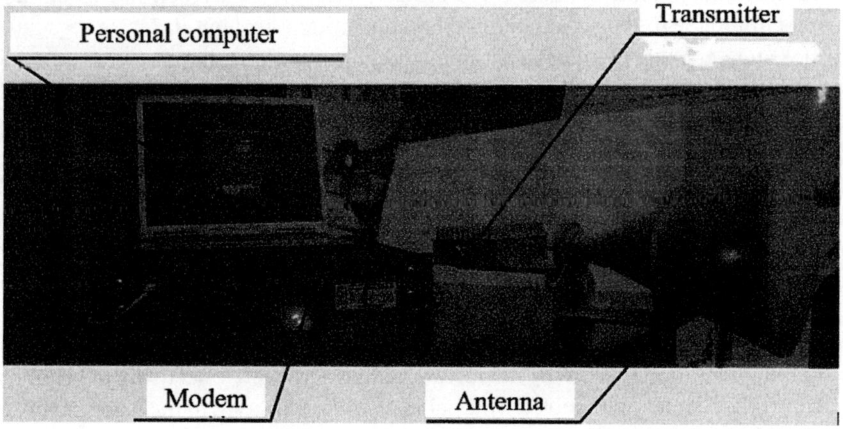

Fig. 19. Exterior view of the digital simplex terahertz radio relay system laboratory testbed

The proposed laboratory testbed of radio relay system was tested according to the block scheme diagram in Fig. 20.

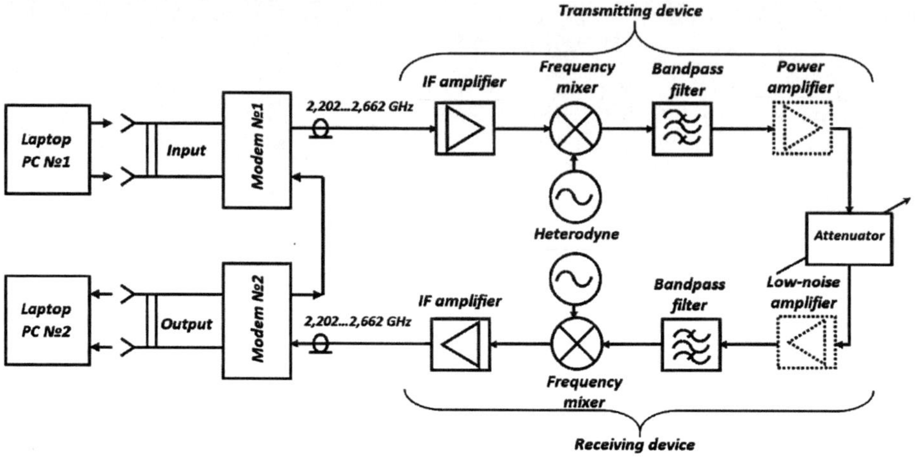

Fig. 20. Block diagram of the terahertz radio relay system laboratory testbed

5 Investigation of the DVB-C Standard Multi-channel Signal Transmission Through the Laboratory Testbed of the Terahertz Transceiver

The aim is to investigate DVB-C standard digital television signal parameters changes when its transmitted through a 130 GHz transmitting and receiving devices of the terahertz transceiver. The reason why DVB-C standard for conducting research was used is the possibility of signals generation with different modulation types (from QAM-64 to QAM-256). It allows to investigate the transceiver laboratory testbed individual nodes parameters influence to the signal parameters transmitted through it.

To investigate DVB-C signals parameters changes when they are transmitted by the receiving and transmitting devices of terahertz range, an experimental testbed of the terahertz transceiver was assembled. The block diagram of this testbed is shown in Fig. 21. As one can see in Fig. 21, DVB-C signal parameters were measured at the input of the transmitting device and at the output of the terahertz receiving device. The transmitting device output is connected to the receiving device input by a rectangular waveguide line.

The following measuring equipment was used in all phases of the research:

- ST-2 ROVER digital TV signal analyzer;
- The output of terahertz receiving device was connected to a TV set using the Homecast DVB-C digital cable TV receiver. This cable receiver was used to measure the DVB-C signal quality in the relative percentage scale.
- The QBox TBS5880 digital terrestrial and cable TV tuner is connected to a computer using a USB cable. This tuner is designed to display the constellation diagrams and numeric demodulation/decoding performance parameters of DVB-C standard TV signals at the transmitter and receiver output of the terahertz radio relay line laboratory testbed.

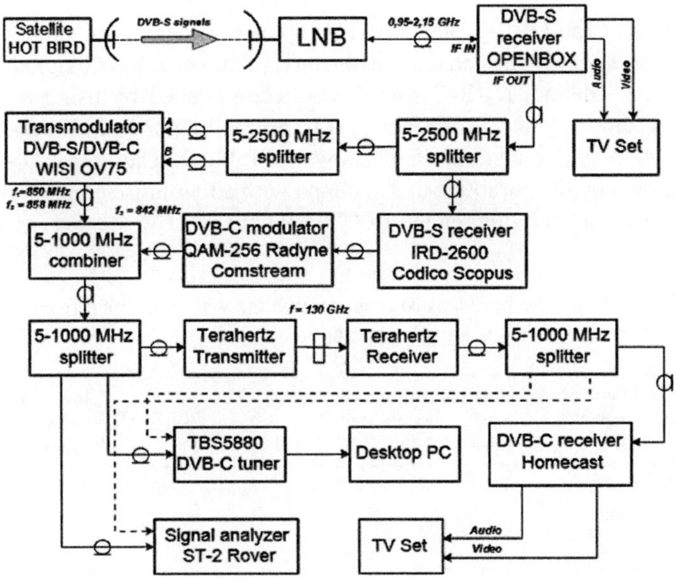

Fig. 21. Block diagram of an experimental laboratory testbed of terahertz transceiver for investigating DVB-C signals parameters

A simplified block diagram of the terahertz transceiver and its photo are shown respectively in Fig. 22a and Fig. 22b.

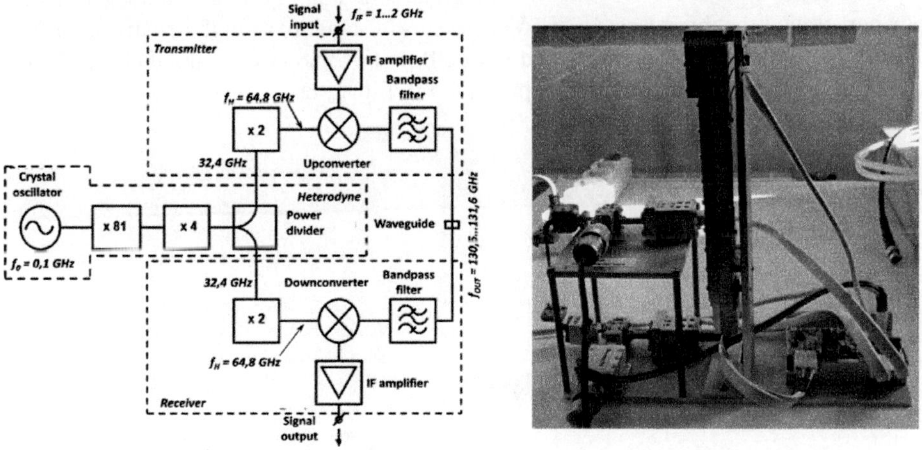

a) b)

Fig. 22. The block diagram (a) and the photo (b) of the terahertz transceiver laboratory testbed

Investigation of the DVB-C standard TV signal parameters for terahertz receiving and transmitting devices was carried out for three-channel DVB-C signal. The generation of the three-channel DVB-C signal was accomplished by using a WISI OV75 transmodulator and Radyne Comstream QAM-256 modulator from DVB-S TV broadcast signals, whose parameters are also shown in Table 2. The above DVB-S signals were received on a by the parabolic dish antenna with a low-noise converter of Ku-band from the geostationary satellite named HOT BIRD (13 E).

Table 2. DVB-S standard TV broadcast signals parameters which received from the HOT BIRD satellite and used for three DVB-C signals generation

Channel number	Input signal, MHz	Intermediate frequency, MHz	Symbol rate, Ksps	Bit rate, Mbps	Modulation	FEC		Information speed, Mbps	Spectrum width, MHz
						Conv.	R.S		
1	11179	1429	27500	55	QPSK	3/4	188/204	38,015	36
2	11137	1387							
3	11334	1584							

The transmodulator and modulator provide the following DVB-C signal parameters to be changed: modulation type (QAM-16, QAM-32, QAM-64, QAM-128, QAM-256), speed (symbol rate, bit rate depending on the modulation type and forward error correction rate), intermediate frequency (in the range 47–862 MHz), MPEG transport stream frame size, DVB-C signal output power level and so on.

Each of the three DVB-C standard TV signals shown in Fig. 23a contains an MPEG transport stream that broadcasts 5–6 TV programs in standard or high density definition and in free-to-air status. The measured with QBox TBS5880 tuner signal constellation diagram for QAM-128 modulation is presented in Fig. 23b.

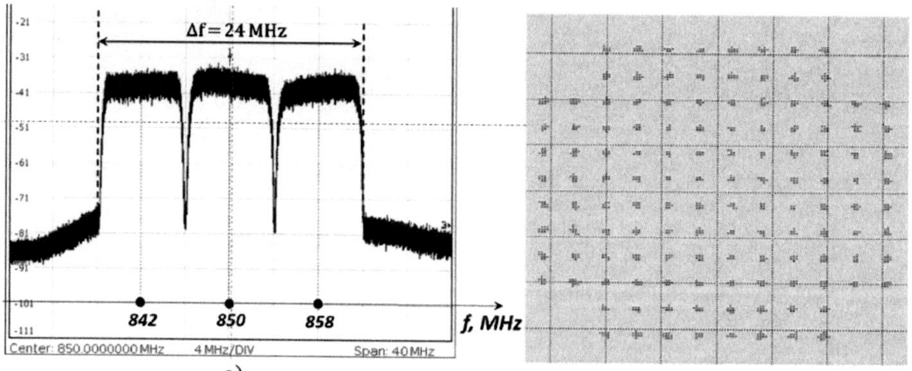

a) b)

Fig. 23. Spectrum of 3-channel TV signal of DVB-C standard (a) and measured signal constellation diagram for QAM-128 modulation (b)

The outputs of QAM-256 Radyne Comstream modulator and WISI OV75 trans-modulator (Fig. 22) are connected to TV combiner. There is a 3-channel TV signal of DVB-C standard formed at the TV combiner output with carrier frequencies 842, 850 and 858 MHz which fed to the input of the transmitting and receiving devices of the terahertz transceiver. The results of 3-channel DVB-C standard TV signal parameters measuring are shown in Table 3.

Table 3. The parameters measuring results of 3-channel DVB-C standard TV signal

Carrier frequency, MHz	Modulation	DVB-C signal parameters at the input of the terahertz transmitting device			DVB-C signal parameters at the output of the terahertz receiving device		
		LEV, dBm	SNR, dB	BER	LEV, dBm	SNR, dB	BER
842	QAM-64	−51	38,6	$<10^{-8}$	−60	33,71	$<10^{-8}$
	QAM-128	−54	42,12	$<10^{-8}$	−57	39,43	$<10^{-8}$
	QAM-256	−32	38,6	$<10^{-8}$	−54	36,8	$<10^{-8}$
850	QAM-64	−38	38,6	10^{-4}	−57	37,6	$7 \cdot 10^{-4}$
	QAM-128	−54	43,43	$<10^{-8}$	−54	39	$<10^{-8}$
	QAM-256	−32	38,6	$<10^{-8}$	−54	37,4	$<10^{-8}$
858	QAM-64	−38	38	$<10^{-8}$	−57	34,57	$<10^{-8}$
	QAM-128	−54	42,12	$<10^{-8}$	−54	39,43	$<10^{-8}$
	QAM-256	−32	38,6	10^{-3}	−54	36,8	$1,5 \cdot 10^{-3}$

6 Simulation of IR-UWB Signal Transmission by Terahertz Radio Line

Nowadays, the maximum distance of UWB communication systems as a rule does not exceed tens of meters between corresponding user terminals. This is due to the fact that the USA Federal Communications Commission has set the permissible spectral density power of the radiated UWB signal at −41,3 dBm/MHz level in the frequencies range of 3.1…10.6 GHz for electromagnetic compatibility ensuring. Thus, dramatically increase the distance of UWB communication systems is possible only through the use of free radio spectrum. The terahertz range (0.1–3 THz) due to its large frequency capacity and inoccupation is promising for the development and deployment of UWB communication systems to explore all their advantages and disadvantages.

It is important to note that the task of generating UWB signal directly in the terahertz range as opposed to generating harmonic oscillations is a very challenging task. So, the most feasible and promising way to build terahertz UWB communication systems is the generation of UWB signals in the decimeter or centimeter frequency range with their next upconvertion to the terahertz range in the transmitting device and downconvertion of the UWB signal from the terahertz range to the decimeter or centimeter range in the receiving device. Therefore, the purpose of this research is to investigate the IR-UWB signal shape change in time and frequency domain (i.e., distortions) during its transmission through an idealized model of terahertz radio link with known parameters.

The Visual System Simulator software package of AWR Microwave Office CAD, which has a user-friendly interface and the necessary functionality to carry out investigations of electric circuits at the schematic and system engineering levels was used for model simulation of IR-UWB signal transmission through terahertz radio link.

Idealized model of terahertz radio link was created based on the parameters and structural scheme of the operational laboratory testbed of terahertz transceiver presented in Fig. 22. This testbed has proven its worth during practical investigations of DVB-S and DVB-C standards TV signals transmission/reception.

The simplified block diagram of IR-UWB terahertz radio communication system, which was built in the AWR Microwave Office used is shown in Fig. 24.

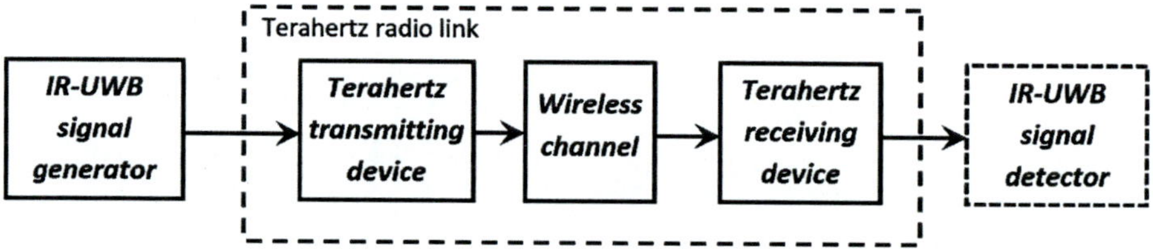

Fig. 24. The block diagram of IR-UWB terahertz radio communication system

6.1 Model of the IR-UWB Signal Generator

The so-called Gaussian monocycle was chosen as the IR-UWB signal, which is widely used as an radiating signal in UWB technology. The model of IR-UWB signal generator, which is built using the Microwave Office CAD tools, is shown in Fig. 25. The model includes a rectangular pulse generator, Gaussian low-pass filter and a derivation circuit.

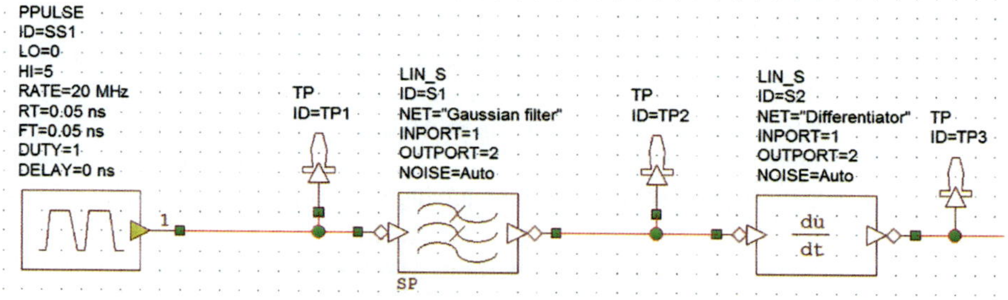

Fig. 25. Schematic diagram of the IR-UWB generator model

Parameters of the ultrashort pulse generator (PPULSE unit):

- repetition rate (RATE): f = 20 MHz;
- pulse repetition period: T = 50 ns;
- pulse edge rise time duration (RT): TR = 50 ps;
- pulse edge fall tine duration (FT): TF = 50 ps;
- Pulse duty cycle (DUTY) = 1% or 0.5 ns = 500 ps.

The iFilter Wizard of Microwave Office CAD was used to synthesize a Gaussian low-pass filter. The electrical scheme of the lumped elements filter is shown in Fig. 26a, and its amplitude-frequency characteristic (AFC) of the transmission coefficient (S_{21} parameter) and the return loss coefficient (parameter S_{11}) are shown in Fig. 26b.

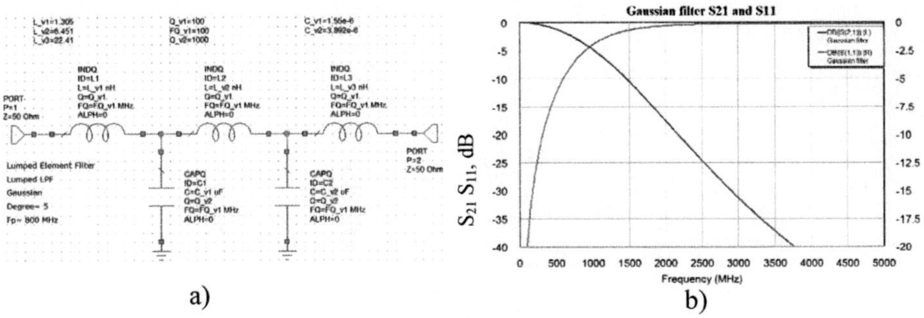

Fig. 26. Electrical circuit of the 5th order Gaussian low-pass filter with cut-off frequency of 800 MHz (a) and frequency response of the Gaussian filter transmission coefficient (parameter S_{21}, blue line) and frequency response of its input return loss coefficient (parameter S_{11}, pink line)

The time response of a data pulse, ultrashort Gaussian pulse and its spectrum is shown in Fig. 27a and Fig. 27b respectively.

To obtain a Gaussian monocycle, it is necessary to "pass" the Gaussian pulse through a derivation circuit, one of the variants of which is shown in Fig. 28a. Also, Fig. 28b shows the AFC of the transmission coefficient of the derivation circuit.

The time response of the Gaussian monocycle at the derivation circuit output is shown in Fig. 29.

6.2 Terahertz Transmitter Model

The received monocycle is fed to the input of the terahertz transmitter which is represented in the Microwave Office CAD model (Fig. 30). According to Fig. 30 the terahertz transmitter model includes: wideband IF amplifier (GALI-5), attenuator (ATTEN), wideband frequency subharmonic mixer (upconverter) model (MIXER_B), heterodyne (TONE), bandpass Chebyshev filter (BPFC).

Figure 31 presents a part of the power spectrum of the radio signal at the mixer output.

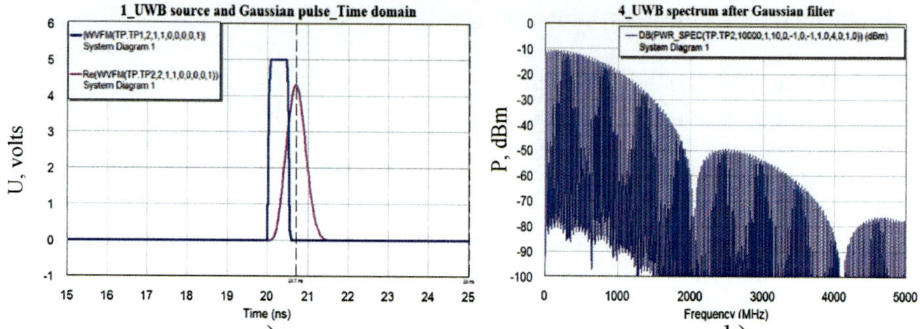

a) b)

Fig. 27. The time response of a data pulse (blue line), ultrashort Gaussian pulse (pink line) (a) and its power spectrum (b)

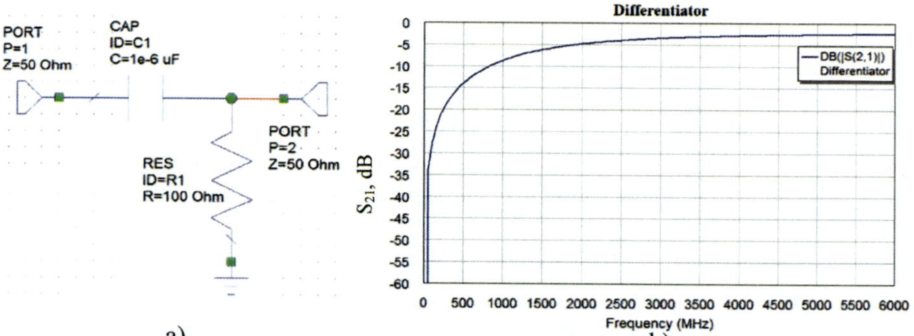

a) b)

Fig. 28. Signal derivation circuit (a) and its amplitude-frequency characteristic (b)

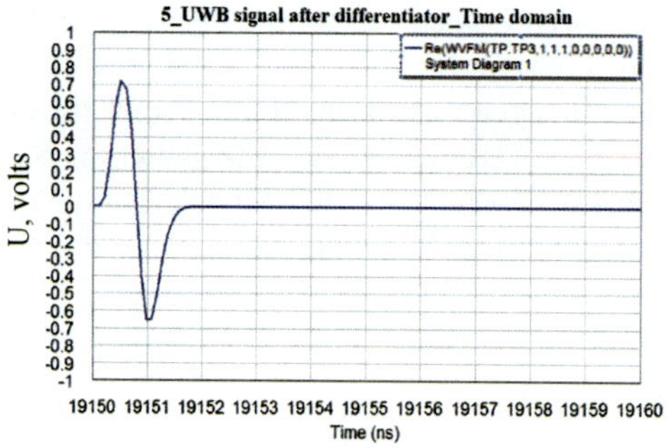

Fig. 29. Gaussian monocycle time response at the derivation circuit output

The AFC of the bandpass filter in the form of 11th order Chebyshev filter model of the transmitting device with a bandwidth of 130.4–131.5 GHz at −3 dB level was simulated and shown in Fig. 32a. The IR-UWB signal spectrum at the output of this bandpass filter is shown in Fig. 32b.

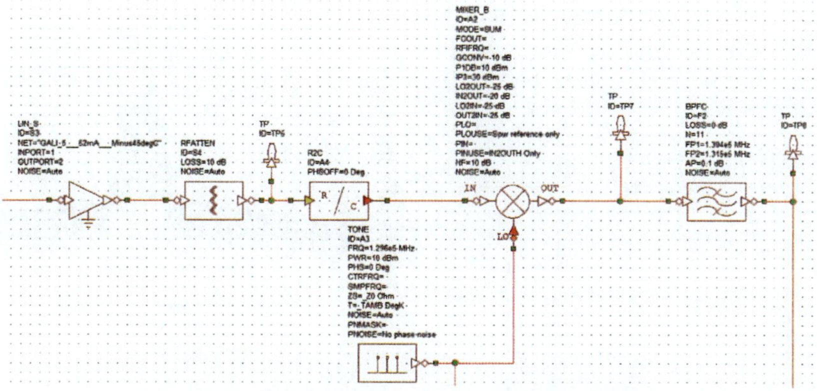

Fig. 30. Simulation model of terahertz transmitter

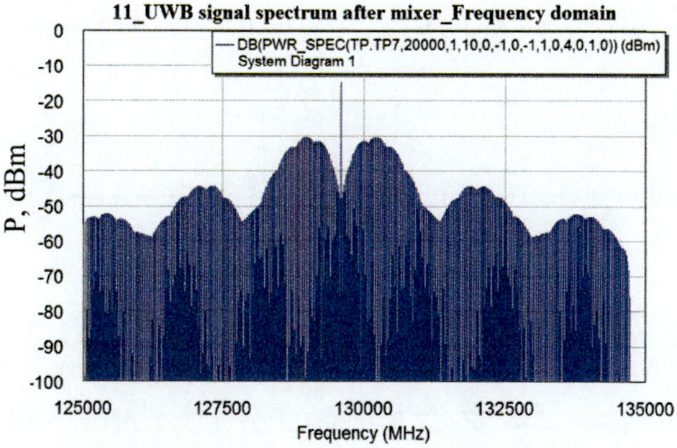

Fig. 31. IR-UWB signal spectrum after its upconvertion in the terahertz transmitting device (heterodyne frequency is $f_0 = 64.8$ GHz)

The time response of the IR-UWB at the output of the bandpass filter of the transmitting device presented in Fig. 33.

6.3 Terahertz Receiver Model

The terahertz receiving device is connected to the terahertz transmitting device via a 20 dB attenuator, which simulates a wireless-free transmission line (free space loss and gains of the transmitting and receiving antennas) without noise. The schematic diagram of the receiving device simulation model is shown in Fig. 34.

The receiving device simulation model in accordance with Fig. 34 consists of:

- bandpass Chebyshev filter (BPFC) with the same parameters as for the transmitting device;
- the 64.8 GHz heterodyne (TONE) with the same parameters as for the transmitting device and with an output power of +10 dBm;

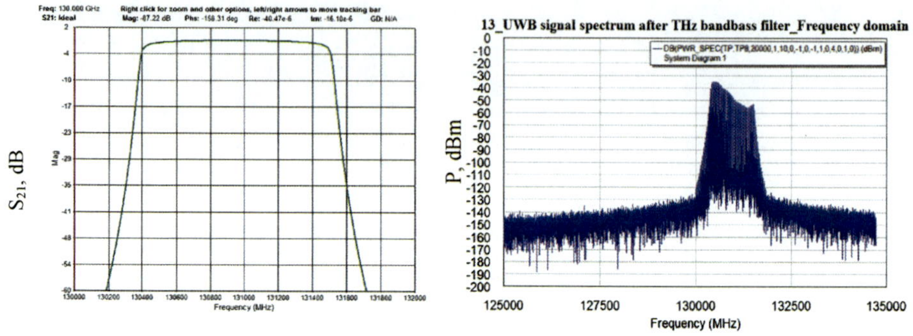

a) b)

Fig. 32. AFC of the bandpass Chebyshev filter (a) and IR-UWB signal spectrum at the output of the this bandpass filter

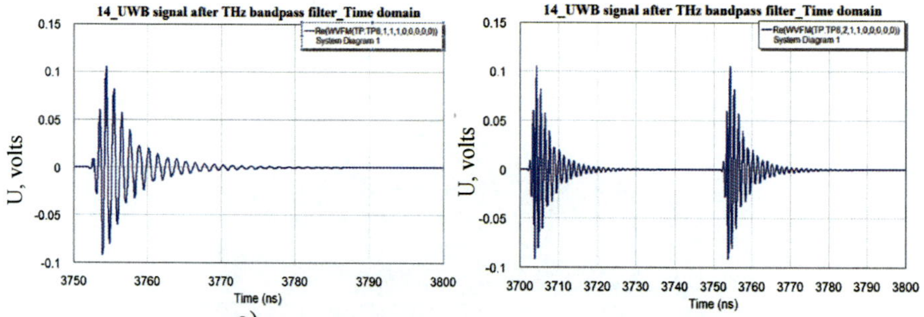

a) b)

Fig. 33. The time response of the IR-UWB at the output of the bandpass filter: 50 ns scale (a), 100 ns scale (b)

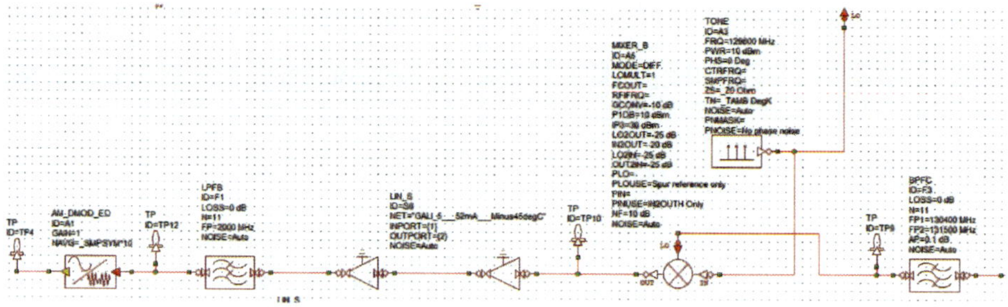

Fig. 34. Simulation model of terahertz receiver

- the frequency subharmonic mixer (MIXER_B) (downconverter). Its parameters are the same as for the transmitting device mixer;
- the wide-band amplifier based on the GALI-5 chip with a 40 dB gain;

- the 11th order low-pass filter LBPF with a Butterworth characteristic and the 2000 MHz bandwidth;
- the envelope detector AM_DMOD_ED.

The spectrum of the IR-UWB signal after its passing through the bandpass Chebyshev filter of the receiving device is shown in Fig. 35a and the time response of the same signal is shown accordingly in Fig. 35b.

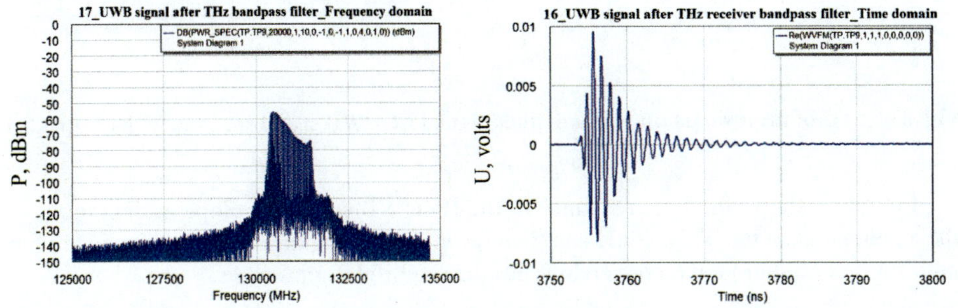

a) b)

Fig. 35. The spectrum of the IR-UWB (a) and its time response (b) at the output of bandpass Chebyshev filter of the receiving device simulation nodel

The spectrum of IR-UWB signal and its time response at the output of the frequency mixer (downconverter) of the transmitting device are shown in Fig. 36a and in Fig. 36b accordingly.

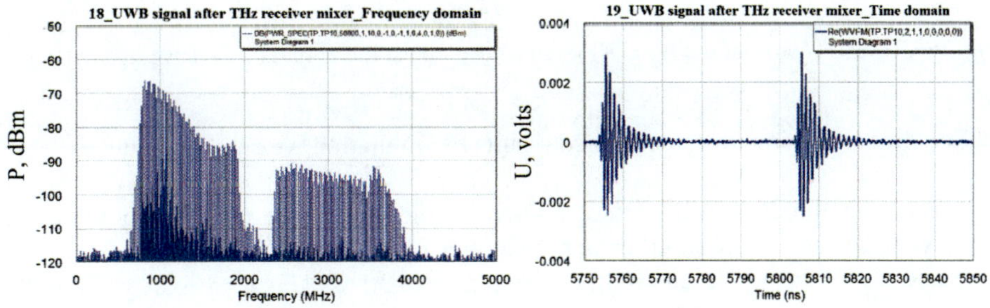

a) b)

Fig. 36. The spectrum of IR-UWB signal (a) and its time response at the output of the frequency mixer (b)

After passing the RF amplifier, the IR-UWB signal is filtered out from the conversion products generated in the mixer by lowpass filter, the AFC of which is shown in Fig. 37a. Accordingly, the IR-UWB spectrum after filtration at the output of lowpass filter is shown in Fig. 37b.

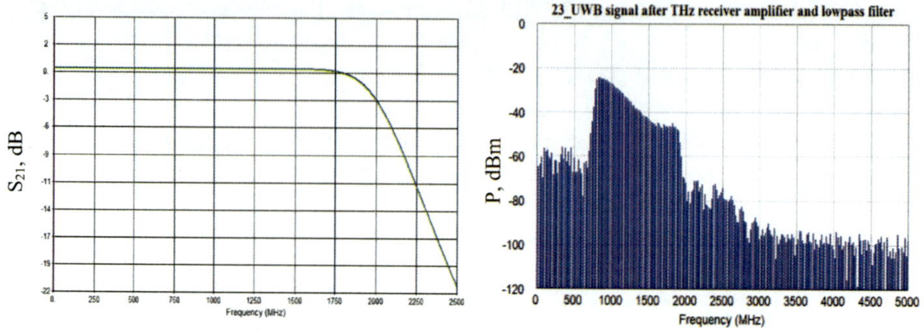

a) b)

Fig. 37. AFC of the lowpass Butterworth filter (a) and IR-UWB signal spectrum at its output (b)

Figure 38 shows the time response of the IR-UWB signal envelope at the output of the envelope detector. This signal can then be applied to the digital part of the receiving path for demodulation and conversion into the required digital code.

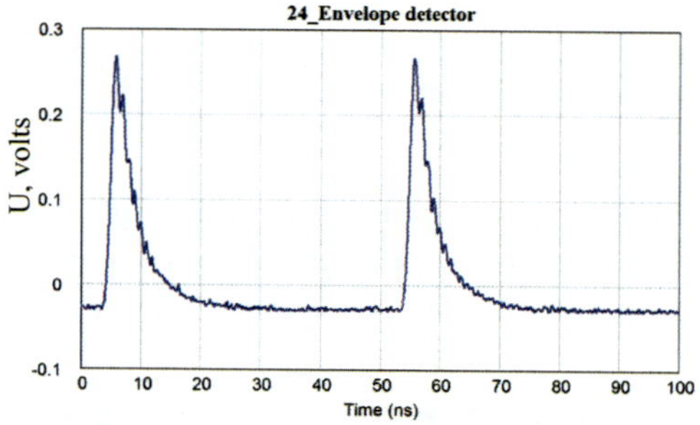

Fig. 38. Time response of the IR-UWB signal envelope at the output of the envelope detector.

So, after analyzing aforementioned results of simulation we can conclude:

1) comparing the IR-UWB at the terahertz radio link input (Fig. 29) and output (Fig. 35b), we see a significant broadening of the IR-UWB signal in the clock range, which is more than 10 times: from 2 ns to 30 ns. This is primarily due to the narrowing of the IR-UWB signal spectrum in the bandpass filters of the transmitting and receiving devices;
2) it is important to note that the radio signal at the output of the terahertz transmitter can not be classified as ultra-wideband signal by the criterion of the relative width of the signal spectrum, because its relative frequency band Δf/f at −10 dB level is less than one 1%. In this case, this means that it is necessary, firstly, to reduce the duration of the Gaussian monocycle, for example, to 50–100 ps, and secondly to significantly expand the frequency band of the bandpass filters of the transmitting and receiving devices approximately from 130 to 160 GHz;

3) the relocation of the IR-UWB signal to terahertz frequencies in the transmitter and the reverse relocation in the receiver sets new challenges in the development of terahertz radio links, namely in the design and manufacturing of the:

- wideband terahertz mixers (upconverter and downconverter);
- bandpass filters with a wide bandwidth;
- wideband power amplifiers;
- wideband low-noise amplifiers,
- heterodynes with high frequency stable.

7 Conclusions

1) Based on the results of research, we can conclude that the most suitable and promising frequency bands for high-speed ultra-wideband wireless telecommunications systems development are 110–150 GHz and 220–270 GHz, where it is possible to use large frequency bands with a width of 40–50 GHz for bandwidth capability increasing, ensure protection against signal detection and unauthorized access to transmitted information.
2) Based on analysis of the existing electronic element base and theoretical investigations, the design of the main modules and the entire terahertz transceiver of the telecommunications system with gigabit data rate in the frequency range of 130–134 GHz was accomplished. The main modules of this system which was designed are frequency converters with subharmonic pumping, heterodyne, which uses a highly stable quartz oscillator and the chain of multiplying and amplifying stages, bandpass septum filter which using a thin metal plate in the E plane of the waveguide channel of 1.6×0.8 mm^2 cross section.
3) For the first time in practical terms, the main modules of the terahertz transceiver were manufactured and experimental studies of a laboratory testbed of a digital simplex radio system in the frequency range 130–134 GHz and a digital modem with bandwidth capability up to 1200 Mbps was accomplished. The technical characteristics of the manufactured terahertz receiving and transmitting devices are investigated. The developed terahertz transceiver can provide a significant breakthrough in the development of the telecommunications industry of Ukraine. Manufactured laboratory testbed of a terahertz radio system with gigabit bandwidth capability could be used for telecommunications systems and networks deployment, such as high-speed radio relay lines for 5G mobile networks to transmit and receive digital information with a data rate at least 1 Gbps in the frequency range 128–134 GHz and distance within 1 km.
4) The results of laboratory tests showed that the use of the lower part of the terahertz frequency range (i.e. 130 GHz band) with a bandwidth of 24 MHz allows transmit three TV signals of DVB-C standard with total data rate of 125 Mbps with high subjective quality of playback of all TV programs through a terahertz transceiver with common local oscillator for the transmitting and receiving devices.
5) The simulation results confirm the possibility of transmitting the IR-UWB signal through the terahertz radio communication system in the frequency range 130.4–131.5 GHz.

Thus, the creation of devices, systems and networks of the terahertz range can radically change the principles and theoretical approaches to the construction of transport distribution networks of 5G and 6G mobile communication, military control systems etc. In addition, there are huge prospects for their use in various other fields of human activity - from aerospace and meteorology to medicine and security.

Further directions for the terahertz range telecommunications development authors see in:

1) the development of power amplifiers with output power of tens mW for the transmitting devices and low-noise amplifiers for the receiving devices. It helps to increase the energy potential of the radio link and thus increase the distance between the corresponding stations more than 1 km;
2) optical technology elements (highly stable lasers, modulators, filters, combiners and power dividers etc.) step-by-step application for the transmitters and receivers design. It will allow, firstly, reduce the dimensions of the transmitting and receiving devices, and secondly, ensure its connection with existing fiber-optic communication lines and thirdly, simultaneously generate several transmission channels with a transmission data rate of up to tens of Gbps in each of them;
3) practical exploration of other parts of the terahertz spectrum, in particular 220–250 GHz, 290–310 GHz, 375–450 GHz frequency bands;
4) practical investigation of IR-UWB signal transmission through the terahertz radio communication system.

References

1. Ilchenko, M.Ye., Narytnik, T.N., Fisun, A.I., Belous, O.I.: Terahertz range telecommunication systems. Telecommun. Radio Eng. **70**(16), 1477–1487 (2011). https://doi.org/10.1615/TelecomRadEng.v70.i16.60
2. Narytnik, T.N.: Possibilities of using THz-band radio communication channels for super high-rate backhaul. Telecommun. Radio Eng. **73**(15), 1361–1371 (2014). https://doi.org/10.1615/TelecomRadEng.v73.i15
3. Ilchenko, M.Ye., Kalinin, V.I., Narytnik, T.N., Didkovsky, R.M.: Potential performance of the communication system using autocorrelation reception of shift-keyed noise signals. Telecommun. Radio Eng. **73**(11), 955–976 (2014). https://doi.org/10.1615/TelecomRadEng.v73.i11
4. Ilchenko, M.Ye., Narytnik, T.M., Idris, A.A.I., Didkowsky, R.M., Turabi, O.: Theoretical and practical aspects of the use of stochastic signals in telecommunication systems. Telecommun. Radio Eng. **73**(20), 1791–1817 (2014). https://doi.org/10.1615/TelecomRadEng.v73.i20
5. Ilchenko, M.Ye., Narytnik, T.N., Fisun, A.I., Belous, O.I.: Conception of development of millimeter and submillimeter wave band radio telecommunication systems. Telecommun. Radio Eng. **67**(17), 1549–156 (2008). https://doi.org/10.1615/TelecomRadEng.v67.i17
6. ECC Recommendation (18)01: Radio frequency channel/block arrangements for Fixed Service systems operating in the bands 130–134 GHz, 141–148.5 GHz, 151.5–164 GHz and 167–174.8 GHz. https://docdb.cept.org/download/a5533f97-5a92/Rec1801.pdf
7. ECC Recommendation (18)02: Radio frequency channel/block arrangements for Fixed Service systems operating in the bands 92–94 GHz, 94.1–100 GHz, 102–109.5 GHz and 111.8–114.25 GHz. https://docdb.cept.org/download/476602d1-6145/ECCRec1802.pdf

8. ECC Report 282: Point-to-Point Radio Links in the Frequency Ranges 92–114.25 GHz and 130–174.8 GHz. https://docdb.cept.org/download/502e984f-9be3/ECCRep282.pdf
9. China puts "6G" test satellite into orbit [electronic resource]. https://www.mobileworldlive.com/featured-content/home-banner/china-puts-6g-test-satellite-into-orbit. Accessed 15 Dec 2020
10. Katayama, K., Takano, K., Amakawa, S., Hara, S., Kasamatsu, A., Mizuno, K., Takahashi, K., Yoshida, T., Fujishima, M.: A 300 GHz CMOS transmitter with 32-QAM 17.5 Gb/s/ch capability over six channels. IEEE J. Solid-State Circuits **51**(12), 3037–3048 (2016). https://doi.org/10.1109/JSSC.2016.2602223
11. Jia, S., Yu, X., Hu, H., Yu, J., Guan, P., Da Ros, F., Galili, M., Morioka, T., Oxenlowe, L.K.: THz photonic wireless links with 16-QAM modulation in the 375–450 GHz band. Opt. Express **24**(21), 23777–23783 (2016). https://doi.org/10.1364/OE.24.023777
12. Dan, I., Ducournau, G., Hisatake, S., Szriftgiser, P., Braun, R.-P., Kallfass, I.: A terahertz wireless communication link using a superheterodyne approach. IEEE Trans. Terahertz Sci. Technol. **10**(1), 32–43 (2020). https://doi.org/10.1109/TTHZ.2019.2953647
13. Nagatsuma, T., Hisatake, S., Pham, H.H.N.: Photonics for millimeter-wave and terahertz sensing and measurement. IEICE Trans. Electron. **E99-C**(2), 173–180 (2016). https://doi.org/10.1587/transele.E99.C.173
14. Yu, X., Asif, R., Piels, M., Zibar, D., Galili, M., Morioka, T., Jensen, P.U., Oxenkowe, L.K.: 400-GHz wireless transmission of 60-Gb/s Nyquist-QPSK signals using UTC-PD and heterodyne mixer. IEEE Trans. Terahertz Sci. Technol. **6**(6), 765–770 (2016). https://doi.org/10.1109/TTHZ.2016.2599077
15. Narytnyk, T.M.: Principles of development of the terahertz band telecommunication system based on the technology of harmonic signal as the information carrier. Telecommun. Radio Eng. **77**(16), 1423–1440 (2018). https://doi.org/10.1615/TelecomRadEng.v77.i16
16. Narytnyk, T.M., Lutchak, O.V., Osypchuk, S.O., Uryvskyi, L.O.: Criteria and algorithms for shaping of the signal-code sequences on the basis of Wi-Fi technology at deployment of the terahertz band telecommunication system. Telecommun. Radio Eng. **75**(20), 1823–1839 (2015). https://doi.org/10.1615/TelecomRadEng.v74.i20.50
17. Ilchenko, M.Ye., Narytnyk, T.M., Radzikhovsky, B.M., Kuzmin, S.E., Lutchak, O.V.: Development of the transmitting and receiving channels for terahertz band relay systems. Telecommun. Radio Eng. **74**(11), 981–998 (2015). https://doi.org/10.1615/TelecomRadEng.v74.i11
18. Ilchenko, M.Ye., Narytnik, T.N., Kuzmin, S.Ye., Fisun, A.I., Belous, O.I., Radzikhovsky, V.N.: Transceiver for 130–134 GHZ band and digital radiorelay system. Telecommun Radio Eng. **72**(17), 1623–1638 (2013). https://doi.org/10.1615/TelecomRadEng.v72.i17.70
19. Ilchenko, M.Ye., Narytnyk, T.M., Denbnovetsky, S.V., May, A.V., Lutchak, O.V.: Simulation of functional units of the terahertz band transmitting and receiving radio systems. Telecommun. Radio Eng. **75**(11), 1027–1039 (2016). https://doi.org/10.1615/TelecomRadEng.v75.i11
20. Ilchenko, M., Denbnovetsky, S., Narytnik, T., Lutchak, O., May, O., Fisun, A., Bilous, O.: Design of the 290…310 GHz frequency range integral receiver. Telecommun. Radio Eng. **76**(15), 1379–1390 (2017). https://doi.org/10.1615/TelecomRadEng.v76.i15
21. Avdeyenko, G.L., Narytnyk, T.N.: Investigation and modeling of transmission of the DVB-C standard television signals and the pulse ultra-wideband signal via the terahertz band radio link . Telecommun. Radio Eng. **77**(17), 1517–1533 (2018). https://doi.org/10.1615/TelecomRadEng.v77.i17

22. Ilchenko, M.Ye., Narytnik, T.N., Cherepenin, V.A., Kalinin, V.I., Chapursky, V.V.: Wireless terahertz communications with spectral modulation of UWB noise signals. Telecommun. Sci. **3**(2(5)), 39–43 (2012). https://doi.org/10.20535/2411-2976.22012
23. Avdeyenko, G., Bunin, S., Narytnik, T.: Transmission and reception of UWB signals based on antenna decoding. Inf. Telecommun. Sci. **9**(2), 27–35 (2018). https://doi.org/10.20535/2411-2976.22018.27-35

5G Slicing and Handover Scenarios: Compulsoriness and Machine Learning

Andriy Luntovskyy[1(✉)], Bohdan Shubyn[2(✉)], Taras Maksymyuk[2(✉)], and Mykhailo Klymash[2(✉)]

[1] BA Dresden University of Cooperative Education (Saxon Study Academy), Hans-Grundig-Str. 25, 01307 Dresden, Germany
Andriy.Luntovskyy@ba-dresden.de

[2] ITRE, Lviv Polytechnic National University, Professorska 2, Lviv 79013, Ukraine
mykhailo.m.klymash@lpnu.ua

Abstract. In the given paper the backgrounds for the deployment of 5G in EU countries are discussed. The most important requirements for 5G networks are examined like slicing, wide network inter-operability (with Wi-Fi 6, IoT devices, LoRa WAN etc.). The operation in 5G micro-cells of small private providers is favoured. With the aim of compulsoriness so-called Blockchained 5G Slices are used. The authors investigated the appropriate Blockchain (BC) architectures, recent BC types, BC applications, platforms as well resource consumption. So-called Smart Contracting for 5G networks and DAO technology can be taken into account for use as 5G Blockchained slices. Intelligent 5G handover and roaming within the 5G Networks can be provided using ML standards. Artificial Intelligence for Handover Optimization is discussed. The demarcation of our research area is given. Recurrent neural networks as well as their modelling and training processes are represented. Modelling with a neural network and prediction of the user number in a micro-cell are investigated. Recurrent neural networks, based on GRU cell, can optimize the handover in 5G cellular structures. The method is extended for the hierarchical roaming in backhauls, mega- and giga-cells.

Keywords: 5G · WSN · RFID · NFC · IoT · Network slicing · Machine learning · Enabling technologies · Hierarchical cell concept and handover · Digital twins · Tele-presence · QoE · NFV · Energy-efficient protocols · Low data unit costs

1 Backgrounds: Deployment of 5G in EU Countries

5G deployment was started last years in multiple countries of EU, in USA and South Korea. In 2019–2020 in Germany after testing, the 5th generation mobile network was launched (Deutsche Telekom, Vodafone, Telefonica Germany) [1–9]. Suisse has assigned in 2019 the 5G frequency bands to three major providers: Swisscom, Sunrise and Salt. In July 2019, 334 antenna towers with 5G frequencies began to operate in the country [1–9]. Through the program "Horizon 2020", the European Commission is investing € 700 million in research and innovation funding, related to 5G nets. The 5G nets provide IoT (Internet of Things), remote industry and events monitoring, plants automation,

P. Vorobiyenko et al. (Eds.): IPF 2020, LNNS 212, pp. 223–255, 2021.
https://doi.org/10.1007/978-3-030-76343-5_12

facility management, safety in the cities, assist health-related care (ambulance, surgery), touristic and more else. The distinguishing features of 5G nets are as follows [1–9]: use of conventional 4G nets; DR for the first time being up to 20 GBit/s; use of higher frequency areas (24.5–29.5 GHz) and still up to 60 GHz; real-time transmission for tele-presence by short latency of approx. 1 ms; QoE and compatibility of multiple machines and devices (so-called ubiquitous connectivity); energy consumption decreasing per transmitted bit (1/1000); 90% lower power consumption per mobile service.

1.1 5G Basic Technologies

The first experiments with 5G nets began in 2014. The main 5G requirements are as follows [1–9]: use of existing 4G-insfrastructure with augmentation via flexible WLAN conform communication everywhere under international voting and conventions; to medium term obtaining of DR = 10 GBit/s; this DR corresponds to up-to-date needs to MM-content download; tiny latencies, real time, inter-operability, services without human intervention; wide use of available frequency bands: mm-Band F = 30 up to 300 GHz (partially and questionable); inter-operability with further mobile and wireless radio networks.

The advanced antenna technique MIMO (Multiple Input – Multiple Output) has been already deployed in diversity network technologies since 2008 like WiMAX 802.16a/d/e/m, WLAN 802.11n/ac/ad/ax, LTE etc. MIMO-antennas allows nowadays communication in3D-mode with NTx = 16–64 transmitting and NRx = 16–64 receiving antennas (Fig. 1). This scenario depicts so-called multi-user MIMO (MU-MIMO) in MxKxM-mode [10–14].

The downlink (DL) with DR = 10GBit/s and uplink (UL) with 1 Gbit/s became available. This DR = 10GBit/s is about ×100 more in contrast to DR = 100 MBit/s of LTE. The prognosis is as follows: in 2025 up to 100 milliards devices will be IPv6-driven working, partially with 5G [1–14]. The architecture and components for 5G Mobile Network are shown in Fig. 2. In the given 5G scenario, the multimodal access and SDN core are represented. Based on the representation the 5G construction points of gratitude are considered as follows. The wide inter-operability to "Wi-Fi world" is foreseen. Furthermore, the 5G architecture is steady based on predecessors: 4G and 3G with appropriate hierarchical cell structure and integrated satellite links. The technology will augment the existing GERAN, UTRAN, SAE and IMS (refer Fig. 2).

1.2 Network Slicing

5G-SDN is based on open interfaces and proven industry standards for hardware like Open Stack, which can be used for GPRS, EDGE, UMTS, HSPA, LTE, Wi-Fi5/6/WLAN/DIDO, NFC, BT, ZigBee, 6LoWPAN too.

The next one is so-called "Network Slicing" (refer Fig. 2). The access and core networks control the creation, orchestration, deployment and operation of 5G and are divided to the so-called "slices" (or partitions). The "Network Slicing" concept means [1–9]: Shorter latency for mobile network access (under 1 ms) for real-time scenarios; Quickly and efficiently creation and providing of differentiated access for different industry requirements (like L3 DiffServ); Better QoE management due to flexible

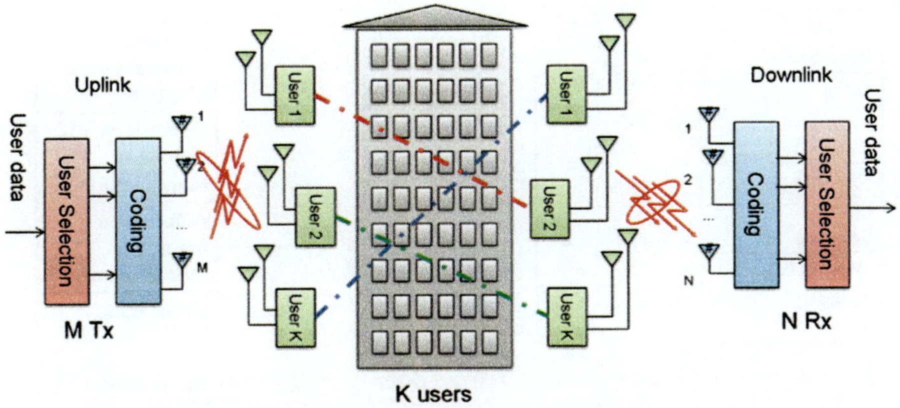

Fig. 1. Up-to-date 3D-MIMO antennas for radio networks in MxKxNmode

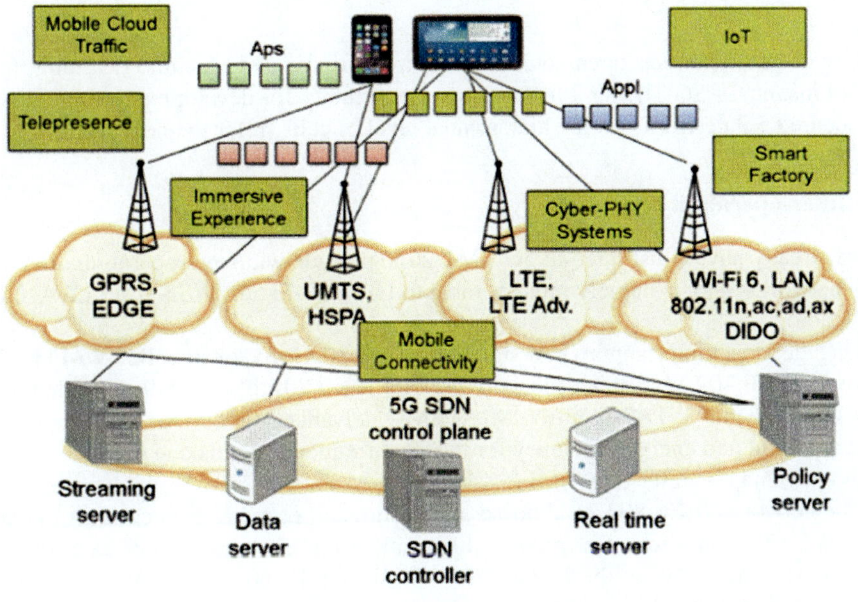

Fig. 2. Legend: SDN – Software-Defined Networking; IoT – Internet of Things

deployment and modification of the necessary "slices" (partitions); Seamless and fast broadband connections. The principle of Network Slicing is shown in Fig. 3.

Network Slicing is guaranteed via the used SDN (Software-Defined Networking) and wideNFV (Network Functions Virtualization).The individually profiled 5G apps can be flexibly deployed under guaranteed QoS/ QoE. For example, the rescue services can start these apps during an avalanche shift or other critical event [1–14].Different apps are supported via the available slices, which are involved taken into account the situation and QoS requirements.

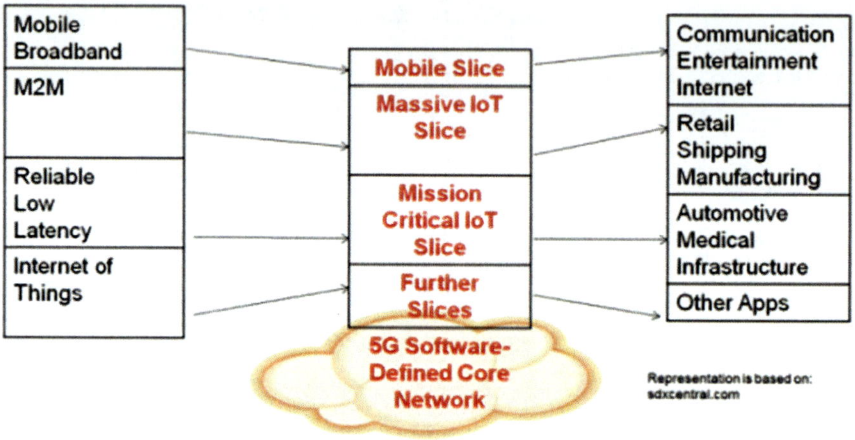

Fig. 3. 5G Network Slicing: example

Use of Blockchain as open-source for compulsoriness and Machine Learning [6, 7, 35, 36] means for 5G Slicing a further great opportunity for development of intelligent and secured 5G networks of any hierarchical level of cells (refer the next sections).

1.3 Inter-operability

As we already have constituted it, 5G/ IMT 2020 provide wide interoperability to other radio networks, and, is namely interoperable to LTE, Wi-Fi 5/6, WSN, LoRa WAN and SAT-Radio.

One the most important is a new standards IEEE 802.11ax, or so-called Wi-Fi 6. The networks IEEE 802.11ac and, especially Wi-Fi6/ IEEE 802.11ax (2019) provide DR = 1–11 Gbit/s, better WPA3-security, MU-MIMO, advanced OFDMA. Wi-Fi 6 possess higher spectral and energy efficiency for the used frequency bands due to use of a set of combo-advantages (refer Fig. 4).

The networks IEEE 802.11ad possess the following properties: in contrast to traditional WLAN is only for a few meters; this results from the high absorption of oxygen at 60 GHz; a large BW to the higher DR = 7 GBit/s; the 60 GHz band is from 57 to 66 GHz, and is divided by a channel spacing of 2160 MHz in four channels having a bandwidth of 1760 MHz. So-called piconets WSN, Bluetooth, 6LoWPAN etc. are also inter-operable to 5G [1–14, 17–21].

The further inter-operability examples are considered below.

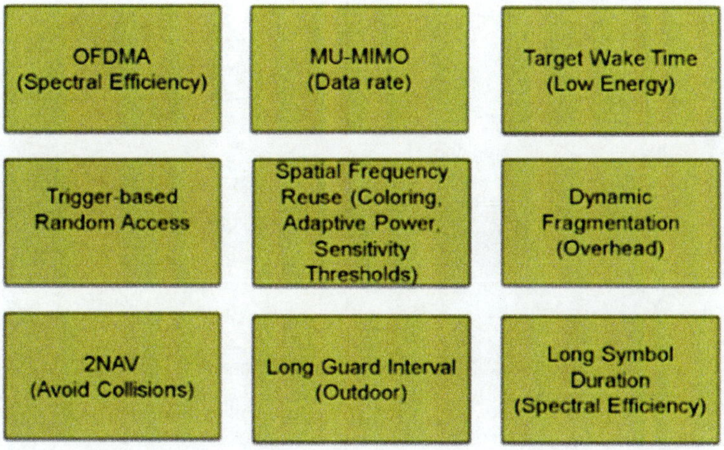

Fig. 4. Advances for IEEE 802.11ax/Wi-Fi 6 in comparison to IEEE 802.11

1.4 Example for 5G and IoT Inter-operability

Which next example for 5G and IoT [17–21] inter-operability do we know? As next example, a smart mobile device can be considered as "All-in-One" (Fig. 5).

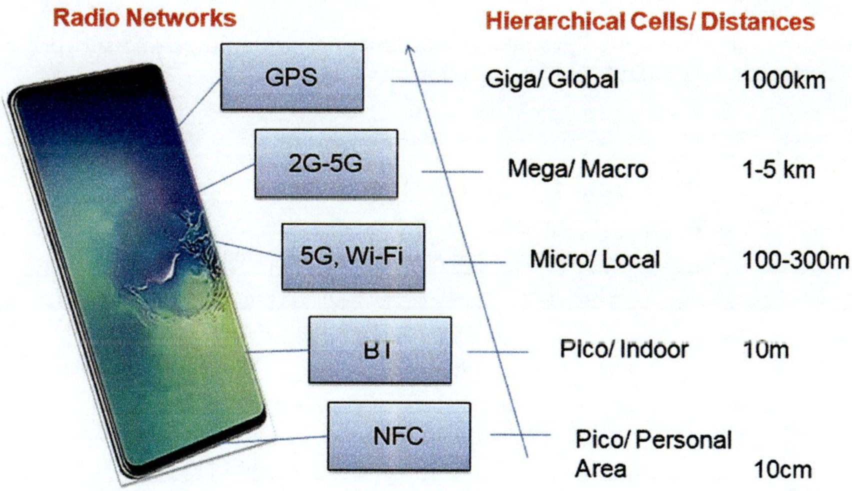

Fig. 5. Smartphone as "All-in-One"

5G networks use the convenient hierarchical cell concept (pico-cells, micro-cells, mega-cells and giga-cells, refer Fig. 6). These networks are completely interoperable (Table 1).

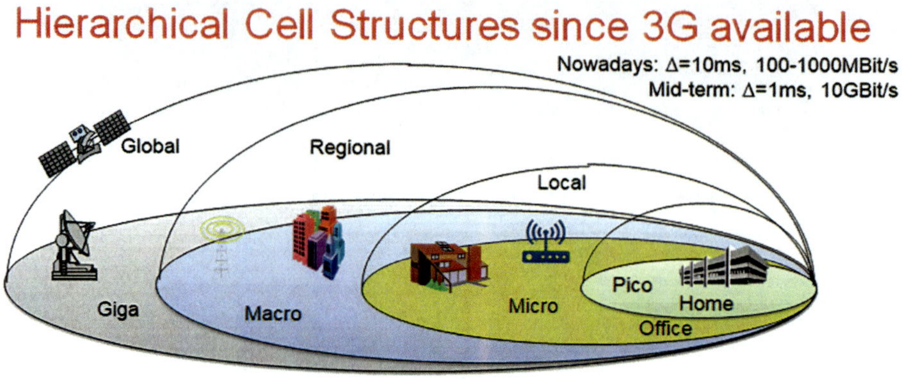

Fig. 6. 5G networks with the hierarchical cell concept

The above-mentioned known 5G Slicing concept can be also realized on the following tiers (refer Sect. 4 and compare with Fig. 6):

1. Macro-provider tier: a central instance, it means a large provider with large cells and backhauls, the central instance controls the cell hierarchy
2. Micro-provider tier: i.e. the small private providers for 5G networks in micro-cells, which are ordered to a central instance (a large provider with large cells and backhauls)
3. Partition tier: slices, modules and virtual functions within a private network with decentralized P2P and M2M organisation [10–14].

Table 1. Overview of hierarchical cells

Type	Distance	Mobility (km/h)	Deployment by …
Giga-cell	~1000 km	~4700 (~1,3 km/s)	International providers, satellite radio
Macro-cell	~1–5 km	up to 500	National and regional providers
Micro-cell	~100–300 m	up to 120	Metropolitan areas, city districts, campus, office area
Pico-cell (Indoor)	~10 m	~10	Hotspots at railway stations, airports, hotels, restaurants, clubs, home area
Pico-Cell (Personal area)	~10 cm	Stationary	PAN, Wearable, Smart Stuff

1.5 Example for 5G Inter-operability with LoRa WAN

LoRa WAN devices are used for so-called low-power wide-area Long Range communication (LPWAN). The Semtech Corporation acts as s pioneer of LoRA Alliance [17]. The LoRa protocol is based on chirp spread spectrum modulation techniques (CSSM). The LPWAN is operated in license-free frequency areas like SRD and ISM: 2.4 GHz, 868/915 MHz, 433 MHz, 169 MHz. The basic modules are available as open source software (OSS). LoRa WAN provides asymmetrical links and energy efficiency for M2M scenarios (Fig. 7).

The uplink (UL) ranges are d ~ 10 km (from the end device to the network). However, low DR by approx. 292 bps to 50 kbps is only available. The middle ranges for LoRa extend from 2 km in urban areas up to 40 km in rural areas. Multiple LoRaWAN networks are deployed Europe- and worldwide: they provide IoT coverage in the Netherlands, Switzerland, South Korea etc. These countries are the first countries with area-wide LoRaWAN coverage. As an example, Swisscom LPN can be mentioned:

- Frequency band from 863 to 870 MHz (SRD Band for Europe).
- Gateway PTx = 500 mW (27 dBm).

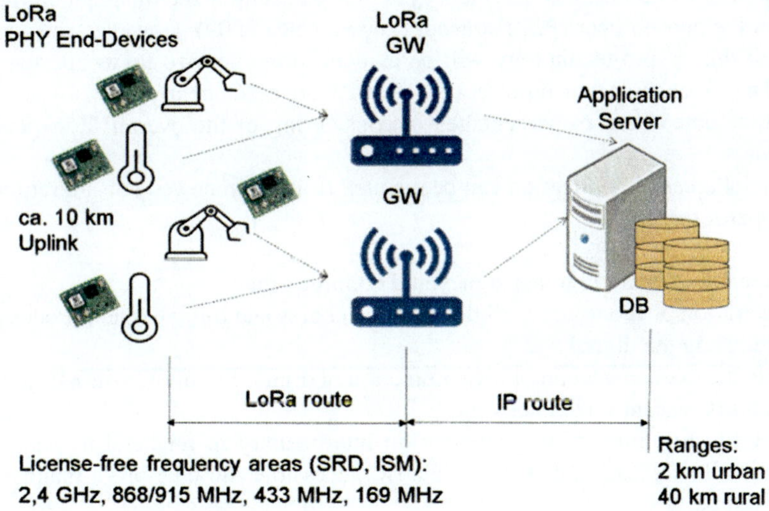

Fig. 7. LoRA Architecture (own representation, based on LoRA Alliance)

1.6 Operation in 5G Micro-cells: Small Private Providers Favoured

5G networks are widely operated in micro-cells with limited supply radius about 150 m. This means a big advantage for use in urban areas with higher user density. The "hotspots" should be supported. Such 5G microcells can be placed on lanterns, light gas columns or house walls and roofs for sparing of the CAPEX. As deployment examples in microcells the festivals, stadiums, event centres and concert halls, theatres, culture sites and institutions can be mentioned.

Considering the new development of 5G apps and software solutions this might be unbeatable argument for the concept. The current application fields in this regard are as follows [1–14]: Robots and "5G digital twins" are supported by 5G; 5G-Agriculture with self-piloted combined harvesters; Self-driving cars (refer Google's Waymo, Tesla Autopilot); Non-stop driving over intersections in the city; Car Navigation and Truck Platooning under shortened distances, control systems for simultaneous acceleration and braking of trucks; "Smart Grid", i.e. combination of energy-efficient IT with so-called "green electricity" under reduced risk of phase losses. The minimal phase shift in the AC is reduced due to a short latency. Compare it: 1 ms means an insignificant phase shift $\varphi = 18°$; 10 ms means a huge phase shift $\varphi = 180°$.

2 Blockchained 5G Slices

2.1 BlockchainedSecurity

Blockchain (BC) is a cryptographically distributed computer network application [22–34] supporting a decentralized payment system and decentralized financial online transactions in the peer-to-peer (P2P) concept (known since 2009). However, the economic success of this crypto-technology will be evident in the next 10 up to 20 years (refer Fig. 8). The Blockchain expenditures (2018–2022) are permanently growing worldwide. They are depicted here in percentages in considering of the overall IT expenditures worldwide.

Potential areas of application can be resulted from the renewed consideration of the main properties of BC:

1. Sustainability, general transparency and commitment
2. Acceleration of economic workflows, value chains and digitalization processes (so-called "IT in the digital age")
3. Completeness and accountability protection in digital Supply-Chain-Management
4. Scalability, decentralized network
5. Possibility to eliminate the need for an intermediary in financial transactions by making all necessary documents via a BC accessible and fake-proof manner for all involved parties
6. Compulsoriness, accountability and liability, especially, for 5G Slicing and IoT solutions.

The deployment of Blockchain technology speaks for a decentralized financial system. The advantages of such a solution are obvious. BC crypto-technology is also well suited to supporting current crypto-currencies (such as Bitcoin, Ethereum, Ripple, Litecoin, ZCash, Monero, Stellar etc.).

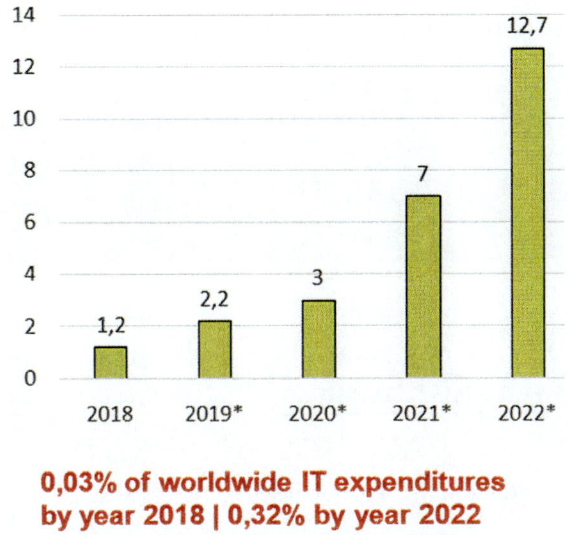

Fig. 8. Prospective Blockchain expenditure growth worldwide (in billions USD)

Digital payment instruments under use of cryptographic principles, which built a fast growing market nowadays, are often created via deployment of the hash algorithms titled SHA256 (FIPS NIST 2008). Such algorithms (refer Table 2) were introduced since 2002. The first representing algorithm was titled SHA-1. The successor was SHA-2 (including the further modifications like SHA-224, SHA-256, SHA-384, SHA-512, SHA-512/224 and SHA-512/256). The hash size is variable between 224, 256, 384 and 512 bits by the parameterized round number 64 or 80.

Table 2. Overview of NIST hash algorithms [21–23]

Hash algorithm	Message Size (bits)	Block Size (bits)	Word Size (bits)	Message Digest Size (bits)
SIIA-1	<264	512	32	160
SHA-224	<264	512	32	224
SHA-256(Blockchain)	<264	512	32	256
SHA-384	<2128	1024	64	384
SHA-512	<2128	1024	64	512

Table 3 contains an overview of current crypto-currencies.

Therefore, there is also multiple BC types, which can be used in public space, by enterprises, for communities or privately (communities, crowd funding). Table 4 depicts the BC types, ordered by their access model with some provided BC implementation examples.

Table 3. Overview of crypto-currencies and market growth (based on: https://coinmarketcap.com/)

#	Currency	Shorthand symbol	Launch	Mining available
1	Bitcoin	BTC	2009	yes, SHA-256
2	Ethereum	ETH	2015	yes, Ethash
3	Ripple	XRP	2013	no
4	Litecoin	LTC	2011	yes, Scrypt
5	Ethereum Classic	ETC	2016	yes, Ethash
6	Monero	XMR	2014	yes, CryptoNight
7	Dash (formerlyDarkcoin)	DASH	2014	yes, X11

Table 4. BC Types, ordered by Access Model

<table>
<tr><th colspan="2">Blockchain Types</th><th>Read OP</th><th>Write OP</th><th>BC example</th></tr>
<tr><td rowspan="2">Open</td><td>No access limitations</td><td>Everyone</td><td>Everyone</td><td>Bitcoin, Ethereum</td></tr>
<tr><td>Limited access</td><td>Everyone</td><td>Authorized</td><td>Sovrin</td></tr>
<tr><td rowspan="2">Limited</td><td>Consortium-limited access</td><td>Authorized</td><td>Authorized</td><td rowspan="2">Corda, Hyperledger, Quorum</td></tr>
<tr><td>Private limited access</td><td>Completely private or authorized</td><td>Network provider</td></tr>
</table>

The following BC types can be chosen: (1) BC with free access; (2) public access limited BC; (3) private access limited BC, or (4) none based on the Wuest-Gervais flowchart for decision-making support [27].

2.2 Constructing a BC

The general "chaining principle" for BC construction is depicted in Fig. 9. Hash and signature algorithms like SHA-256 and RSA are commonly used for chaining, authentication and integrity verification.

The financial transactions are used as a rule as blocked transactions. Therefore, the BCs possess in the practice the following detailed structures.

The central element of a BC is a block, which contains the transactions that are carried out between the participants within a P2P network. According to Distributed Ledger Technology, the transactions are saved in chronological order. A block has only a limited storage capacity at a time; when the maximum number of transactions is reached, the current block is closed and a new one is created. In addition to the transactions, a block contains a header area with various attributes. The attributes are usually the

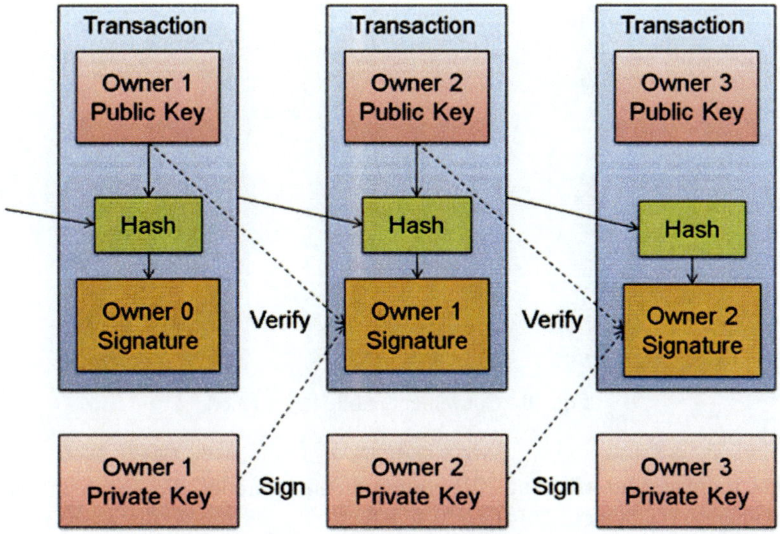

Fig. 9. General principle BC

block number, a time stamp, the previous block key value (hash) and, depending on the technology, a sample and a nonce optionally.

When a block is completed, a unique key value (hash) is created from the combination of the transactions and the header attributes. This hash is used for validation and is reused in the following block as part of the header area. The individual key values (hashes) of the blocks are interdependent as well as the blocks are cryptographically chained.

The key values for the block: previous key, common key of the transactions and current key are generated from the considered block and called the "Hash Values" according to thc SHA-256 algorithm from NIST.

By definition, so-called SHA-256 hash algorithm delivers a hash, i.e. an output with a fixed size, from an input of any size and shape. The hash functions are unidirectional, so that conclusions about the origin are not available.

It means that each change for each character in a string in the initial data (i.e. transactions) creates a completely new hash value.

In addition to the SHA-256 hash algorithm, the public key method (Fig. 10) is another important component (e.g. RSA) for integrity control of the data (transactions) within a BC. The method can be used both to verify the sender and to encrypt a transaction. A transaction is the sending of data via the BC network, for example, the digital payment of an invoice under use of crypto-currency (BTC, ETH, ZEC, XMR).

2.3 Blockchain Architecture

Decentralized, cryptographically secured and unified blocks, their chains and transactions are grouped under a general, global public ledger (account), the structure of which is as follows. The Blockchain, as a networked Public Ledger, consists of participating

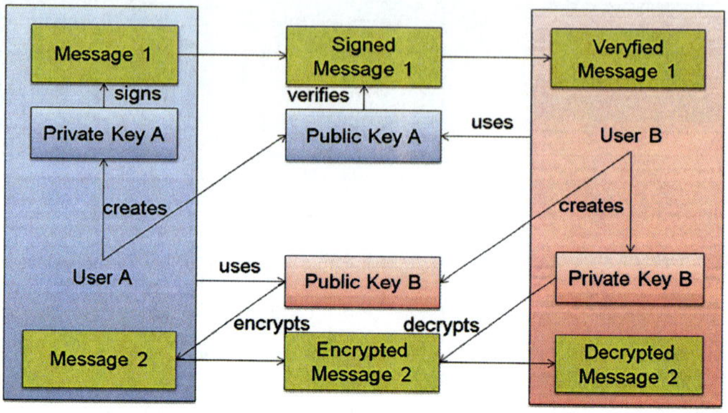

Fig. 10. Convenient public key method

nodes that represent an efficient P2P communication model. Typical features of the Blockchain are as follows [22–34]:

- Redundancy and synchronization
- Cryptographic hash procedures for integrity assurance and attack safety
- Decentralized management and control of the Blockchain
- Network subscribers are also referred to as Nodes (Full-Nodes, Miners, Validators) and run redundantly with mutual synchronization
- In addition, large block volumes can cause the "Big Data" problem.

Figure 11 depicts the structure for an exemplary BC. The defining block chain (green colour) consists of the longest sequence of secured blocks from the origin (genesis, red colour) to the current block. Alternative chains (orange colour) became orphan as soon as they are shorter than another chain.

Within the Blockchain architecture between the following basic components can be distinguished (Fig. 12): the simple Nodes, the Full-Nodes, and Miner/Validator [22–34]:

1. **Nodes:**
   Each Blockchain participant (computer, smartphones, tablets, or even clusters) is qualified as Node, if he has installed the corresponding software, which runs based on the Bitcoin protocol or the program code of Bitcoin.
2. **Full-Nodes:**
   - A Node with full local copy of the Blockchain
   - Checking for so-called "consensus rules" (consensus method).
3. **Miner/Validator:**
   - The individual participants or mining pool (high resource requirements regarding hardware and energy consumption)
   - Finalising of blocks (Miner – block generation, Validator – proving)

- Externally they act each like a large participant, but in fact, many small blocks are generated for payment in fractions of the crypto-currency units.

However, the following problems occur during the Blockchain operation:

- Enormous energy consumption due to "mining" of crypto-currencies (processing of the hash blocks via its algorithmic complexity).
- Exponential memory growth (including capacity migration between USB media, smartphones, PC, storage media such as SAN / NAS, as well as cloud storages)
- Cryptographic data security is guaranteed, but privacy issues may arise. One-way out is as follows: no processing the complete Blockchain with all the transactions, but only use of excerpts of the Blockchain without a prehistory.

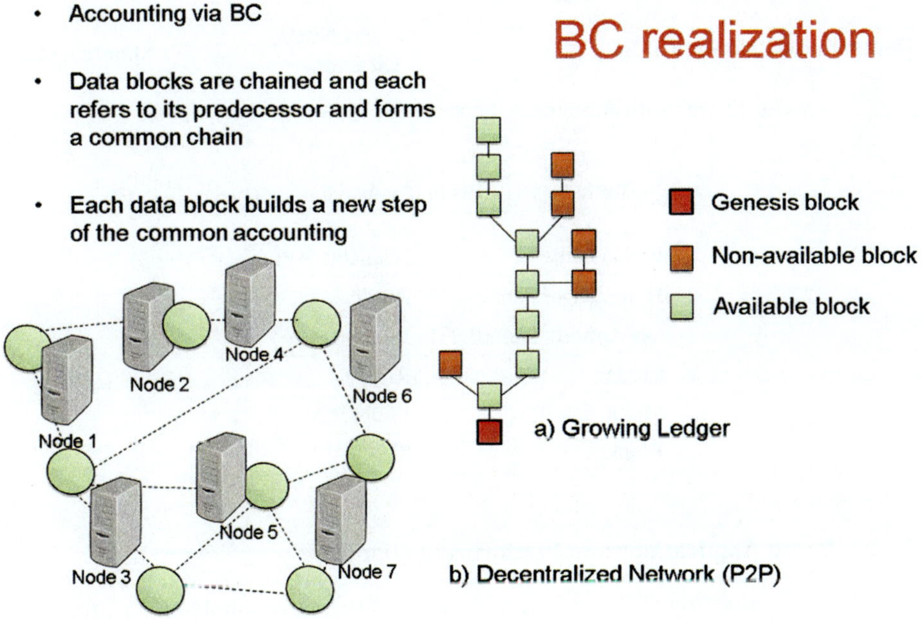

Fig. 11. Structure for an exemplary BC: a) growing ledger; b) decentralized network (P2P)

2.4 Recent BC Types

Multiple crypto-platform scan be used for BC integration within different applications: Ethereum Classic, Codius for Ripple, Hyperledger Sawtooth [22–34] etc. The following APIs are used nowadays for these multiple BC types (s. Table 5). A detailed comparison between some recent open source BCs and their advantages are given in Table 6.

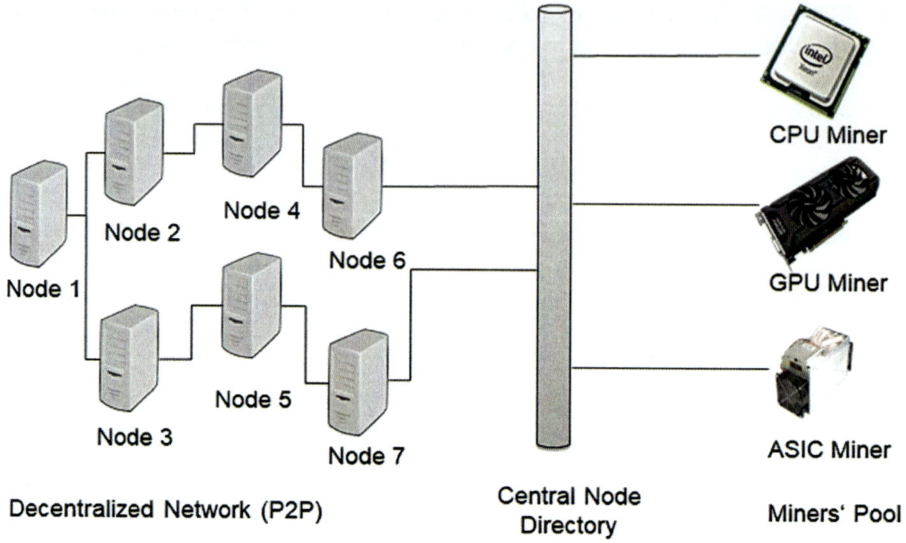

Fig. 12. BC with decentralized nodes and Crypto-Currency Mining

Table 5. BC APIs in Comparison

Blockchains	Available API
Hyperledger Fabric	REST
Quorum/Ethereum/ETH	JSON-RPC
Corda	RPC
MultiChain/BTC	JSON-RPC
Ripple	REST

2.5 BC Based Applications and Platforms at Glance

In the Smart Contracting (SC), in the logistics industry, entire supply chains, from the production of raw materials to the ready-on-market products for the end user can be represented via an appropriate BC. Every participant within the logistics chain has insight into the good conditions and WF steps. Subsequent manipulation of data afterwards is completely excluded. Therefore, the trustiness within a supply chain grows between the participants.

There are some pilot projects of a kind of micro-grids, in which the residents can sell on excess self-generated solar power directly to their neighbours. The use of BC technology can also bring some advantages in this area via the SC.

However, a main disadvantage of the BC technology, of the BC-based platforms, tools, APIs is resource-intensity and enormous energy consumption. Especially, with a view to the large energy and consumption for BC it makes sense for further research and integration consultants to develop further skills in this area. Multiple projects are

Table 6. Comparison of recent BC types (based on [22])

Blockchain Type	Hyperledger Fabric 2016	MultiChain 2015–2019	Quorum 2016
BaaS in a cloud	Suitable	Suitable	Suitable
Private or public P2P network?	Private, participation is only by prior invitation, subject to approval	Designed for private networks, subject to approval, both publicly and privately operated	Both publicly and privately operated
Open source	Open source	Open source	Open source
Modularity	Available	Available	Available
Suitability as a solution in the corporate environment	Suitable	Suitable, a modified BC version of BTC	Quorum is the opposite of MultiChain: Like MultiChain is an enterprise version of the public BTC-BC, is Quorum as an enterprise version of the public ETH
Transactions und blocks	All transactions within the BC run via previously defined, encrypted channels Programs are created in the form of "chain code" Chain code management	Public key encryption Since v2.0 SW for so-called "Smart Filters" Two consensus mechanisms: - "Proof-of-Work" (like BTC) - resource optimized Round-Robin	BASED on Ethereum/ETH Multiple alternative consensus mechanisms Programs, which are based on Quorum-BC are called analogously the ETH like "Smart Contracts"
Support via PL	Go, Java, JavaScript	Python, JavaScript, Ruby, PHP, C#	Proprietary PL: Solidity as JavaScript-Dialect
System examples	Collection of container weights by Kuehne + Nagel and Capgemini platform: monitoring and transparency of the container weights for all parties involved in a supply chain	"Air-Quality-Chain " for measured environmental data of the city of Vienna are saved in a MultiChain and made publicly available	This technology has not yet been used productively Bank and Business BC in mid-term

still in a pilot phase, but with increasing of maturity degree can be expected so that Blockchained 5G Slices can already be provided in mid-term.

A further important requirement for 5G and its applications is the data security and trust especially by risks of ransomware and other crypto-Trojans like Petya, WannaCry or GandCrab.

In opposite to standard solutions, based on PKI and combined symmetric-asymmetric encryption (RSA) and digital signatures, BC provides its own security decentralized infrastructure, which distinguish from centralized PKI or bilateral Web-of-Trust incorporated in convenient distributed apps.

In Germany, the BC cloud initiatives came from the Telekom.de and Siemens too. So-called augmented BaaS (BC as a Service) among other XaaS (Everything as a Service) from the cloud can reduce the investment risks for the enterprises in the digital age. IoT can be successfully combined with a correct digitization solution like a BC is.

MindSphere from Siemens as well as further Siemens solutions for Blockchained IoT can be used for food and beverage industry aimed to complete ensure traceability from the farm and field up to the consumers.

Furthermore, Siemens relies on the BC, which secures the tool management during repair of power plants Siemens brings transparency to the supply chain.

The increasing recognition of BC technology in the enterprise cloud environments by "the digital age" has also prompted multiple giants of IT industry, i.e. companies such as MS and SAP to offer their own BaaS among other XaaS. The following kinds of BC are integrated within these software platforms [22–34]: Hyperledger Fabric, MultiChain, Quorum, and Ripple.

A further successful BC example is based on the use of Hyperledger Fabric BC from the Hyperledger Foundation and concerns to the collection of container weights. The worldwide known logistics company Kuehne+ Nagel and the consulting company Capgemini have developed a platform for monitoring and transparency of the container weights for all parties involved in a supply chain.

MultiChain from CoinSciences Ltd. is a modified version of the BTC BC, which is specially designed for private P2P networks and can also be adapted to the respective requirements. The proven public key method is used.

Quorum by J. P. Morgan is an open source BC based on Ethereum. It can be operated both publicly and privately and offers several alternative consensus procedures. The software on the Quorum BC offers analogously Ethereum SC network. Thus, the software has some legal restrictions and cannot yet be directly deployed; probably, in mid-term as bank and business BC.

2.6 Cryptocurrency Mining and Resource Consumption by Blockchain

At least five factors are necessary and must be considered when calculating the profitability of the mining [22–34]:

- Investment (basis device costs): hardware investment for a Mining Rig
- Electricity costs: in EUR per kWh
- Energy consumption: electrical power of Mining Rig in KW
- Hash-rate: how much hash values can be computed each second?
- Mining difficulty: factor has always a new actual value!
- Pool fees: how much in % belongs to the joined pool
- Unit "reward" per Block: Bitcoin amount for each new computed block (or for other units)

- Unit price: exchange course for the crypto-currency BTC (or other units like XMR, ETH, ZEC, LTC).

There are different multipurpose and specialized devices for mining available, so-called Mining Rigs. The old good PCs or plain smartphones can be used too but under considering of the energy consumption problems. The following types of devices can be deployed to generate the hash values for mining process:

- CPU mining: powerful processor is required
- GPU mining: powerful graphics card is required
- ASIC mining (Application-Specific Integrated Circuit, s. Table 7).

Table 7. Features of a Mining Device (source: amazon.de)

Features of AntMiner S15	High-Performance Mode	Low-Energy Mode
Hash rate	28 TH/s	18TH/s
El. power	1596 W	900 W
Power Efficiency	57 J/TH	50 J/TH
NW connection	Ethernet	Ethernet
Weight and dimensions	7 kg, 240 mm × 178 mm × 296 mm	

The practical experience has shown that in a lot of cases the Mining of the crypto-currencies like BTC, ETH, ZEC, XMR etc. leads unfortunately to "no reward" cases due to a large energy consumption as well as essential CAPEX + OPEX.

2.7 Smart Contracting in 5G

One of the mostly important Blockchain applications after the mining of the crypto-currencies are so-called Smart Contracts [10–16]. Historically, Smart Contracts (SC) does not require exceptionally Blockchain, but certain consensus algorithms (protocols), which are cryptographically conditioned via hashes, private and public keys and signatures. A possible solution is given in Fig. 13.

The Blockchain must guarantee the secured transitions between the interlinked Web Services, Micro-Services and further modules within the slicing structures and provide their compulsoriness. In the best way, the solution is possible based on Blockchain. In opposite to standard solutions, oriented to PKI and combined symmetric-asymmetric encryption (RSA) and digital signatures Blockchain provide its own security decentralized infrastructure, which distinguish from centralized PKI or bilateral Web-of-Trust, incorporated in convenient distributed apps [10–14]. The main goal is to transfer service data between 5G Network Slicing using a Blockchain that will introduce obligatorily and irreversibility by organizing linear chained structure. In the paper [9] this idea realized by using Key Distribution Center (KDC), which is co-operating with the Authentication

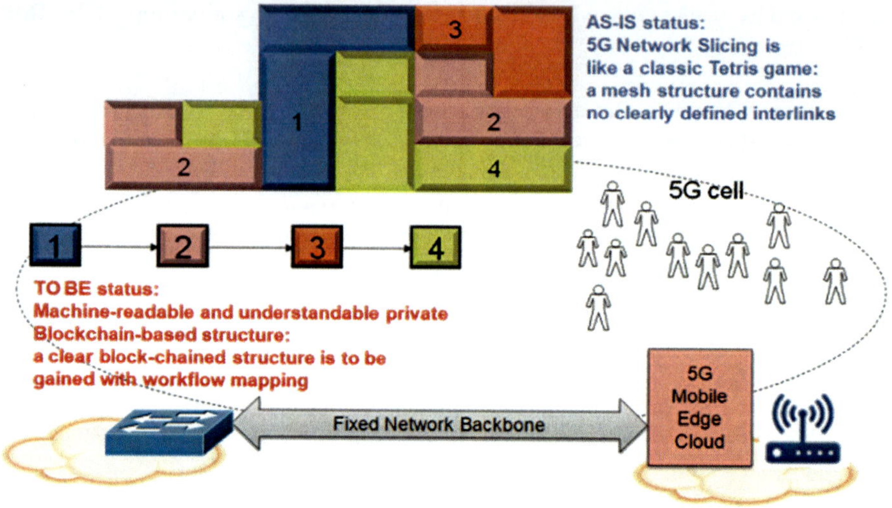

Fig. 13. Multivariate "Tetris" vs. a block-chained clearness

Server Function (AUSF) and for a given network slice that serves n devices (or n distinctive use cases), KDC generates a key-pair (d,e) for the El-Gamal cryptosystem: d and e are private and public keys. In our opinion, this method will not be effective enough. Using Blockchain we can win in terms of security and productivity even unless more complicated and resource consuming character of hash computing. Let us to compare the both status and the advantages of the offered solution (refer Fig. 13):

- **AS-IS status:** 5G Network Slicing is like a classical Tetris game. There is a mesh structure with no clearly defined interlinks. The workflow is not mapped.
- **TO-BE status:** Machine-readable and understandable privateBlockchain-based structure must be constructed instead of them. A linear chained structure must be gained. The compulsoriness of the steps and algorithm blocks is guaranteed as well as the further security aims: confidentiality and authentication.

2.8 DAO as a Prospective BC Application for 5G

DAO means a "Digital Decentralized Autonomous Organization" and simultaneously is signifying in Chinese the "way", "path", a holistic philosophical doctrine. DAO operates via crowd-funding (group financing) by interested donors. DAO operates as a SC-based application. Such organization works without a conventional management structure (CEO, CTO, and CIO) or board of directors and represents an instance of Ethereum Blockchain.

The above-mentioned application is related to ETH, Ethereum Classic and other Ethereum BCs as well as to recent crypto-currencies.

DAO code is open-source that means the great opportunity and relevance to use at different platforms as well as by development of so-called intelligent and secured 5G

Slicing (refer the previous sections). DAO can be a prospective BC Application for 5G networks.

An essential disadvantage for this solution for 5G micro-providers and macro-providers can be the law conformity. DAO has, in fact, no nationality, is stateless. It means a problematic deployment because of existing EU legislation and state regulatory authorities.

2.9 Consensus protocols for Blockchain

Consensus protocols in simple terms are concerning to so-called "rules to reach an agreement". There are a lot of different ways, how Blockchains can organize communication between any two IoT devices while they are can be located in a different part of the world without any third party (centre), which can control their communication [22–34].

This solution certainly has many advantages, but with such a M2M-communication, the BC can be vulnerable in cases where attackers on intermediate nodes have not falsified information.

For these cases, we need consensus protocols, which reduce the chances of failure of distributed information and forgery of information at intermediate nodes. The consensus protocol largely serves two purposes: to maintain a single version of the truth and to prevent sick actors from interfering with data already stored in blocks (refer Table 8).

Table 8. Comparison of different Consensus Protocols [8]

Consensus algorithm/blockchains	Number of transactions per second	Confirmation latency	Features
Proof-of-Work (PoW)/Bitcoin, Ethereum	Tens	6–60 min	High security Low throughput Low scalability
Proof-of-Stake (PoS)/Peercoin	Tens	10–60 min	High security Low throughput Low scalability
Proof of Authority (PoA)	60	10–15 s	Medium security Medium throughput Low scalability
Delegated Proof-of-Stake (dPoS)/EOS	Thousands	<1 s	Low security High throughput High scalability

Below you find the short property descriptions for the above discussed consensus protocols: **{PoW, PoS, PoA, dPoS}.**

1. **Proof of Work (PoW)**

The “Proof of Work” consensus is exactly what supports bitcoin blockchain and the security of bitcoin currency.

The Proof of Work consensus is one of the most decentralized protocols. However, it has certain disadvantages that can greatly affect the feasibility of its use in mobile networks:

1. This is the time, because miners have to randomly look for the right solution, checking a number of possible solutions.
2. is energy-intensive, as miners use energy-intensive equipment, such as GPUs, to find a solution as quickly as possible.

2. **Proof of State (PoS)**

The next most used consensus protocol is the «Proof of Stake protocol» (PoS). Network nodes that use the PoS consensus mechanism buy a stake (crypto tokens or coins) in the network that acts as collateral and gives them reliability to confirm transactions. The algorithm selects validators on the PoS blockchain based on the size of their collateral, thus allowing stakeholders with a larger amount on validation to verify more transactions.

After the block is created by the interested party, the other validators sign the block for approval, and the block is added to the chain after receiving the signature from most miners.

3. **Delegated Proof of State (dPoS)**

The “DelegatedPoof of Stake protocol” is similar to PoS, but here decentralization is significantly reduced, and the number of validators required to process each transaction by consensus dPoS is relatively lower than in PoS.

Like the PoW consensus, dPoS is also energy efficient and less time-consuming. In addition, it is also more scalable, ie it can process more transactions per second. This is because fewer validators are responsible for verifying transactions.

The most famous example of the dPoS consensus is the EOS blockchain, which has a record speed of 4,000 transactions per second.

4. **Proof of Authority (PoA)**

In PoA-based networks, transactions and blocks are validated by approved accounts, known as validators. Validators run software allowing them to put transactions in blocks. The process is automated and does not require validators to be constantly monitoring their computers. It, however, does require maintaining the computer (the authority node) uncompromised.

For each individual task, it is necessary to clearly understand what are the requirements for the operation of the blockchain, and which of the consensus will be the best in a particular situation.

3 Intelligent 5G Handover and Roaming

The AI (Artificial Intelligence) deployment allows exploring the environment and can use the ever-growing massive datasets available on up-to-date radio or wireless systems, inter alia, 5G systems. This knowledge can be used to solve complex and previously unsolvable goals. Many problems in wireless systems, such as decision-making, resource optimization, and network management, can be presented in a form that is suitable for artificial intelligence.

AI, mobile networks and wireless systems are becoming an important social infrastructure that mobilizing our lives. However, wireless networks and artificial intelligence should be considered as one area of research because they have huge potential for the development of future mobile networks. AI can be used to optimize radio communication through intelligent adaptive learning and decision making. The next generation of mobile and wireless technologies also require system optimization to maximize the uses of network resources, which can be provided by AI.

3.1 Using ML and AI Standards

The last development on the area of AI (Artificial Intelligence) is mainly based on ML (Machine Learning) processes, which deal with so-called adaptive algorithms. Such algorithms can be systematically improved over the time through the learned experiences (training data). ML can provide the following main approaches: 1) (partially) supervised, 2) unsupervised and 3) reinforcement learning [35, 36].

In mid-term, the standards for ML and AI will accompany the industries, digital economy and everyday life over the world and for each institution. Surely, they will find it deployment in 5G networking, and surely, in 5G Slicing. Therefore, let us to give a more detailed overview below. The main purpose of ML is to look for learning algorithms, performed without human intervention, where the time and amount of data required for training is one of the most important indicators of performance.

The main ML approaches are traditionally divided into Supervised Learning (SL), Unsupervised Learning (UL), and Reinforcement Learning (RL). Some authors distinguish this categorization, for instance, by adding of partially supervised learning as a combination of supervised and unsupervised approaches (so-called evolutionary learning, refer Fig. 14):

In supervised learning (Fig. 15), the algorithm collects a series of training data with corresponding labels. The algorithm learns independently a context between training data and labelling i.e. finds patterns to match input and label and then can generalize this to any input data.

The strategy is different with unsupervised learning (Fig. 16): the algorithm only collects unlabeled input data. The algorithm identifies similarities and patterns in the input data in order to classify the data or to identify the clusters and exceptions (founded outliers).

Reinforcement learning (Fig. 17) is a strategy of ML or usual paradigm, inspired by psychology: the agent learns through reward and punishment. He learns to master a given task autonomously by interacting with his environment, trying to maximize the possible reward.

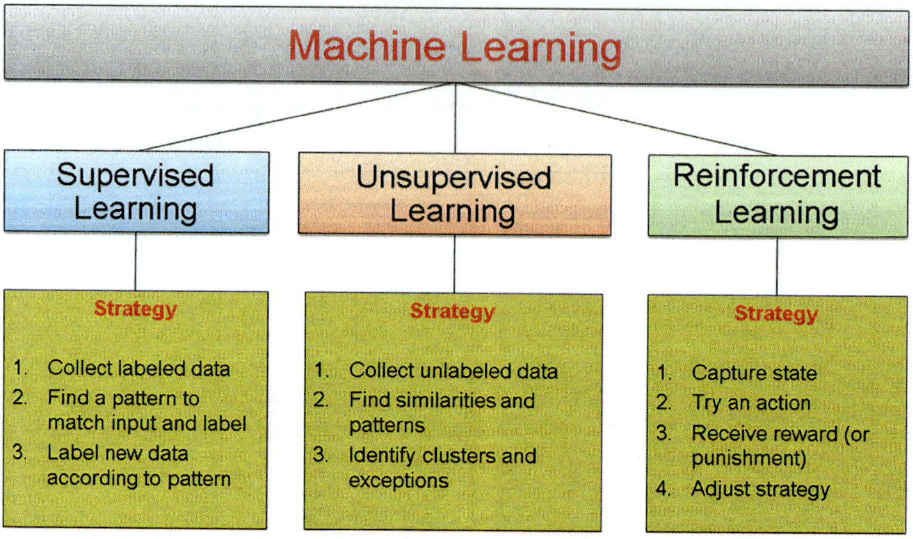

Fig. 14. ML paradigms in comparison

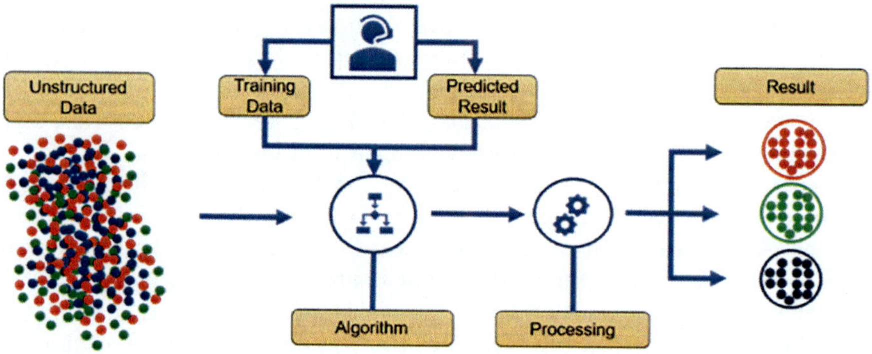

Fig. 15. Supervised Learning for 5G traffic classes

3.2 Artificial Intelligence for Handover Optimization

Nowadays, with the current 5G development and large traffic amount caused by multiple subscribers, it became increasingly difficult to provide the required QoS parameters, especially for users with high mobility. One of the main problems for these users is the proper initiation of handover. The main handovers problems are too-early, too-late and so-called "ping-pong" handover, when users move e.g. on the edge of two cells or two providers' areas.

In this work, we consider two types of handover. The first one is within one provider when there is a handover between two microcells. The second type is provided when a user moves from the microcell of one provider to the megacell of another provider (refer Fig. 6). Asit was shown in Fig. 18, it's impossible to define handover offsets statically

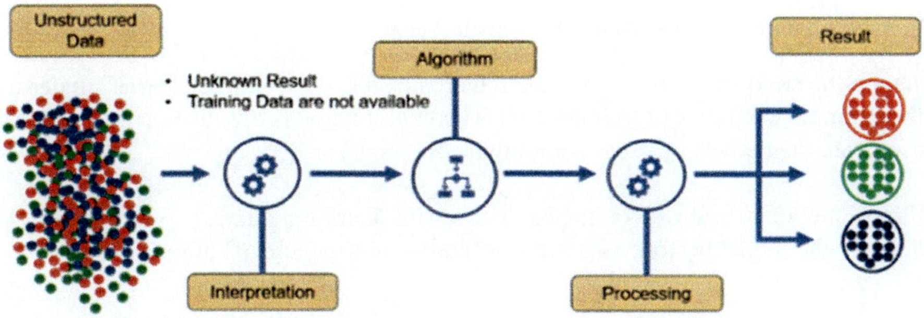

Fig. 16. Unsupervised Learning for 5G traffic classes

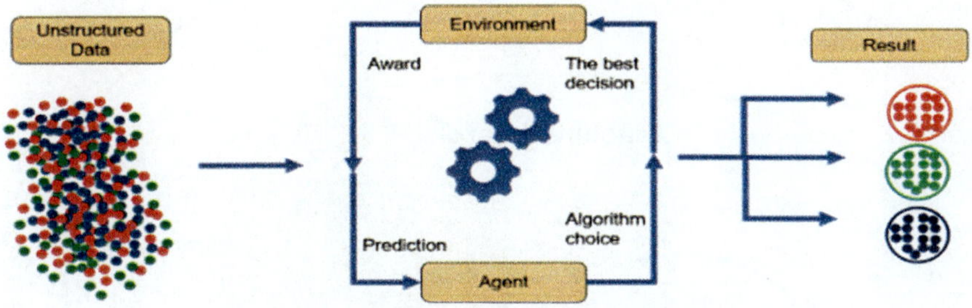

Fig. 17. Reinforcement Learning Algorithms aimed to 5G

and select an average value of offsets for all the time that network is running because users have, as a rule, high mobility and throughout the day they are located in different places. Therefore, it is suggested to use intelligent approaches to optimize the network and solve this problem.

Fig. 18. Mobility of users to different timespans of the day

3.3 To the Demarcation of Our Research Area

Herewith the short facts about our research area (refer Fig. 19). ML "learns" under use of the acquired examples or training data. The neural networks we us together with ML but separate the modelling process into two-part processes [35, 36]:

- First, a model is "trained" on the basis of available training data.
- Second, the model performs later the ML tasks and specific 5G algorithms.

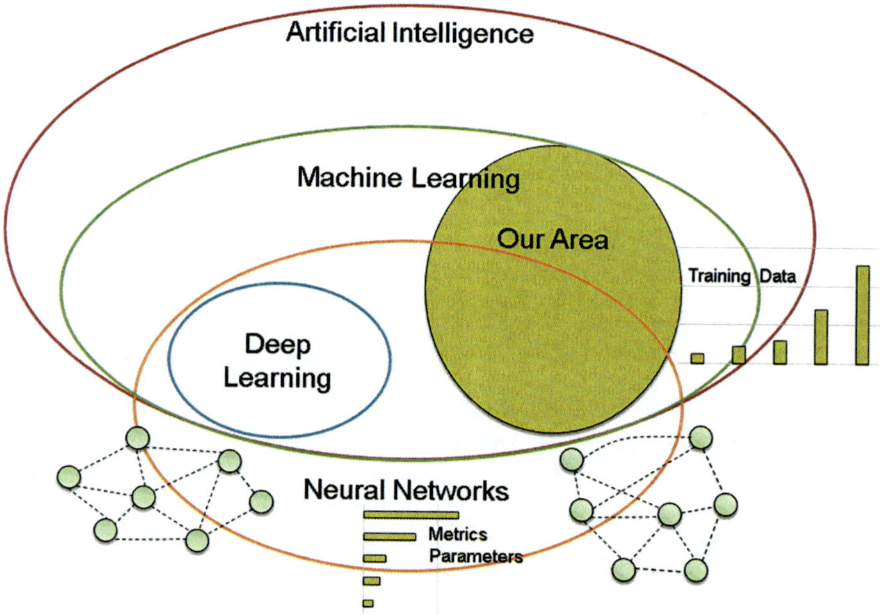

Fig. 19. Demarcation of the research area

For the modelling of 5G Slicing we use some metrics. The usage purpose for so-called metrics are as follows:

- Understandable parameters/ criteria for comparing processes and products within a discussed model can be provided.
- Quality and model properties can be quantitatively described.

The metrics are the model variables and numerical values used to measure the quality of the used models for ML and neural networks. A metric can be identified under use of so-called Confusion Matrix (Fig. 20) for the relevance. This matrix records how many errors and hits the discussed model has, when the model is executed.

The following metrics can be e.g. identified in the context: handover, roaming, DR, delay Δ, jitter.

Confusion Matrix	Actually Positive (1)	Actually Negative (0)
Predicted Positive (1)	True Positives	False Positives
Predicted Negative (0)	False Negatives	True Negatives

Fig. 20. Confusion Matrix for relevance of the metrics

A method of secured and intelligent handover process between micro-cells of small 5G micro-providers as well as roaming to the integrative macro-cells of the big providers in 5G hierarchical cellular structures under use of ML, neural networks as well as Smart Contracting via Blockchain technology is offered.

The method guarantees adaptive control of QoS for the 5G users independently from the type of provider/cell as well as safety of billing and compulsoriness of workflow within 5G slicing structures (Fig. 21). Multi-Layered Handover and Roaming based on 5G Cell Hierarchy is implemented within three strata:

- 5G Gateway Stratum
- Megacell Backhaul Stratum
- Microcell Basic Stratum.

The simulations and performance analysis based on a recurrent neural network are discussed below.

3.4 Recurrent Neural Network: Modelling and Training

In general, recurrent neural networks (Fig. 2) are not very different from classic neural networks. The main aim of recurrent neural networks is that they can relate prior information that was in the past to current tasks, such as moving a group of subscribers can help to understand current user mobility. However, they are not always appropriate for such purposes, as we can't control the amount of past information that needs to be considered.

One of the most important problems of the recurrent neural network is that they have a very short memory. This limitation is caused by the vanishing gradient problem during back propagation. So, we can't use them if we want consider information that was recently available.

To solve this problem, it was decided to use two different modified recurrent neural network architectures based on Long Short-Term Memory (LSTM) and Gated Recurrent Unit (GRU) cell architecture.

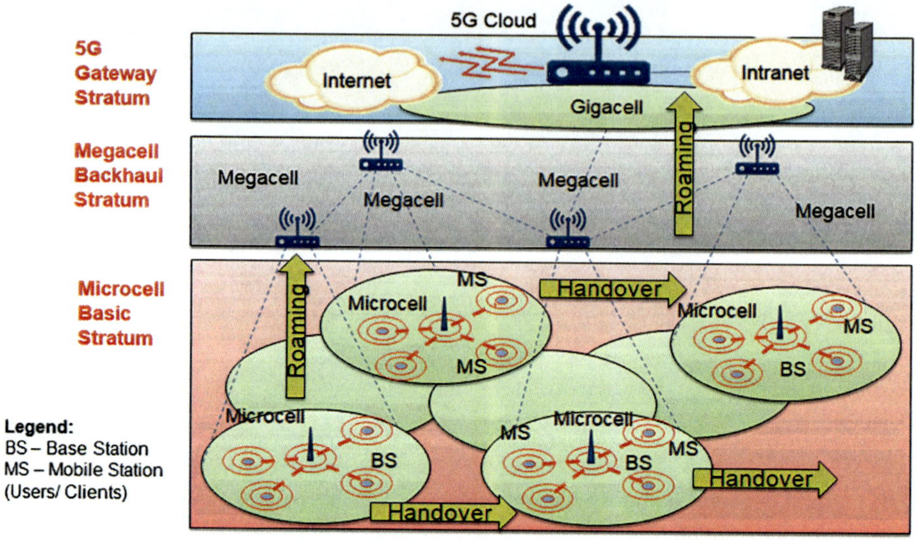

Fig. 21. A method of secured and intelligent Handover and Roaming

Recurrent neural network architectures based on LSTM and GRU cells have been used to create a neural network that allows predicting the number of users in a particular cell, this approach will allow experimentally see which of these architectures will be better to optimize the performance of the 5G network.

The next step is to collect user mobility information [6–8]. All training data were divided into training and test data (Fig. 22). The training data is used to train the neural network, this data collected from user's mobility over two months at intervals of 5 min. And the test of the neural network is performed on test data, which contains data about the subscribers for 5 days at intervals of 5 min.

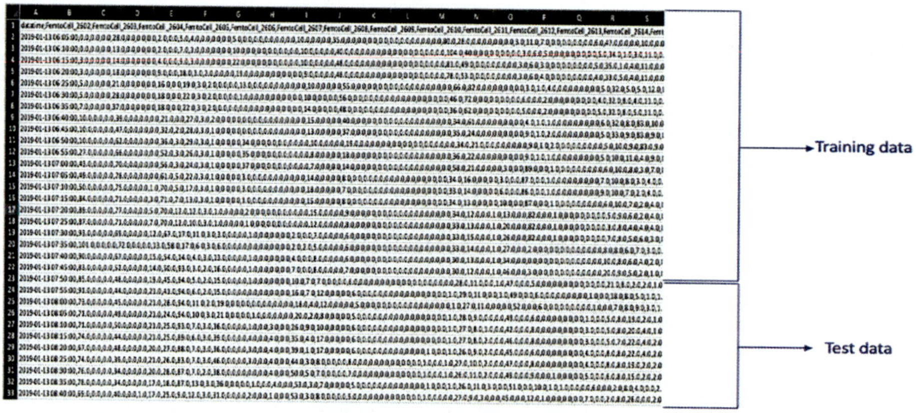

Fig. 22. Example of the distinguishing of the data into training data and test data

3.5 Modelling with a Neural Network and Prediction of the User Number in a Cell

During the experiment, several simulations were performed for recurrent neural network architectures based on LSTM and GRU cells. The main purpose of the experiment is to allow the simulated neural network to predict how many subscribers will be in a particular cell in a given time of day or several days. Consider the obtained experimental values of the two architectures and compare the effectiveness of their work at different amounts of learning epochs.

Modelling in the implementation of five epochs (Fig. 23):

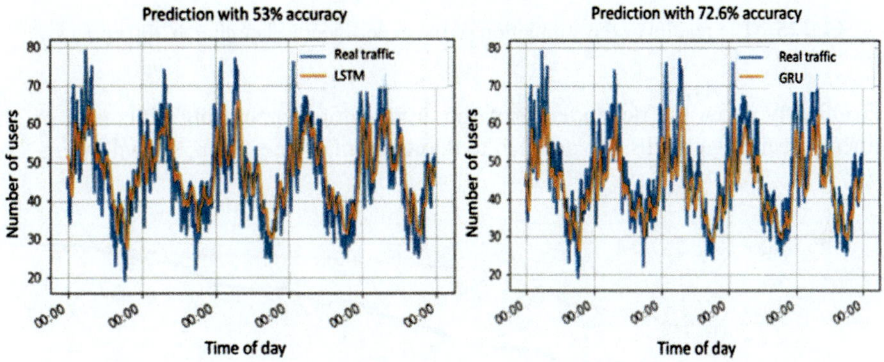

Fig. 23. The results of the simulation of the experiment in learning at 5 epochs

Modeling in the implementation of 100 learning epochs (Fig. 24):

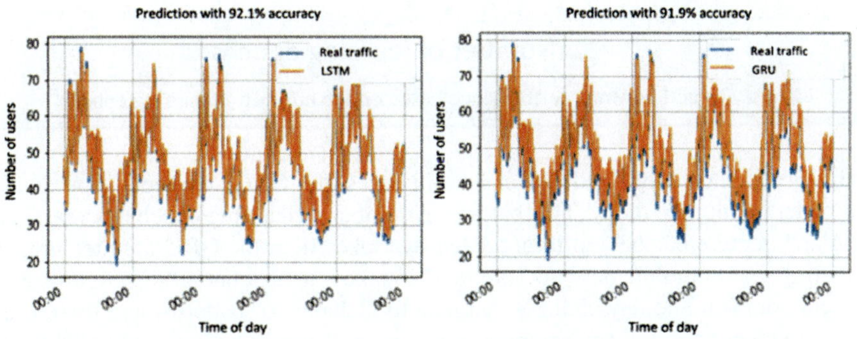

Fig. 24. The results of the simulation of the experiment in learning at 200 epochs

Modelling in the implementation of 200 learning epochs (Fig. 25):

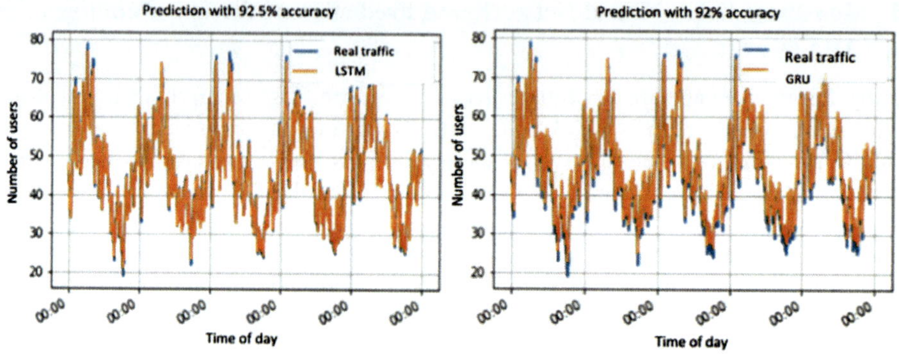

Fig. 25. The results of the simulation of the experiment in learning at 200 epochs

To clearly show the dependence of the architecture forecasting efficiency on the number of learning epochs, we decided to depict them in one graph, shown in Fig. 26.

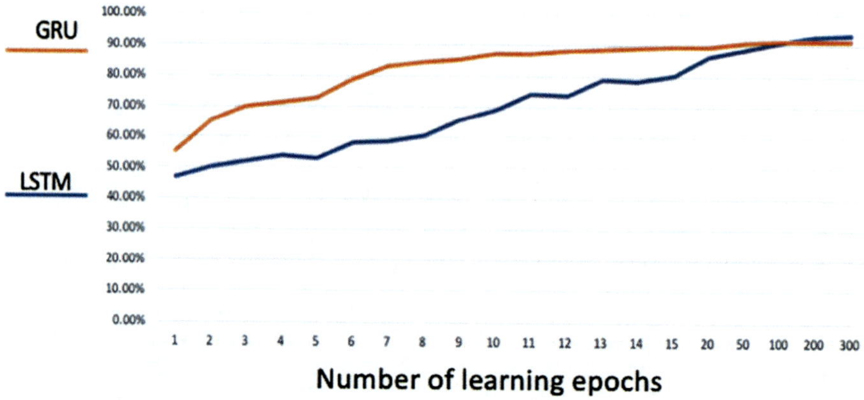

Fig. 26. Neural network performance based on the number of learning epochs

This graph shows that with fewer learning eras, the GRU architecture produces a much better result than the LSTM, but as the number of learning epochs increases, the LSTM architecture catches up to and even overtakes then the GRU architecture. But considering that in mobile networks we need to take into account all the circumstances in the environment and any changes can lead to a change of patterns, and therefore to the need to respond to the environment in this case, it would be better to use the GRU architecture, which with a minimum time delay will be able to provide us with the desired result of predictive cell performance.

3.6 Recurrent Neural Networks Based on GRU Cell to Optimize Handover in 5G Networks

At the time of initiation of the handover, it is necessary to take into account the value of MLB and MRO offsets, because in some case when the cell is overloaded it needs to use MLB offsets, but in other cases when there isn't overloaded cell it's better to use MRO offsets. So, we propose to introduce a new coefficient k, which will show which offset is more relevant in each situation. This decision will be taken by the recurrent neural network, based on knowledge of the mobility of subscribers and on the load capacity of the cells, before thatin previous sections we showed how neural network can predict the mobility of users. So, we introduce a new coefficient k = [0…100].

If the current cell 1 is overloaded (Fig. 27), the neural network considers coefficient K close to 80–90%, this means that MLB offsets consider near 80–90%. Therefore, the power shifts to the side of cell 1, which in turn visually reduces the size of the cell, and to make a handover in it more difficult, but it becomes easier to make a handover into the cell 2.

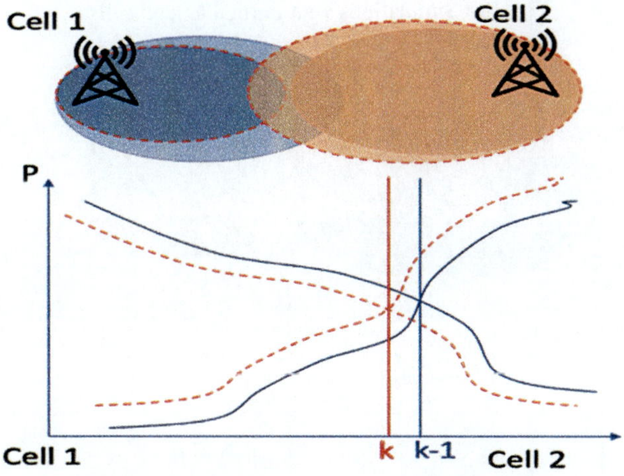

Fig. 27. Simulations with overloaded cell 1

When we have normally loaded cells (Fig. 28), the neural network takes the average, when the MRO and MLB are considered the same and the k factor is about 50%, the handover initiation location does not change.

If overloaded cell 2 (Fig. 29), the neural network considers coefficient K close to 10–20%, and the values of the MRO offset will be taken as 80–90%. Therefore, the power shifts to the side of cell 2, which in turn visually reduces the size of the cell, and to make a handover in it harder, but it becomes easier to make a handover in the cell 1.

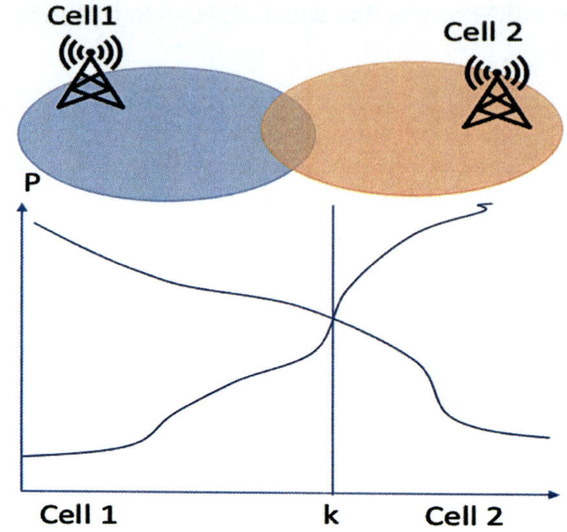

Fig. 28. Simulations with normal loaded cells

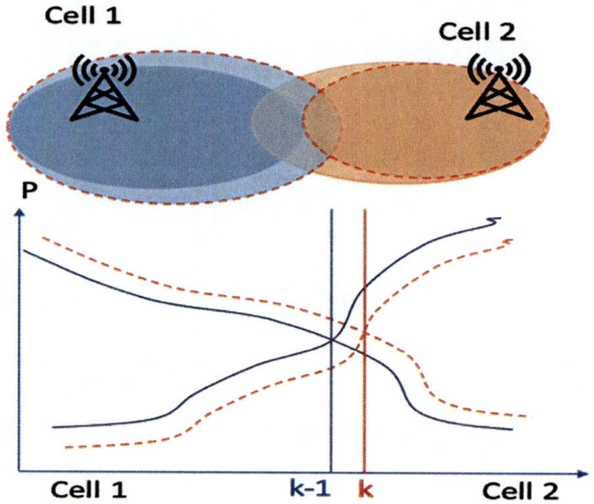

Fig. 29. Simulations with overloaded cell 2

4 Conclusions and Outlook

1. The given paper represents an overview on 5G challenges and deployment scenarios. The 5G mobile radio promises "a new technological breakthrough" in communication, industry, entertainment and education. Among the deployment scenarios on 5G networks so-called "digital twins" play a standing increasing role.
2. The case studies on optimizing of slicing and handover within the hierarchical 5G networks based on ML and neural networks are offered.

3. The recent 5G slicing scenarios are nowadays frequently implemented with security and compulsoriness via Blockchain and with intelligence via Machine Learning (refer Fig. 30).
4. The following main results were reached and can be established:

- 5G Deployment Scenarios with Cell Hierarchy and Interoperability are examined.
- 5G Slicing via Blockchain in Cell Hierarchy is discussed.
- 5G Intelligent Handover and Roaming in Cell Hierarchy is provided under use of ML paradigms and recurrent neural networks.

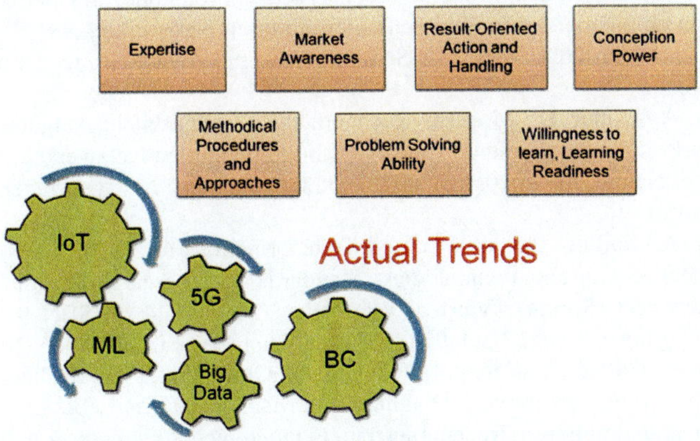

Fig. 30. New fields of competence for 5G Slicing Scenarios for 5G micro-providers

Acknowledgements. Authors' great acknowledgement to the colleagues from BA Dresden (Saxon Academy of Studies), Lviv National Polytechnic University and Odessa Telco Academy, especially to Prof. I. Strelkovska, Mr. E. Zumpe, Mr. M. Stoll, Mr. F. Franke, Mr. T. Zobjack for inspiration and challenges by fulfilling of this work.

References

1. 5G Whitepaper of Vodafone. https://www.vodafone.de/media/downloads/pdf/5G_Whitepaper
2. 5G Suisscom. https://www.swisscom.ch/de/about/unternehmen/portraet/netz/5g.html
3. IEEE 5G World Forum in Dresden. https://ieee-wf-5g.org/
4. Fettweis, G., Franchi, N.: FlächendeckendeVersorgungmit 5G-Mobilfunk in Deutschland (Online, in German). https://5glab.de/wp-content/uploads/2018-10-22_Diskussionspapier_5GLab_final.pdf
5. Globa, L., Prokopets, V., Gvozdetska, N.: Prognostic-reactive NFV resource allocation method for implementation in virtualized mobile network EPC of Ukraine. In: BlackSeaCom-2018, Batumi, Georgia (2018). 8 p.

6. Maksymyuk, T., Han, L., Larionov, S., Shubyn, B., Luntovskyy, A., Klymash, M.: Intelligent spectrum management in 5G mobile networks based on recurrent neural networks. In: IEEE CADSM 2019. IEEE Xplore (2019). https://ieeexplore.ieee.org/document/8779301
7. Maksymyuk, T., et al.: Deep learning based massive MIMO beamforming for 5G mobile network. In: IEEE International Symposium on Wireless Systems within the International Conferences on Intelligent Data Acquisition and Advanced Computing Systems, Lviv, Ukraine, pp. 241–244, September 2018
8. Maksymyuk, T., Gazda, J., Volosin, M., Bugar, G., Horvath, D., Klymash, M., Dohler, M.: Blockchain-empowered framework for decentralized network management in 6G. IEEE Commun. Mag. **58**(9), 86–92 (2020). https://doi.org/10.1109/MCOM.001.2000175
9. Fekade, et al.: Clustering hypervisors to minimize failures in mobile cloud computing. Wirel. Commun. Mob. Comput. **16**(18), 3455–3465 (2016)
10. Luntovskyy, A., Gütter, D.: Moderne Rechnernetze: Protokolle, Standards und Apps in kombinierten drahtgebundenen, mobilen und drahtlosen Netzwerken, 481 Seiten, 263 Abb. Springer Nature, Juli 2020. ISBN 9783658256166, https://www.springer.com/gp/book/9783658256166
11. Luntovskyy, A., Gütter, D.: Moderne Rechnernetze - Übungsbuch: Aufgaben und Musterlösungen zu Protokollen, Standards und Apps in kombinierten Netzwerken, 145 Seiten, 44 Abb. Springer Nature, Juli 2020. ISBN 9783658256180. https://www.springer.com/gp/book/9783658256180
12. Luntovskyy, A., Spillner, J.: Architectural Transformations in Network Services and Distributed Systems: Current Technologies, Standards and Research Results in Advanced (Mobile) Networks. Springer Vieweg, Wiesbaden (2017). 344 p. ISBN 9783658148409, https://www.springer.com/gp/book/9783658148409#otherversion=9783658148423
13. Luntovskyy, A., Gütter, D., Melnik, I.: Planung und Optimierung von Rechnernetzen. Methoden, Modelle, Tools für Entwurf, Diagnose und Management im Lebenszyklus von drahtgebundenen und drahtlosen Rechnernetzen: Planung von Rechnernetzwerke theoretisch anspruchsvoll und praxisnah, 415 Seiten, 245 Abb. Springer/Vieweg+Teubner, Wiesbaden (2012). ISBN 9783834814586, https://www.springer.com/gp/book/9783834814586#otherversion=9783834882424
14. Luntovskyy, A.: Integration Concepts for Computer-Aided Design Tools for Wired and Wireless Local-Area Networks, Handbook. Shaker Publishing House, Aachen (2008). 196 p.
15. Schill, A., Springer, T.: Verteilte Systeme: Grundlagen und Basistechnologien: Kompakte Darstellung der Grundlagen und Techniken Verteilter Systeme, 2 Ausgabe. Springer, Berlin (2012). ISBN 9783642257957, https://www.springer.com/gp/book/9783642257957#otherversion=9783642257964
16. BSI – Federal Office for Information Security (2020). https://www.bsi.bund.de/
17. LoRa Alliance (2020). https://lora-alliance.org/
18. Khan, J.Y., Yuce, M.R. (eds.): Internet of Things (IoT): Systems and Applications. Jenny Stanford Publishing, New York (2019). ISBN 9780429399084, 366 p.
19. Lueth, K.L.: The 25 best IoT Platforms 2019 - Based on Customer Review (2020). https://iot-analytics.com/the-25-best-iot-platforms-2019/
20. SAP Cloud Integration (2020). https://www.sap.com/
21. SAP Leonardo for IoT (2020). https://blogs.sap.com/2019/05/06/sap-leonardo-iot-for-smbs/
22. Survey on Crypto-platforms (2002). https://hackernoon.com/top-blockchain-platforms-to-watch-out-in-2019-aa80e336a426/
23. Luntovskyy, A., Guetter, D.: Cryptographic technology blockchain and its applications. In: Ilchenko, M., Globa, L., et al. (eds.) Advances in Information and Communication Technologies. Processing and Control in Information and Communication Systems (International Conference on UkrMiCo-2019). LNCS, pp. 14–33. Springer (2019). ISBN 978-3-030-16769-1, https://link.springer.com/book/10.1007/978-3-030-16770-7

24. Antonopoulos, A., Wood, G.: Mastering Ethereum: Building Smart Contracts and Dapps. O'Reilly Media (2019). 345 p. ISBN 978-1491971-949
25. Smart Contracts (2020). https://www.icertis.com/
26. MIT Blockchain Course (2020). https://executive-education.mit.edu/MIT-Blockchain/Online-Course/
27. Wuest, K., Gervais, A.: Do you need a Blockchain? ETH Zurich & Imperial College London (2020). https://eprint.iacr.org/2017/375.pdf
28. MS Bletchley (2020). https://github.com/Azure/azure-blockchain-projects/blob/master/bletchley/bletchley-whitepaper.md/
29. Leske, C., Göbel, A., Joswig, S.: Blockchain mit SAP, 1. Aufl. (2020)
30. Codius: Open-source Hosting Platform for Smart Programs (2020). https://codius.org/
31. Hyperledger Sawtooth (2020). https://www.hyperledger.org/projects/sawtooth/
32. Blockchain as a Service: Teamwork ohne Datengrenzen (2020). https://www.t-systems.com/blockchain/
33. Siemens Blockchain IoT (2020). https://new.siemens.com/
34. Siemens Blockchain (2020). https://www.plm.automation.siemens.com/
35. Seegerer, S., Michaeli, T., Romeike, R.: So lernen Maschinen, pp. 27–31 (2020). (in German)
36. Mitchell, T.M.: Machine Learning. McGraw-Hill, Boston (1997)

Method for Constructing an Adaptive Model for Optimizing Service Platforms of Information and Communication Networks

L. A. Nikityuk(✉) and R. Y. Tsaryov(✉)

O.S. Popov Odessa National Academy of Telecommunications, Odessa, Ukraine

Abstract. The work proposes a method for constructing an adaptation model for optimizing service platforms in infocommunications. The current stage of development of infocommunications is characterized by the introduction of concepts such as "Internet of Things" (IoT), "Internet of Everything" (IoE), which undoubtedly stimulates the development of new information technologies and telecommunication means. However, the demand for new infocommunication services and their quality is growing at a faster pace. According this, there is the problem of creating effective service platforms (a set of network resources that used in the production and provision of certain services) based on the resources of the network infrastructure and the possibility of optimization their technical characteristics during maintenance. A feature of modern service platforms is that the requirements for their technical characteristics change dynamically depending on the conditions of both external and internal environment. Take to account this fact, the process of optimization service platform should be continuous that is adapted to the changing situations. The proposed method for constructing a mathematical model allows solving the problems of structural and parametric optimization of the service platform during the life cycle from the moment of its design and further during maintenance at the stages of reconfiguration and reconstruction, taking into account the influence unforeseen factors of the environment and object itself. Solving the problems of reconfiguration and reconstruction is a cyclical process during all time maintenance until the service platform will be disposal. The proposed method for constructing an adaptive mathematical model is based on models of the morphological and functional description of the service platform, which allow to formalize the problems of finding the optimal values of its parameters. The effectiveness of the proposed method is demonstrated on specific examples of constructing adaptive models for optimizing service platforms for IPTV, telemedicine, etc.

Keywords: Service platform · Optimization · Adaptive model · Life cycle · IPTV

1 Introduction

The ever-growing need for optimizing service platforms for information and communication networks, which at the present stage are called "infocommunications" [1], poses

P. Vorobiyenko et al. (Eds.): IPF 2020, LNNS 212, pp. 256–271, 2021.
https://doi.org/10.1007/978-3-030-76343-5_13

a number of challenges for providers related to both organizing new network services and ensuring the high quality of the services provided over a long period of operation. It is not cost-effective to implement a new network service to provide information services without taking into account the capabilities of the existing network infrastructure. A segment approach is expedient, which allows to increase the technological capacity of the corresponding network segment in an optimal way, both by adding individual network components and by increasing the technical characteristics of the existing ones.

This situation actualizes the development of new approaches to determining the optimal set of network resources and methods for their dynamic reconfiguration during operation.

The set of network resources involved in organizing a network service to provide a specific service is called a "service platform". However, the term "service platform" currently does not have a unified interpretation in scientific works. Different sources [2–4] give different definitions of this term.

In this work, a service platform will be understood as a set of components of network resources that provide the implementation of network technology that allows to meet the requirements of users in relation to the quality of the information transfer service provided.

The work of many domestic and foreign scientists is devoted to the problems of creating effective service platforms. These are such scientists as L. N. Berkman., P. P. Vorobienko, V. A. Kaptur, M. M. Klymash, L. Globa., O. V. Lemeshko, B. S. Goldstein, Hong J. W., Yu C. Z. Eli Weintraub, Yuval Cohen, Christine Strauss, Xiaohu Ge, Jo M., Aramudhan M., Schmidt H., Walter F. Witt, Bloomers J. and others.

The authors explored aspects related to the optimization the structure of the service platforms and question of management the service platforms. While recognizing the scientific and practical value of the scientific works of the above authors, it should be noted that they mainly consider the service platform from only one side or from one phase of the life cycle, such as: design, management, operation, etc.

Life cycle stages (LC) can be conventionally considered as independent projects, each of which has its own purpose, characteristics, limitations and results [5]. However, the life cycle (LC) of a service platform is a continuous sequential process consisting of several iterations, starting from the moment of making a decision to create it and ending with the moment of its complete reconstruction or disposal. Formalization of the description of this process requires taking into account all possible changes in the characteristics of the state of the object in each iteration, and therefore the service platform life cycle model should be adaptive.

For the object "service platform", which belongs to the objects of the info-communication network and is a complex system, the following stages of LC are distinguished:

- system research
- design and implementation
- technical maintenance
- reconfiguration (change of resource parameters);
- reconstruction (structural-parametric optimization);
- disposal.

The purpose of this work is to develop an adaptive model of service platform, which allows to consistently solving the problem of optimizing its parameters and characteristics in according to the requirements for their changes during all stages of its life cycle under the influence of environmental changes and requirements of service users.

Adaptation in this case can be defined as the process of purposefully changing the parameters and structure of the service platform, based on the sequential determination of the criteria for its functioning and ways to achieve these criteria. As you can see, there is no clearly defined optimization requirement for a given criterion and complex systems (such as a service platform), as a rule, do not have a single criterion of functioning.

The proposed method for constructing a mathematical model of a service platform, adaptive in relation to its life cycle, is based on models of its morphological and functional description.

2 Models of Morphological and Functional Description of the Service Platform

The main problem of exploring the architecture of a service platform as a complex system is the need to take into account its many elements, and the relationships between them, their properties, changes characteristics. This, by turn, requires the formation of a multilayered family of models of morphological and functional description of the object of research. The problem is compounded by the need to take into account the impact of environmental changes and customer service requirements over time. That is, creating an effective service platform involves building an adaptive model that allows you to optimize it at all stages of its life cycle.

The solution to this problem can be reached based on a stratified approach [6], the essence of which is that the object of study is presented in the form of a set of stratum, each of which reveals the behavior of the system from the point of view of different levels of abstraction. Thus, stratum is the level of abstraction in terms of which the system is described.

Considering the service platform SP at the black box level (the inputs and outputs of the object are defined), it can be represented as:

$$SP : F(I) \rightarrow H, \tag{1}$$

then stratification of the system is possible if the sets of inputs I and outputs H form Cartesian product, that is, some input I_i and output H_i form two independent bases:

$$I = (I_1 \times I_i \ldots I_n); H = (H_1 \times H_i \ldots H_m), \tag{2}$$

where n is the number of inputs and m is the number of outputs. The *SP* service platform can be represented as subsystems by the following formulas:

$$\begin{aligned} &\mathrm{SPS1} : \mathrm{I_1} \times \mathrm{V_1} \times \mathrm{H_1}; \\ &\mathrm{SPSi} : \mathrm{I_i} \times \mathrm{D_i} \times \mathrm{V_i} \times \mathrm{H_i}; \\ &\mathrm{SPSn} : \mathrm{I_n} \times \mathrm{D_n} \times \mathrm{H_n}; \end{aligned}$$

$$n = \min(n, m) \tag{3}$$

where D, V are respectively the downstream and upstream flows that provide the interaction between the stratums.

During the development of the morphological description model, the question arises as to the number of stratums it is necessary to assign. M. Mesarovich, the founder of the stratified approach, states that the choice of stratums in terms of which the object of research will be depends on the researcher himself, his knowledge and the goals of the research [6], that is, the expert method of solving this problem is the main one.

It is proposed that the consideration of service platforms in infocommunications be carried out in five aspects, that is should be select the following five stratums.

The first stratum. The service platform is considered in terms of its resource base, that is, the set of *R* resources used in the production and provisioning of services:

$$R = \{r_1, r_2, \ldots r_k\}, \tag{4}$$

where r_i – is some resource that belongs to the service platform (communication channel, computing processor), as well as input resources (energy, financial resource, etc.).

The second stratum. The service platform is considered in terms of technological process of production and provision of modern information and communication services, which consists of many technological operations:

$$TP = \{TO_1, TO_2, \ldots TO_m\}, \tag{5}$$

where TO_i - some technological operation performed within the technological process.

Third stratum. The service platform is considered in the economic aspect, that is, it determines the set of its financial indicators (capital costs for implementation, operating costs, etc.).

Fourth stratum. The service platform is considered from the point of view of information security, where the indicators are information security and reliability assessments (security factor, security policy, etc.) of both the service platform as a whole and its individual elements.

Fifth stratum. We consider the service platform in terms of quality of service (QoS), where many indicators are estimates of service quality (delay, jitter, packets lost, etc.).

To create models of morphological description of service platforms, it is enough to consider them on the first and second stratums. This task can be formalized in terms of graph theory as follows [7, 8].

The technological process can be present in the form of an oriented graph *G (N, E)*, where the set of vertices *N* is corresponded the set of technological operations $\{TO_t\}$, and the arcs *E* reflect the transitions from one technological operation to another (Fig. 1).

Accord to each technological operation, on the first stratum, we determine the composition of the resources involved in its execution, which can be represented as a hierarchical tiered undirected graph $G^*(N_i^*, E^*)$ [8], which directly reflects the model of the morphological description of the service platform *SP*, that is, the composition of its elements and the relationships between them (Fig. 2).

The set of vertices $\{N_i^*\}$ of the graph $G^*(N_i^*, E^*)$ are correspond the elements of the stratums.

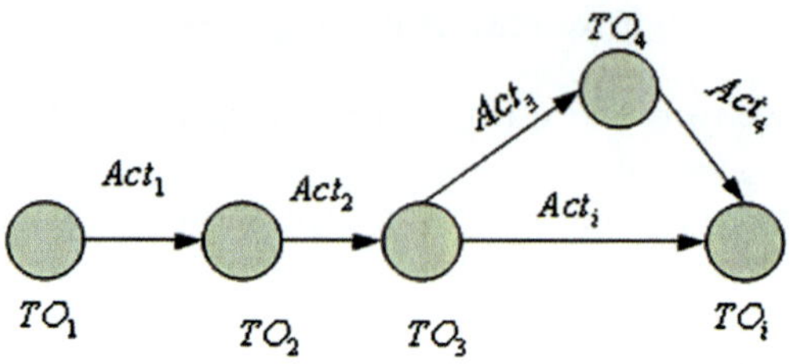

Fig. 1. The graph of technological process.

In the first layer of the hierarchy, the set N_1^*, with a power k, correspond to the elements of the stratum, which considers the service platform from the point of view of the resource base $\{r_i\}$, $i = \overline{1, k}$.

On the second layer of the hierarchy, the set N_2^*, of power m, correspond to the elements of the stratum, which considers the service platform from a technological point of view $\{TO_t\}$, $t = \overline{1, m}$.

In the third layer of the hierarchy, the set N_3^* correspond to *{SPi}* service platforms that provide individual services. In this case, set vertices N_3^* are represented by a single vertex that reflects the $SP*$ platform itself, but there can be several vertices.

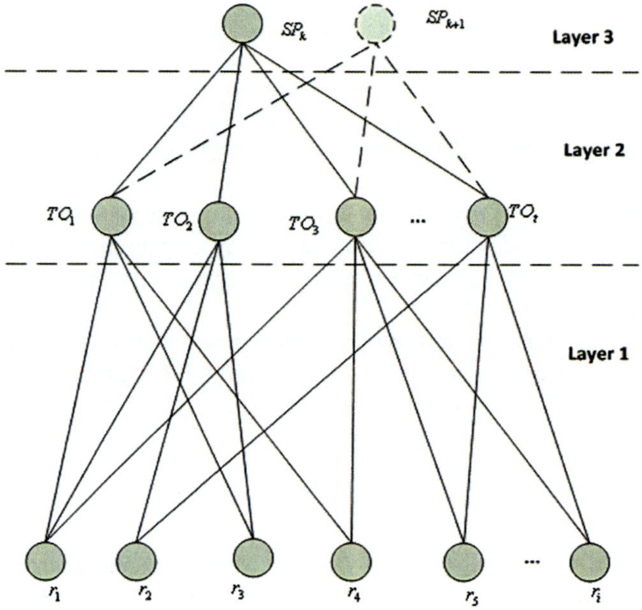

Fig. 2. The model of morphological description of service platform in the form of multilayer graph

The set of edges of an $E*$ graph reflects the set of connections between its vertices, that is, the elements of the $SP*$ platform.

The elements of the stratums that describe the service platform in terms of financial, information security, and quality of service indicators can be used to determine the corresponding weight characteristics of the vertices of G^*.

The next step is to create a model for a functional description of the service platform. Building a model of a functional description involves the creation of a hierarchy of epistemological levels, which forms the taxonomy framework of any complex system [9]. A rational way to form a functional description of a research object is to apply a hierarchy of functional descriptions, in which higher-level elements will depend on generalized and factorial lower-level variables.

All elements of the functional description of the service platform, as a complex system, can be divided into three groups: target functions; criterion-parametric functions; base variables.

To create a functional description of the service platform, it is proposed to apply the five epistemological levels.

At zero epistemological level, the system is defined by a set of variables $\overline{C} = (C_1, C_2, ..., C_k)$ that are assigned to the characteristics of the elements of the service platform, their properties or possible states.

$$\overline{\mathrm{C}}\colon \begin{cases} c_{11}, c_{12}, \ldots, c_{1n} \\ c_{21}, c_{22}, \ldots, c_{2n} \\ \ldots \\ c_{k1}, c_{k2}, \ldots, c_{kn} \end{cases}, \tag{6}$$

where $c_{i1}, c_{i2}, ..., c_{in}$, $n \in N-$ are the variables that describe the i-th vertex of the first layer.

At the first epistemological level, the system is complemented by values that variables can take. Depending on the type of task being solved at the appropriate stage in the service platform lifecycle, these values can be obtained by measuring, analyzing statistics, or defining them as desired.

At the second epistemological level, restrictions are placed on the set of the first level values:

$$\Phi(C) \leq 0 \tag{7}$$

At the third epistemological level, a description of the service platform is displayed, based on interrelated structural elements, which are linked by some common variables. In other words, the criteria $\overline{Y} = (Y_1, Y_2, ...Y_p)$ and their functional dependencies on the C variables are defined:

$$\overline{\mathrm{Y}}\colon \begin{cases} f\colon C_1 \to Y_1 \\ f\colon C_2 \to Y_2 \\ \ldots \\ f\colon C_p \to Y_p \end{cases} \tag{8}$$

At the fourth epistemological level, the limitations that are appropriate for the criterion parameters, depending on the stage of the service platform lifecycle are defined:

$$\Phi_Y(C) \leq 0 \tag{9}$$

At the fifth epistemological level, goal functions are formed that reflect the desired level of performance of the service platform at the relevant stage of its life cycle, for example:

$$\Upsilon = F(\overline{Y}) \to extr \tag{10}$$

The functional description model of the service platform provides the opportunity to formalize its optimization tasks. At the initial stage of the life cycle, this corresponds to the construction of a formalized mathematical model of optimal design, which involves finding the optimal structure and parameter values of the target object. However, at the stage of technical operation of the service platform under the influence of unpredictable environmental factors and the object itself, maintaining the optimal state of the object is not possible without the influence of the control system. The impact of the control system in this case involves the use of coefficient matrices that determine the requirements and the nature of the change in the parameters of the object. The specified matrixes of coefficients should be formed by the control system on the basis of data received from the object state monitoring system [11, 12]. Consideration of this effect in a formalized mathematical description of the optimal state of an object involves ensuring the possibility of its parametric and structural optimization throughout the entire life cycle. At the same time, maintaining the optimal state due to reconfiguration of resources, which in our case corresponds to parametric optimization, will be correlated with the stage of technical operation, and the need to simultaneously change the structure of the object with the stage of its reconstruction. Reconfiguration and reconstruction is a cyclic process that proceeds in time until the service platform is decommissioned.

Summarizing the above, it can be stated that the service platform is a complex object, the search optimization of which in the context of changing its effects is equivalent to the adaptation process, for the implementation of which it is necessary to have an adequate model of the object.

3 Creating an Adaptive Mathematical Model of a Service Platform

The optimization model of the service platform, as the target of adaptation within its life cycle T, will be called the adaptive model [10].

At the design stage, the optimal state of the object is achieved by finding the values of the basic variables that satisfy the given constraints and provide the extremum of the objective function, i.e.:

$$\begin{aligned} &\Upsilon = \mathrm{F}(\overline{\mathrm{Y}}) \to \mathrm{extr} \\ &\mathrm{F}(\mathrm{X,Y}) \cdot \left\| \mathrm{a_{ij}} \right\| \\ &\Phi_{\mathrm{Y}}(\mathrm{X}) \le 0 \\ &\Phi_{\mathrm{X}}(\mathrm{X}) \le 0 \end{aligned} \tag{11}$$

where X – is the set of basic variables.

The operator F determines the structure ST and the corresponding parameters C of the object, i.e. $F = \langle ST, C \rangle$, $\|a_{ij}\|$ - matrix coefficient, where:

$$\mathrm{a_{ij}} = \begin{cases} 1, & \text{if } \mathrm{x_{ij}} \text{ is included the coupling equation} \\ 0, & \text{if not} \end{cases}$$

$$i = 1, 2, \ldots, k;\ j = 1, 2, \ldots, m \tag{12}$$

Since the object is static at this stage, we obtain the problem of mathematical programming.

At the reconfiguration stage, the object undergoes adaptation, i.e. the influence of the control system $\overline{U} = U(t)(0 \leq t \leq T)$ is taken into account, as a result of which there is a change in the parameters of the object:

$$\begin{aligned}
&\Upsilon = \mathrm{F}(\overline{\mathrm{Y}}) \Rightarrow \psi(\mathrm{X} \cdot \mathrm{U(t)}) \to \mathrm{extr}\\
&\mathrm{F}(\mathrm{X} \cdot \mathrm{U(t)},\mathrm{Y}) \cdot \left\| \mathrm{a_{ij}} \right\|\\
&\Phi_{\mathrm{Y}}(\mathrm{X} \cdot \mathrm{U(t)}) \leq 0\\
&\Phi_{\mathrm{X}}(\mathrm{X} \cdot \mathrm{U(t)}) \leq 0\\
&0 \leq \mathrm{t} \leq \mathrm{T}
\end{aligned} \tag{13}$$

At the reconstruction stage, the problems of parametric and structural optimization are solved, i.e. under the influence of U (t), a new structure is formed and the corresponding object parameters are determined.

$$\begin{aligned}
&\Upsilon = \mathrm{F}(\overline{\mathrm{Y}}) \Rightarrow \psi(\mathrm{X} \cdot \mathrm{U(t)}) \to \mathrm{extr}\\
&\Omega(\mathrm{X} \cdot \mathrm{U(t)}, \mathrm{Y}) \cdot \left\| \mathrm{a_{ij}} \right\|\\
&\Phi_{\mathrm{Y}}(\mathrm{X} \cdot \mathrm{U(t)}) \leq 0\\
&\Phi_{\mathrm{X}}(\mathrm{X} \cdot \mathrm{U(t)}) \leq 0\\
&0 \leq \mathrm{t} \leq \mathrm{T}
\end{aligned} \tag{14}$$

where Ω – the operator determines the structure *ST(t)* and the corresponding parameters *C(t)* of the object, i.e. $\Omega = \langle ST, C, U \rangle$.

Reconfiguration and reconstruction tasks are performed cyclically throughout the *T* life cycle and can be implemented based on the genetic algorithm. The decision on the need for reconfiguration or reconstruction of the service platform is made based on the results of work the management system. The management system controls the indicators of quality the service platform and on the basis of cognitive monitoring [12] make decides of the need for reconfiguration or reconstruction the service platforms.

4 Adaptive Model for Optimizing Service Platforms IPTV

Let us illustrate the application of the proposed method by the example of constructing an adaptive model for a service platform for the provision of IPTV services.

Formulation of the problem. It is required to design a service platform for IPTV that provides reliable and high-quality transmission of video content in volume L_k for servicing q subscribers, with an intensity λ requests of service and an average volume of request L_r. The level of costs for the creation and operation of the service platform should not exceed the permissible value of C_{max}.

The main components of the IPTV platform are (Fig. 3) [12, 13]:

- Server of Head Station IPTV
- Service delivery network

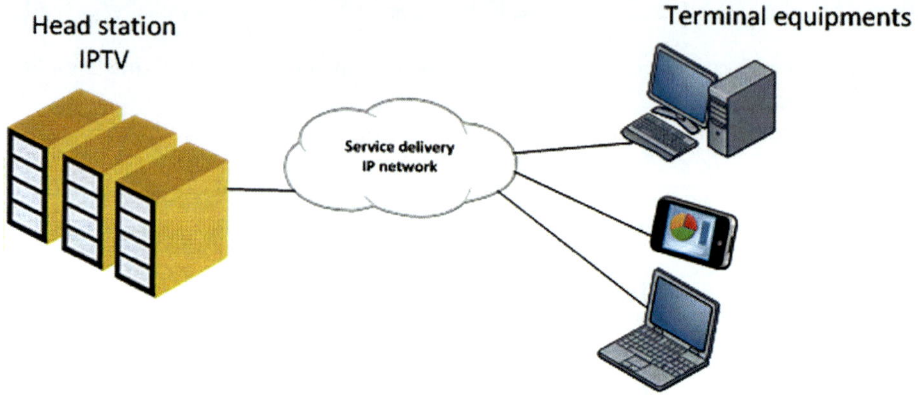

Fig. 3. The IPTV service platform

- Terminal equipment of clients.

In this case as the independent internal variables of the problem are a set of technical characteristics *TTN*, a set of operational characteristics *TTE*, and a set of cost characteristics of components *C* that can be used to create an IPTV service platform, namely, servers and channels of an IPTV service delivery network.

As the technical characteristics of the server, we will consider the power of the processor - *CPU* and the amount of memory size - *RAM*. Then the set of technical characteristics TTH^{Ser} of *n* servers that can be used to organize the Head Station IPTV can be represented as:

$$TTH^{Ser} = \left\{ <CPU^{Ser}(i), RAM^{Ser}(i)> \right\} = \{<CPU_1^{Ser}, RAM_1^{Ser}>, <CPU_2^{Ser}, RAM_2^{Ser}>, \ldots, <CPU_n^{Ser}, RAM_n^{Ser}>\},\ i = \overline{1,n} \quad (15)$$

The operational characteristics of the server *i* are *MTBF* (Mean Time Between Failure) and *MTTR* (Mean Time To Repair) [14]. The set of operational characteristics TTE^{Ser} of *n* servers can be represented as:

$$TTE^{Ser} = \left\{ <MTBF^{Ser}(i), MTTR^{Ser}(i)> \right\} = \{<MTBF_1^{Ser}, MTTR_1^{Ser}>, <MTBF_2^{Ser}, MTTR_2^{Ser}>, \ldots, <MTBF_n^{Ser}, MTTR_n^{Ser}>\},\ i = \overline{1,n} \quad (16)$$

The set of cost characteristics of *n* servers can be represented as:

$$C^{Ser} = \left\{ c^{Ser}(i) \right\}, i = \overline{1,n} \quad (17)$$

The components of the IPTV service delivery network can be represented by a set of TTH^{Ch} technological characteristics, namely, by a set of speeds *k* of network channels:

$$\overline{V} = \{v(i)\}, i = \overline{1,k} \quad (18)$$

As the set TTE^{Ch} of the operational characteristics channels of the IPTV service delivery network, we will consider the mean *MTBF* and the mean *MTTR*. These characteristics depend on the length of communication channels, the provided transmission rate and defines in the recommendation G.826:

$$TTE^{Ch} = \left\{ <MTBF^{Ch}(i), MTTR^{Ch}(i)> \right\} = \{<MTBF_1^{Ch}, MTTR_1^{Ch}>, \\ <MTBF_2^{Ch}, MTTR_2^{Ch}>, \ldots, <MTBF_k^{Ch}, MTTR_k^{Ch}>\},\ i = \overline{1,k} \quad (19)$$

The set of cost characteristics of k channels of the IPTV service delivery network can be represented as:

$$C^{Ch} = \left\{ c^{Ch}(i) \right\}, i = \overline{1,k} \quad (20)$$

The vector $\overline{X}$ of internal variables of the system in general form can be represented by a set of: input data of the problem, characteristics of n servers and characteristics of k channels of the delivery network, namely:

$$\begin{aligned} \overline{X} &= [\overline{X^{In}}, \{\overline{X_i^{Ser}}\}, \{\overline{X_j^{Ch}}\}] \\ \overline{X^{In}} &= \{x_z\} = \{L_k, q, \lambda, L_r\},\ z = \overline{1,4}; \\ \overline{X_i^{Ser}} &= \{x_{ir}\},\ i = \overline{1,n};\ r = \overline{1,m} \\ \overline{X_j^{Ch}} &= \{x_{jp}\},\ j = \overline{1,k};\ p = \overline{1,l} \end{aligned} \quad (21)$$

where z – number of input data of the problem; m – number of variables that describe the entire set of characteristics of n servers involved in organizing the service platform; l - number of variables describing the entire set of characteristics of k channels of the IPTV service delivery network.

Or else:

$$\overline{X} = (x_1, x_2, x_3, \ldots, x_h) \quad (22)$$

where h – number of all internal variables of the problem, i.e. $h = z + n \cdot m + k \cdot l$.
The following restrictions are imposed on internal variables:

$$x_i \geq 0,\ \ \forall x_i \in \overline{X},\ \ i = \overline{1,n} \quad (23)$$

As criterion parameters characterizing the external variables of the task, it is advisable to consider such characteristics as: the average response time T^{avr} of the platform to user requests, which in our case characterizes the transmission quality; availability factor K (characterizes the reliability of content transmission) of the service platform and reduced costs G, that is:

$$Y_1 = T^{avr},\ \ Y_2 = K,\ \ Y_3 = G. \quad (24)$$

In the conditions of the problem, the following restrictions can be imposed on external parameters:

$$Y_1 = T^{avr} \leq T_{\max}$$

$$
\begin{aligned}
&Y_2 = K = K_{in} \\
&Y_3 = G \le G_{\max} \\
&Y_i > 0, \forall Y_i \in Y\ i = \overline{1, 3}
\end{aligned}
\tag{25}
$$

where $T_{\max} = 400$ ms is the maximum possible platform response time to user requests, as defined by ITU recommendations [15, 16]; K_{in}- the required level of availability; $G_{\max}$- the maximum admissible reduced costs.

Let compose the constraint equations, thus expressing the external criterion parameters as functions of the internal variables of the problem:

$$Y_k = f(\overline{X}), \forall Y_k \in \overline{Y} \tag{26}$$

Let the average response time T^{avr} to user requests be defined as:

$$T^{avr} = T_r^{avr} + T_w^{avr} + T_s^{avr} + T_c^{avr} \tag{27}$$

where T_r^{avr} is the average time of request transmission over the network; T_w^{avr} - the average time a request is in the queue for server servicing; T_s^{avr} - average time of service of the request by the server; T_c^{avr} - is the average delivery time of the requested content.

Determination of these parameters requires consideration the network component of the service platform, which can be represented by a typical tree-like structure of the *SN* network, which ensures the connection of a given number of q subscribers to the Head station IPTV.

Let represent the original service delivery network in the form of a weighted graph *G (N, E)*, where *N* is the set of vertices that correspond to network nodes, and *E* is the set of edges that correspond to communication lines. The weight characteristics of the nodes are the delay times of messages in the nodes $\{t_i\}, i \in N, i = \overline{1, n}$. The weight characteristics of the edges are the transmission rate $\{v_{ij}\}$, where $(i, j) \in E, i, j \in N$. Let *M* be the set of transmission paths in the network $M = M' \cup M'' = \{\mu_{is}\} \cup \{\mu_{si}\}$, where $\{\mu_{is}\}$ is the set of transmission paths for a request from the *i-th* user to the server *s*, and $\{\mu_{si}\}$ is the set ways of transferring content from server *s* to the *i-th* user (Fig. 4).

Obviously, the values T_r^{avr} and T_c^{avr} depend on the transmission rate of the user's request and the transmission rate of the requested content, respectively, along a some path $\mu \in$ M and on the delay time t_i in the communication nodes.

The average transmission time of a user's request is defined as: [17]

$$
\begin{gathered}
T_r^{avr} = \frac{1}{q}\left[\sum_i t_i + \left(\sum_i \sum_j \frac{1}{v_{ij}} \sum_k L_{rk}^{avr}\right)\right] \\
i \in N(\mu_{is}),\ i, j \in N(\mu_{is}),\ k : i, j \in N(\mu_{ks})
\end{gathered}
\tag{28}
$$

and the average delivery time of the requested content, respectively:

$$
\begin{gathered}
T_c^{avr} = \frac{1}{q}\left[\sum_i t_i + \left(\sum_i \sum_j \frac{1}{v_{ij}} \sum_k L_{ck}^{avr}\right)\right] \\
i \in N(\mu_{si}),\ i, j \in N(\mu_{si}),\ k : i, j \in N(\mu_{ki})
\end{gathered}
\tag{29}
$$

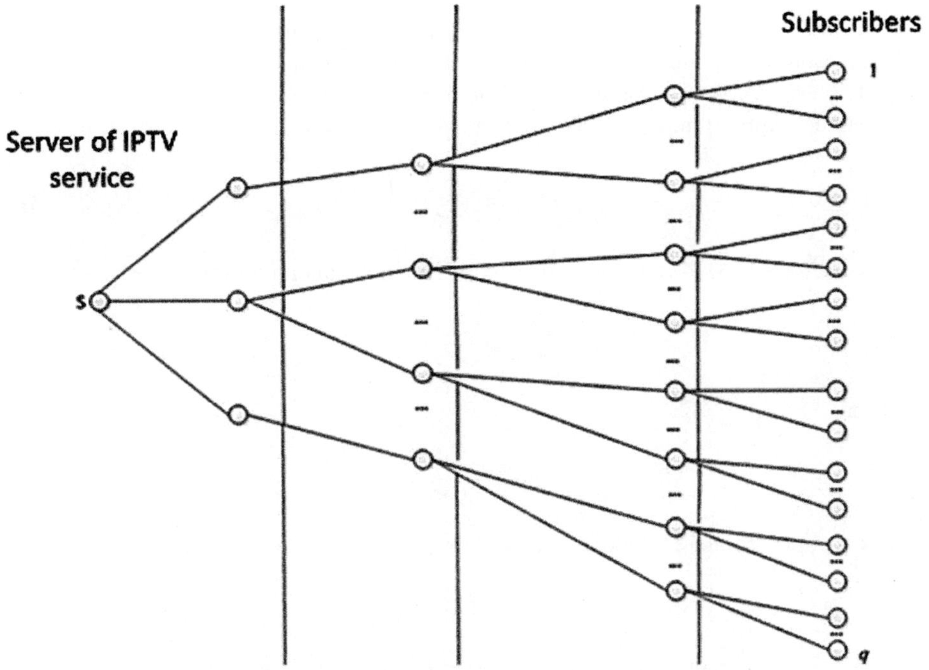

Fig. 4. The graph of service delivery network

where L_{rk}^{avr} – the average volume of a user's request in bytes; L_{ck}^{avr} - the average volume of content requested by the user in bytes; v_{ij}- data transmission channel speed; q – the number of subscribers; t_i – delay in the i -th node of the network.

The value t_i depends on the functional purpose of the node (access node, gateway, etc.) and can take a number of integer values, while any case $t_i \leq 10$ ms [17].

Suppose that the incoming flow of requests is a simple Poisson process with the flow rate of requests λ (initial data), and the server load (utilization) ρ does not exceed 80%. Let's define the performance of the *RPS* server as follows:

$$RPS = \min\{RPS^{CPU}, RPS^{MEM}\}$$
$$RPS^{CPU} = \lambda^{CPU} = \frac{CPU^{\Sigma} - CPU^{SYS}}{CPU^{q}}$$
$$RPS^{MEM} = \lambda^{MEM} = \frac{MEM^{\Sigma} - MEM^{SYS}}{MEM^{q}} \quad (30)$$

where, RPS^{CPU} – server performance in terms of processor capabilities; RPS^{MEM} – server performance in terms of the capabilities of the memory subsystem; λ^{CPU} – the maximum intensity of requests that the server can handle, taking into account the capabilities of the processor; λ^{MEM} – the maximum intensity of requests that the server can process, taking into account the capabilities of the memory subsystem; CPU^{Σ} – server processor power; CPU^{SYS} – processor power required to perform in-system tasks; CPU^q – processor power allocated to process one request; MEM^{Σ} – the capacity of the server memory subsystem; MEM^{SYS} – the volume of memory required to perform in-system tasks; MEM^q – the volume of memory allocated for processing one request.

Then the required number of N^{Ser} servers is determined as:

$$N^{Ser} = \left\lceil \frac{\lambda}{RPS} \right\rceil \tag{31}$$

According to [18] the average time a request is in the queue for service by the server:

$$T_w^{avr} = \frac{\rho \cdot T_s^{avr}}{1-\rho} \tag{32}$$

where

$$\rho = \frac{\lambda \cdot T_s^{avr}}{N^{Ser}} \tag{33}$$

Taking into account the above, the expression for Y_1 will be as follows:

$$\begin{gathered} Y_1 = T^{avr} = \tfrac{\rho}{\lambda}\left(N^{Ser} + \tfrac{\rho}{1-\rho}\right) + \\ + \tfrac{1}{q}\left[2\sum_i t_i + \left(\sum_i\sum_j \tfrac{1}{v_{ij}} \sum_k L_{rk}^{avr}\right) + \left(\sum_i\sum_j \tfrac{1}{v_{ij}} \sum_k L_{ck}^{avr}\right)\right] \\ i \in N(\mu_{is}),\ i,j \in N(\mu_{is}),\ k : i,j \in N(\mu_{ks}),\ i,j \in N(\mu_{si}), \\ k : i,j \in N(\mu_{ki}) \end{gathered} \tag{34}$$

The reliability of a service platform can be expressed through the availability factors of its elements. According to [14], the availability factor of the *i-th* element K_i is defined as:

$$K_i = \frac{MTBF_i}{MTBF_i + MTTR_i} \tag{35}$$

Using expression (35), the coefficient Y_2 availability of the service platform can be defined as:

$$Y_2 = K^{SP} = \prod_{i=1}^{n} K_i = \prod_{i=1}^{n} \frac{MTBF_i}{MTBF_i + MTTR_i} \tag{36}$$

The reduced costs G for the service platform are defined as:

$$G = K_{TE} + K_n Q \tag{37}$$

where, K_{TE} – annual operating costs; K_n – standard coefficient of efficiency of capital investments; Q – capital costs of the service platform:

$$Q = N^{Ser} \cdot C^{ser} + \sum_{i=1}^{k} C_i^{Ch} \tag{38}$$

Therefore:

$$Y_3 = G = K_{TE} + K_n(N^{Ser} \cdot C^{ser} + \sum_{i=1}^{k} C_i^{Ch}) \tag{39}$$

The solution of the system of communication equations will allow to obtain a set of admissible values $\overline{X_{PS}} \in \overline{X}$ for a set of admissible values $\overline{Y_{PS}} \in \overline{Y}$.

To find the optimal, from the point of view of the goal set in the task, the option of a set of components of the IPTV service platform among the set of admissible ones, it is proposed to use the objective function γ of the form:

To find the optimal, from the point of view of the goal set in the task, the option of a set of components of the IPTV service platform among the set of admissible ones, it is proposed to use the objective function γ of the form:

$$\Upsilon = \frac{T^{avr}}{T_{\max}} + \frac{G}{G_{\max}} \rightarrow \min \tag{40}$$

Where $T^{avr} = Y_2 = f(\overline{X})$; $G = Y_3 = f(\overline{X})$; - criterion parameters of the system to which the function is most sensitive.

Since the optimum in this formulation of the problem is conditional, its search must be performed subject to the constraints:

$$\begin{aligned} &Y_1 = f(\overline{X}) \leq T_{\max} \\ &Y_2 = f(\overline{X}) = K_{in} \\ &Y_3 = f(\overline{X}) \leq G_{\max} \\ &x_i \geq 0, \ \forall x_i \in \overline{X} i = \overline{1, n} \end{aligned} \tag{41}$$

The solution of the task is a vector $\overline{X_0} \in \overline{X}$, the values of which determine the required parameters of the components for building a service platform.

The adaptive nature of the optimization model of the service platform, which takes into account the impact of the control system on the facility during operation during the stages of its life cycle, is taken into account in this task using coefficients a_i reflecting the required changes in the values of internal variables, tasks, namely:

$$\overline{X}' = \overline{X} \cdot \{a_i\} = \{x_i \cdot a_i\}, \ i = \overline{1, h} \tag{42}$$

where

$$\begin{aligned} a_i = 1 \ \forall x_i \neq 0 - \text{for design stage;} \\ a_i \geq 0, \ \ \forall x_i \neq 0 - \text{for reconfiguration or reconstruction stage} \end{aligned} \tag{43}$$

The adaptive optimization model of the IPTV service platform can be represented as:

$$\begin{aligned} &\gamma = F(\overline{Y}, \overline{X}') \rightarrow \min \\ &\Phi_Y(\overline{X}') \leq \overline{Q_Y} \\ &\Phi_{X'} \geq 0 \end{aligned} \tag{44}$$

where $\Phi_Y(\overline{X})$ – the system of communication equations for the external and internal variables of the problem; $\overline{Q_Y}$ – constraints on external variables; Φ_X – restrictions on internal variables.

The decision on the need to move to the stage of reconstruction is made by the control system based on the analysis of changes in the initial data of the task, as well as data from the monitoring system of the state of the technical and operational characteristics of the object. At the same time, the formation of coefficients reflecting the required changes in the values of the internal variables of the problem is preceded by a decision to change the structure of the *SN* network.

5 Conclusion

The paper proposes a method for creating adaptive models for optimizing service platforms for information and communications networks throughout their life cycle. The construction of the model is based on the principles of a stratified approach and adaptation of complex systems. The adaptive model makes it possible to formalize the procedure for optimizing the structure and parameters of the object, taking into account the control action of the control system, which, in turn, makes it possible to automate the process of adapting the state of the object to internal and external changes. Further work will be aimed at developing genetic algorithms for solving problems of optimizing service platforms for information and communications networks.

References

1. Vorobienko, P.P., Nikityuk, L.A.: Infokommunikaczii: terminy i opredeleniya [Infocommunication: terms and definitions]. Eastern-Eur. J. Enterp. Technol. **6/2**(54), 4–7 2011. (in Russian)
2. Vorobiienko, P.P., Nikitiuk, L.A., Reznichenko, P.I.: Telekomunikatsiini ta informatsiini merezhi [Telecommunication and information networks]. Ukraine:SAMMIT-KNYHA, Kyiv (2010). (in Ukrainian)
3. Uvarov, R.V., Skliarenko, V.K., Zuev, S.I., Gulian, V.G., Uleev, A.P., Grin, A.A.: Telekommunikatsionnye seti i tekhnologii. Ch1 Telekommunikatsionnye i informatsionnye seti [Telecommunication networks and technologies. Part 1. Telecommunication and information networks], Kharkov, Ukraine (2006). (in Russian)
4. Tsifrovaia platforma [Digital platform]. Regional seminar ITU (2017). https://www.itu.int/en/ITU-D/Regional-Presence/CIS/Documents/Events/2017/09_Tashkent/Presentations/ITU%20Workshop%2019.09%20-%20Nikolay%20Kovtun%20presentation%203.pdf. (in Russian)
5. Lysetskyi, Yu.M.: Informatsiini systemy i tekhnolohii v menedzhmenti [Information systems and technologies in management], Lohos, Kyiv, Ukraine (2014). (in Ukrainian)
6. Mesarovic, M., Takahara, Y.: General Systems Theory. Academic Press, New York (1975)
7. Ray, S.S.: Graph Theory with Algorithms and Its Applications. Springer, New Delhi (2013)
8. Diestel, R.: Graph Theory. Springer, New York (2006)
9. Klir, G., Elias, D.: Architecture of Systems Problem Solving. Springer US, Boston (2003)
10. Chernikov, B., Antonchikov, S.: Adaptivny'e organizaczionnye sistemy i ikh modelirovanie [Adaptive organizational systems and their Modeling]. Fundamental'nye issledovaniya, no. 9–1, pp. 243–252 (2017). (in Russian)

11. Nikitjuk, L.A., Babich, Y.O.: Imitacionnoe modelirovanie predikativnogo monitoringa verojatnosti poteri ciklovoj sinhronizacii v konvergentnyh setjah [Simulation of predictive monitoring of frame alignment losses probability within convergent networks]. Zbirnik naukovih prac' ONAZ im. O. S. Popova, no. 2, pp. 169–175 (2014). (in Russian)
12. ITU-T Recommendation Y.1910. IPTV functional architecture, September 2010
13. ITU-T Recommendation Y.1910. Terms and definitions for IPTV, March 2010
14. ITU-T Recommendation G.827. Availability performance parameters and objectives for end-to-end international constant bit-rate digital paths, September 2003
15. ITU-T Recommendation Y.1540. IP Packet Transfer and Availability Performance Parameters, December 2002
16. ITU-T Recommendation Y.1541. Network Performance Objectives for IP-Based Services, May 2002
17. ITU-T Recommendation Y.1542. Framework for achieving end-to-end IP performance objectives. December 2010
18. Kul'gin, M.: Tehnologii korporativnyh setej. [Technology of corporative networks]. SPb.: «Piter» (1999). 704 p. (in Russian)

Research of Ways to Increase the Efficiency of Functioning Between Firewalls in the Protection of Information Web-Portals in Telecommunications Networks

Zoya Dudar(✉), Ihor Shubin(✉), Victoria Skovorodnikova(✉), and Svitlana Litvin(✉)

Department of Software Engineering, Kharkiv National University of Radio Electronics, Kharkiv, Ukraine
{zoia.dudar,igor.shubin,viktoriia.skovorodnikova}@nure.ua

Abstract. The results of a full-factor experiment and analysis of experimental data to increase the performance between network screens in information security systems of telecommunications networks are presented. A number of factors that significantly affect the functioning between firewalls are considered, namely: fragmentation of network packets, security policy, network structure, auxiliary functions between firewalls. Empirical dependences characterizing the influence of the adopted security policy on the performance between network screens are obtained. The possibility of increasing the level of information security of the network by redistribution of security rules is proved. A technique for studying the performance of firewalls is described, which involves the primary processing of experimental data, and their secondary processing to ensure satisfactory accuracy of the results. The results of the study can be used in information security systems of telecommunications networks to increase the security of web-portals.

Keywords: Telecommunication networks · Information security · Firewalls · Information web-portals · Security policy · Security rules · Experiment planning · Linear regression analysis

1 Goal and Objectives of the Research

In connection with the rapid development of the Internet and related services, the problems of ensuring network information security are becoming increasingly important. Websites of all kinds also require security attention, as evidenced by recurrent web attacks on the sites of major companies such as Microsoft and Yahoo. One of the most common types of web sites is information web portals, which are publicly accessible repositories of information of a wide variety of topics. Such portals are characterized by a large number of users and, accordingly, high traffic intensity. In this case, the main means of protection today is the firewall system. However, the use of such a system can be the main reason for the slowdown in the response rate of the portal to user requests.

P. Vorobiyenko et al. (Eds.): IPF 2020, LNNS 212, pp. 272–292, 2021.
https://doi.org/10.1007/978-3-030-76343-5_14

This problem is discussed in [1]. Because all traffic passes through a firewall (FW), understanding its performance characteristics is a key element in predicting and ensuring overall system performance. Comparison of the performance characteristics of MEs with different architectures, such as classical and distributed screens, is presented in [2, 3].

Some software methods for increasing the performance of the class of software under consideration are described in publications [4–6]. In addition, many ME manufacturers publish specific recommendations to ensure high performance of their products. An example is a fairly detailed guide published by Microsoft for their ME ISA Server.

Many companies that own information web portals in one way or another face the problem under discussion, but most of the proposed solutions require significant financial costs to improve the hardware. Thus, the study of the performance of software ME is a relevant and insufficiently disclosed topic. This is especially true of the degree of influence of certain factors on the problem under consideration and the possibility of reducing their influence.

The object of research in this article is network information security as a property of information web portals in Ethernet type networks, the subject is models and methods of ensuring network information security of information web portals in ethernet type networks. The ultimate goal of the study is to develop recommendations for improving the performance of MEs used to protect information web portals in ethernet networks. To achieve this goal, the following tasks were set:

1) Analyze the conditions for the operation of the ME used to protect information web portals in ethernet networks.
2) Develop a network model of the information web portal and build a test network corresponding to this model.
3) Prepare and conduct an experiment to determine the performance of the ME system included in the network according to the standard protection scheme for information web portals.
4) Describe the methodology for researching the performance of software firewalls in telecommunication networks.

2 Analysis of the Current State of the Problem of Achieving High Performance of Software Tools to Ensure Network Information Security

The development of network technologies, the rapid qualitative and quantitative growth of networks around the world, the expansion of the global Internet make the issues of secure network operation more and more urgent. The safety and confidentiality of data located on computers connected to large networks, as well as public websites, are at great risk due to the real possibility of an attack with the aim of stealing or destroying information. According to the data given in [7], 63% of websites have critical vulnerabilities, and in 93 cases out of 100 the software (software) of a web application contains medium-risk vulnerabilities.

Among the most common types of vulnerabilities are:

1) Lack of parameter checking in http requests. As a result, using special parameters, an attacker can gain access to server resources via a web application.
2) Failure to comply with resource access control policies allows an attacker to use private resources or gain access to other user accounts.
3) Failure to comply with the rules for managing accounts and user sessions. This vulnerability is primarily related to the lack of reliable protection of user data and session identifiers such as cookies. This allows hackers to intercept other users' data and use the system on their behalf [8].
4) Vulnerabilities related to errors in the Cross-Site Scripting mechanism used to redirect the user to other sites. In this case, the attack could lead to an attacker gaining access to user data or compromising the local computer.
5) Buffer overflow errors found in many software products (PP) - from scripts and drivers to operating systems and server software. Failure to check some parameters can lead to a buffer overflow, and an attacker can seize control of the computer. Overflow error messages appear extremely frequently.
6) Lack of proper control over the parameters transmitted by computers when accessing external resources. If a hacker manages to enter his commands into these parameters, the consequences can be very serious.
7) Bad use of cryptography. Often, tools for encrypting information have their own vulnerabilities, which makes it meaningless to use even the most effective cryptographic tools.
8) Lack of proper protection of remote administration subsystems. Although the presence of a web interface is convenient, since it allows the administrator to control the system from any computer connected to the Internet, in the absence of reliable protection, an attacker can do the same [9–11].

Specialized software systems for ensuring network information security are able to reliably protect a website from threats caused by these and other types of vulnerabilities. However, this is only possible with proper configuration and maintenance of such tools: timely installation of the latest updates, monitoring of suspicious network activity, etc.

The main types of software tools (PS) that ensure network information security include network monitors, network protocol analyzers and ME.

Network monitors are used to constantly monitor the corporate network for detecting attacks, that is, they are designed for "active" protection. These systems detect attacks on the nodes of the corporate network and respond to them in the manner specified by the administrator. At the same time, modern monitors support many functions, for example, maintaining network statistics, which includes data such as segment utilization, collision level, error rate and broadcast traffic level, and determining the signal propagation speed. This also includes checking the legitimacy of network adapters if a "suspicious" one suddenly appears. As you know, monitors are subdivided into software and hardware. However, hardware monitors are mainly used for statistical functions and do not provide the required level of information protection.

Protocol analyzers are a separate class of software, although they are essentially part of networked program monitors. Each monitor has at least several protocol analyzers built into it. They are used separately from network monitors in cases where it is impractical to use a powerful and, therefore, more expensive monitor. Protocol analyzers put the network adapter into promiscuous mode and collect all network traffic for analysis. The Protocol Analyzer is a very useful tool to help you troubleshoot, remove bandwidth bottlenecks, and detect intrusion of computer intruders in a timely manner.

Firewalls are the most common firewalls (for example, ME is a standard element of the Windows operating system). They are widely used both for protecting home computers and for complex shielding of large corporate networks.

ME is a system or a combination of systems that allows you to divide a network into two or more parts and implement a set of rules that determine the conditions for the passage of packets from one part to another. ME, passes all traffic through itself. For each passing packet, it decides to let it pass or discard it. There is a set of rules by which the DOE decides to skip a packet.

ME are used everywhere in the protection of local networks of small businesses, large corporate networks, web services, home computers connected to global networks. And if, when used in small networks, you can not pay much attention to performance, then when protecting large corporate networks or popular web portals connected to the Internet via high-speed connections, it is ME that becomes the main factor that reduces the speed of Internet access.

When deploying ME in networks with high performance requirements, they are configured in such a way as to reduce the sources of bottleneck formation to computer hardware resources. In the case of a single entry-level computer, the potential bottleneck is not the processor, but the bandwidth of the Internet channel.

From a cost perspective, it is most efficient to design a system with an emphasis on central processing unit (CPU) resources, since this is the most expensive resource to upgrade.

As is the case with most applications, ME performance increases with CPU speed. At the same time, this increase does not lead to a linear increase in productivity. Due to frequent and large memory accesses, increased CPU performance can result in some of the CPU cycles being idle while waiting for a response from memory. When not more than 60% of CPU resources are consumed, the "hyper-threading" technology helps to reduce the load on the processor [12]. With a higher level of resource consumption, turning it on will have no effect.

Also, when processing large amounts of data, the memory is accessed frequently. Level 2 and 3 caches improve performance on large memory samples.

Since the operation of the ME is associated with the passage of significant amounts of data through the network devices, memory and CPU, the performance of the ME is influenced not only by the processor, but also by the characteristics of other elements of the system. The overall throughput increases as the speed of the external memory bus and I/O buses increases.

The performance of the ME can also be increased by adjusting it accordingly.

One part of this setup involves configuring application-level filters. For example, an HTTP filter might be enabled to scan the HTTP payload for specific signatures. When this function is enabled, the amount of processing increases, which means that the computer on which the ME is installed begins to consume more resources. Therefore, if possible, replace the application layer filters with simple network and transport layer filtering.

When working with HTTP traffic, caching is also widely used, but it is possible only when using a web proxy, which in itself somewhat slows down performance. However, if the bulk of the traffic passing through the ME is HTTP traffic, then the use of a properly configured web proxy and caching gives a very large performance gain for both the ME and web servers, reducing the consumption of the Internet channel bandwidth.

Filtering rules should also be placed in the list of rules in the order corresponding to the highest performance. Since most of the ME works on the principle of “first match” when deciding whether to send or reject a packet, the most commonly used rules should be placed as close to the top of the list as possible. Thus, if 90% of the traffic matches the first rule in the list, then only one check will be performed for this traffic.

Productivity can also be improved by configuring the ME event logging [13]. For example, ISA Server provides the ability to log ME activity as a file. At the same time, the efficiency of consumption of processor resources increases by 10–20% due to a decrease in functionality.

Despite the fact that ME manufacturers often give recommendations on how to improve their efficiency, the final solution to the problem under discussion has not yet been obtained.

3 Planning an Experiment to Investigate the Performance of Software Firewalls

In general, the ME is not a single device or program. As a rule, it is better to think of it as a system of components that are designed to control access to the internal local and external networks based on the chosen security policy. The functioning of this system of components is influenced by many factors, the most significant of which will be discussed below.

Not all network segments have the same data frame size. Since the Internet is an interconnection of many networks, the size of IP packets transmitted in any segment may be larger than allowed in it [14]. IP allows a packet to be split into several smaller packets, which are reassembled after reaching their destination. Severely fragmented packets can slow down the operation of the ME, since fragmentation complicates the process of analyzing packets for their compliance with the rules set for the ME. In a real web portal, packet fragmentation is random. During the experiment, the test program simulates packet fragmentation, but this factor is uncontrollable.

The main components of ME as a system are packet filters and proxy servers. These components are implemented in different ways and provide different levels of protection. There are typical schemes for connecting components to a network in order to form a ME. The simplest of these is to install a packet filter at the edge of the protected subnet. The packet filter simply discards packets from the hosts specified by the administrator and does not allow them to enter the protected subnet. Likewise, packets from the protected subnet that do not match the rules are not allowed to the external subnet. Packet analysis is done only based on the contents of the IP packet header. However, the header mainly contains addresses and other data required by the protocol to deliver the packet to its destination. The packet header does not include details that would help the filter decide whether to pass the packet through the screen. For example, a packet filter can determine that the packet is of some protocol, say FTP [15]. At the same time, the packet filter is not able to recognize which request it is - get (receive) or put (transmit). At the same time, it is possible that requests of one type will be quite acceptable, and requests of a different type for a node will have to be blocked. For the purpose of analyzing the contents of an IP packet, proxy servers (or application layer gateways) are used. You can build a ME by simply turning on such a proxy server at the boundary of the protected subnet. However, more efficient methods involve a combination of packet filters and proxy servers. An example of such a combination is a circuit for connecting a ME into a network in the form of a shielded node. In this case, the ME uses the protective capabilities of both the packet filter and the proxy server. At the same time, a packet filter is used to connect to an external network, and a proxy server is used to ensure the security of various services. The shielded subnetting scheme is a further development of the shielded node network. According to this scheme, two packet filters are installed at the border of the protected subnet, between which one or more proxy servers are connected, providing the operation of one or several network services [16]. Thus, the latter scheme at least doubles the number of checks that a packet needs to pass in order to get from the external network to the protected subnet as compared to the scheme with a single packet filter at the subnet border. The time it takes for a packet to pass through the ME system also increases, but the latter scheme, if properly configured, is much more reliable than the first.

Due to these circumstances, during the experiment, it is necessary to use a fixed scheme for connecting the ME system to the network, which corresponds to the standard scheme used in organizing the protection of the web portal.

As a result of competition among ME manufacturers and their attempts to improve their product, firewalls were endowed with new features. Since the ME is located on the subnet border, it must perform many tasks, including those not related to security [17]:

- caching. This property is especially true for networks containing Web servers with a large amount of information available from the Internet. By storing frequently requested data locally, a caching server can improve response times to user requests and save bandwidth that would be required to reload data;
- address translation. A properly configured ME allows you to use any IP addresses for the internal network. In this case, only the address of the ME is visible from the outside;
- content filtering (content restriction). An increasing number of products provide restriction of information received by users from the Internet by blocking access to URLs containing unwanted content, or by searching for specified keywords in incoming IP packets [3, c 150];
- forwarding (address vectoring). This feature provides the firewall with the ability to modify requests so that they are directed to the server with a different IP address than the IP address specified in the request packet. In this way, it will be possible to distribute the load among several servers, which for an external user looks like a single server.

It should be noted that when conducting experiments, additional functions of the ME system should be disabled in order to exclude their impact on the performance of firewalls.

Passing through the firewall system, IP packets are checked for their compliance with the specified security rules. The presence of a large number of such rules on packet filters and ME proxy servers increases the number of checks and, accordingly, increases the packet transit time. Thus, the set of security policy rules also affects the speed of the ME.

Security policy rules and rule groups are used as controllable factors in experiments. To develop such a policy and the corresponding security rules, it is necessary to select the categories of users of the web portal and define the rights for each category. As a rule, an information web portal has the following categories of users:

1) Public users - the most numerous category of users, the web portal itself is focused on serving information needs.
2) Unwanted users who are seen spreading spam or other malicious activity. The use of the web portal by such users may be restricted.
3) Security administrator, whose functions include monitoring and managing the portal's security system, as well as responding to attempts to breach the portal's security.
4) The database administrator, whose functions include maintaining the full volume of operational and accumulated information in an up-to-date working condition, as well as protecting information from unauthorized access.
5) The content manager defines the policies, attributes and rules for the use of the various types of content in the web portal.

Let's define the IP addresses and ports that will correspond to the above groups in the information portal network model:

1) 192.168.0.1 ports 5000–65535 - public users;
2) 192.168.0.1 port 4444 - database administrator;
3) 192.168.0.1 port 5555 - security administrator;
4) 192.168.0.1 port 6666 - content manager;
5) 192.168.0.2 port 80 - portal web server;
6) 192.168.0.2 port 25 - mail server;
7) 192.168.0.4 port 2222 - database server;
8) 192.168.0.4 port 2224 - application server;
9) 192.168.0.4 port 2225 - authentication and authorization server;
10) 192.168.0.4 port 1111 - Trust Center server.

During the experiment, the ports used by the test computer software for service purposes are excluded from the 5000–65535 port range.

On the basis of these roles for the protection of the portal formed set of filtering rules, which are presented in Table 1.

Table 1. Basic terms filtering external firewall

Rules №	senderIP	senderPort	IPDestination	Destination Port	Status
1	192.168.0.1	Any	192.168.0.2	Service port http, https, smtp, dns, imapi, imapis, ldap, pop3, pop3s, smtp, telnet, minger	Allowed
2	192.168.0.2	Service ports http, https, smtp, dns, imapi, imapis, ldap, pop3, pop3s, smtp, telnet, minger	192.168.0.1	Any	Allowed
3	192.168.0.1	Any	192.168.0.2	Ports for msa and odmr services	Allowed
4	192.168.0.1	Any	192.168.0.4	Service ports http proxy, https, telnet, 1111, 2222, 2224, 2225	Allowed
5	192.168.0.4	Ports of services http proxy, https, telnet, 1111, 2222, 2224, 2225	192.168.0.1	Any	Allowed
	Any	Any	Any	Any	Forbidden

Table 2 shows a set of packet filtering rules at the network and transport layers. The first two rules define permissions for inbound and outbound traffic for a specific set of protocols by which users can interact with the portal. The third rule sets permissions only for incoming traffic via the msa and odmr protocols. These protocols are applied by the mail server used in the model. The next two rules determine the permission for interaction of users from the external network directly with the servers of the control segment and the segment of service servers using certain protocols. These rules are required to ensure remote administration of the portal.

Internal firewall, unlike the external one, filters only at the transport and network layers. The first two rules allow working with servers in the service server segment from the external network. The following two rules allow work with the servers of the control segment from the external network. Thus, the screen supports remote control of the web portal. The following two rules define the permission to use the application server on the service server segment for the public segment servers. The last rule is the default, and according to the “forbid anything that is not explicitly allowed” policy, all other traffic between the management and service segments of the portal and the rest of the network is prohibited.

Table 2. Internal firewall filtering

rulesrule number	Sender IPSender	Port	Recipient IPRecipient	portport	Status
6	192.168.0.1	Any	192.168.0.4	2222, 2224, 1111	Connection establishment allowed
7	192.168.0.4	2223, 2224, 1111	192.168.0.1	Any	Allowed
8	192.168.0.1	Any	192.168.0.5		Allowed to establish connection
9	192.168.0.5	1111	192.168.0.1	Any	Alowed
10	192.168.0.2	80, 8080, 25, 110	192.168.0.4	2224	Allowed
11	192.168.0.4	2224	192.168.0.2	80, 8080, 25, 110	Allowed
	Any Any	Any	Any	Any	Forbidden The

Table 3. Rules for filtering e-mail by an external firewall

Rule no.	Direction	Attribute	Status
12	Incoming	e-mail addresses from the list of unwanted is	Prohibited
13	Incoming	Message contains attachment of the "message/partial" type is	Prohibited
14	Incoming	Message contains the IFRAME tag is	Prohibited
15	Incoming	Message contains the SCRIPT tag	Forbidden
16	Incoming	Message size $\geq$ maximum allowed	Forbidden
17	Incoming	Message identified as unwanted by subject, attachment name or text	Prohibited
18	Incoming	Message contains a compressed password-protected file	Forbidden

Table 3 shows a set of rules for filtering mail messages for an external FW. Such filtering is carried out on the principle of eliminating known vulnerabilities and removing messages with unwanted content, therefore all rules are prohibited. The first rule blocks mail from unwanted addresses, such as spam. The next three rules block messages that are potentially unsafe for the portal. Fragmented, deformed, encrypted and messages in the subject or content of which words from the prohibited list are found are also discarded by the external screen.

Each security rule may or may not be used by the system. Thus, the number of factor levels was taken to be two.

The total number of safety rules under study when measuring the performance of the first ME is 16, and of the second 6. If we single out each rule as a separate factor of a full-factor experiment, then the total number of measurements in the first case will be 65536, and in the second case - 64. Thus, it makes sense combine rules that are similar in meaning and, presumably, have an equally small effect on system performance, into one factor of the experiment. Table 4 presents a set of experimental factors and the corresponding safety rules for studying the performance of an external ME.

Table 4. Factors of the experiment to measure the performance of the external ME

Factor number	1	2	3	4	5
Rule number	1, 2, 3	4, 5	13, 16, 18	12, 14, 15	17

The first three rules handle traffic between the demilitarized zone (DMZ) and an external subnet, and the greatest impact on performance has rules 1 and 2, which check the traffic compliance with a large number of network protocols through which it is possible to interact with the portal. The third rule handles traffic that performs administrative functions for the administration of the ME itself and, presumably, has a minimal impact on performance.

The fourth and fifth rules handle traffic between the external subnet and internal segments of the portal network. Since the share of such traffic in the total mass is small, then, presumably, the impact of these rules on performance is small and it makes sense to present them as a single factor in a full-factor experiment.

Rules 13, 16, and 18 handle mail traffic and deal with the external attributes of the message, such as the size and composition of attachments. Analyzing these attributes presumably takes relatively little time, so it makes sense to also present these rules as a separate factor.

Rules 12, 14, and 15 also deal with mail traffic. Presumably, all of them are quite resource-intensive and can be considered as one significant factor in the experiment. The first rule matches the sender's address to the list of unwanted addresses. Since such a list is usually quite lengthy, this analysis can be quite time consuming. The fourteenth and fifteenth rules are very similar in that they analyze the message for the presence of dangerous html content, which is also a time-consuming task.

The seventeenth rule is presented as a separate factor since, presumably, the longest period of time is executed, since it analyzes both the header and the content of the mail message for the presence of words from the forbidden list.

Table 5 presents a set of experimental factors and the corresponding safety rules for studying the performance of an internal ME.

All pairs of rules presented in the table allow incoming and outgoing traffic of the same type, so it is logical to present them as one factor.

Thus, when measuring the performance of an external ME, the influence of five factors is analyzed. Each factor has two levels of significance. The number of full-factor

Table 5. Factors of the experiment to measure the performance of the internal firewall

Factor #	1	2	3
Rule #	6.7	8.9	10.11

experiments in this case is 25 = 32. When measuring the performance of the internal screen, 3 factors are used, and the number of experiments is 23 = 8.

The model of the information web portal during the research included four computers connected to a local network. Each computer in the network will simulate a separate segment or network element of a real portal.

The first computer simulates an external subnet with respect to the portal - the Internet. All free ports on this computer are used to simulate the addresses of portal users.

The second computer simulates the public portal server segment. It has an external ME portal installed and configured accordingly. In addition, a mail server is installed on the computer, which is integrated with the screen in order to provide mail filtering at the application level. The model will also use two ports of this computer - one to simulate the mail server, the other to simulate the http portal web server.

On the third computer, an internal ME is installed and configured accordingly, whose tasks include protecting the service server segment and the security management segment of the web portal.

The fourth computer simulates the segment of the portal service servers. The experiment uses three ports of this computer to simulate a database server (DB), an application server, and an authentication and authorization server.

Figure 1 shows the logical model of the web information portal network.

In the network model, during the experiment, the following PSs were used:

1) Kerio WinRoute Firewall 6 - cross-platform enterprise-level ME. In the model, it functions as an external firewall.
2) MDaemon 10.1 is a mail server.
3) Ipfirewall - cross-platform ME of the basic level. In the model, it functions as an internal firewall.
4) A specialized program for measuring network performance.

During the experiments, the load on the ME will be created by sending a large number of test packets of different sizes via the TCP protocol. The server of the test program creates connections with clients on the other side of the ME, simulating the interaction of users and administrators with the portal. After the connection is established, the server generates a large amount of traffic. The arrival time of each package is fixed by the clients, as well as the start and end times of the test. Then this information is sent to the server for further processing.

Figures 2 and 3 show the layouts of the server and clients of the test program in the network model of the information web portal when carrying out experiments on the processing speed of incoming and outgoing traffic, respectively.

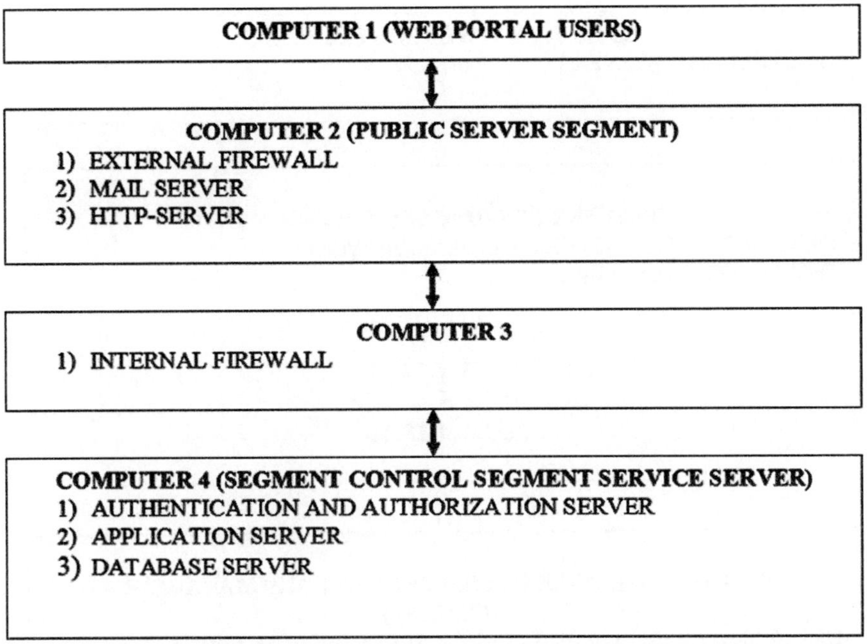

Fig. 1. Logical model of the network of the information web portal used in the research

Based on the results of measurements on the above metrics for different categories of users using the method of linear regression analysis, empirical formulas were built for the dependence of the speed of packet transmission through the TCP/IP protocol through the ME system on the security rules used.

The process of preparing for experiments involves a number of the following actions: connecting computers to the test network; setting by disabling all unnecessary operating system services. The tested ME and the mail server are configured to implement the specified security rules, all their additional functions are disabled.

As part of this study, an experiment was carried out on a model of an information portal network. At the same time, to clarify the performance characteristics of the studied types of MEs, we used an approach to measurements by various direct metrics:

1) The average time for an IP packet to travel to a segment of the demilitarized zone via the HTTP protocol.
2) Average transit time of an IP packet to the DMZ segment via postal protocols.
3) Average transit time of an IP packet from the DMZ segment to the external network via the HTTP protocol.
4) Average transit time of an IP packet from the DMZ segment to the external network via postal protocols.
5) Average transit time of an IP packet to the service segment.
6) The speed of processing incoming traffic by external FU is the number of IP packets checked for compliance with the specified security policy per unit of time (1 s).

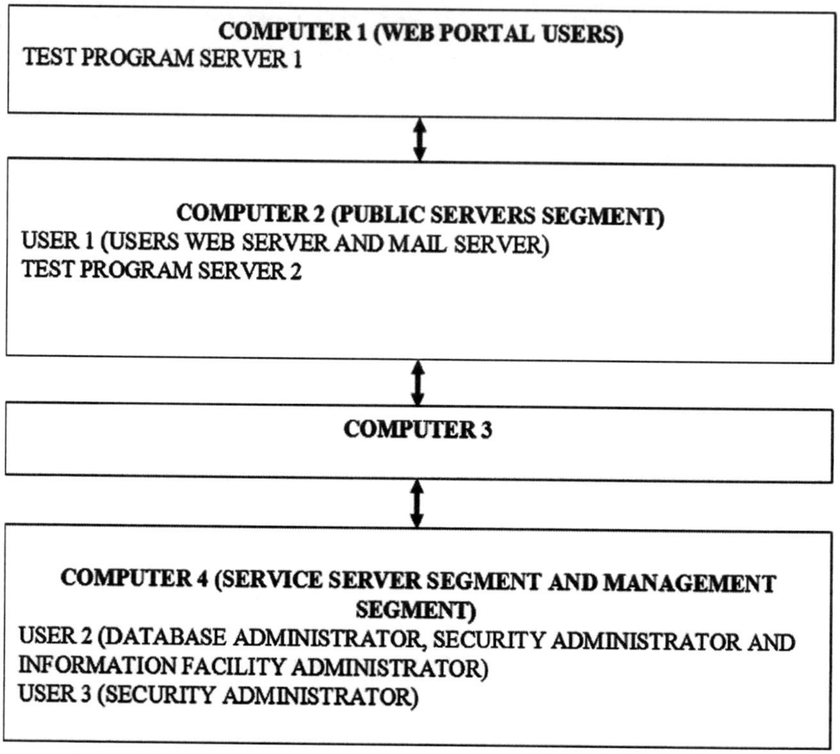

Fig. 2. Layout of test program components in the network model of the information web portal during the experiment on the processing speed of incoming traffic

7) The rate of processing of outgoing traffic by external ME is the number of IP packets checked for compliance with the specified security policy per unit of time (1 s).
8) The rate of traffic processing by the internal ME is the number of IP packets checked for compliance with the specified security policy per unit of time (1 s).

After the operating systems of the computers participating in the testing and the tested PS are prepared, using the test program, two measurements are created to measure the speed of incoming and outgoing traffic. The meter for measuring incoming traffic contains 153 test connections. These are presented in Table 6. The metering for the outbound traffic measurement includes 150 test connections. Their description is presented in Table 7.

The beginning of the experiment coincides in time with the end of the test measurements. In this case, the processes of the test program were assigned the "Realnoe Vremya" priority. The very process of the experiment consisted in the alternate execution of incoming and outgoing test measurements on the corresponding computers of the test network. The safety rules for the tested ME were preliminarily established in accordance with the experimental plan presented in Tables 8 and 9. The nomenclature of rules was corrected before performing each next pair of measurements. The scope of

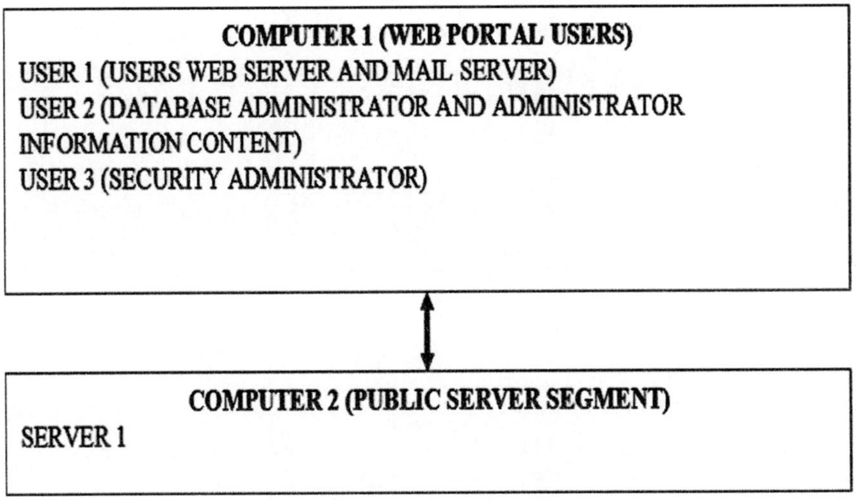

Fig. 3. Layout of the test program components in the network model of the information web portal during the experiment on the processing speed of outgoing traffic.

Table 6. Description of test connections for measuring the speed of work with incoming traffic

Number of connections of the	IPhost Host	port	Target IP	Target port
65	192.168.0.1	5000–5064	192.168.0.2	80
65	192.168.0.1	5065–5129	192.168. 0.2	8080
10	192.168.0.1	5130–5139	192.168.0.2	25
10	192.168.0.1	5140–5149	192.168.0.2	110
1	192.168.0.1	4444	192.168.0.4	2222
1	192.168.0.1	5555	192.168.0.4	2224
1	192.168.0.1	6666	192.168.0.4	1111

Table 7. Description of test connections for measuring the speed of work with outgoing traffic

Number of connections of the	IPhost Host	port	Target IP	Target port
65	192.168.0.2	80	192.168.0.1	5000–5064
65	192.168.0.2	8080	192.168.0.1	5065–5129
10	192.168.0.2	25	192.168.0.1	5130–5139
10	192.168.0.2	110	192.168.0.1	5140–5149

the experiment was 60 test measurements in accordance with the experiment plan - 30 measurements of incoming traffic and 30 measurements of outgoing traffic.

Table 8. Experiment plan for measuring the performance of an external ME

n	x1	x2	x3	x4	x5
1	1	1	1	1	1
2	−1	1	1	1	1
3	1	−1	1	1	1
4	−1	−1	1	1	1
5	1	1	−1	1	1
6	−1	1	−1	1	1
7	1	−1	−1	1	1
8	−1	−1	−1	1	1
9	1	1	1	−1	1
10	−1	1	1	−1	1
11	1	−1	1	−1	1
12	−1	−1	1	−1	1
13	1	1	−1	−1	1
14	−1	1	−1	−1	1
15	1	−1	−1	−1	1
16	−1	−1	−1	−1	1
17	1	1	1	1	− 1
18	−1	1	1	1	−1
19	1	−1	1	1	−1
20	−1	−1	1	1	−1
21	1	1	−1	1	−1
22	−1	1	−1	1	−1
23	1	−1	−1	1	−1
24	−1	−1	−1	1	−1
25	1	1	1	−1	−1
26	−1	1	1	−1	−1
27	1	−1	1	−1	−1
28	−1	−1	1	−1	−1
29	1	1	−1	−1	−1
30	−1	1	−1	−1	−1
31	1	−1	−1	−1	−1
32	−1	−1	−1	−1	−1

Table 9. Experiment plan for measuring the performance of internal ME

n	x1	x2	x3
1	1	1	1
2	−1	1	1
3	1	−1	1
4	−1	−1	1
5	1	1	−1
6	−1	1	−1
7	1	−1	−1
8	−1	−1	−1

4 Development of Experimental Studies of the Performance of Software Firewalls Used to Protect Information Web Portals in Ethernet Networks

In the course of experimental measurements, it is assumed that the test pro gram saves the average, maximum and minimum packet transit times for each test connection in special report files in XML format. Initial processing of reports based on the results of the experiment consists in calculating the average, minimum and maximum intervals of packet transit times for various groups of connections, for example, connections via the http protocol, based on the data on each test http connection. To fulfill the specified condition, all compounds belonging to a certain group were selected, after which the average values of their maximum, minimum and average time intervals are found.

In the next stage calculation capacity was conducted physical connection for each host by the formula:

$$N = \frac{n}{t}$$

where

n - the total number of test packets transmitted by the host;

t is the total testing time.

All calculated values are entered into the general table of the experiment, which must be edited before secondary processing.

The next step was to combine the results obtained for different ports of the http protocol, the test results for the smtp and pop3 protocols, as well as the test results for connections that simulate different portal administrators into one category. As a result of this combination, the final table will contain the average measurement results of the packet transit time over the test network and the throughput readings for its different sections. Since in the real Internet it is impossible to predict the transit time of packets through the network, and, accordingly, the throughput, then for secondary analysis it is necessary to select the share that the ME system brings to these indicators. This requires subtracting the value of this indicator from the packet transit time in the case when the influence of all factors is minimal (that is, all security rules are disabled). For throughput, on the contrary, it is necessary to subtract the results of all other rows from the row of the table in which the results in which the influence of ME is removed. After carrying out the calculations described above, Table 10 of the results of the experiment on measuring the performance of the external ME will contain 6 columns: two columns reflect the results of testing the throughput of incoming and outgoing traffic between the DMZ and the external subnet, and one column is reserved for the measured values of the average transmission rate of incoming packets. and outgoing traffic via http and mail protocols.

Table 11 contains the experimental results for performance measurement of the internal FW and consists of two columns: one column with the results of measuring the throughput between the DMZ and a column with the results of measurements of the average transmission rate of incoming traffic packets for connections simulating portal administrators. Since the internal shield security rules are the same for inbound and outbound traffic, measurements are only allowed for outbound or inbound traffic.

Secondary processing results of the experiment is to perform linear regression analysis of the experimental results and empirical formulas receipt reflecting the dependence of the packets speed and bandwidth the network model of the information web portal from the security policy rules used.

Within the framework of the described technique, the method of least squares is applied, in accordance with which the sum of the squares of deviations of actually observed Y from their estimates is minimized (we mean estimates obtained on the basis of the analysis of the values of the linear function reflecting the sought regression dependence).

Obviously, to obtain reliable results with a certain confidence level p, it is necessary to carry out at least a certain number of observations n. The problem of determining the required number of experiments, n, was solved using an interval estimate of the mathematical expectation of this value and its normalized form

$$u = \frac{\overline{y} - M\{y\}}{\sigma/\sqrt{n}},$$

where y is the average value of a random variable over a sample.

The interval estimate for M {y} is represented by the inequality

$$\overline{y} - \frac{u_p \sigma_y}{\sqrt{n}} < M\{y\} < \overline{y} + \frac{u_p \sigma_h}{\sqrt{n}}, \qquad (1)$$

where Up is the tabular quantile of the standard value corresponding to the probability p;

M {y} - mathematical expectation;

The sample has a certain scatter of values from the left border q1 to the right border q2. Then the length of the interval of values is L = q2 − q1. The larger the range of values of the quantity, the less reliable and less accurate the sample estimates [16].

The value ε was taken as an estimate of the accuracy

$$\varepsilon = L/2\sigma.$$

Table 10. Results of the experiment to measure the performance of the external firewall after preprocessing

n	Avg http	Avg mail	192.168.0.2	Avg http	Avg mail	192.168.0.1
1	29.371	114.823	728.765	33.933	90.890	619.687
2	26.027	77.006	709.252	24.443	81.690	574.801
3	34.500	79.476	902.919	28.751	69.292	758.505
4	23.068	92.875	719.684	20.703	90.345	665.115
5	30.310	83.572	558.371	37.649	66.848	494.123
6	16.808	51.465	530.985	18.804	75.867	329.424
7	15.560	63.547	604.970	13.736	62.075	347.099
8	11.417	37.908	497.690	10.627	42.518	445.581
9	28.665	85.838	514.468	36.190	62.305	485.684
10	20.185	68.922	423.689	15.549	58.860	447.796
11	22.976	68.506	519.041	27.867	60.253	403.633
12	10.526	60.247	324.199	11.849	43.185	358.073
13	32.171	29.680	287.778	14.890	25.359	327.178
14	11.535	27.586	74.716	11.669	37.969	74.371
15	22.147	30.489	315.798	16.468	30.514	200.150
16	5.187	22.044	52.251	8.640	9.741	119.355
17	35.679	87.978	682.803	21.529	67.904	562.515
18	24.530	67.363	569.790	29.002	65.557	528.072
19	27.308	76.732	627.038	23.447	78.278	538.739
20	21.819	83.786	622.487	21.066	59.619	480.479
21	19.879	54.711	414.383	29.875	39.439	355.122
22	9.758	29.162	372.225	12.391	44.693	219.335
23	24.549	41.311	431.940	17.298	39.250	308.794
24	6.955	28.459	345.191	9.106	16.755	185.311
25	19.342	57.131	368.018	28.921	40.447	378.292
26	17.221	44.799	270.616	16.325	30.784	289.577
27	22.138	47.475	337.338	27.565	40.676	332.001
28	7.582	27.666	277.442	10.166	27.665	265.912
29	12.618	22.843	120.796	21.489	12.457	124.422
30	8.316	0.500	37.226	1.480	3.952	34.920
31	10.373	11.043	85.669	14.489	7.962	74.585

Table 11. Results of the experiment on measuring the performance of the internal firewall after primary processing

n	Avg внутр	192.168.0.3
1	22.398	189.770
2	13.633	125.151
3	14.869	144.996
4	7.056	74.720
5	12.989	128.524
6	6.010	52.340
7	6.551	75.574

Here the denominator is a constant - the larger the interval of L values, the lower the accuracy and the greater the relative deviation ε, i.e. the value of this characteristic is inverse to the accuracy.

The left and right sides of expression (2) will be considered as the boundaries of q1 and q2, then

$$L = 2\frac{u_p \sigma_y}{\sqrt{n}},$$

and the relative error ε will be $\varepsilon = u/\sqrt{n}$, whence it follows

$$n \geq \left(\frac{u_p}{\varepsilon}\right)^2$$

This formula can be written in the following form:

$$n \geq \left(\frac{u_p}{u/\sqrt{n}}\right)^2$$

from where we derive

$$n = \left(2u_p\right)^2 \tag{2}$$

Let us assume the confidence level is 0.95 and the significance level is 0.01. Tabular quantile $U_{0.95}$ while it is equal to 2.629. Then, in accordance with (2), the required minimum number of observations n will be

$$n = (2 * 2.629)^2 = 28$$

Thus, the permissible measurement accuracy will be achieved when conducting twenty-eight experiments.

5 Conclusion

1) The analysis of the features of the functioning of firewalls used to protect information web portals in ethernet networks. The analysis showed that when examining the performance of such screens, the parameters for connecting screens to the network and their additional functions are fixed. Managed parameters are sets of security policy rules.
2) A model of the information web portal network has been developed, which contains only the elements necessary for researching the performance of ME; a test network was built corresponding to this model.
3) An experiment was planned and carried out to determine the performance of the ME system included in the network according to the standard protection scheme for information web portals.
4) It has been established that the rational number of factors when measuring the performance characteristics of the external ME is five, and when measuring the performance characteristics of the internal ME - three. The number of tests required should be 32 and 8, respectively.
5) A technique for studying the performance of ME is described, which provides the necessary measurement accuracy due to the sequential implementation of the primary processing of the experimental results based on the calculation of the average values of the measured quantities, and the secondary processing of these quantities using linear regression analysis.

References

1. Зорин, В.: Архитектура защищённого портала. www.elvis.ru/files/prot_portal.pdf
2. Качинский, И.: Программные межсетевые экраны. www.osp.ru/lan/2002/03/135923/
3. Зорин, В.: Система защищённого доступа в Internet. www.elvis.ru/files/Internet_Sec.pdf
4. Сердюк, В.: Как защитить Web-портал от информационных атак "Журнал сетевых решений LAN", №. 9 (2004)
5. Равичев, Л.В., Ломакина, И.А.: СТАТИСТИКА: РХТУ им. Д.И.Менделеева. Москва (2007)
6. Максимов, В.: Защита сети: комплексный подход. www.ellib.gpntb.ru/?journal=ap&year=2006&num=7&art=8
7. Биячуев, Т.А., под ред, Л.Г.: Осовецкого Безопасность корпоративных сетей. – СПб: СПб ГУ ИТМО, 161 p (2004)
8. Wool, A.: How to Use firewall management solutions to improve firewall performance and security. https://www.eweek.com/c/a/Security/How-to-Use-Firewall-Management-Solutions-to-Improve-Firewall-Performance-and-Security/
9. Twedt, L.: A Comparison of firewall performance in distributed systems. Saint Mary's University of Minnesota (2005)
10. Research and Markets Ltd. Firewalls are critical components of every network - from Small-Office/Home-Office (SOHO) through the largest Enterprises and Service Providers. https://www.download3k.com/Press-Firewalls-are-critical-components-of-every.html

11. Albert Greenberg. Simulation Study of Firewalls to Aid Improved Performance. http://www2.computer.org/portal/web/csdl/doi/10.1109/ANSS.2006.42.
12. IEEE 1471, Recommended Practice for Architecture Description of Software-Intensive Systems
13. Вэк, Д.: Содержание сети вашей организации в безопасности при работе с Интернетом (Введение в межсетевые экраны (брандмауэры)). https://www.citforum.ru/internet/nist/.
14. CERT Coordination Center. Packet Filtering for Firewall Systems. https://www.cert.org/tech_tips/packet_filtering.html
15. Мэдисон, А.: Руководство по обеспечению безопасности ISA Server 2006. https://technet.microsoft.com/ru-ru/magazine/2008.09.isahardening.aspx
16. Сердюк, В.: Комплексный подход к обеспечению безопасности на основе многоуровневой защиты от вирусных угроз. https://www.antivir.ru/main.phtml?/press-center/press
17. Давлетханов, М.: Система защиты веб-портала. https://hostinfo.ru/articles/568

The Approach to Network Planning Process Improvement

L. S. Globa(✉) and V. A. Prokopets(✉)

National Technical University of Ukraine "Igor Sikorsky Kyiv Polytechnic Institute", 37, Prosp. Peremohy, Kyiv 03056, Ukraine
lgloba@its.kpi.ua

Abstract. The rapid growth of modern communication technologies, moving to the 5G ideology requires a rethinking of approaches related to the networks planning and the services they provide. Current trends like energy efficiency, planning with traffic uncertainty considerations requires a significant change in existing network planning technologies. The main idea of this research is to provide a pipeline that will process the network planning process to focus on additional designing stage as marketing forecast, B2B, B2C, rollout plan, target cells, equipment in stock. The pre-planning algorithms for the definition of the outputs (required equipment, budget spending, others), the algorithms of equipment-based optimization, problem detection (solution type selection) are suggested. The suggested approach takes into account all new key features and technologies and focuses on improving the pre-planning stage.

Keywords: Planning · Cellular · Mobile · Network · Optimization

1 Urgency of the Problem

The development of new technologies requires new demands. New technologies such as massive MIMO, fractional frequency reuse, Small Cells, Coordinated Multipoint (CoMP) should be reflected in the cell planning approach as ways to reach capacity and coverage targets. A new conception of Self Optimization Networks (SON), which aims at Total Cost Ownership (TCO) reduction, could be introduced in LTE networks. Other improvements as performance improvement and complexity management, which SON provides, could have much effort on the network deployment [1]. Not less important are current eco-trends and, in particular – energy efficiency or a combination of self-sufficiency in energy with power management [2].

Future 5G networks will bring additional features such as widespread Software Defined Networks (SDN), Network Function Virtualization (NFV), millimeter-wave spectrum, network ultra-densification, M2M connections with high mobility, and new radio access technologies [3]. All these features and trends should be reflected in the network planning approach.

P. Vorobiyenko et al. (Eds.): IPF 2020, LNNS 212, pp. 293–307, 2021.
https://doi.org/10.1007/978-3-030-76343-5_15

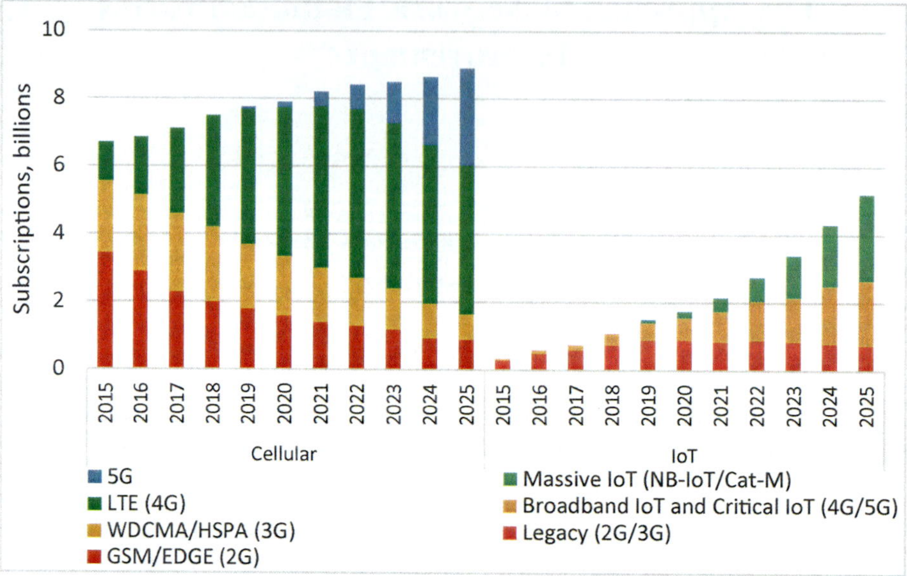

Fig. 1. Mobile sybscriptions by technology (Ericsson mobility report 2019)

Figure 1 shows that there is a significant growth of 4G/5G subscriptions in the next five years. Also, IoT will add extra network load due to exponential connection growth.

One more important thing is the response speed on force majeure circumstances. Coronacrisis in 2020 forced many people to observe self-isolation, leading to increased load in cellular networks [4]. The majority of cellular network operators reported significant traffic consumption increasing. For example, AT&T statistic shows 20% core traffic growth and rise of voice traffic up to 40% [5].

All statistics above show that every kind of network (GSM, UMTS, LTE) should be taken into account during the pre-planning stage because there is a need to serve both voice and data traffic. Also, a lot of new input data appears every year. New technologies and the unpredictable growth of traffic usage caused by these technologies make the existing state of the art planning useless without accurate pre-planning.

2 A Brief Tutorial on the Planning Process

The network planning process consists of three steps [6]:

- Pre-planning
- Detailed planning
- Optimization

All these steps go sequentially and iteratively. Usually, the whole iteration (the period between two sequential pre-planning actions) takes from 3 months to 1 year.

The pre-planning stage is the first step in the planning process to get approximate BS locations and the number of equipment necessary for meeting current requirements in network development. This stage is heavily coupled with budget planning because it defines current CAPEX. Usually, high-level calculations of this stage repeat until the result will meet the budget spending plan. List with the ordered equipment amount per location is a final result of this step. Authors in [7] propose a statistical approach that uses derivations of probability density functions, which are used for capturing tradeoff trends between coverage, capacity, and QoS in LTE. Later, they modified it in [8] to the Radio Network Dimensioning step of the planning process. The results show that this approach has better planning results (lower level of an unjustified number of radio sites and related costs). This method's drawbacks are focus on a single Radio Access Network type and lack of optimization algorithms.

Detailed planning is held after equipment ordering. The purpose of this stage is to get exact BS locations with tuned technical parameters. Most of the researches are dedicated to cellular network planning concerns this stage as most time and labor-intensive. For example, Ant Colony Optimization Algorithm for Network Planning in Heterogeneous Cellular Networks [9], Frequency Plan Optimization Based on Genetic Algorithms for Cellular Networks [10], Multi-Objective Optimization for Heterogeneous Cellular Network Planning [11], Artificial intelligence-based novel scheme for location area planning in cellular networks [12]. These current researches are dedicated to the usage of state of the art technologies in detailed planning. Orders on BS creation and rollout are the final results of this step.

The optimization stage comes after network commissioning and statistic collection. The purpose is to find flaws and correct them. There are two ways how to do it:

- Parameter tuning. In case of interference or minor lack of coverage, antenna tilt and transmission power could be tuned to better performance. Cellular network operators could perform this approach without any investments in a short time.
- Equipment – based. This approach has a priority to use while there is a significant lack of coverage or capacity. The essence is to install new equipment (multi-beam antennas, new BS, carriers, etc.), which requires investments. This approach is heavily coupled with the pre-planning stage as equipment ordering is held after it.

3 The Proposed Approach

The purpose of this research is to provide a pipeline that will process the network planning process described in Fig. 2. The focus is on blocks highlighted with orange – new approaches that will reflect is are future work. Other parts like Parameter tuning and Detailed planning tool could be introduced with any detailed planning software (Forsk Atoll, Infovista Planet, etc.) or other planning or optimization algorithms, respectively.

Figure 2 displays the proposed cellular network planning process pipeline. Several additional components define all timings between stages. Marketing analysis usually is done every quarter or year and triggers the pre-planning stage. Equipment ordering is a trigger of a detailed-planning stage – there is no sense to do this without assurance that equipment will be installed in time. BS rollout allows operators to collect relevant statistics and to perform optimization.

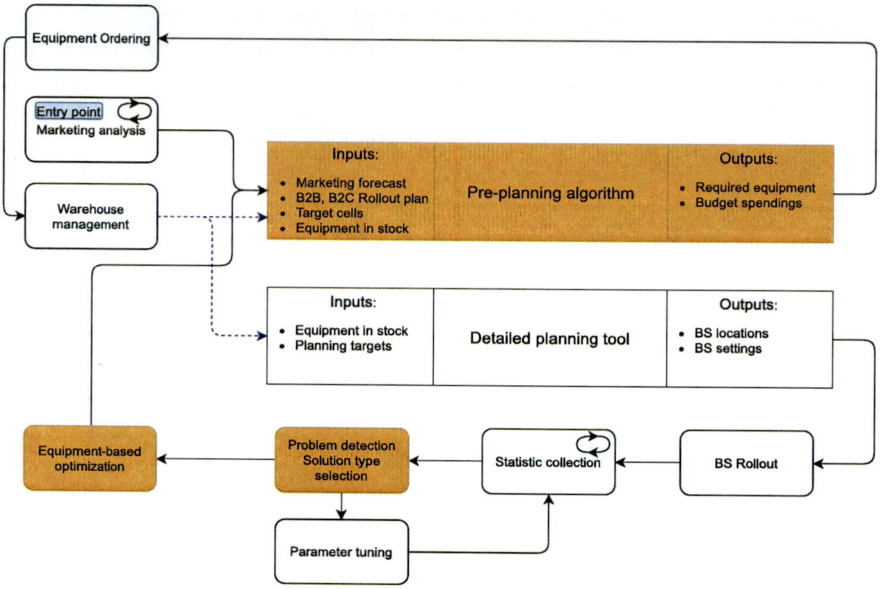

Fig. 2. Network planning process pipeline

The Problem detection and Solution type selection block will analyze the collected statistic and check whether it is possible to solve drawbacks using only parameter tuning without additional equipment.

The Equipment-based optimization block will analyze performance statistics and propose, if necessary, some types of equipment that could be used to fix capacity and coverage issues like multi-beam antennas, additional spectrum RRU (Remote Radio Unit), etc.

The pre-planning algorithm is the research topic. It will use marketing forecasts, existing rollout plans, and suggestions from Equipment-based optimization blocks as inputs. The purpose of this block is to provide a list of equipment that should be purchased to meet current trends. Since the equipment purchase process lasts much time and the detailed planning process is highly dependent on the equipment list, there is a need to create a new pipeline that will consider all input data and features described above.

One more research question is to analyze the efficiency of modern prediction and optimization techniques like Neural Networks, Genetic Algorithms, Simulated Annealing in the network planning process. The resulting algorithm should automatically decide what optimization method to use according to the statistical data and results from previous predictions.

4 Detailed Description of Proposed Approach

In this section more detailed algorithm of Network planning process is presented. Each algorithm step is categorized and described.

The proposed algorithm is shown in Fig. 3. The blocks highlighted in green represent modules in which it is recommended to reuse existing approaches. The blocks highlighted

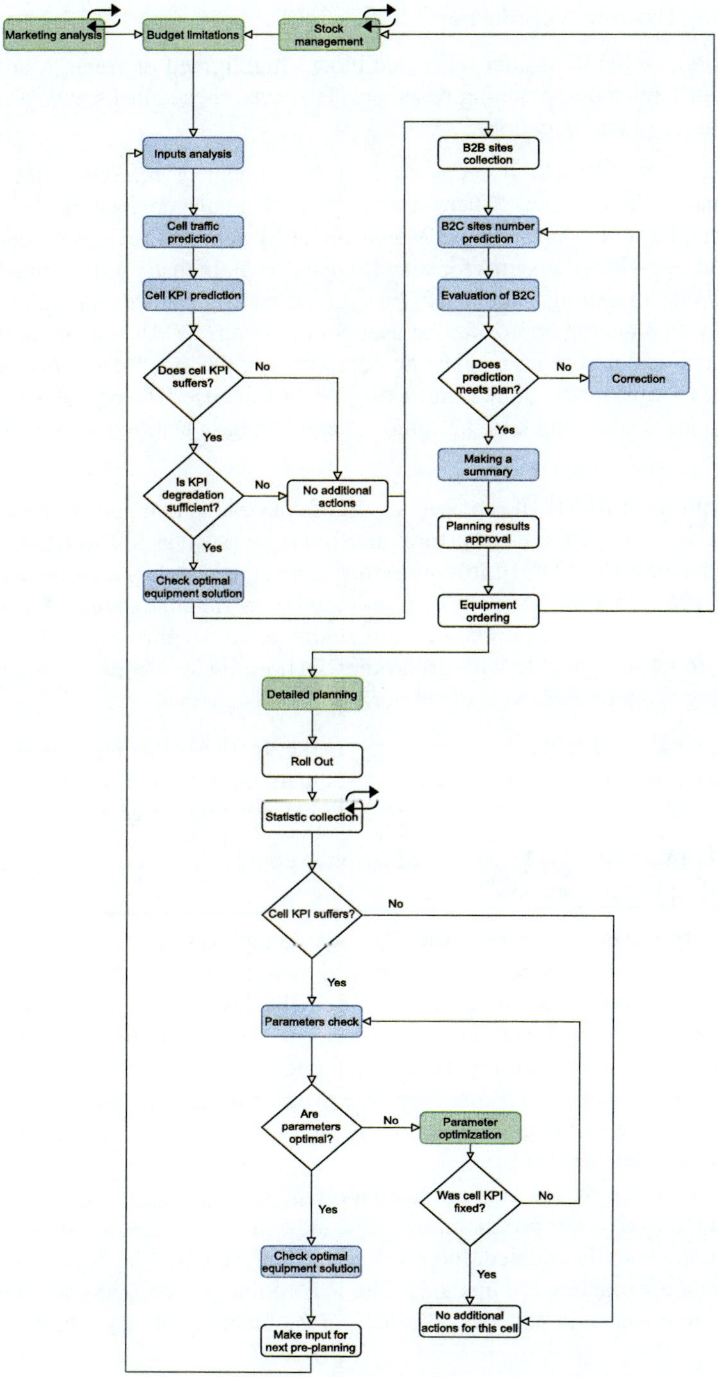

Fig. 3. Detailed network planning process algorithm

in blue contains completely new realization. All other blocks are not the subject of this research.

4.1 Reusing Existing Approaches

In this section, we will consider borrowed blocks highlighted in green. Many of them are not directly related to planning processes. However, these blocks have a significant influence on the planning course.

Stock Management Block. It keeps track of the amount of equipment present, their incomes, and their outcomes. There are many stock management models as manual so as automatic (WMS – Warehouse Management System). WMS's last trends are the Internet of things – based systems for smart logistics [13], Industry 4.0 automation trend: usage of Cloud Computing, Blockchain-based systems, Big Data technologies, etc. [14]. In the proposed planning approach, the user could use any WMS. The main demand is to provide the actual number of existing equipment at the stock and solve equipment distribution's simple task if there is more than one warehouse. The equipment should be delivered to the nearest stock to the future or existing base station to save the CAPEX and OPEX.

Budget Limitation Block. It represents a set of constraints that need to be satisfied to keep main economic KPI's (Key Performance Indicator) such as ROMI (Return on marketing investment), EBITDA (Earnings before interest, taxes, depreciation, and amortization), etc. at a high level. These limitations could be as valuable (budget for equipment ordering) so as invaluable (equipment manufacturer, usually equipment of different manufacturers are not compatible with each other). All limitations could be defined using some existing economic analysis of network [15].

Marketing Analysis Block. It contains any market forecasts that try to predict market changes during the next period. This analysis's main output is the trend of user profile changes and subscriptions amount change due to existing coverage, price policy, etc.

Detailed Planning Block. This block description could be found in the section "A brief tutorial on the planning process".

Parameter Optimization Block. Aims to find optimal parameters set of existing base stations to fix minor drawbacks. A lot of these drawbacks are caused by coverage overlapping, which leads to interference. In this case, CINR (Carrier to interference + noise ratio) parameter suffers, worsening user experience. This could be fixed with antenna tilt and transmit power. The current trend in parameter optimization is SON (Self Organizing Network) - automation technology of network management and optimization. An example of the state of the art research is the auto-tuning self-optimization algorithm for mobility management [16].

Blocks described in this section are important steps in the planning process. As this work is focused on the pre-planning stage and equipment-based optimization solutions, which are heavily coupled, blocks described above could be borrowed from the other planning approaches and methods. The main point of such integration is keeping consistent interconnection between all blocks in the planning process chain.

4.2 New Blocks of the Enhanced Planning Process

This section is dedicated to new and rethought blocks. They are all related to the pre-planning stage and equipment-based optimization, which are not enough covered in current research.

Input Analysis. This block is the entry point of the enhanced planning process. The main purpose of this block is to collect all input data:

- Marketing forecast;
- Budget and other constraints;
- Warehouse status report;
- Set of optimal equipment selected by this approach (will be described later).

One other function is to transform all the data into the appropriate format. As most of the inputs are produced by different blocks, they could have a different format. One way to fix it is to use a transformer with a pre-defined set of transformation rules. One of the most widely used unification formats is JSON (JavaScript Object Notation), which is supported by most programming languages and tools. However, the JSON string is hardly understandable by humans, especially if there are many data encapsulated. Another popular format is CSV (Comma Separated Values), which has much higher readability for human (via spreadsheet programs such as Microsoft Excel, LibreOffice Calc, etc.) than JSON. However, CSV could not represent data structure in the case of hierarchical and other relational data. Formats comparison is presented in Table 1.

Table 1. Comparison of JSON and CSV data formats.

Evaluation criteria	JSON	CSV
Versatility	Higher	Lower
Scalability	Higher	Lower
Support of hierarchical data	Yes	No
File size	Higher	Lower
Human readability	Lower	Higher
Memory consumption	Higher	Lower

The most important properties listed in Table 1 are versatility, scalability, and support of hierarchical data. The higher versatility of JSON means that it is more unified and could be used for describing more types of systems. Also, JSON has more scalability in terms of content management (adding, change, movement). Support of hierarchical data allows describing more complex systems and processes.

All other properties have a lower priority; however, they also need to be supported. Smaller file size and memory consumption of CSV format allow the parse of a larger amount of data during a defined time period. Also, CSV compatibility with spreadsheets allows easy and understandable processing of data for humans.

The preferred data transmission format is JSON, as it much easier for data processing. However, support of CSV should be present as it also widely used in current business processes.

Cell Traffic Prediction. This is the block aimed at traffic volume prediction on each cell of the network. It uses Marketing forecast and current statistical data for this. In the simplest case, traffic volume is calculated as the multiplication of current traffic volume and growth multiplicator:

$$V_{predicted} = V_{actual} * K_{growth} \tag{1}$$

There are also more advanced traffic prediction models as Temporal Graph Convolutional Networks [17], Deep Learning approaches [18] and other forecasting models [19].

The one more important function of the Cell traffic prediction block is evaluating the impact of the user profile changes due to long-time trends (video traffic consumption growth, mobile gaming popularization, etc.) and short-period influence (commissioning of new tariffs, event service, etc.) The key approach is user profiling – building a comprehensive user profile. A user profile summarizes the user's interests, characteristics, behaviors, and preferences, while user profiling collects, organizes, and inferring the user profile information [20]. In terms of cellular networks, the main indicators are the QCI (Quality of Service Class Identifier) proportions table, representing user interests in different kinds of traffic, user mobility indicator, and others. User profile changes significantly influence traffic change trends, so it needs to be evaluated among other key marketing factors.

The Cell KPI Prediction. Is the block is heavily coupled with the Cell traffic prediction block. The results of the last one are used as inputs in cell KPI prediction. All traffic changes cause load level changes and/or service level demands. This leads to main KPI changes (traffic growth causes a higher level of interference, lack of frequency resources, etc.) and to KPI thresholds change (for example, video call requires much higher downlink and uplink speed than audio call). We need to be able to predict all the KPI changes due to traffic changes.

Check Optimal Equipment Solution. The block is responsible for equipment solution choice when parameter tuning, described in the section "Reusing existing approaches" will not help. These problems could be joined into two main groups:

- Lack of capacity;
- Major lack of coverage.

Lack of capacity is when the number of active users is higher than the threshold defined according to SLA (Service Level Agreement). This agreement approves key

KPI that should be provided to each user in the target cell. These KPI could be reached if giving some resources (time, frequency, and codes). Resources volume is limited in each cell. So, the target threshold could be defined as the division of the total amount of resources that could provide a maximal level of KPI on target KPI:

$$Capacity\ threshold\ (users) = \frac{total\ resources \Rightarrow \max KPI}{target\ KPI} \tag{2}$$

However, the possibility that all users will require the full amount of resources granted by SLA is quite low. This means that the Capacity threshold could be increased. This increasing factor is an object of future work. Correspondingly, the capacity threshold equation could be redefined:

$$Capacity\ threshold\ (users) = \frac{total\ resources \Rightarrow \max KPI}{target\ KPI} * increasing\ factor \tag{3}$$

In some cases, the capacity problem could be solved via parameter tuning. Such kind of problem usually is caused by mistakes in coverage planning. For example, in some highly loaded areas, two neighbor cells serve this area with different load proportions. In some cases, cell reconfiguration could solve this problem and uniformly distribute all load between neighbor cells. This result could be reached by coverage area changing using antenna tilt and transmit power reconfiguration. An example of such an approach is described in Fig. 4. In this case, Site 1 serves most users in the load area while having the same coverage square as Site 2, which serves only 20% of users. We need to migrate some users from Site 1 to Site 2 to make the load satisfactory. To make this, we could change the electrical tilt of Site 1 antennas and/or reduce transmission power, while at Site 2, we need to do opposite actions. As a result, both sites' coverage area will change, and all the users will be distributed more fairly between sites.

The only solution is to install some additional equipment to provide more resources amount in all other cases.

The main ways to increase cell capacity are:

- Usage of multi-beam antennas;
- New site construction;
- Broadening the spectrum.

Multi-beam antennas provide a higher level of capacity without an additional space requirement in antenna stands. A Multi-beam antenna is an analog of several single beam antennas placed together. The difference is that all beams are placed in a single body. To reduce mutual interference between beams, their width at level −3 dBm is divided into the number of beams. Also, all beams are shifted to each other (Fig. 5).

Like a set of separate single-beam antennas, each beam of multi-beam antenna has its own RRU (Remote Radio Unit) – transceiver used to provide wireless data transfer. Because each RRU has its own resource volume and each beam covers its own area, in ideal conditions profit of multi-beam antenna installation is linear and equals the number of beams, so resulting capacity could be calculated as:

$$Resulting\ capacity = RRU\ capacity * number\ of\ beams \tag{4}$$

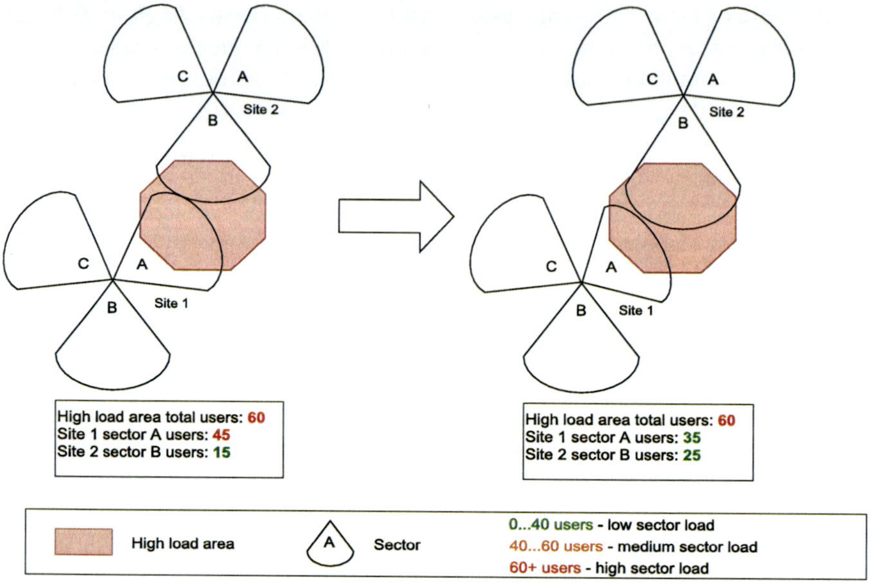

Fig. 4. Cell reconfiguration to fit load area.

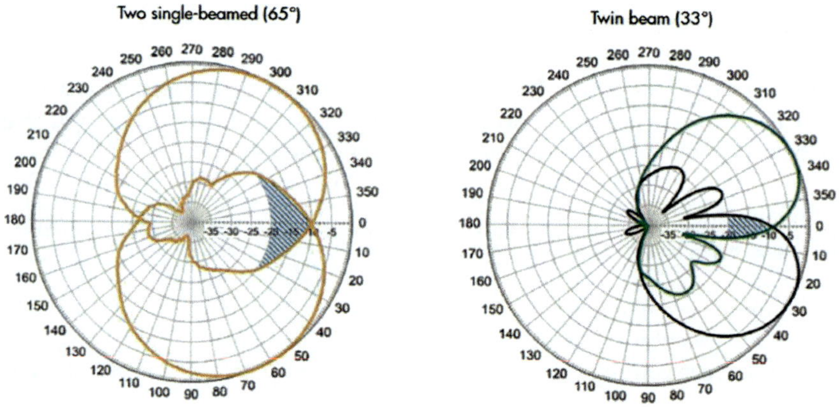

Fig. 5. Left – two single-beamed antennas radiation pattern. Right – dual beam antenna radiation pattern [21].

There is no possibility to avoid mutual interference between beams in real conditions completely so that the resulting capacity will be lower. Usually, the multi-beam antenna solution is used when traffic hot-spot is placed completely inside the single sector and has a quite large overload coefficient (Figs. 6 and 7).

New site construction. The solution described in Fig. 4 is applicable only if the hot-spot load is lower than the neighbor cells' combined capacity. In the opposite case, there are 2 possible solutions: to use multi-beam antennas or to build a new site. To make a decision on which solution to use, a detailed situation analysis is required. In

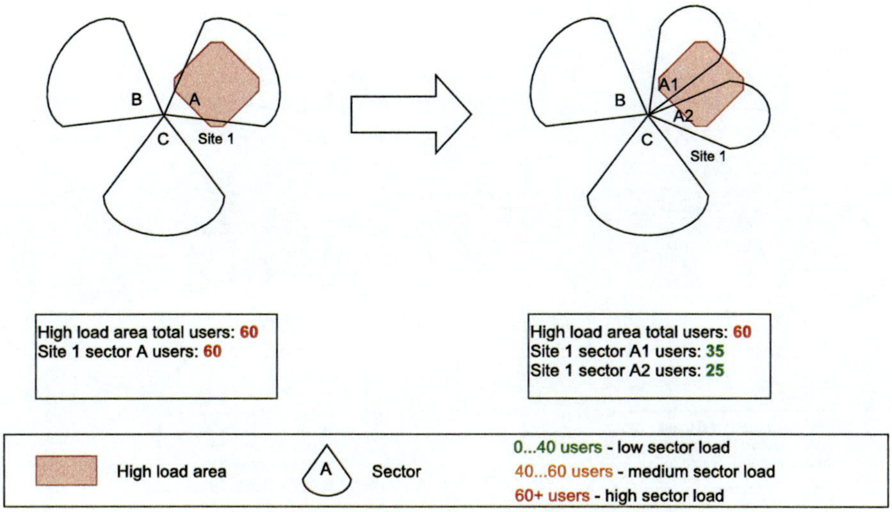

Fig. 6. Usage of dual-beam antenna solution.

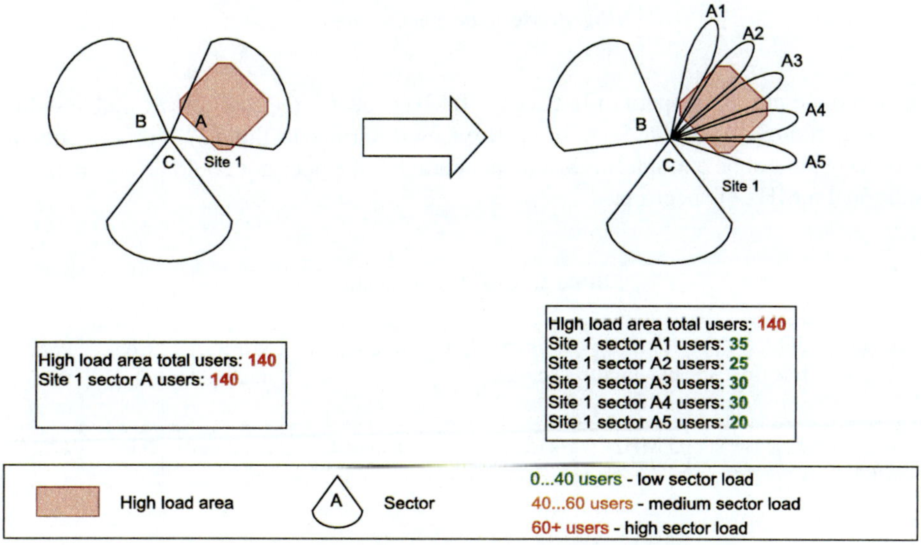

Fig. 7. Usage of penta-beam antenna solution.

many cases, overloading areas with multi-beam antennas will lead to high interference, frequent handovers, etc., which is not the best option. The other way to deal with such load is to build a new site near the hot-spot area with parameter tuning on the neighboring cells to distribute the load. This solution is described in Fig. 8.

Broadening the spectrum is used only in case of part utilization of frequency resources. For example, Mobile Network Operator has a license on 15 MHz spectrum for the 3G network. The single carrier in 3G has a bandwidth of 5 MHz. On several sites

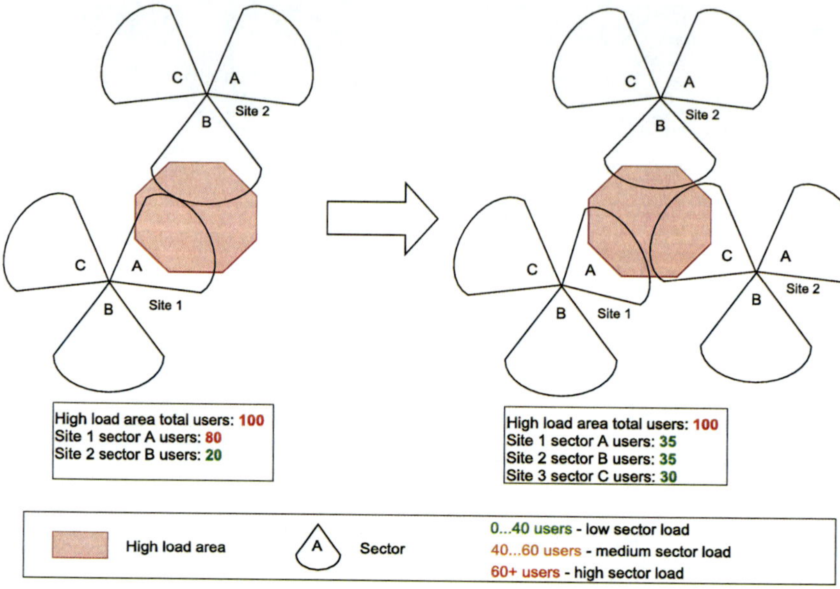

Fig. 8. New site construction.

for economic purposes operator uses only 10 MHz spectrum (2 carriers available – this amount is enough to provide SLA for 3G services, described in Table 2). If this amount of resources will not be enough for the current demand, the operator could always activate additional 5 MHz of spectrum.

Table 2. UMTS specification

Protocol	3GPP release	Carrier bandwidth	HS-DSCH codes	Modulation	MIMO, dual-cell	Code rate	Peak DL rate
Dual-cell HSDPA	Release 8, Category 24	5 MHz	15	QAM-64	Dual-cell	0.98	42.2

Another case is the common usage of some frequency bands by several networks, like Band 3 (1800 MHz), for 2G and 4G networks. In case of lack of resources in the 4G network, while the number of resources in 2G will be excess, the operator could relocate additional spectrum from 2G to 4G—Table 3 shows possible 4G spectrum allocation options.

The solution selection algorithm should carefully analyze the situation at each problem cell. It needs to take into account:

- Load at the target cell and its neighbors;

Table 3. LTE specifications

Channel bandwidth	1.4 MHz	3 MHz	5 MHz	10 MHz	15 MHz	20 MHz
RBs	6	15	25	50	75	100
Subcarriers	72	180	300	600	900	1200
FFT size	128	256	512	1024	1536	2048
Peak DL rate	4.4	11.1	18.3	36.7	55.1	75.4
SC spacing	15 kHz					
OFDM samples per subframe	7 for normal CP					

- KPI which influence user experience at the target cell and its neighbors;
- Traffic distribution area, calculated from previous knowledge;
- An effort from each type of solution;
- Parameter configuration at the target cell and its neighbors;

The steps of solution selection algorithm to make a decision:

- Analyze load and other KPI at influenced cells;
- Build traffic distribution area;
- Calculate the required effort from the solution;
- Analyze budget and other restrictions;
- Make a decision;
- Tune neighbor cells to fit their parameters according to a new solution.

B2C Sites Number Prediction Block. It makes a prediction based on the following inputs:

- Number of sites targeting coverage expansion;
- Results obtained from Check optimal solution block.

The number of coverage-targeted sites could be evaluated using marketing defined targets (focus-spots) and detailed planning tools. The first one defines some areas where coverage should be provided, while the second one, with the help of optimization techniques [22], could find the number of sites required.

Then, both inputs being combined, and constraint solvent is executed. This is held by block Evaluation of B2C. If some of the constraints are not met, then the correction block tries to change solution configuration in some areas, trying to meet restrictions with minimal sacrifices.

Parameters check block. It analyzes whether the site configuration (technical parameters, like antenna tilt, transmit power) is optimal. This could be done by analyzing such key parameters as noise level, the traffic distribution function of distance to the site, etc. However, as a result of planning or configuration mistakes, any parameter could have a wrong or forbidden value. If the cell has a high level of noise caused by interference, some anomalies in traffic distribution, or wrong parameter values, we need to fix parameters and find a better cell configuration before applying the equipment-based solution.

All blocks presented in this section will help to make pre-planning output results more precise. Flexible adjustment possibilities allow adapting planning results to a set of existing constraints, while the main target – solution efficiency maximization remains close to optimal.

5 Conclusions

There is a need to improve the existing planning approach (which consists of all three planning stages). In the state of art literature, there are no approaches that will meet current requirements in network planning. Almost all modern researches are dedicated to the detailed planning stage. There are no researches that will take into account all three planning stages linking technical and marketing inputs and targets into one holistic system. Also, no algorithm could evaluate the impact of new services commissioning or other changes in user behavior. The purpose of our research is to investigate a new planning approach that will fill in all the gaps in the planning process, eliminating manual planning processes. The pipeline that consists of new blocks of Equipment-based optimization and Pre-planning algorithms and custom blocks of Parameter optimization and Detailed-planning tool is presented in this paper. Future work is related to the research of new blocks' functionality.

References

1. Cisco: SON and the LTE Challenge: How to Get More for Less. C11–733194–00 01/15 (2015)
2. D'Amours, M., Girard, A., Sanso, B.: Planning solar in energy-managed cellular networks. IEEE Access. 1 (2018). https://doi.org/10.1109/ACCESS.2018.2877040
3. Aguilar-Garcia, A., Barco, R., Fortes, S.: Coordinated location-based self-optimization for indoor femtocell networks. Comput. Netw. **106** (2016). https://doi.org/10.1016/j.comnet.2016.06.009
4. Rajiullah, M., Khatouni, A.S., Midoglu, C., Alay, O., Brunstrom, A.: Mobile network performance during the COVID-19 outbreak from a testbed perspective (2020)https://doi.org/10.1145/3411276.3412194.
5. AT&T. 6 Months. In: Unyielding Connections in the Age of COVID-19/AT&T – Resource access mode. https://about.att.com/innovationblog/2020/09/fn_covid_19_six_months.html
6. Taufique, A., Jaber, M., Imran, A., Dawy, Z., Yaacoub, E.: Planning wireless cellular networks of future: outlook, challenges, and opportunities. IEEE Access 1 (2017). https://doi.org/10.1109/ACCESS.2017.2680318
7. Jaber, M., Akl, N., Dawy, Z., Yaacoub, E.: Statistical link budget analysis approach for LTE cellular network dimensioning. In: European Wireless 2014; 20th European Wireless Conference, Barcelona, Spain, pp. 1–6 (2014)
8. Jaber, M., Dawy, Z., Akl, N., Yaacoub, E.: Tutorial on LTE/LTE-a cellular network dimensioning using iterative statistical analysis. IEEE Commun. Surv. Tutor. **18**(2), 1355–1383 (2016). https://doi.org/10.1109/COMST.2015.2513440

9. Tseng, F.-H., Kao, F.-Y., Liang, T.-T.: Ant colony optimization algorithm for network planning in heterogeneous cellular networks (2020). https://doi.org/10.1007/978-981-15-3308-2_2
10. Megnafi, H.: Frequency plan optimization based on genetic algorithms for cellular networks. J. Commun. Softw. Syst. **16** (2020). https://doi.org/10.24138/jcomss.v16i3.1012
11. Chen, C.-Y., Tseng, F.-H.: Multi-objective optimization for heterogeneous cellular network planning. IET Commun. **13**, 322–330 (2019). https://doi.org/10.1049/iet-com.2018.5332
12. Vimal, V., Singh, T., Qamar, S., Nautiyal, B., Singh, K., Kumar, A.: An artificial intelligence-based novel scheme for location area planning in cellular networks. Comput. Intell. (2020). https://doi.org/10.1111/coin.12371
13. Lee, C., Lv, Y., Ng, K., Ho, W., Choy, K.L.: Design and application of internet of things-based warehouse management system for smart logistics. Int. J. Prod. Res. 1–16 (2017). https://doi.org/10.1080/00207543.2017.1394592
14. Fernández-Caramés, T., Blanco-Novoa, Ó., Froiz-Míguez, I., Fraga-Lamas, P.: Towards an autonomous industry 4.0 warehouse: a UAV and blockchain-based system for inventory and traceability applications in big data-driven supply chain management. Sensors **19**, 2394 (2019). https://doi.org/10.3390/s19102394
15. Hoikkanen, A.: Economics of 3G long-term evolution: the business case for the mobile operator, pp. 1–5 (2007). https://doi.org/10.1109/WOCN.2007.4284154.
16. Alhammadi, A., Mardeni, R., Alias, M.Y., Shayea, I., Alraih, S., Mohamed, K.: Auto tuning self-optimization algorithm for mobility management in LTE-A and 5G HetNets. IEEE Access 1 (2019). https://doi.org/10.1109/ACCESS.2019.2961186.
17. Zhao, L., Song, Y., Zhang, C., Liu, Y., Wang, P., Lin, T., Deng, M., Li, H.: T-GCN: a temporal graph convolutional network for traffic prediction. IEEE Trans. Intell. Transp. Syst. 1–11 (2019). https://doi.org/10.1109/TITS.2019.2935152
18. Yisheng, L., Duan, Y., Kang, W., Li, Z.: Traffic flow prediction with big data: a deep learning approach. IEEE Trans. Intell. Transp. Syst. **16**, 865–873 (2014). https://doi.org/10.1109/TITS.2014.2345663
19. Bastos, J.: Forecasting the capacity of mobile networks. Telecommun. Syst. **72**, 1–12 (2019). https://doi.org/10.1007/s11235-019-00556-w
20. Wang, Z., Tu, L., Guo, Z., Yang, L., Huang, B.: Analysis of user behaviors by mining large network data sets. Future Gener. Comput. Syst. **37**, 429–437 (2014). https://doi.org/10.1016/j.future.2014.02.015
21. CommScope: Dr. Mohamed Nadder Hamdy, Multibeam antennas planning - limitations and solutions. AN-109688-EN (01/16) (2016)
22. Wang, S., Ran, C.: Rethinking cellular network planning and optimization. IEEE Wirel. Commun. **23**, 118–125 (2016). https://doi.org/10.1109/MWC.2016.7462493

Research of Public and Special Applications of Information Technologies

Fundamentals of Functioning of Global Contact Monitoring Systems in the Context of COVID-19 Spread Prevention

Alexandr Kuznetsov(✉), Nikolay Poluyanenko(✉), and Anastasiia Kiian(✉)

V. N. Karazin Kharkiv National University, Svobody sq., 4, Kharkiv 61022, Ukraine
kuznetsov@karazin.ua

Abstract. Coronavirus disease 2019 (COVID-19) has a clear potential for a prolonged global pandemic, high mortality and overload of health systems. Until vaccines become widely available, the only available approaches to infection prevention are case isolation, contact and quarantine tracking, physical distancing, decontamination, and hygiene measures. The paper considers the conceptual principles of combining the bases of transparency and confidentiality of personal life and the need to track personal contacts, substantiates the need for such a combination and provides a possible technical solution using blockchain technology. The practical development and deployment of the system based on such protocols will allow each citizen to track the personal history of probable contact with infected persons or other potential risks (including deciding on the need for self-isolation or additional examination). In this case, the decentralized system guarantees the privacy of such information both for each participant (due to the inability to determine the personal data of a potentially infected person) and for the system as a whole.

Keywords: Global monitoring system · Epidemiological situation · Transparency and confidentiality of personal life · Blockchain technology

1 Introduction

The main in the comprehensive strategy to prevent COVID-19 is a number of public health priorities aimed at disrupting the chain of transmission, including (1) detection, isolation, testing and treatment of all patients, (2) tracking and quarantine. contact persons, as well as (3) assistance in maintaining a safe distance of at least 1 m, as well as the rules of regular hygienic treatment of hands and respiratory etiquette. These three components play a leading role in national COVID-19 threat response plans in each country [1].

Quarantine measures to control the spread of infectious diseases have been around for centuries. Today, many countries have the legal authority to introduce quarantine, which must be carried out with full respect for human rights and fundamental freedoms according to Article 3 of the International Health Regulations (2005) [2]. Under the conditions of the spread of COVID-19 quarantine of contact persons means separation

P. Vorobiyenko et al. (Eds.): IPF 2020, LNNS 212, pp. 311–328, 2021.
https://doi.org/10.1007/978-3-030-76343-5_16

from others or restriction of activities of people who are not sick, but could come into contact with patients [3].

One of the elements of the approach to control COVID-19, adopted by China and a number of other countries in East and Southeast Asia, which has been very successful for reducing the incidence, is the use of mobile phone data in combination with intensive testing programs [4]. Effectiveness of Digital Proximity Tracking to Facilitate Contact Tracking Remains Unknown Digital tracking technologies have already been implemented in several countries to respond to COVID-19. Meanwhile, governments, universities and companies in other countries are developing technologies and scaling them at an unprecedented rate.

The issue of combining the possibility of monitoring personal contacts and privacy is considered at the level of the World Health Organization. Thus, the Interim Guide "Ethical Considerations for the Use of Digital Contactless Tracking Technologies for Tracking COVID-19 Contacts" [5] defines the following.

Contact tracking is a process of identifying, evaluating, and managing people who have been affected by the disease to prevent further spread. When used systematically, contact tracing breaks the chain of spreading of infectious diseases and, therefore, is an important health tool to combat outbreaks of infectious diseases.

Among other things, the generalized goals of contact tracking are [6]:

- both the "direct" definition of possible spreading, which follows from the confirmed case, and the "reverse" definition (search for sources of infection);
- identification of an already confirmed or potential outbreak (cell, cluster) that will take precedence over individual cases;
- informing contact persons about the risk of the disease, about reducing contact and about what to do if symptoms appear;
- early detection of diseases among contact persons.

For contact tracking to be effective, sufficient capacity, including human resources, is needed to verify suspected cases in a timely manner [7]. Digital technology can play a role in state-of-the-art contact tracking programs.

States are required by the International Health Regulations to develop health surveillance systems [8] that collect important data for their response to COVID-19, while ensuring the transparency of such systems, responding to community concerns and not imposing unnecessary encumbrances, such as breach of confidentiality [9]. Failure to implement effective surveillance systems can impede effective health care and clinical response [10].

One form of digital surveillance technology that many countries facing the COVID-19 epidemic have been paying is contact tracking. Contact tracking measures signal strength to determine if two devices (such as smartphones) are close enough to allow users to spread the virus from an infected person to an uninfected person. If one user is infected, it is possible to notify others who have been identified as being in close contact to such a person, and thus take appropriate measures to reduce the health risks to themselves and others [11]. Tracking is a broad area of healthcare, and closeness tracking is a new technique to help keep track of contacts.

Health authorities could use digital contact tracking tools to alert a person to an increased risk of contact with another who has been infected by COVID-19. Such person notification who may have been in close contact with a COVID-19-positive person may prompt testing or precautionary measures to limit potential transmission, such as self-isolation and physical removal, even before any symptoms appear [12]. Early public health responses can significantly change the control and spread of COVID-19. In addition, data from digital contact tracking technologies can be useful for researchers in preparing for future COVID-19 outbreaks and promoting overall preparedness for future epidemics and pandemics.

However, such use of data may also threaten fundamental human rights and freedoms during and after the COVID-19 pandemic. Surveillance can quickly bridge the line between disease surveillance and population surveillance. Thus, there is a need for legislation, policies and oversight mechanisms to severely restrict the use of digital contact tracking technologies and any research that uses data obtained from such technologies.

The state can achieve its goal in the field of health care, while protecting fundamental rights such as privacy. Moreover, human rights laws and documents provide for the use of personal data that are in the public interest while preventing unnecessary intrusion or commercial exploitation. The integration of these and other ethical considerations into the development of new technologies can ensure that its technical specifications preserve and promote certain values, such as transparency and confidentiality.

Unfortunately, the current virus has great epidemic potential because mild or asymptomatic manifestations of infection are difficult to track. As treatment is not currently available, one of the effective tools which we can use to stop the epidemic is a global contact tracking system [13].

Persons who have been in contact with the person with confirmed COVID-19 during the infectious period are referred to as "contact persons". According to data published in October 2020 by the Federal Institute for Infectious Diseases of Germany - the Robert Koch Institute [6, 14], the following assumptions are implemented to determine the likelihood of infection:

- In most cases, the incubation period lasts a maximum of 14 days.
- The average incubation period is 5–6 days.
- The duration of the infectious period is about 12 days and begins with

- in asymptomatic case (contact is known): 3 days after contact;
- in asymptomatic case (only the date of sampling is known): on the 2nd day before sampling;
- in case of symptoms: 2 days before the onset of symptoms.

High to low infection priorities for "contacts" are determined. Here are some criteria that can be determined by digital contact tracking. High priority (increased risk of infection):

- close contact (<1.5 m);
- people with a total personal contact of at least 15 min with the source of infection, for example, during a conversation.

Low priority (lower risk of infection) if the person was at close range (<1.5 m) for less than 15 min.

2 World Practice of Tracking Contacts in the Fight Against COVID-19

2.1 «Corona-Datenspende» Germany

With the participation of the Robert Koch Institute (RKI), E-Health Thryve (a company specializing in digital health care) and with the assistance of the Federal Data Protection Agency, a mobile application called "Corona-Datenspende" was developed [15]. This application works in conjunction with smart watches or fitness trackers designed to collect additional information about the location and rate of spread of coronavirus (SARS-CoV-2) in Germany.

The idea behind the development is based on the analysis of vital indicators of the body, which are collected and processed by Corona-Datenspende:

1. The actions of the fitness tracker are automatically and manually registered, as:

   - Sports (eg cycling, running).
   - Sleep and sleep phases.
   - Active state of life (eg, walking, activity).
   - Periods of rest.

2. Vitally important data of the fitness tracker, such as:

   - Pulse.
   - Heart rate variability.
   - Stress.
   - Temperature.
   - Weight.
   - Blood pressure.

3. Socio-demographic data, if they were previously entered by users in the fitness tracker program:

   - Age (rounded up to 5 years).
   - Size (rounded to 5 cm).
   - Sex.
   - Weight (rounded up to 5 kg).

Socio-demographic data as well as fitness tracker data are needed to identify possible symptoms. Zip code is required to record spatial location. This data does not allow the user to be identified.

In acute respiratory disease, some of the indicators in most cases deviate significantly from normal. The application shows various symptoms associated with coronavirus infection. These include, for example, a fast heartbeat at rest (indicating fever), as well as changes in sleep and activity. Accordingly, the program can identify typical symptoms of COVID-19, such as fever.

"Corona-Datenspende" will allow specialists to more accurately identify foci of infection, as well as analyze the effectiveness of antiviral measures. At the same time, the new development is not designed to track the contacts of infected people - a separate mobile application has been developed for this purpose.

After installing Corona-Datenspende, the user must agree that the indicators collected through the application will be sent to the Robert Koch Institute. TLS/SSL encryption interfaces are used for data transmission. Use of the program is based on an individual user ID, which is assigned to the user personally, a so-called alias, which is generated by the server. Thus, "Corona-Datenspende" is not anonymous, but pseudo-anonymous. Your personal health information is associated with an alias. The developers ask users to keep this nickname secret as a regular password for security reasons.

Apple Health and Samsung Health retrieves data directly from the smartphone, and all other vendors retrieve data from the server of the respective manufacturer. In the case of Google Fit, the manufacturer sends the developers an image of profile and ID in addition to the fitness tracker data. The date of birth is also transmitted from Amazfit. This information is processed together with other data in main memory and is reset when saving data.

All incoming data are processed by specialists of the institute, and then displayed in the form of the map of Germany, which captures the location of potentially infected with coronavirus by region.

2.2 «Vocalis Health» Israel

Vocalis Health is developing an artificial intelligence platform that uses voice analysis to assess human health [16].

The company's platform allows healthcare providers to use remote voice communication through a call center or smart device to passively monitor, index, and track millions of patients living with a range of voice-related conditions such as chronic respiratory or heart disease and depression. Early clinical data from Vocalis Health demonstrate its ability to effectively index and track chronic patients.

Vocalis Health was formed as a result of the merger of two medical companies engaged in artificial intelligence technologies, Beyond Verbal and Healthymize.

It is believed that by analyzing the voice beyond what the human ear can hear, it is possible to detect special voice biomarkers that will allow the medical community to get an idea of the symptoms and possibly the appearance of the COVID-19 virus.

The developers are implementing modern methods and techniques of artificial intelligence to match the voice with the symptoms of COVID-19. This will warn of early symptoms only with a smartphone.

Vocalis Health products and solutions may be subject to regulatory approval and are not currently available for sale.

2.3 COVIDSafe Australia

COVIDSafe [17] is a contact tracking application announced by the Australian Government on April 14, 2020 to prevent the COVID-19 pandemic. The application is based on the BlueTrace protocol, developed by the Government of Singapore, and was first released on April 26, 2020.

The application works with Bluetooth technology, which allows you to capture cases where two users of the program spend more than 15 min at a distance of 1.5 m from each other. If a user of the program is diagnosed with a coronavirus, he or she may consent to the transfer of information about his or her recent contacts to health authorities. COVIDSafe collects the following information:

- Registration information consists of name or nickname, mobile phone number, age range and zip code. It is stored at the COVIDSafe National Data Warehouse in Australia.
- Digital handshake information is an encrypted control code, date and time, Bluetooth signal strength and phone model. This is stored on the user's phone. If the user has a positive result and uploads the information to the COVIDSafe National Data Warehouse, the algorithm will use the Bluetooth signal strength to determine the proximity and duration of contact with other COVIDSafe users.
- Diagnostic information is an operating system on the phone, version of COVIDSafe installed on the phone, language, application entries with any contacts with other COVIDSafe users for the last seven days, whether Bluetooth is enabled on the phone, whether battery optimization is enabled on the phone and whether phone location service (Android / Google devices only).
- COVIDSafe stores information as follows:

  - COVIDSafe encrypts a name or nickname, verified mobile phone number, age range and zip code that are registered in the COVIDSafe National Data Warehouse.
  - COVIDSafe provides information in the form of an encrypted hash code. This is the only information passed on to other users as part of a Bluetooth digital handshake.
  - COVIDSafe stores the collected digital handshakes on your phone. The information is encrypted, apparently the user does not have the opportunity to view it.
  - COVIDSafe automatically removes digital handshakes older than 21 days from your phone.

The 21-day COVIDSafe window includes a 14-day incubation period plus the time required to confirm a positive test result. The sliding 21-day window allows the application to constantly monitor only those user contacts that occur during the coronavirus incubation window. COVIDSafe automatically deletes from the user's phone contacts that took place outside the 21-day window.

Information about the digital handshake leaves your phone only in case of a positive test result on COVID-19 and consent to download. If you agree, the information will be uploaded to the COVIDSafe National Data Warehouse. When accessing and using downloaded information, healthcare professionals must comply with the obligations set out in Part VIIIA of the Privacy Act 1988 and the Australian Privacy Principles. The information is intended to be used to alert people if they have come into contact with a person who has tested positive for COVID-19.

Registration information will be stored in encrypted form in the COVIDSafe National Data Warehouse. COVIDSafe generates a temporary unique identifier of the user's mobile phone number (unique identifier), which is stored in the application [17].

After the user provides his registration information, he will be prompted to enable push notifications for the program. The app will create a daily push message for users to remind them to use the app when they are away. If the user does not turn on push notifications or the user's device is in "do not disturb" mode, the application will continue to run. COVIDSafe automatically generates new unique identifiers for each user every two hours and sends these new unique identifiers to the user's application.

The application will only accept new unique identifiers if it is open and running. If the application successfully accepts a new unique identifier, an automatic message will be generated and sent back to the COVIDSafe National Data Warehouse. This message will only effectively indicate the answer "yes" (new unique identifier successfully delivered) to the COVIDSafe National Data Warehouse. If the application is not open and does not work, it will not be able to accept the new unique ID. It will continue to store the previous unique identifier and use it when opening the program until a new unique identifier is generated and accepted [18].

COVIDSafe will regularly compile and store reports (unique identifier reports) based on whether the application has adopted a new unique user ID, which will determine the proportion of users who downloaded, opened and launched the application.

Unique identity reports will be sent to the National Data Warehouse COVIDSafe. The Health Service will use the information contained in the unique identifier reports to assess the appropriate use of the program across Australia and to determine whether further communication with the public is necessary (for example, if the level of use is very low, even though the level of downloads are high, the health service may decide to implement additional communication strategies for the use of the program).

Professor Dali Kaafar is the Executive Director of the Center for Cyber Security at OPTUS Macquarie University. In early April 2020, he and a group of Australian researchers assessed some problems with the Singaporean application on which the Australian program is based.

Researchers have found that good work has been done to protect data from other users of the program and from anyone who tries to hack the user's phone, but the program does not provide very good privacy from the central authority where your data is sent - in this case, the government.

Professor Kaafar pointed out that the problem stemmed from the huge amount of information available to the central authority that receives data from the application. Over time, this central body will be able to create a very detailed "social schedule" for thousands of people: a photo of everyone they spend time with and how much time they spend with them.

"... The only information that is not required is the location. We may not know where they met, but we know they met".

"Thus, the general concern is that this technology can be used later as a tool for mass surveillance".

2.4 «Corona-Warn-App» Germany

The Corona-Warn-App [19, 20] allows people to track the risk of personal infection via mobile phones. The Corona-Warn-App uses a new framework provided by Apple and Google, called the Exposure Notification Framework. The structure uses Bluetooth Low Energy (BLE) technique. BLE allows individual mobile phones to act as beacons, which means that they constantly broadcast a temporary identifier called a Rolling Proximity Identifier (RPI), which is stored, and at the same time allows the mobile phone to scan the identifiers of other mobile phones. Identifiers are identification numbers sent by mobile phones. To ensure confidentiality and prevent tracking of user application traffic patterns, these translation IDs are temporary and constantly changing. The new identifiers are derived from the Temporary Exposure Key (TEK), which is replaced at midnight (UTC) daily by cryptography. Once the FEC is associated with a positive test result, it remains technically unchanged, but is then called the diagnostic key.

Collected IDs from other users, as well as their own keys, which can then be used to obtain IDs, are stored locally on the phone in a secure repository of the framework provided by Apple and Google. The application cannot access this secure repository directly, but only through the interfaces provided by the Exposure Notification Framework. To prevent misuse, some of these interfaces are subject to speed limits. If users of the application pass the SARS-CoV-2 test, they can update their status in the application by providing confirmation of their test and choose to send their latest keys 14 days ago.

On the Corona-Warn-App server, all the keys of positively tested people are combined and then made available to all mobile phones on which the application is installed. In addition, the configuration parameters for the framework are available for download, so you can make adjustments to the calculation of the risk assessment. After downloading the keys and establishing the fact of infection, the data is transmitted to the notification structure, which analyzes whether one of the identifiers collected by the mobile phone corresponds to the identifiers of the positive test. In addition, metadata that was broadcast along with identifiers, such as transmission power, can now be decrypted and used. Based on the data collected, the risk alert framework provided by Apple and Google calculates a risk assessment for each individual exposure as well as for the overall situation. The risk of infection is defined as the sum of all meetings with another person during one calendar day (UTC time zone). For reasons of confidentiality, it is not possible to track meetings with others for several days.

It is important to note that those who have passed the positive test are not informed by the central authority, but the risk of infection is calculated locally on their phones. Information about the possibility of infection remains on the user's mobile phone and is not transmitted.

The Corona-Warn-App has two goals:

1. Help people find out if they have been exposed to human influence, which was later tested positive.
2. Get the test result for SARS-CoV-2 on the user's mobile phone through the online system. This helps reduce the time to take the necessary precautions, such as reducing contact and testing.

To prevent misuse, individuals must provide proof that they have passed a positive test before they can download their keys. With this integrated approach, the verification required to download diagnostic keys requires no further action from users. They only need to confirm in the appendix and within the exposure notification system that they agree to share their diagnostic keys. Manual verification is also possible if the testing laboratory does not support direct electronic transmission of test results to users' mobile phones or if users have opted against electronic transmission of test results.

However, the application does not provide complete anonymity of users. It starts with the developers of the project [21], as soon as temporary access keys are uploaded to the server (in case of a positive test result), anonymity becomes pseudonymous.

A more detailed overview of digital solutions and specific health initiatives (programs) used in the EU and around the world in response to COVID-19 can be found in [22].

3 Trust Issues and the Effectiveness of the Contact Tracking System

According to the survey conducted in June 2020 in the United States on the question of whether people plan to download the application for tracking contacts, the vast majority (71%) said no. This looks worse than the previous year. In April, a poll found that every second American would probably or definitely not use the app to track contacts [23].

Respondents who answered "no" to the first question were then asked to explain this decision through a multi-choice survey. The most common reason was privacy concerns. Overall, 44% of those who said "no" to the contact tracking app said they would not trust technology to protect their digital privacy. But almost as many (39%) said they thought the applications created a false sense of security, and 37% said they thought the programs would not work to slow the pandemic. 35% also indicated a lack of trust in application providers.

According to Censuswide surveys of 1,000 UK citizens, 48% of respondents believe that the application being developed to track the spread of coronavirus will be used by cybercriminals [24]. Of the 43% surveyed said that the contact tracking application could be used for phishing cyberattacks, but only 52% said that they could identify a fake email or text.

A group of 173 scientists from 27 countries working on privacy and security issues published an open letter in April 2020 [25]. The statement said that the current COVID-19 crisis was unprecedented, that innovative ways of overcoming the current constraints were needed, and expressed concern that some "solutions" to the crisis could lead to systems that would allow unprecedented oversight of society as a whole. It is important to ensure that solutions maintain the privacy of their users, thus protecting them from many other problems, and that such programs can be redesigned to provide unjustified discrimination and oversight.

The letter further states that GPS-based solutions for contact detection are not sufficiently accurate and carry the risk of confidentiality because GPS data is sent to a centralized location. For this reason, Bluetooth-based solutions for automated contact tracking are highly desirable when they are available.

Some proposals based on Bluetooth respect the individual's right to privacy, while others will allow a form of public or private oversight that would catastrophically impede public trust and decision-making. It is very important that citizens trust decisions to ensure a sufficient level of decision-making to resolve the crisis. It is vital that, given the current crisis, no tool be created to allow large-scale population data collection, either now or later. Thus, decisions to reconstruct invasive population information should be discarded without further discussion. Such information may include a "social graph" of someone you have physically met over a period of time.

Having gained access to the social graph, an attacker (state, private sector or hacker) can spy on the real activities of citizens. Some countries are seeking to build systems that could allow them to access and process this social graph. On the other hand, highly decentralized systems do not have a clearly defined entity that can learn something about the social graph.

The letter called on all countries to rely only on decisions that are subject to public scrutiny and protect the confidentiality of their architecture (rather than hoping to be governed by a credible party) as a means of enforcing citizens' data protection rights. principles:

- Contact tracking programs should only be used to support health measures which are needed to combat COVID-19. The system should not be able to collect, process or transmit more data than is necessary to achieve this goal.
- Any solution under consideration must be completely transparent. Protocols and their implementation, including any subcomponents provided by companies, should be available for general analysis. The processed data, as well as information on how, where and for how long the data is stored, must be clearly documented. Such data should be kept to a minimum for this purpose.
- When there are several possible options for implementing a particular component or functionality of the program, then you should choose the most secure option. Deviations from this principle are permissible only if it is necessary to achieve the goal of the program more effectively and must be clearly justified.
- The use of contact tracking programs and systems that support them should be voluntary, used with the express consent of the user, and systems should be designed so that they can be turned off and all data deleted when the current crisis is over.

The following documents regulate the foundations of human rights and freedoms with regard to the fight against COVID-19, which are given in [26].

Article 12 of the International Covenant on Economic, Social and Cultural Rights (ICESCR) guarantees [27] everyone the "right to the highest attainable standard of health". ICESCR member states (including all EU member states) are obliged to take effective measures for the prevention and treatment and control of epidemic, endemic, occupational and other diseases. The need to protect public health may justify restricting certain rights "in emergencies that threaten the lives of people and the nation". However, such measures must be lawful, necessary and proportionate (Syracuse Principles [28]).

Article 8 of the European Convention for the Protection of Human Rights and Fundamental Freedoms provides for the right to respect for private and family life [29]. Any interference by a public authority in the exercise of this right must be "in accordance with the law" and "necessary in a democratic society" to protect the interests of national security, including public safety and "health care".

Article 7 of the Charter of Fundamental Rights of the European Union [30] provides for the right to respect for 'private and family life, home and communication", while Article 8 provides for the right to the protection of personal data. According to Article 35, "a high level of human health shall be ensured in the definition and implementation of all policies and activities of the Union". Any restriction on the exercise of rights and freedoms must be lawful, proportionate, "respect the essence of those rights and freedoms" and apply "only if they are necessary and genuinely meet the objectives of the general interest recognized by the Union" (Article 52).

The Header of the Committee of the Convention 108 [31] and the Data Commissioner of the Europe Council [32] issued a joint statement on 30 March 2020 stating that "large-scale processing of personal data can only take place when, on the basis of scientific evidence, the potential public benefits digital epidemic surveillance (such as contact tracking) for health, including their accuracy, outweighs the benefits of other alternative solutions that are less intrusive." On 7 April, the Secretary General of the Council of Europe [33] issued a set of tools for European governments on respect for human rights, democracy and the rule of law during the COVID-19 crisis.

Regulation (EU) 2016/679 [34] (GDPR) establishes rules and principles for the processing of personal data and on the free movement of such data within the EU. As defined in Article 4 of the GDPR, personal data is any information relating to the identification of an individual (eg names, dates of birth, photographs, e-mail addresses, IP addresses). "Location data" is directly mentioned as a potential identifier of an individual. If the location data may be related to an identified or identifiable individual, this data is considered personal data and is processed by the GDPR.

Mobile applications used to prevent the spread of COVID-19 can potentially be classified as medical devices, in which case they must comply with certain safety regulations. According to the provisions of Regulation (EU) 2017/745 on medical devices (MDR) [35], an application may be considered a medical device if it is intended by the manufacturer for use, inter alia, for diagnosis, prevention, monitoring, prognosis, prognosis, treatment or relief. disease.

Article 52 of the European Parliament resolution of 21 April 2020 on the use of location and contact tracking tools in the context of the COVID-19 outbreak (2020/2616 (RSP)) [36] states: "Takes note of the emergence of mobile contact tracking applications to warn people if they are near an infected person and the Commission's recommendation to develop a common EU approach to the use of such programs; Notes that any use of programs developed by national and EU authorities may not be mandatory and that the data created should not be stored in centralized databases that are at potential risk of abuse and loss of trust and may threaten to spread throughout Unions; calls for the decentralization of all data storage, full transparency for the commercial interests of the developers of these applications (non-EU) and the demonstration of clear forecasts of how part of the tracking programs, combined with specific other measures, will lead to significantly fewer infected people; Requests that the Commission and the Member States be fully transparent about the operation of contact tracking applications so that people can check both the basic protocol on security and confidentiality and the code itself to verify that the program works as required by the authorities; recommends that termination provisions be established and that data protection principles be fully adhered to by designing and minimizing data.

In order to achieve the ultimate goal of combating COVID-19 and respecting the above-mentioned human rights and freedoms, applications that implement contact tracking, as indicated in [37], are divided into two large groups according to their communication protocols. These groups differ mainly in the place of storage and/or the method of processing arbitrary COVID-19 identifiers and contact data. Contact tracking programs can be distinguished between:

1) Decentralized applications: Arbitrary ephemeral identifiers of all phones in contact with another user are generated, stored and processed on the user's device (ie mobile phone), which calculates risk assessments for all users and stores all identifiers at risk of infection. When a person receives a positive COVID-19 test from a health authority, they upload their public contact information to the server. Examples of such systems are the DP-3T [38, 39] and TCN [46] protocols and the Google-Apple Exposure Notification [40] programming interface.
2) Centralized applications: random ephemeral identifiers of all telephones located near the user are generated, stored and processed on a central server operated by public health authorities, which calculates updated risk assessments for all relevant users and decides which affected users inform. When a person receives a positive COVID-19 test from a health authority, they upload their public contact information to the server. Examples of such systems are PEPPPT [41, 42] and OpenTrace/BlueTrace/TraceTogether [43].

4 A Model Proposed for the Design of System for Global Monitoring of the Epidemiological Situation

Given the above global practice of creating such models and the existing legal framework, it is proposed to implement global monitoring of the epidemiological situation by tracking contacts without any trusted party using, inter alia, methods of building distributed decentralized systems. This will ensure the processing of data with a guarantee of information security and its availability in conditions of complete distrust of the parties. The Global Monitoring Protocol sets out the basic guidelines of the Interim Guidelines on Ethical Considerations for the Use of Digital Contactless Tracking Technologies for Tracking COVID-19 Contacts, issued by WHO [5], which set out 17 principles by which the government, health authorities and non-governmental projects can be controlled using digital contact tracking technologies to combat COVID-19. The document recommends that the contact tracking program be based on the voluntary and informed consent of users, and that the collection and processing of data be transparent.

Leading world companies and individual countries have already developed and put into trial operation elements of the global monitoring system, but the emergence of new system technologies always poses additional threats. For example, as part of a joint effort to combat the COVID-19 pandemic, Apple Inc. and Google Inc. developed contact tracking technology to help governments and healthcare reduce the spread of the virus.

Using the Stopp Corona, the Austrian Red Cross (ÖRK) provides a contact tracking program to control new SARS-CoV-2 infections in Austria.

It is planned to use Bluetooth Low Energy and/or StreetPass contact tracking for passive Wi-Fi communication according to the following construction principles (and/or other models using GPS systems to locate the mobile device):

1. The device installed on the phone generates a secret master key, known only to this device, which always remains on it (see Fig. 1). Every day, the application generates a new daily key as a hash according to the algorithm of cryptographic one-way hash function from a secret master key and some salt (serial number, current date, external data, etc.). Together with the daily key, the device generates session IDs (ephemeral ID, in Apple & Google terminology - Rolling Proximity ID). These are 24 * 60 session IDs per minute of the day used to exchange contacts. ECB encryption mode is used to generate session IDs, where the daily key is used as the encryption key, and the current session ID number is used as an input (a total of 1440 per day).

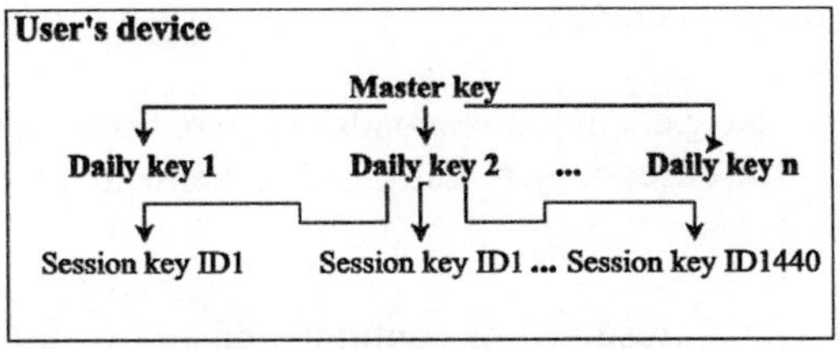

Fig. 1. Creating session IDs.

2. By analogy with Bluetooth interaction with a headset that uses a Bluetooth service with a standard headset code assignment (characteristic UUID), in this case session ID is used, which determines the protocol for tracking contacts on phones with the installed application. Phones periodically search via Bluetooth for other devices with a fixed UUID and when they find out they understand that a contact has taken place. The task of the protocol is to pass anonymous device identifiers to each other (see Fig. 2).
   That is, if a user is near a phone with such a BLE service, he sees one of his 1440 session IDs, which change every minute, making it impossible to track the user.
   Given that the identifiers are generated using the daily key, then receiving the day key, any user will be able to generate all session IDs.

Fig. 2. Interaction of mobile devices.

3. When the devices are in contact, they remember each other's session IDs and RSSI (Received Signal Strength Indicator) values, which calculate the distance between

the devices in their local contact database (see Fig. 3). The local contact database is private and never leaves the user's mobile device.

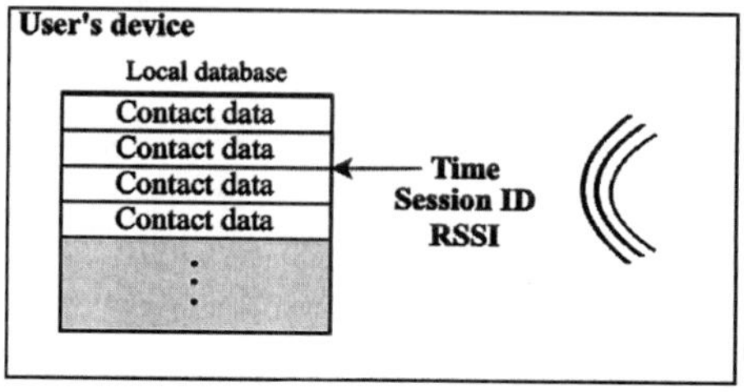

Fig. 3. Creating a local contact database.

4. According to WHO recommendations, all persons who have been in contact with patients with a confirmed or probable diagnosis of COVID-19 should be quarantined for 14 days from the last contact.

The definition of contact persons corresponds to persons who in the period from two days to or 14 days after the onset of symptoms in a patient with a confirmed or probable diagnosis of COVID-19 [44]:

- talked to a person with a presumed or confirmed diagnosis of COVID-19, at close range within 1 m for more than 15 min;
- have been in direct physical contact with a person with a presumed or confirmed diagnosis of COVID-19;
- directly assisted a person with a presumed or confirmed diagnosis of COVID-19, without the use of appropriate personal protective equipment [45];
- was in other types of interaction specified in the risk assessment at the local level.

Given the above and in case of infection, the user, on a voluntary basis, adds data to the blockchain register using the appropriate transaction (Tx). The data consists of the latest 14-day keys, current time, diagnosis and qualified data signature. The procedure for transferring key data to the blockchain register, in accordance with the principles of the GDPR and the degree of reliability, can be performed in three ways:

1. Reliable. When the diagnosis of the patient is confirmed by the medical institution and the user on a voluntary basis transmits to the medical institution in a blinded state the formed body of the transaction. The medical institution returns to the user the body of the transaction with its own signature (EDSmed.inst). The user's software removes blindness and transmits an already anonymous transaction to the blockchain.
2. Medium trust. The user signs the transaction on his own initiative and sends it to the blockchain network.
3. Low trust. The transaction is transmitted to the blockchain network without a signature.

The second and third methods can be useful when the user considers himself infected, but for some reason (mild disease, self-isolation, etc.) does not seek medical attention, but for his own reasons considers it necessary to warn other contacts about the possible danger (Fig. 4).

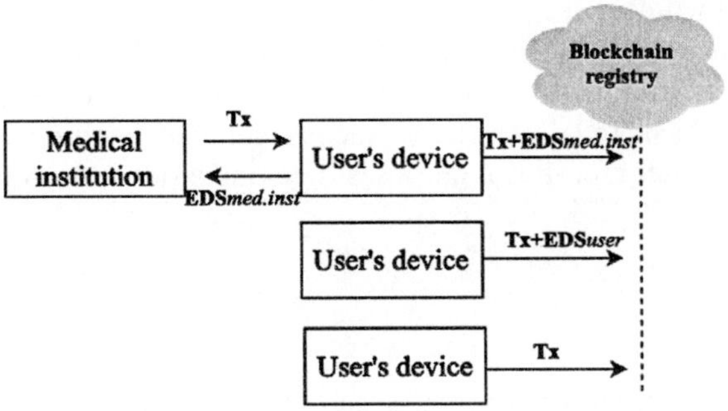

Fig. 4. Three ways to transfer daily keys to the blockchain registry.

5. Other users of the global monitoring of the epidemiological situation periodically receive new data from the blockchain register. The application on the smartphone generates all 24 * 60 session IDs of the infected person and compares them with their locally stored contacts (see Fig. 5). If there is compliance, the user receives a message about contact with the infected, its duration and approximate distance to the patient, on the basis of which a conclusion is made about the likelihood of infection and appropriate recommendations are given.

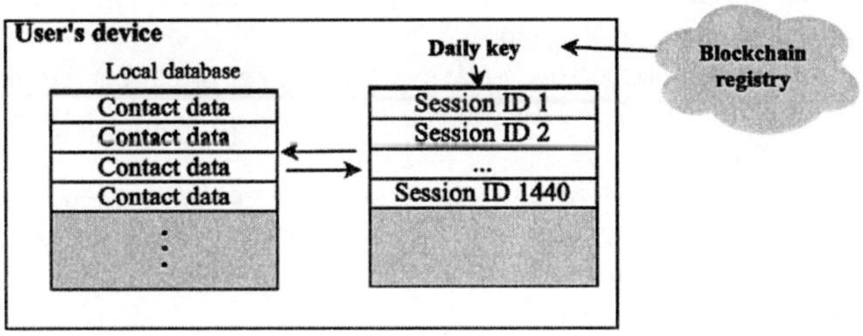

Fig. 5. Check for dangerous contacts.

5 Conclusion

It is proposed to implement global monitoring of the epidemiological situation by tracking contacts without the presence of any trusted party using, inter alia, methods of building distributed decentralized systems. The practical development and deployment of the system based on such protocols will allow each citizen to track the personal history of probable contact with infected persons or other potential risks (including deciding on the need for self-isolation or additional examination). At the same time, the decentralized system guarantees the privacy of such information both for each participant (due to the impossibility of determining the personal data of a potentially infected person) and for the system as a whole.

The development of such a method will allow using the already deployed information infrastructure to effectively prevent the spread of infectious diseases, without creating risks to the basic freedoms of civil society, completely eliminating the possibility of using such a system for illegal purposes (tracking or identification) by any person (organization).

References

1. Core measures to prepare for and respond to COVID-19. Geneva: World Health Organization 2020. https://apps.who.int/iris/bitstream/handle/10665/331826/WHO-2019-nCoVCommunity_actions-2020.3-rus.pdf. Accessed 11 Aug 2020
2. Key considerations: quarantine in the context of COVID19. In: Social Science in Humanitarian Action: A Communication for Development Platform. UNICEF, Institute of Development Studies, New York (2020). https://www.socialscienceinaction.org/resources/february2020-social-science-humanitarian-action-platform/. Accessed 11 Aug 2020
3. International Health Regulations (2005), 3rd edition. 2016. World Health Organization, Geneva (2020). https://www.who.int/ihr/publications/9789241580496/ru/. Accessed 11 Aug 2020
4. Parker, M.J., Fraser, C., Abeler-Dörner, L., et al.: Ethics of instantaneous contact tracing using mobile phone apps in the control of the COVID-19 pandemic. J. Med. Ethics **46**, 427–431 (2020)
5. Ethical considerations to guide the use of digital proximity tracking technologies for COVID-19 contact tracing. WHO reference number. https://www.who.int/publications/i/item/WHO-2019-nCoV-Ethics_Contact_tracing_apps-2020.1
6. Robert Koch-Institut: Kontaktpersonen-Nachverfolgung bei Infektionen durch SARS-CoV-2 19.10.2020 https://www.rki.de/DE/Content/InfAZ/N/Neuartiges_Coronavirus/Kontaktperson/Management.html
7. Contact tracing in the context of COVID-19: interim guidance. Geneva; World Health Organization (2020). https://apps.who.int/iris/handle/10665/332049. Accessed 21 May 2020
8. International health regulations – 2nd ed. World Health Organization, Geneva (2005). https://apps.who.int/iris/bitstream/handle/10665/43883/9789241580410_eng.pdf?sequence=1. Accessed 26 May 2020
9. Surveillance strategies for COVID-19 human infection: interim guidance. Geneva; World Health Organization (2020). https://apps.who.int/iris/handle/10665/332051. Accessed 21 May 2020
10. WHO guidelines on ethical issues in public health surveillance. Geneva; World Health Organization (2017). https://apps.who.int/iris/bitstream/handle/10665/255721/9789241512657-eng.pdf?sequence=1. Accessed 7 May 2020
11. Crocker, A., Opsahl, K., Cyphers, B.: The challenge of proximity apps for COVID-19 contact tracing. Electronic Frontier Foundation (2020). https://www.eff.org/deeplinks/2020/04/challenge-proximity-apps-covid-19-contact-tracing. Accessed 7 May 2020

12. Parker, M.J., Fraser, C., Abeler-Dörner, L., et al.: Ethics of instantaneous contact tracing using mobile phone apps in the control of the COVID-19 pandemic. J. Med. Ethics. https://jme.bmj.com/content/early/2020/05/05/medethics-2020-106314. Accessed 10 May 2020
13. Ferretti, L., Wymant, C., Kendall, M., Zhao, L., Nurtay, A., Abeler-Dörner, L., Parker, M., Bonsall, D., Fraser, C.: Quantifying SARS-CoV-2 transmission suggests epidemic control with digital contact tracing (2020). https://science.sciencemag.org/content/368/6491/eabb6936/tab-article-info
14. Robert Koch-Institut: SARS-CoV-2 Steckbrief zur Coronavirus-Krankheit-2019 (COVID-19), 16October 2020. https://www.rki.de/DE/Content/InfAZ/N/Neuartiges_Coronavirus/Steckbrief.html
15. Corona-Datenspende. https://corona-datenspende.de/
16. Vocalis Health. https://www.vocalishealth.com/
17. Department of Health. The COVIDSafe application. Privacy impact assessment https://www.health.gov.au/sites/default/files/documents/2020/04/covidsafe-application-privacy-impact-assessment-covidsafe-application-privacy-impact-assessment.pdf
18. Privacy experts are concerned about the government's coronavirus tracing app. Here's why. https://www.sbs.com.au/news/the-feed/privacy-experts-are-concerned-about-the-government-s-coronavirus-tracing-app-here-s-why
19. Corona-Warn-App: Documentation. GitHub. https://github.com/corona-warn-app/cwa-documentation
20. Corona-Warn-App Open Source Project. https://www.coronawarn.app/en/
21. GitHub: Criteria for the Evaluation of Contact Tracing Apps. https://github.com/corona-warn-app/cwa-documentation/blob/master/pruefsteine.md
22. Annex, I.V.: Inventory mobile solutions against COVID-19. https://ec.europa.eu/health/sites/health/files/ehealth/docs/covid-19_apps_annex_en.pdf
23. ArsTechnica: More than 7 in 10 Americans won't use contact-tracing apps, data shows. https://arstechnica.com/science/2020/06/more-than-7-in-10-americans-dont-want-contact-tracing-data-shows/
24. Contact tracing app will be exploited for cyberattacks, half of UK citizens fear. https://www.verdict.co.uk/contact-tracing-app-cyberattacks/
25. Joint Statement on Contact Tracing, 19 April 2020. https://drive.google.com/file/d/1OQg2dxPu-x-RZzETlpV3lFa259Nrpk1J/view
26. Tracking mobile devices to fight coronavirus. https://www.europarl.europa.eu/RegData/etudes/BRIE/2020/649384/EPRS_BRI(2020)649384_EN.pdf
27. International Covenant on Economic, Social and Cultural Rights. https://www.ohchr.org/en/professionalinterest/pages/cescr.aspx
28. The Siracusa Principles on the Limitation and Derogation Provisions in the International Covenant on Civil and Political Rights. https://www.refworld.org/docid/4672bc122.html
29. European Convention on Human Rights. https://www.echr.coe.int/Documents/Convention_ENG.pdf
30. Charter of fundamental rights of the European Union. https://eur-lex.europa.eu/eli/treaty/char_2012/oj
31. Convention for the Protection of Individuals with regard to Automatic Processing of Personal Data. https://www.coe.int/ru/web/conventions/full-list/-/conventions/rms/0900001680078b37
32. Statement by Alessandra Pierucci and Jean-Philippe Walter. The Chair of the Committee of Convention 108 and the Data Protection Commissioner of the Council of Europe recall the principles of data protection in these times of fight against the COVID-19 pandemic. https://www.coe.int/en/web/data-protection/statement-by-alessandra-pierucci-and-jean-philippe-walter

33. Respecting democracy, rule of law and human rights in the framework of the COVID-19 sanitary crisis. A toolkit for member states. https://rm.coe.int/sg-inf-2020-11-respecting-democracy-rule-of-law-and-human-rights-in-th/16809e1f40
34. Regulation (EU) 2016/679 of the European parliament and of the council of 27 April 2016 on the protection of natural persons with regard to the processing of personal data and on the free movement of such data, and repealing Directive 95/46/EC (General Data Protection Regulation). https://eur-lex.europa.eu/eli/reg/2016/679/oj
35. Regulation (EU) 2017/745 of the European parliament and of the council of 5 April 2017 on medical devices, amending Directive 2001/83/EC, Regulation (EC) No 178/2002 and Regulation (EC) No 1223/2009 and repealing Council Directives 90/385/EEC and 93/42/EEC. https://eur-lex.europa.eu/eli/reg/2017/745/oj
36. European Parliament resolution of 17 April 2020 on EU coordinated action to combat the COVID-19 pandemic and its consequences (2020/2616 (RSP)). https://www.europarl.europa.eu/doceo/document/TA-9-2020-0054_EN.pdf
37. National COVID-19 contact tracing apps. https://www.europarl.europa.eu/RegData/etudes/BRIE/2020/652711/IPOL_BRI(2020)652711_EN.pdf
38. GitHub: DP^3T Decentralized Privacy-Preserving Proximity Tracing. https://github.com/DP-3T
39. Wikipedia: Decentralized Privacy-Preserving Proximity Tracing. https://en.wikipedia.org/wiki/Decentralized_Privacy-Preserving_Proximity_Tracing
40. Exposure Notification. Cryptography Specification. Preliminary – Subject to Modification and Extension. https://covid19-static.cdn-apple.com/applications/covid19/current/static/contact-tracing/pdf/ExposureNotification-CryptographySpecificationv1.2.pdf
41. GitHub: PEPP-PT – Pan-European Privacy-Preserving Proximity Tracing. https://github.com/pepp-pt
42. Wikipedia: Pan-European Privacy-Preserving Proximity Tracing. https://en.wikipedia.org/wiki/Pan-European_Privacy-Preserving_Proximity_Tracing
43. Bay, J., Kek, J., Tan, A., Hau, C.S., Yongquan, L., Tan, J., Quy, T.A.: BlueTrace: a privacy-preserving protocol for community-driven contact tracing across borders. Government Technology Agency, Singapore. https://bluetrace.io/static/bluetrace_whitepaper-938063656596c104632def383eb33b3c.pdf
44. Sanitary and epidemiological surveillance of COVID-19. World Health Organization, Geneva (2020). https://apps.who.int/iris/bitstream/handle/10665/333752/
45. WHO-2019-nCoV-SurveillanceGuidance2020. 7-rus.pdf. Accessed 11Aug 2020
46. GitHub: TCN Protocol. https://github.com/TCNCoalition/TCN

A Three-Aspects Approach for Technical Systems Quality Evaluation

S. Volkov(✉), L. Kolomiets(✉), and O. Hrabovskyi(✉)

Odessa State Academy Regulation and Quality, Kuznechnaya Street, 15, Odessa 65029, Ukraine
greyw@ukr.net

Abstract. A conceptual approach to technical, including cyberphysical, systems quality evaluation at the stage of exploitation in three aspects: functional, technical and process is presented in the article. The dominance of the reductionist approach to the evaluation of technical systems, which is impractical in terms of management decisions at the current level of complexity of systems, is shown. A general approach, which allows considering and researching functional, technical and process approaches to evaluation, as a holistic evaluation system based on their interaction, within a single information structure based on project specifications and system information links, is offered. The presentation of the informational structure in the form of the sensory infrastructure is considered, the model and the scale of the quality state evaluation, based on relative value of the amount of information, are given. A model, that combines technical, functional and process aspects into the quality state evaluation unified system, is developed.

Introduction. The main idea of the fourth industrial revolution (Industry 4.0), which was started at the 2011 Hannover Fair, is the automation of all spheres of life through the widespread introduction of intelligent technologies. By definition, the driving force behind Industry 4.0 is cyber-physical systems - technical systems (TS), operating in an automatic mode under the control of artificial intelligence, which, depending on the user's production needs, is able to provide an autonomous change in the behavioral logic and work patterns of the system's components. In practice this means that information about the necessary technological processes is embedded in the product (service) as we move through the stages of creation, which allows us to implement fully flexible "individual production or service to order", moreover, the decision to change the functional behavior of the components and the system in generally accepted by artificial intelligence and not controlled or partially controlled by specialists. Of particular concern is autonomous operation and appropriate decision making in the event of abnormal situations, for example, in case of failures in technological processes or failures of equipment or software. What decision on system management will be made by artificial intelligence and which, as a result of the decision, will be his behavior, is perhaps the most difficult and uncertain question.

In this context, the definition of the qualitative state and, accordingly, the quality of work and TS behavior and their components at the Operation stage acquires great weight what determines not only the release of products or the provision of services, but also the safety of human life and environmental safety.

Problem statement. An analytical review of literary sources showed the presence of two main approaches to the analysis and determination of the qualitative

P. Vorobiyenko et al. (Eds.): IPF 2020, LNNS 212, pp. 329–339, 2021.
https://doi.org/10.1007/978-3-030-76343-5_17

state of artificial, including technical, systems - functional and process. The main thesis of the publications is a direct comparison of these approaches, which consists in a dualistic view of their interaction from direct opposition to determining them parallel to the current ones with similarities in basic premises. For example, turn to take ISO/IEC 25010: 2011 standard [1], where the definition of the qualitative state of the system begins with measuring and assessing the quality of processes, as the basis for further definition of the quality of the system under study. Moreover, the separation of problem-oriented functional properties (system or software functional specifications) and quality properties is clearly stands out, the latter, by standard definition, have nothing to do with the former. But the standard introduces a quality characteristic - functional suitability, criticizing which, it is indicated in [2] that a software application can be suitable for a certain production process, be useful, but not perform, or partially not perform some of its functions.

Another approach to assessing systems is to assess their technical condition (monitoring), which is determined by the value of the monitored parameters during technical diagnostics. According to DSTU 2389-94 [3], determining the state of the system under study is carried out at a certain point in time, under certain environmental conditions, by the values of the parameters established by the technical documentation. Conformity or non-conformity assessment takes place according to the criteria of malfunction, defect, degree of deterioration, regulation deviation, and the like.

The above allows us to conclude that there is reductionism in assessing the state of technical systems, which is expressed in a differentiated approach to assessing and the absence of research publications that combine all three approaches when assessing the TS state.

A separate assessment of the state of technical systems arose according to their evolutionary development, from simple to System systems and, accordingly, from determining the quality of the functional state, the technical state to the process state. The process approach began to be applied when systems reached such complexity that testing all their functions and technical condition in all modes became almost impossible. Its essence is quite simple: if the life cycle processes of the studied (target) system correspond to the best world practices, then, most likely, its work will be of high quality [4]. The key phrase here is "probably", which means that the quality of the work may NOT meet the necessary requirements. It is possible to check this only at the Utilization Stage, assessing the general condition of the TS, taking into account the functional, technical and process aspects.

The aim of the presented study is to determine an approach that combines technical, functional and process aspects into a single system for assessing the quality state of a technical system at the Utilization Stage.

The object of the research work is methods and models of a unified system for assessing the qualitative state of a technical system at the Utilization Stage.

The subject of the research is the assessment of the qualitative state of the technical system at the Utilization Stage.

The scientific novelty of the work consists in the definition and development of an approach and model of a unified system for assessing the qualitative state of technical systems, taking into account the functional, technical and process aspects of its work.

Keywords: Technical systems · Quality evaluation functional · Technical and process aspects · Informational structure · Quality state evaluation unified system

The differentiation of the above approaches is based on the separation of technical, functional and process structures within one TS, and this separation is fundamental, since at this stage of development, these structures are generated by a set of components and connections that differ in relations, properties and characteristics. This division is one of the manifestations of the reductionist approach, which doesn't see a systematic approach to assessing the state, thereby entering into a contradiction with the real state of affairs. The principle of consistency is shown schematically in Fig. 1. The embedded quality of the TS functions determines the requirements for the processes that implement them and the requirements for the technical components in which they (functions) are implemented. In its turn, both the flow of processes and the performance of system functions depend on the quality of technical implementation, and the quality of processes will determine the quality of functions and the behavior of hardware and software.

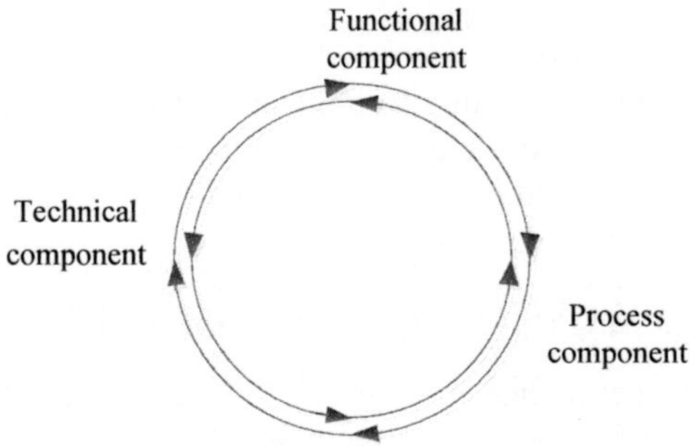

Fig. 1. The principle of consistency in assessing the state

The system solution proposed in the article is based on certain systems according to ISO/IEC/IEEE 15288:2015, as a combination of interacting elements organized to achieve one or more goals [5]. Since, as noted above, there is a fundamental structural separation of the functional, process and technical aspects of assessment, the organization of system connections, in order to ensure their interaction on structural or other grounds, is quite problematic. In order to solve this problem a concept was adopted based on the representation of the target system in the form of a system information structure (SIS), formed in the general information space for the life cycle of the target system, which unites these structures on the basis of system-wide information links.

The proposed information approach is based on the following provisions.

1. Information belongs only to the system information structure and is invariant with respect to the properties of technical, functional or process structures:
   At the Utilization Stage, information about the current state of the target system does not exist outside the SIS.
   - One and the same information can be applied, processed and transmitted by any of the structures included in the SIS.

2. The hierarchical SIS configuration coincides with the hierarchy of the hardware-software implementation of the target system:

   - The hierarchical SIS levels correspond to the hierarchical hardware and software levels of the studied TS (connected acyclic graph "tree" according to ISO/IEC/IEEE 15288:2015 [5]).
   - Each of the structures included in the SIS has its own hierarchical structure within the levels of the latter.

The links between SIS components are embodied in the form of an organized set of information channels.

Information in information channels is generated by sensors (physical or virtual sensors) and its quantitative value corresponds to their state.

We represent the information space of the life cycle (LC) of the target system as a set:

$$I_{LC} = \{I_1, \ldots, I_j, \ldots, I_k\} \tag{1}$$

where I_j – stage system information structure $(1, \ldots, j, \ldots, k)$ of the life cycle, subset of set I_{LC}. For example, according to the standards of the series ISO/IEC 24748, the life cycle of artificial systems consists of six stages: Concept Stage, Development Stage, Production Stage, Utilization Stage, Support Stage, Retirement Stage. For this formulation, expression (1) can be rewritten as:

$$I_{LC} = \{I_{CS}, I_{DS}, I_{PS}, I_{US}, I_{SS}, I_{RS}\}. \tag{2}$$

For each of the LC stages of the target system, SIS is built according to the established approaches and methods for determining its condition. So, for the Utilization Stage, based on the provisions of the triaspect approach and the definition of SIS and taking into account (2), we obtain:

$$I_{US} = \{I_F, I_T, I_P\}, \tag{3}$$

where I_F, I_T, I_P – information structures, the configuration of which coincides with the functional, technical and process structures, respectively. The essence of coincidence and construction I_T, which consists in associating elements of a multiple definition of a technical structure with information about their state, discussed in detail in [6, 7]. Since the construction of functional and process structures, according to the definition of the system and position, is similar to the construction of a technical structure, I_F and I_P are constructed by analogy with I_T.

According to the definition of the system and [5–7], we represent the structure of the hardware-software implementation of the target system in the form of a graph, in which the levels correspond to the hierarchical hardware implementation, nodes - hardware and software components of the corresponding levels, and the branches - to the backbone links that were laid down at the Development stage. The general plural definition of the structure of the hardware-software implementation matches to the expression:

$$Sys = \{Lev_1\{\ldots\{Lev_\mu\{\ldots\{Lev_\alpha\}\}\}\}\} = \left\{\bigcup_{i^1=1}^{m^1} Con_{i^1}\left\{\ldots\bigcup_{i^\mu=1}^{m^\mu} Con_{i^\mu}\left\{\ldots\left\{\bigcup_{i^\alpha=1}^{m^\alpha} Con_{i^\alpha}\right\}\right\}\right\}\right\}, \tag{4}$$

where $Lev\langle\cdot\rangle$ – hardware structural levels of the target system (graph levels), $Con\langle\cdot\rangle$ – hardware and software components of the appropriate level (graph nodes), $\mu = \overline{(1, \alpha)}$ – arbitrary level index, $i = \overline{(1, m)}$ – arbitrary index of the level component.

The belonging of the sets (4) that make up the levels to the sets of the highest level is determined by the characteristic vector (binary relations between the components of the – branch of the graph), which is built on the basis of system-forming connections:

$$Con_{i^{\mu}} \subset Con_{i^{\mu-1}} \left| \forall Con_{i^{\mu}} \in Lev_{\mu},\ b_i = 1, \right. \tag{5}$$

where $Con_{i^{\mu}}$ – component of an arbitrary level, $B^{(\mu-1)} = \{b_1, \ldots, b_{m^{(\mu-1)}}\}$ – characteristic level vector $Lev_{\mu-1}$.

By replacing in (4) the structural identification of hardware and software components with identification by their information state, we get:

$$I_{Sys} = \left\{ \bigcup_{i^1=1}^{m^1} I_{Con_{i^1}} \left\{ \ldots \bigcup_{i^{\mu}=1}^{m^{\mu}} I_{Con_{i^{\mu}}} \left\{ \ldots \left\{ \bigcup_{i^{\alpha}=1}^{m^{\alpha}} I_{Con_{i^{\alpha}}} \right\} \right\} \right\} \right\}, \tag{6}$$

Figure 2 that is, the nodes of the structural graph of the target system are identifiers of the information state of the corresponding components. The connections between the components of different levels are determined according to (5).

Based on the provisions and in accordance with [6–8], define the concepts: a sensor, like any source of information with an expected quantitative value and probability distribution, the current quantitative value of which can be measured; sensor infrastructure, as a configuration of information channels formed by sensors, where the information channel is a system of sensors that makes it possible to receive, transform and transmit information received in a message from the object of research.

The sensory infrastructure has a hierarchical structure, which has been studied in detail in [7], here we just note that the information state of the sensor s defined as the ratio between the received and expected amount of information from an object

$$I_s^{\Delta} = \frac{I(s_{meas})}{I(s_{exp})}, \tag{7}$$

where $I\langle\cdot\rangle$ – the amount of the sensor's own information according to K. Shannon,s_{meas}, s_{exp} – current (measured) and expected state of the sensor, respectively.

The information state of the set of sensors, based on the additivity of their own information, is determined by the expression:

$$I_{\Sigma}^{\Delta} = \frac{\sum_{n} I_{s_n}(s_{meas})}{\sum_{n} I_{s_n}(s_{exp})}, \tag{8}$$

Let's consider an arbitrary node with (4) in the context of identifying its state by functional, technical and process aspects. To do this, let's turn to the definition of the specifications: Technical specification - a document that establishes the technical requirements that a product, process or service must satisfy [9]. Requirements are established regarding technical and quality characteristics, as well as compliance with the standards of hardware and software components and the system as a whole.

Functional specification - a document that describes the characteristics of a technical object that are required [10]. Here, first of all, we are talking about formal specifications, which are a mathematical description of what the object under study (the system or its component) should do, and how it does it is not specified.

Process specification - a document that describes the properties of a process in the form of a mathematical object [11].

Each description in certain documents is a structure, with respect to requirements, characteristics and properties, they are given. In addition, as follows from the definitions, each structural unit of any description can be represented in the form of a mathematical model, that is, its state can be quantified. Typically, the evaluation is carried out by comparing the measured values with the expected values obtained from the simulation results. Thus, the state of each hardware-software component (4) can be determined by the quantitative values of its specifications.

$$Sys = \left\{ \bigcup_{i^1=1}^{m^1} \{F, T, P\}_{i^1} \left\{ \ldots \bigcup_{i^\mu=1}^{m^\mu} \{F, T, P\}_{i^\mu} \left\{ \ldots \left\{ \bigcup_{i^\alpha=1}^{m^\alpha} \{F, T, P\}_{i^\alpha} \right\} \right\} \right\} \right\}, \tag{9}$$

where F, T, P - structures of functional, technical and process specifications of components, respectively.

Since the hardware and software components are the nodes of the graph (6), we have:

$$I_{Sys} = \left\{ \bigcup_{i^1=1}^{m^1} \{I_F, I_T, I_P\}_{i^1} \left\{ \ldots \bigcup_{i^\mu=1}^{m^\mu} \{I_F, I_T, I_P\}_{i^\mu} \left\{ \ldots \left\{ \bigcup_{i^\alpha=1}^{m^\alpha} \{I_F, I_T, I_P\}_{i^\alpha} \right\} \right\} \right\} \right\}, \tag{10}$$

which corresponds to expression (3) and is a multiple SIS definition.

Automatic (autonomous) mode of operation assumes the absence or limited control, management and, accordingly, assessment of the TS state by dispatchers and operators. But the existing standards imply an assessment of the state of systems and components at the operational stage, specifically by users, for example, assessing quality indicators when used in the standards of the series ISO/IEC 25000. In automatic systems, the evaluation functions are performed by expert diagnostic systems [12], which, in relation to the target system, can be both external and built-in. The only source of information for expert systems is the data transmitted through the information channels of the sensor infrastructure. It should be noted that according to the above definition, which source of information belongs to the concept of a sensor, this means that sensors can be both traditional physical and virtual (software) sensors, and specifications of lower hierarchical levels (objects of research) in relation to the specifications of higher levels, according to the backbone connections. Based on the foregoing and (7, 8), expression (10) can be represented in the form that forms the structure and determines the quantitative values of the relative information state of the target system:

$$I_{Sys}^{\Delta} = \left\{ \bigcup_{i^1=1}^{m^1} \left\{ I_F^{\Delta}, I_T^{\Delta}, I_P^{\Delta}, I_{\Sigma}^{\Delta} \right\}_{i^1} \left\{ \ldots \left\{ \bigcup_{i^\alpha=1}^{m^\alpha} \left\{ I_F^{\Delta}, I_T^{\Delta}, I_P^{\Delta}, I_{\Sigma}^{\Delta} \right\}_{i^\alpha} \right\} \right\} \right\}, \tag{11}$$

where $\left\{I^{\Delta}_{\langle\cdot\rangle}\right\}_{i^{\langle\cdot\rangle}}$ – relative informational state of level components; I^{Δ}_{F}, I^{Δ}_{T}, I^{Δ}_{P} – relative informational state of specifications, $I^{\Delta}_{\sum}$ – relative informational state of physical or software sensors corresponding to the component.

A general example of the formation of the structure of the information state of the hardware component of the target system is shown in Fig. 2. It is necessary to pay attention to the hierarchical construction of the given structures, for example, in the technical structure specifications 3 and 4 are at a lower level, relative to specification 2 and according to system information links, they can be sources of information for it. Specification 1 is at the highest hierarchical level, the quantitative value of its state determines the state of the entire technical structure and is assessed taking into account the states of all technical specifications.

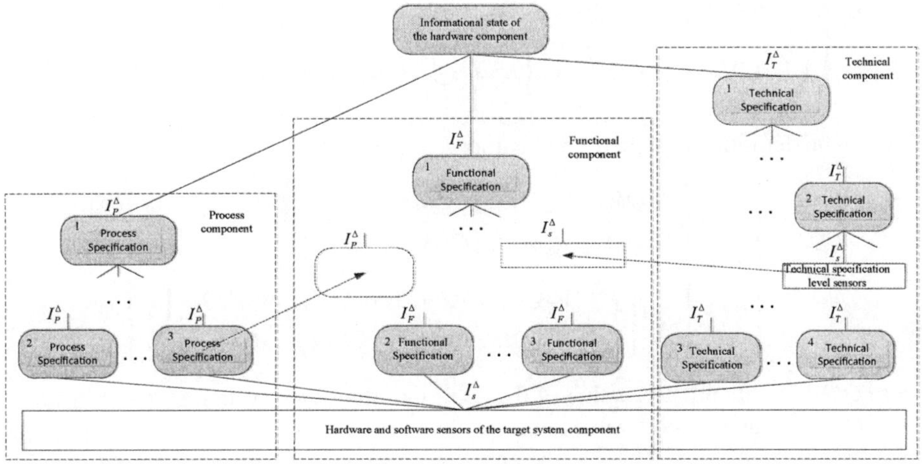

Fig. 2. General example of the information state formation the hardware component of the target system

In addition to visualizing the use case of the expression (11), an example (Fig. 2) shows a possible implementation of one of the provisions, namely, that the same information can be applied, processed and transmitted by any of the structures included in the SIS. The dashed line shows how information about the state of the process specification and state of the sensor of a separate level of the technical component can be used to assess the state of the functional specifications. That is, for functional specifications, for example, these sources become their own virtual sensors, providing data invariance with respect to the properties of structures.

The transition to the definition of the qualitative state of the target system and its individual components is carried out according to [7, 13], where the quality assessment function is introduced as the inverse value of the relative information state of the components TS (in our case, physical and virtual sensors, specifications, hardware and software components). Based on (7, 8), we get:

– to quantify simple and quasi-simple quality indicators:

$$SQM = \frac{I(s_{exp})}{I(s_{meas})}, \quad (12)$$

– to determine the quantitative value of complex quality indicators:

$$CQM = \frac{\sum_n I_{s_n}(s_{exp})}{\sum_n I_{s_n}(s_{meas})}. \quad (13)$$

According to (11, 12, 13), a quantitative assessment of the qualitative state of the target system and its individual components can be obtained by the expression:

$$\begin{aligned} QM_{Sys} &= \{QM_F, QM_T, QM_P\} \\ &= \left\{ \bigcup_{i^1=1}^{m^1} \{QM_F, QM_T, QM_P\}_{i^1} \left\{ \ldots \left\{ \bigcup_{i^\alpha=1}^{m^\alpha} \{QM_F, QM_T, QM_P\}_{i^\alpha} \right\}\right\}\right\}, \end{aligned} \quad (14)$$

or in the definitions of quality indicators:

$$QM_{Sys} = \left\{ \bigcup_{i^1=1}^{m^1} \begin{array}{l} \left\{ \begin{array}{l} \{CQM_F, SQM_F\}, \{CQM_T, SQM_T\}, \\ \{CQM_P, SQM_P\}, \{CQM_{\sum}, SQM_s\} \end{array} \right\}_{i^1} \\ \left\{ \ldots \left\{ \bigcup_{i^\alpha=1}^{m^\alpha} \begin{array}{l} \{CQM_F, SQM_F\}, \{CQM_T, SQM_T\}, \\ \{CQM_P, SQM_P\}, \{CQM_{\sum}, SQM_s\}_{i^\alpha} \end{array} \right\}\right\} \end{array} \right\}, \quad (15)$$

where QM_{Sys} – quantitative value of the quality of the target system at the Utilization Stage, $QM\langle\cdot\rangle$ – quantitative value of the quality of the corresponding components of the target system, $SQM\langle\cdot\rangle$, $CQM\langle\cdot\rangle$ – the quantitative value of simple and complex quality indicators of the corresponding components of the target system.

The expected quantitative value of quality indicators is determined at the Development stage based on the simulation results, in accordance with the design requirements for the Utilization Stage of the target system.

The quantitative value of the quality indicators obtained by expression (15) doesn't carry information about the qualitative state of the object under study. The process of determining the quality consists in assessing the belonging of the obtained value to a certain rating scale. If we consider the scale, which consists of three subbands - "good", "satisfactory", "bad", we can use the dynamic scale given in [7, 13], which in terms of the article will take the following form. For simple quality metrics:

$SQM(s_{exp}, s_{meas}) \in [SQM(s_{exp}, x))$ – range "good",

$SQM(s_{exp}, s_{meas}) \in [SQM(s_{exp}, y))$ – range "satisfactory",

$SQM(s_{exp}, s_{meas}) \notin \{[SQM(s_{exp}, x)] \cup [SQM(s_{exp}, y)]\}$ – range "bad".

For complex quality indicators:

$[(\forall SQM \notin [SQM(s_{exp} + y)]) \wedge (\forall SQM \notin [SQM < SQM(s_{exp} + y)])|\forall SQM \in CQM)$ – range "good",

$[(\exists SQM \in [SQM(s_{exp} + y)]) \wedge (\forall SQM \notin [SQM < SQM(s_{exp} + y)])|\forall SQM \in CQM)$ – range "satisfactory",

$\left[\exists SQM \in \left[SQM < SQM\left(s_{exp} + y\right)\right] | \forall SQM \in CQM\right]$ – range "bad".

Where $\left(s_{exp} - b \leq x \leq s_{exp} + a\right)$, (a, b) – permissible maximum and minimum deviation from the expected value, respectively; $\begin{cases} s_{exp} + a < y \leq s_{exp} + a + c \\ s_{exp} - b - d \leq y < s_{exp} - b \end{cases}$, (c, d) – maximum and minimum deviation from the permissible deviations (a, b), respectively.

The following method is proposed to build a triaspect model:

1. We take the quality of the technical, functional and process specifications of the target system or its component as spatial dimensions.
2. Using expressions (14, 15), we calculate the quantitative value of the quality indicators of specifications, as identifiers of the object under study.
3. We present the calculated quantitative value of complex quality indicators QM_F, QM_T, QM_P specifications in the form of vectors in space, each of the dimensions of which corresponds to the specified indicator.
4. We calculate the resulting vector in three-dimensional space, where the numerical values of the axial coordinates correspond to the quantitative values of the quality indicators of the specifications.

The evaluation space is limited to a rectangular parallelepiped with side lengths $QM_F = QM_T = QM_P = QM_{max} = 1$, де $QM_{max} = 1$ – the maximum possible quantitative value of the quality indicator (Fig. 3).

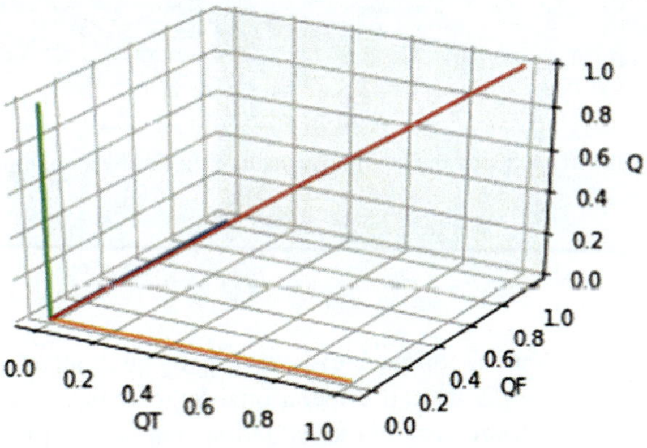

Fig. 3. Tri-aspect model for technical systems quality evaluation

The quantitative determination of the qualitative state of the technical system or its component, taking into account the rationing to the unit, is carried out according to the expression:

$$Q\langle\cdot\rangle = \frac{D}{\sup(D)} = \frac{\sqrt{QM_T + QM_F + QM_P}}{\sqrt{3}}, \tag{16}$$

where D – the length of the resulting vector, $\sup(D)$ – the length of the maximum possible result vector.

In Fig. 4 shows an example of assessing the qualitative state of the hardware component of TS according to the information structure shown in Fig. 2. Evaluation is carried out while ensuring the hardware component of the function according to specification 3 during the execution of the process corresponding to specification 2, while the work of the technical component corresponds to specification 3. All specified specifications and specifications of the highest hierarchical levels to which they refer are noted in the quality assessment chart. Quantitative value of quality $QM\langle\cdot\rangle$ each of the specifications is calculated by expression (15). Quantitative value of the qualitative state of the investigated hardware component is calculated by expression (16). The quality assessment is carried out according to the scale proposed in the article.

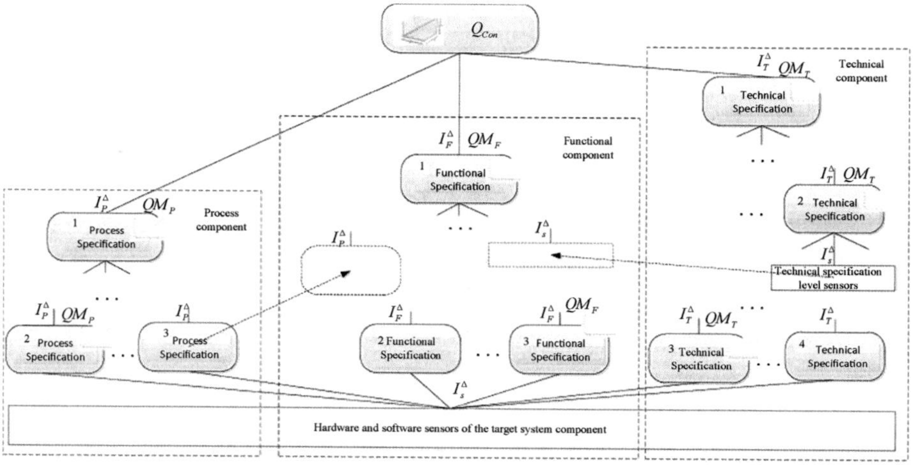

Fig. 4. General example of assessing the qualitative state of the hardware component of the target system

Conclusion

The article presents a conceptual approach to assessing the qualitative state of technical, in particular cyber-physical systems at the operational stage, in three aspects: functional, technical and process. The dominance of the reductionist approach to the assessment of technical systems is shown, which is inappropriate in the context of making managerial decisions at the modern level of systems complexity. A general approach and provisions are proposed that make it possible to consider and study the functional, technical and process approaches to assessment, as an integral assessment system based on their interaction within a single systemic information structure based on design specifications and system information links. The presentation of the information structure of a technical system in the form of a sensor infrastructure is considered. A mathematical model has been developed and a qualitative state assessment scale has been proposed, based on determining the relative value of the amount of specification information. A construction method and a model have been developed that combines technical, functional and process aspects into a single system for assessing the quality state of technical systems. Examples are given.

Further research is the development of models: quality indicators of specifications, a system for collecting measurement information in a system information structure, an expert assessment of the state of the technical system under study and its components, with the subsequent integration of all models into a single monitoring and diagnostics system based on a triaspect approach to quality assessment.

References

1. ДСТУ ISO/IEC 25010:2016: Інженерія систем і програмних засобів. Вимоги до якості систем і програмних засобів та її оцінювання (SQuaRE). Моделі якості системи та програмних засобів (ISO/IEC 25010:2011, IDT)
2. Лукин, В.Н.: Проблемы качества программного обеспечения. Моделирование и анализ данных. **10**(1), 140–156 (2020). https://doi.org/10.17759/mda
3. ДСТУ 2389-94: Технічне діагностування та контроль технічного стану. Терміни та визначення
4. ДСТУ ISO 9000:2015: Системи управління якістю. Основні положення та словник термінів (ISO 9000:2015, IDT)
5. ДСТУ ISO/IEC/IEEE 15288:2016: Інженерія систем і програмного забезпечення. Процеси життєвого циклу систем (ISO/IEC/IEEE 15288:2015, IDT)
6. Volkov, S.: Set-theoretic model of the information state of the industrial cyberphysical system. Sci. J. Ternopil Ivan Pul'uj Natl. Tech. Univ. № 1(89), 132–138 (2018)
7. Волков, С.Л.: Теоретичні засади моніторингу якісного стану технічних систем: монографія. Одеса: Видавництво ТОВ «Лерадрук», 112 с (2018)
8. Волков, С.Л., Кисельова, О.І.: Філософський аспект інформаційного стану штучних систем. Метрологія та прилади № 5, 119–122 (2017)
9. Закон України про технічні регламенти та оцінку відповідності Відомості Верховної Ради (ВВР), № 14, ст. 96 (2015)
10. Lightsey, B.: Systems Engineering Fundamentals; Technical Report, Department of Defense-Systems Management College, Fort Belvoir, VA, USA (2001)
11. Миронов, А.М.: Теория процессов. Переславль: Издательство НОУ Институт программных систем. Университет г. Переславля, им. А.К, Айламазяна Переславль-Залесский, 346 с (2008)
12. Волков, С.Л., Казакова, Н.Ф., Асабашвілі, С.Д.: Модель експертного оцінювання якісного стану технічної системи. Вісник НТУ "ХПІ", № 44, pp. 157–161 (2017)
13. Волков, С.Л., Волков, К.С.: Функція і шкала ентропійної оцінки якісного стану датчика. Збірник наукових праць Одеської державної академії технічного регулювання та якості. Випуск **1**(12), 10–15 (2018)

A Method for Diagnosing SCADA Failures Based on Predicate Logic Within the Expert System

Olena Syrotkina(✉), Mykhailo Aleksieiev(✉), Borys Moroz(✉), Iryna Udovyk(✉), and Andrii Martynenko(✉)

Dnipro University of Technology, 19 D. Yavornytskoho Ave., Dnipro 49005, Ukraine
{syrotkina.o.i,alekseev.m.o,moroz.b.i,udovyk.i.m, martynenko.a.a}@nmu.one

Abstract. This paper addresses the issue of detecting and localizing failures in the distributed SCADA system based on the analysis of changes in the information flow states when passing through structural modules and levels of SCADA hierarchy. The structural and logical model presented describes SCADA of any topology. We formed diagnostic features to detect independent failures based on the characteristic functions of three-valued logic. We put forward a methodology of expert diagnostic systems to analyze changes in the state of SCADA information flows. We also developed a predicate system for knowledge representation in the expert system to implement a method for diagnosing SCADA operability when having incomplete or unreliable data. This includes a predicate system for describing SCADA topology and detecting the location of any failures. The configuration diagram of SCADA topology is presented using the parameters of the structural and logical model for diagnosing failures. This model is used by the predicate system within the expert system knowledge base. We calculate the instability coefficients to evaluate the operability of each system's structural module in real-time based on the analysis of changes in the state of SCADA information flows within the structural and logical model developed. The proposed method for detecting and localizing failures in the SCADA system can be applied both for current diagnostics and for predicting the operability of the structural modules of the SCADA system. The system operability prediction will be formed on the basis of statistical processing and analysis of the retrospective array containing the instability coefficients of the SCADA structural modules' operation.

Keywords: SCADA system · Expert diagnostic system · Predicate system · Structural and logical model · Structural SCADA module · Information flow · Instability coefficient for the SCADA structural module

1 Introduction

At the current stage of development, which includes the implementation of information technologies in the field of industrial automation, the problems of diagnosing the SCADA

P. Vorobiyenko et al. (Eds.): IPF 2020, LNNS 212, pp. 340–358, 2021.
https://doi.org/10.1007/978-3-030-76343-5_18

operability for a manufacturing enterprise remains relevant [1, 2]. In the event of a system failure, or when detecting an unusual situation, large volumes of semi-structured low-level primary diagnostic information are generated within seconds. This requires further processing and analysis to restore the system's performance. Various factors affect the quality of diagnostics and the recovery time of the system. These include a lack of time as well as the usage of computing and information resources during data processing. These all affect the methods used for data analysis and the need for prompt decision-making in real-time. Thus, the most pressing tasks are the creation and application of mathematical methods for automatic diagnostics of failures in a distributed SCADA system. Fulfilling these requirements will allow us to minimize the resources we use to a real-time scale [3].

2 Aim of the Research

The aim of the research is to increase the efficiency of the automated control system of an industrial enterprise. This is done by providing automatic self-diagnostics of SCADA operability in real-time. In order to achieve this, it is necessary to conduct a study of the processes concerning functional diagnostics of SCADA operability based on a diagnostic model specifically created for use by the expert system [4, 5]. Within the framework of this model, it is necessary to form criteria for the detection and localization of independent failures in the system. It can be performed by analyzing the patterns of changes in the state of information flows during their passage through the structural modules and levels of SCADA hierarchy.

3 Problem Statement in General Terms

In order to solve the problem of automatic self-diagnostics of the SCADA system operability with the possibility of auto-recovery after reversible failures, we propose a scheme of interaction between SCADA and the expert system as shown in Fig. 1.

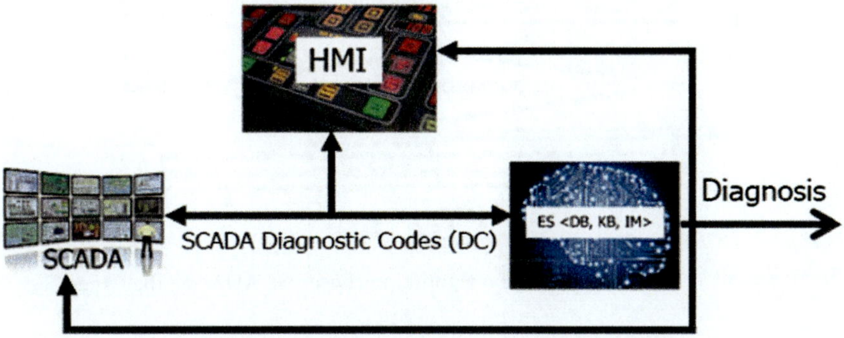

Fig. 1. An interaction scheme between SCADA and the Expert System.

In Fig. 1, the following designations are employed:

ES is an expert diagnostic system working in real-time; DB is a database (basic components of the subject area, their properties and connections between them); KB is a knowledge base (rules for constructing variants of structures from database components); IM is an inference machine; DC are SCADA diagnostic codes; HMI is a human-machine interface.

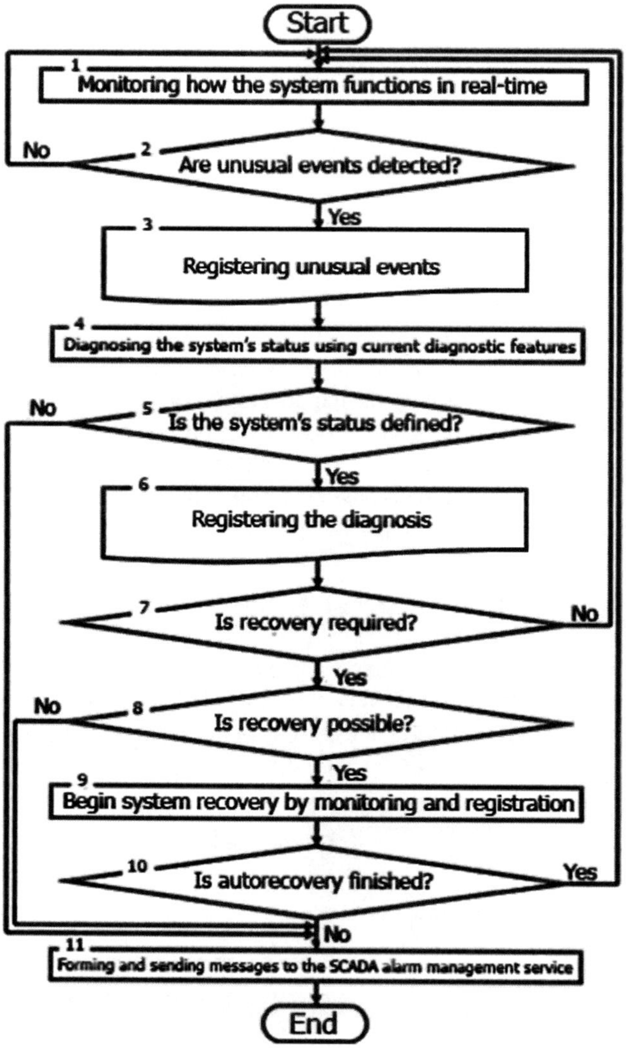

Fig. 2. Flowchart showing the interaction algorithm between SCADA and the Expert System.

The expert system draws conclusions on the technical state of SCADA, including its functional components and processes. It is based on the analysis of input data it receives (system diagnostic codes) from the SCADA system in real-time. SCADA then stores the self-recovery algorithms in the system. When the expert system sends a diagnosis for reversible failures to SCADA, the relevant procedure associated with backbone node self-recovery is executed. A flowchart of an interaction algorithm between SCADA and the Expert System is shown in Fig. 2.

4 Main Part

The method to automatically detect and localize failures in real-time is based on the analysis of changes in the state of information flows. This is grounded on the development of a knowledge representation system (KRS) for the expert system knowledge base. In our case, KRS consists of a predicate system and is intended as a universal description, such as:

- SCADA structure of any topology;
- distribution of diagnostic indicators for detecting independent failures among structural elements of different hierarchies;
- determining instability coefficients for the operation of SCADA structural modules as well as effectively searching the diagnosis in real-time.

4.1 Example of SCADA Topology

In order to describe SCADA topology and localize failures, we consider an example of a SCADA topology as shown in Fig. 3.

At time t, the set of controlled parameters of the technological control object (CP TCO) is as follows:

$$X(t) = \{x_1(t), x_2(t), \ldots, x_i(t), \ldots x_{n(X)}(t)\}. \tag{1}$$

These control parameters are measured by Sensors (SR), and are then sent to Remote Terminal Units (RTU) through Data Transmission Channels (DTC1). After being received by RTUs, the control parameters are sent to Servers through Data Transmission Channels (DTC2).

The following designations are used in Fig. 3:

HL is a SCADA Hierarchy Level; RTU is a Remote Terminal Unit; DTC is a Data Transmission Channel; SR are Sensors; CP TCO is a Controlled Parameter of the Technological Control Object; DTC1 is a Data Transmission Channel between Sensors and RTUs; DTC2 is a Data Transmission Channel between RTUs and Servers; SR_i is a Sensor index number; n_CP is the number of controlled parameters measured by the appropriate sensor; Ch1_i is the index number of DTC1; n_SR is the number of sensors connected to the appropriate data transmission channel; RTU_i is the RTU index number; n_ Ch1 is the number of data transmission channels from sensors connected to the appropriate RTU; Ch2_i is the index number of DTC2; n_ RTU is the number of RTUs connected to the server through the appropriate data transmission channel; Srv_i is the index number of the server; n_Ch2 is the number of data transmission channels from the RTUs connected to the appropriate server.

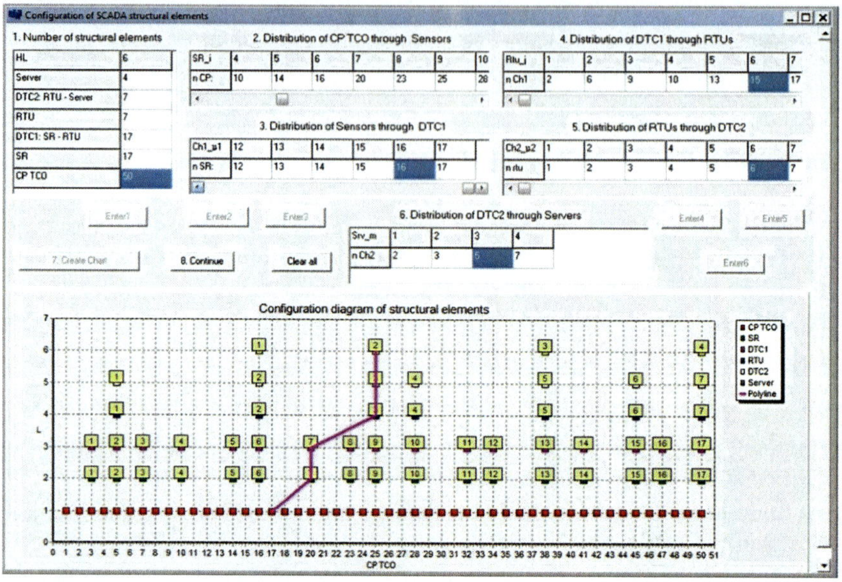

Fig. 3. Software interface example for the configuration of SCADA structure.

4.2 SCADA Structural and Logical Model

We define the controlled parameter distribution through RTUs (see Fig. 3) using a non-decreasing sequence of positive integers I_x:

$$I_x = i_1,\ i_2,\ \ldots,\ i_j,\ \ldots i_N,$$

where j is the position of the element in the sequence and corresponds to the RTU sequence number; N is the number of RTUs; $i_N = n(X)$ is the number of controlled parameters; $i_j - i_{j-1}$ is the number of controlled parameters connected to RTU_j.

Another non-decreasing sequence of positive integers M_k defines the distribution of Data Transmission Channels through RTUs from sensors (see Fig. 3),

$$M_k = m_1,\ m_2,\ \ldots,\ m_j,\ \ldots m_N,$$

where j is the position of the element in the sequence member and corresponds to the RTU sequence number; N is the number of RTUs; m_N is the number of Data Transmission Channels from sensors to RTUs; $m_j = m_{j-1}$ is the number of Data Transmission Channels from sensors connected to RTU_j.

The distribution of controlled parameters through Data Transfer Channels from sensors to RTUs is defined using a non-decreasing sequence of positive integers K_x:

$$K_x = k_1,\ k_2,\ \ldots,\ k_\mu,\ \ldots i_{mN},$$

where μ is the position of the element in the sequence and corresponds to sequence number of the Data Transfer Channels from sensors to RTUs Ch_μ; N is the number of RTUs; m_N is the number of Data Transmission Channels from sensors to RTUs; $k_\mu - k_{\mu-1}$ is the number of controlled parameters. These are configured to transfer data through RTU Ch_μ.

The configuration diagram of the CP TCO X through the Data Transmission Channels Ch and Remote Terminal Units RTU is shown in Fig. 4.

We define certain predicates using the parameters of the structural and logical model of failure diagnostics (see Fig. 4).

Predicate $H_1(i,j)$ to represent the connection between CP TCO x_i and SR_j is as follows:

$$H_1(i,j) := ((i \le k_{m_N})\&(j \le m_N))?((j > 1)?((i > k_{m_{j-1}})\&(i \le k_{m_j})) : (i \le k_{m_j})) : 0, \quad (2)$$

where $m_j \in K_X$.

Predicate $H_2(i,j)$ to represent the connection between CP TCO x_i and RTU_j is as follows:

$$H_2(i,j) := ((i \le i_N)\&(j \le N))?((j > 1)?((i > i_{j-1})\&(i \le i_j)) : (i \le i_j)) : 0, \quad (3)$$

where $i_j \in I_X$.

Predicate $H_3(i,j)$ to represent the connection between SR_i/Ch_i and RTU_j is as follows:

$$H_3(i,j) := ((i \le m_N)\&(j \le N))?((j > 1)?((i > m_{j-1})\&(i \le m_j)) : (i \le m_j)) : 0. \quad (4)$$

where $m_j \in M_K$.

The above predicates fully describe the SCADA topology, taking into account all the connections between the structural elements of various hierarchy levels according to the configuration shown in the diagram (see Fig. 4).

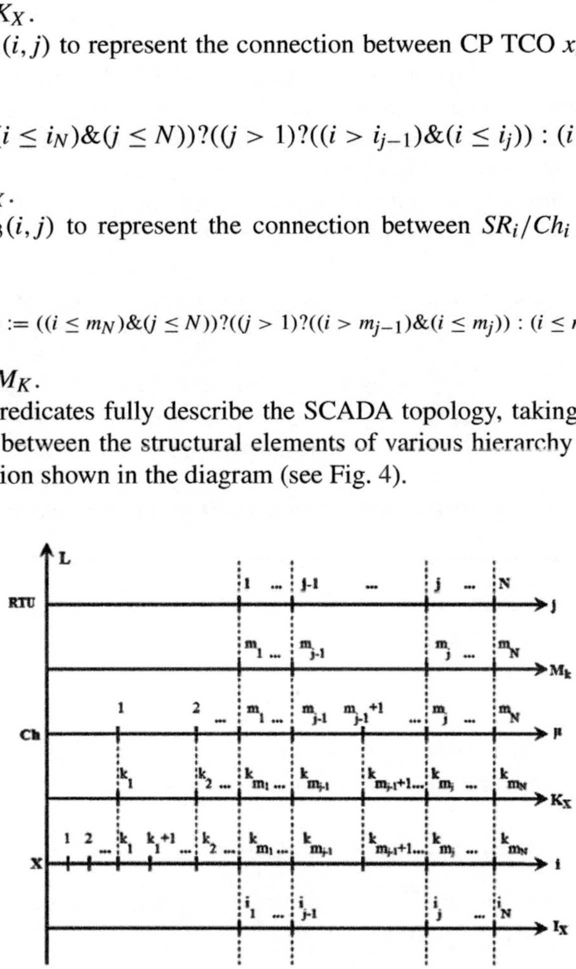

Fig. 4. The configuration diagram showing the distribution of CP TCO through Data Transmission Channels and Remote Terminal Units of SCADA.

4.3 Formation of Diagnostic Features for Failure Detection in SCADA

The method of automatic failure detection and localization in a SCADA system is based on the analysis of changes in the state of information flows. The input data are diagnostic matrices $D(t)$ which represent the information flow state at time t. Such a matrix is formed with the frequency of obtaining the data from sensors. In this case, the number of rows of the diagnostic matrix corresponds to the number of SCADA hierarchy levels and the number of columns corresponds to the number of controlled parameters.

$$\begin{cases} D(t) = [d_{iL,iC}(t)] \\ d_{il,iC}(t) \in E_3, E_3 = \{0, 1, 2\} \\ iL = l(S_1) + 1 - l, \quad 1 \le l \le l(S_{1)} \\ 1 \le iC \le n(X) \end{cases}, \tag{5}$$

where iL is the matrix $D(t)$ row index corresponding to SCADA hierarchy level l; iC is the matrix $D(t)$ column index corresponding to the controlled parameter index $x_{iC}(t)$; $l(S_1)$ is SCADA server hierarchy level; $n(X)$ is the number of CP TCO.

At each SCADA hierarchy level, the controlled parameter can be in one of three states: the value of the parameter is "Missing", "Unreliable" or "Reliable". These states can be described using the three-valued set $E_3 = \{0, 1, 2\}$.

In order to analyze the diagnostic matrix, we use an elementary function of three-valued logic called the characteristic function of the first kind, which is defined as follows:

$$\varphi_e(x) = \begin{cases} 1, & x = e, \quad e \in E_3 \\ 0, & x \neq e, \quad e \in E_3 \\ & E_3 = \{0, 1, 2\} \end{cases}. \tag{6}$$

We define the various diagnostic features (DF) of failure detection as described below:

a) a sufficient DF of failure absence in the SCADA structural element with coordinates (iL, iC) at time t is as follows:

$$\varphi_2(d_{iL,iC}(t)) = 1; \tag{7}$$

b) A necessary, but insufficient DF of failure detection for the SCADA structural element with coordinates (iL, iC) at time t is as follows:

$$\neg\varphi_2(d_{iL,iC}(t)) = 1; \tag{8}$$

c) a necessary, but insufficient DF of failure detection in the SCADA structural element with coordinates (iL, iC) at time t due to.
   - either the absence of the CP TCO on the system's hierarchy level corresponding to the backbone nodes;
   - or the absence of the process responsible for receiving and transmitting data. This is located in a system's hierarchy level and corresponds to the data transmission

channels in the SCADA structural element with coordinates (iL, iC) at time t. It is as follows:

$$\varphi_0(d_{iL,iC}(t)) = 1; \tag{9}$$

d) a necessary, but insufficient DF of failure detection can be due to.
   - either the unreliability of the CP TCO at a SCADA hierarchy level corresponding to the backbone nodes;
   - or the unreliability of the process responsible for receiving and transmitting data at a system's hierarchy level corresponding to the data transmission channels in the SCADA structural element with coordinates (iL, iC) at time t. It is as follows:

$$\varphi_1(d_{iL,iC}(t)) = 1. \tag{10}$$

When analyzing diagnostic matrix $D(t)$, it is possible to assert that for all SCADA structural modules and functional components involved in the transmission of the CP TCO from sensors to servers, no failures were found at time t, if for the first row ($iL = 1$) of matrix $D(t)$ corresponding to the server's hierarchy level $l(S_1)$, we have the following:

$$\&_{iC=1}^{i_N} \varphi_2(d_{1,iC}(t)) = 1. \tag{11}$$

In general, the function of failure detection in the system based on the analysis of diagnostic matrix $D(t)$ is written as follows:

$$g_2(iL, \alpha, \beta, t) = \neg(\&_{iC=\alpha}^{\beta} \varphi_2(d_{iL,iC}(t))), \tag{12}$$

where iL is the row index of matrix $D(t)$ corresponding to SCADA hierarchy level l; α, β are the initial and final serial numbers of the CP TCO passing through the system's structural elements at a given level of hierarchy l.

In order to form the criteria for the distribution of DF for independent failures, taking into account the characteristic features for each SCADA hierarchy level, we consider predicate $S(i, y, l)$ for determining state y for CP TCO x_i at hierarchy level l is as follows:

$$S(i, y, l) := (y = d_{l(S1)+1-l,i}). \tag{13}$$

Therefore, the diagnostic feature of failure detection is described by predicate $S(i, y, l)$ as follows:

a) $S(i, 2, l)$;
b) $\neg S(i, 2, l)$;
c) $S(i, 0, l)$;
d) $S(i, 1, l)$.

4.4 The Analysis of Diagnostic Features for Failure Detection in SCADA

It should be noted that both diagnostic features of failure detection $\neg\varphi_2(d_{iL,iC}(t))$ and function $g_2(iL, \alpha, \beta, t)$ of failure detection in the SCADA system do not distinguish between independent and secondary failures. We assume that, for the process, when moving the CP TCO up through the system's hierarchy levels, at the lower SCADA hierarchy level l_{min} all failures are considered independent of the hierarchy level $l < l_{min}$ and all diagnostic features for detecting failures belong to the diagnostic features of detecting independent failures. We also accept that at hierarchy levels corresponding to the data transmission channels, the diagnostic features for detecting failures refer to the diagnostic features for detecting failures independent of the state of the CP TCO.

Therefore, at this stage of the analysis, we can assert the following:

- the absence of a failure detection diagnostic feature at a certain SCADA hierarchy level is a sufficient condition. This means that no failure was detected at this hierarchy level;
- the presence of a failure detection diagnostic feature at a certain SCADA hierarchy level l_{min} is a sufficient condition, which means that independent failures appeared at this hierarchy level l_{min} and all failure detection diagnostic features refer to independent failures;
- the number of independent failures at a sensor hierarchy level ($l_{min} = 2$) is equal to the number of failure detection diagnostic features;
- In order to determine the number of failures at a certain lower hierarchy level ($l_{min} = 2$), an additional analysis of diagnostic matrix $D(t)$ is required, since different diagnostic features may refer to the same failure;
- the presence of diagnostic features for detecting a failure at a certain hierarchy level $l > l_{min}$ for $l \in L_1$ (L_1 is a subset of the SCADA hierarchy levels related to the backbone nodes). It is a necessary, yet insufficient condition for a failure to appear which does not depend on lower hierarchy levels. Additional diagnostic criteria are required to classify diagnostic features as a secondary or independent failure;
- the absence of diagnostic features at a certain hierarchy level $l > l_{min}$ to detect independent failures is a sufficient condition for the fact that there are no independent failures at this hierarchy level;
- the additional analysis of diagnostic matrix $D(t)$ is required in order to determine the number of independent failures when having diagnostic features for detecting independent failures at a certain hierarchy level $l > l_{min}$, since different diagnostic features may refer to the same failure.

We define the lower SCADA hierarchy level of failure detection l_{min} using the example of SCADA structure shown in Fig. 3:

$$\begin{aligned} iL_{min} &:= (g_2(1, 1, i_N, t))?((g_2(2, 1, i_N, t))?((g_2(3, 1, i_N, t))? \\ &((g_2(4, 1, i_N, t))?((g_2(5, 1, i_N, t))?5 : 4) : 3) : 2) : 1) : 0. \end{aligned} \tag{14}$$

If $iL_{min} = 0$, then no failures were found in the system at time t, otherwise, it is the lower hierarchy level for detecting failures in the SCADA system:

$$l_{min} = l(S_1) + 1 - iL_{min}. \tag{15}$$

4.5 Criteria for Detecting Independent Failures in SCADA

Due to the structural and logical model taken into consideration for the failure detection and localization in SCADA, we assume that at the lower hierarchy level l_{min} all diagnostic features for detecting failures $\neg\varphi_2(d_{iL,iC}(t))$ refer to independent failures. For hierarchy level l_{min} we form a matrix of independent failure markers $\Lambda(t)$:

$$\lambda_{iL_{min},iC}(t) = \neg\varphi_2(d_{iL,iC}(t)). \tag{16}$$

If $l_{min} < l(S_1)$, then it is necessary to determine other SCADA hierarchy levels $l_{min} < l \leq l(S_1)$ for which, when analyzing the current diagnostic matrix $D(t)$, it is possible to detect independent failures. Thus, it takes into account possible changes in the state of the CP TCO when moving up the various SCADA hierarchy levels.

We consider an algorithm for detecting independent failures if the lower SCADA hierarchy level belongs to backbone nodes.

For row iL of diagnostic matrix $D(t)$, the number of diagnostic features $\varphi_0(d_{iL,iC}(t))$ and the number of diagnostic features $\varphi_1(d_{iL,iC}(t))$ are determined by the following formulas:

$$n_{\varphi_0}(iL, 1, i_N, t) = \sum\nolimits_{iC=1}^{i_N} \varphi_0(d_{iL,iC}(t)), \tag{17}$$

$$n_{\varphi_1}(iL, 1, i_N, t) = \sum\nolimits_{iC=1}^{i_N} \varphi_1(d_{iL,iC}(t)). \tag{18}$$

The total number of diagnostic features for failure detection $\neg\varphi_2(d_{iL,iC}(t))$ for row iL of diagnostic matrix $D(t)$ is as follows:

$$n_{\overline{\varphi_2}}(iL, 1, i_N, t) = n_{\varphi_0}(iL, 1, i_N, t) + n_{\varphi_1}(iL, 1, i_N, t). \tag{19}$$

In accordance with the logic of SCADA functioning at higher hierarchy levels $l_{min} < l_{h+1}$ which also belong to the backbone nodes, a necessary, yet insufficient criterion for the presence of independent failures η is the increase in the number of diagnostic features for detecting failures in comparison with the system's lower hierarchy level $l_{min} < l_h$.

$$\begin{cases} l_h = l_{min} + 2 \times h \leq l(S_1), h = 1, 2, \ldots \\ iL_h = l(S_1) + 1 - l_h \\ n\neg\varphi_2(iL_h, 1, i_N, t) = n_{\varphi_0}(iL_h, 1, i_N, t) + n_{\varphi_1}(iL_h, 1, i_N, t) \end{cases} \tag{20}$$

If at a higher system's hierarchy level l_{h+1} the number of diagnostic features for detecting failures has increased in comparison with the lower system's hierarchy level l_h, then we conclude that the necessary condition for the presence of independent failures is fulfilled at the SCADA hierarchy level l_{h+1}.

$$\begin{aligned} \eta(iL_{h+1}) = (n_0(iL_{h+1}, 1, i_N, t) - n_0(iL_h, 1, i_N, t) > 0) \vee (n_1(iL_{h+1}, 1, i_N, t) \\ - n_1(iL_h, 1, i_N, t) > 0). \end{aligned} \tag{21}$$

If $\eta(iL_{h+1}) = 1$, then we calculate the number of diagnostic features for detecting independent failures at SCADA hierarchy level $l_{h+1} \in L_1$.

Within the method framework of detecting a failure in SCADA, the function distinguishing between independent and secondary failures was derived when moving the CP TCO through the system's hierarchy levels. It took into account the type of process completion responsible for reception and transmission of data between the neighboring system's hierarchy levels. This function has the following form:

$$f(x, y, z) = (2xyz - 2x^2yz - 2xy^2z - xyz^2 + x^2y^2 + 2x^2y^2z + x^2yz^2 \\ +xy^2z^2 - 2x^2y^2z^2)(mod3), \tag{22}$$

where x is the parameter state at the transmitting hierarchy level; y is the type of process completion responsible for receiving and transmitting the parameter at the hierarchy level; z is the state of the controlled parameter at the receiving hierarchy level; $f(x, y, z) = 1$ is a necessary, yet insufficient condition for the presence of a diagnostic feature required to detect an independent failure; $f(x, y, z) = 0$ is a sufficient condition for the absence of diagnostic features needed to detect an independent failure.

We determine the number of diagnostic features for independent failures at a hierarchy level $l_{min} < l_{h+1} \leq l(S_1)$ using formula 22.

$$n_f(iL, 1, i_N, t) = \sum\nolimits_{iC=1}^{i_N} f(d_{iL+2,iC}(t), d_{iL+1,iC}(t), d_{iL,iC}(t)). \tag{23}$$

If $n_f(iL, 1, i_N, t) = 0$, then this criterion is a sufficient condition for the fact that independent failures are not found at the current hierarchy level.

If $n_f(iL, 1, i_N, t) > 0$, then this criterion is a sufficient condition for detecting independent failures.

We form a row for the matrix of independent failure markers $\Lambda(t)$:

$$\lambda_{iL,iC}(t) = f(d_{iL+2,iC}(t), d_{iL+1,iC}(t), d_{iL,iC}(t)). \tag{24}$$

Accordingly, predicate $M(i, l)$ showing a marker of independent failure for CP TCO x_i at hierarchy level l is as follows:

$$M(i, l) := \lambda_{(S_1)+1-l,i}. \tag{25}$$

4.6 Predicate System of Expert System Knowledge Base Describing the Distribution of Diagnostic Features Through SCADA Structural Elements

We determine the predicates for the distribution of diagnostic features over SCADA structural elements taking into account the characteristic peculiarities of each hierarchy level.

Predicate $\exists iP_1(i, l)$ for the existence of at least one CP TCO x_i at hierarchy level l having diagnostic features to detect an independent failure is as follows:

$$\exists iP_1(i, l) := \exists i(M(i, l)\&\neg S(i, 2, l)). \tag{26}$$

Predicate $\exists iP_{11}(i, y, l)$ for the existence of at least one CP TCO x_i at hierarchy level l having diagnostic features to detect an independent failure with value y is as follows:

$$\exists iP_{11}(i, y, l) := \exists i(M(i, l)\&S(i, y, l)). \tag{27}$$

Predicate $\exists i \exists j P_2(i, j, l)$ for the existence of at least two different CP TCO x_i and x_j at hierarchy level l having diagnostic features to detect independent failures is as follows:

$$\exists i \exists j P_2(i, j, l) := \exists i \exists j((i \neq j)\&P_1(i, l)\&P_1(j, l)). \tag{28}$$

Predicate $\exists i \exists j P_3(i, j, l)$ for the existence of at least two different CP TCO x_i and x_j at hierarchy level l having different diagnostic feature types to detect independent failures is as follows:

$$\begin{aligned}\exists i \exists j P_3(i, j, l) := \exists i \exists j((i \neq j)\&M(i, l)\&M(j, l)\&((S(i, 0, 1) \oplus S(j, 1, l))\vee \\ \vee(S(i, 1, l) \oplus S(j, 0, l)))).\end{aligned} \tag{29}$$

Predicate $\exists k \exists i \exists j P_4(i, j, k, l)$ for the existence of at least one SR_k/Ch_k at hierarchy level referring to sensors ($l = 2$)/data transmission channels from sensors to RTUs ($l = 3$), for which there are at least two CP TCO x_i and x_j having different diagnostic feature types to detect independent failures is as follows:

$$\exists k \exists i \exists j P_4(i, j, k, l) := \exists k \exists i \exists j((i \neq j)\&H_1(i, k)\&H_1(j, k)\&P_3(i, j, l)). \tag{30}$$

Predicate $\exists k \exists i \exists j P_5(i, j, k, l)$ for the existence of at least one RTU_k/Ch_k at hierarchy level referring to RTU ($l = 4$)/data transmission channels from RTUs to the server ($l = 5$), for which there are at least two CP TCO x_i and x_j having different diagnostic feature types to detect independent failures is as follows:

$$\exists k \exists i \exists j P_5(i, j, k, l) := \exists k \exists i \exists j((i \neq j)\&H_2(i, k)\&H_2(j, k)\&P_3(i, j, l)). \tag{31}$$

Predicate $\exists i Q_1(i, j, l)$ for the existence of at least one CP TCO x_i at the hierarchy level referring to sensors ($l = 2$)/data transmission channels from sensors to RTUs ($l = 3$) having diagnostic features for detecting independent failures and which is registered at SR_j/transmitted over channel Ch_j is as follows:

$$\exists i Q_1(i, j, l) := \exists i(H_1(i, j)\&P_1(i, l)). \tag{32}$$

Predicate $\exists i Q_{11}(i, y, j, l)$ for the existence of at least one CP TCO x_i at the hierarchy level referring to sensors ($l = 2$)/data transmission channels from sensors to RTUs ($l = 3$) having diagnostic features for detecting an independent failure with value y and which is registered at SR_j/transmitted over channel Ch_j is as follows:

$$\exists i Q_{11}(i, y, j, l) := \exists i(H_1(i, j)\&P_{11}(i, y, l)). \tag{33}$$

Predicate $\exists i Q_2(i, j, l)$ for the existence of at least one CP TCO x_i at the hierarchy level referring to RTUs ($l = 4$)/data transmission channels from RTUs to the server ($l = 5$) having diagnostic features for detecting an independent failure and which is registered at RTU_j/ transmitted over channel Ch_j is as follows:

$$\exists i Q_2(i, j, l) := \exists i(H_2(i, j)\&P_1(i, l)). \tag{34}$$

Predicate $\exists i Q_{21}(i, y, j, l)$ for the existence of at least one CP TCO x_i at the hierarchy level referring to RTUs ($l = 4$)/data transmission channels from RTUs to the server (l

= 5) having diagnostic features for detecting an independent failure with value y and which is registered at RTU_j/ transmitted over channel Ch_j is as follows:

$$\exists i Q_{21}(i, y, j, l) := \exists i(H_2(i, j) \& P_{11}(i, y, l)). \quad (35)$$

Predicate $\exists i Q_3(i, j, k, l)$ for the existence of at least one CP TCO x_i belonging to SR_j at the hierarchy level referring to RTUs ($l = 4$)/data transmission channels from RTUs to the server ($l = 5$) having diagnostic features for detecting an independent failure and which is registered at RTU_k/ transmitted over channel Ch_k is as follows:

$$\exists i Q_3(i, j, k, l) := \exists i(H_3(i, k) \& P_1(i, l)). \quad (36)$$

Predicate $\exists i Q_{31}(i, y, j, k, l)$ for the existence of at least one CP TCO x_i belonging to SR_j at the hierarchy level referring to RTUs ($l = 4$)/data transmission channels from RTUs to the server ($l = 5$) having diagnostic features for detecting an independent failure with value y and which is registered at RTU_k/ transmitted over channel Ch_k is as follows:

$$\exists i Q_{31}(i, y, j, k, l) := \exists i(H_3(i, k) \& P_{11}(i, y, l)). \quad (37)$$

Predicate $\exists i \exists j Q_4(i, j, k, l)$ for the existence of at least two CP TCO x_i and x_j at the hierarchy level referring to sensors ($l = 2$)/data transmission channels from sensors to RTUs ($l = 3$) having different diagnostic feature types to detect independent failures and which are registered at SR_k/transmitted over channel Ch_k is as follows:

$$\exists i \exists j Q_4(i, j, k, l) := \exists i \exists j((i \neq j) \& H_1(i, k) \& H_1(j, k) \& P_3(i, j, l)). \quad (38)$$

Predicate $\exists i \exists j Q_5(i, j, k, l)$ for the existence of at least two CP TCO x_i and x_j at the hierarchy level referring to RTUs ($l = 4$)/data transmission channels from RTUs to the server ($l = 5$) having different diagnostic feature types to detect independent failures and which are registered at RTU_k/ transmitted over channel Ch_k is as follows:

$$\exists i \exists j Q_5(i, j, k, l) := \exists i \exists j((i \neq j) \& H_2(i, k) \& H_2(j, k) \& P_3(i, j, l)). \quad (39)$$

Predicate $\exists i \exists j Q_{51}(i, j, m, k, l)$ for the existence of at least two CP TCO x_i and x_j belonging to SR_m at the hierarchy level referring to RTUs ($l = 4$)/data transmission channels from RTUs to the server ($l = 5$) having different diagnostic feature types to detect independent failures and which are registered at RTU_k/ transmitted over channel Ch_k is as follows:

$$\begin{aligned} \exists i \exists j Q_{51}(i, j, m, k, l) := \exists i \exists j((i \neq j) \& H_1(i, m) \\ \& H_1(j, m) \& H_3(m, k) \& P_3(i, j, l)). \end{aligned} \quad (40)$$

4.7 Instability Coefficient for the SCADA Structural Module

We determine the number of independent failures for each SCADA hierarchy level using the predicate logic formulas described above.

Therefore, for the structural and logical model of the SCADA system in question, and the hierarchy level referring to sensors/data transmission channels from sensors to RTUs, the number of independent failures $n_F(l = 2 \vee l = 3)$ for a CP TCO with timestamp t is defined as:

a. the number of independent failures at a given hierarchy level is equal to the number of structural modules for which there is at least one diagnostic feature to detect an independent failure which meets the following conditions:
   - there are at least two CP TCO x_i and x_j, which have diagnostic features for detecting an independent failure at the hierarchy level referring to sensors/data transmission channels from sensors to RTUs;
   - there is not at least one sensor/data transmission channel for which there are at least two CP TCO x_i and x_j, which have different diagnostic feature types to detect independent failures;
b. the number of independent failures at a given hierarchy level is determined by the number of different diagnostic feature types to detect an independent failure per structural module in case the following condition is met:
   - if there is at least one sensor/data transmission channel for which there are at least two CP TCO x_i and x_j, which have different diagnostic feature types to detect independent failures.

$$n_F((l=2) \vee (l=3)) := (\exists i(P_1(i,l)))?((\exists i \exists j(P_2(i,j,l)))? \tag{41}$$

$$((\exists k \exists i \exists j(P_4(i,j,k,l)))?(\sum_{\mu=1}^{m_N}(\exists i \exists j(Q_1(i,\mu,l)$$
$$+Q_4(i,j,\mu,l)))) : (\textstyle\sum_{\mu=1}^{m_N} \exists i(Q_1(i,\mu,l))))) : 1) : 0.$$

By analogy, for the SCADA structural and logical model in question, the number of independent failures $n_F(l=4 \vee l=5)$ at the hierarchy level referring to RTUs/data transmission channels from RTUs to the server for a CP TCO with timestamp t is defined as:

$$n_F(l=4 \vee l=5) := (\exists i(P_1(i,l)))?((\exists i \exists j(P_2(i,j,l)))? \tag{42}$$

$$((\exists i \exists j \exists k(P_5(i,j,k,l)))?(\sum_{\mu=1}^{N}(\exists i \exists j(Q_2(i,\mu,l)$$
$$+Q_5(i,j,\mu,l)))) : \left(\sum_{\mu=1}^{N} \exists i(Q_1(i,\mu,l)\right)))) : 1) : 0.$$

Finally, the number of independent failures at the server's hierarchy level employs the following formula:

$$n_F(l=6) := (\exists i(P_1(i,l)))?((\exists i \exists j(P_3(i,j,l)))? : 2 : 1) : 0. \tag{43}$$

We define the instability coefficient characterizing the work (from the point of the data flow view) of structural module j for each SCADA hierarchy level l using the predicate logic formulas described above.

In the structural and logical models of failure diagnostics under consideration, we accept the following weight coefficients to determine the instability coefficients of the system's structural module: $a_{y=0} = 1$; $a_{y=1} = 0,5$; $a_{y=2} = 0$.

The number of diagnostic features for independent failures with value y at the hierarchy level referring to sensors ($l = 2$) for SR_j/data transmission channels from sensors to RTU ($l = 3$) for Ch_j is as follows:

$$n_{F_1}(y, j, l) := (\exists i(Q_{11}(i, y, j, l)))?(\sum\nolimits_{i=1}^{i_N}(H_1(i, j)\&(P_{11}(i, y, l)))) : 0. \tag{44}$$

The instability coefficient for structural module j at hierarchy level ($l = 2$) referring to data transmission channels from sensors to RTUs ($l = 3$) is defined by using the following formula:

$$K_w(j, l) = \frac{a_0 \cdot n_{F_1}(0, j, l) + a_1 \cdot n_{F_1}(1, j, l)}{k_{m_j} - k_{m_{j-1}}}. \tag{45}$$

The number of diagnostic features for independent failures with value y at the hierarchy level referring to RTUs ($l = 4$) for RTU_j/data transmission channels from RTUs to the server ($l = 5$) for Ch_j is defined by using the following formula:

$$n_{F_2}(y, j, l) := (\exists i(Q_{21}(i, y, j, l)))?(\sum\nolimits_{i=1}^{i_N}(H_2(i, j)\&(P_{11}(i, y, l)))) : 0. \tag{46}$$

The instability coefficient describing the operation of structural module j at the hierarchy level referring to RTU ($l = 4$)/data transmission channels from the RTUs to the server ($l = 5$) is defined as follows:

$$K_w(j, l) = \frac{a_0 \cdot n_{F_2}(0, j, l) + a_1 \cdot n_{F_2}(1, j, l)}{i_j - i_{j-1}}. \tag{47}$$

However, when calculating the instability coefficient of the structural module for the higher SCADA hierarchy level using formula (47), the data source reliability for this module is not taken into account. Therefore, for the hierarchy level referring to RTUs ($l = 4$)/data transmission channels from RTUs to the server ($l = 5$), the instability coefficient for structural module j, taking into account the data source reliability for this module, is calculated by the following formula:

$$\begin{cases} K_w(j, l) = \sum\nolimits_{i=m_{j-1}+1}^{m_j} K_{w1}(i, l) \\ K_{w1}(i, l) = \frac{a_0 \cdot n_{F_1}(0, j, l) + a_1 \cdot n_{F_1}(1, j, l)}{k_i - k_{i-l}} \end{cases} \cdot \tag{48}$$

5 Description of the Experiment

We developed the DgnMethod software in order to implement a method of automatic failure detection and localization in real-time. This method is based on the analysis of changes in the state of information flows. The DgnMethod uses the predicate system described above and is part of the knowledge representation system. Diagnostic matrix $D(t)$ is the input data of the DgnMethod software. The SCADA system is responsible for the formation of diagnostic matrix $D(t)$ with the frequency of obtaining the data from sensors as well as the provision of an access interface to it.

The reliability, unreliability or absence of the controlled parameter values at the SCADA backbone nodes or the process state which is responsible for receiving and transmitting data via SCADA data transmission channels at each hierarchy level are a separate task and should be automatically determined and recorded by the SCADA system itself.

For example, the parameter value is unreliable in the following cases:

- the parameter value registered in the system is physically impossible (it is outside the range of permissible values/sensor measurement scale);
- the combination of parameter values is physically impossible for a given operating mode of the TCO or in connection with the specifics of the controlled technological process;
- there was an impossibly large jump in the value of the controlled parameter for an impossibly small period of time.

In addition, all analog-to-digital converters (ADC), when converting the signal from the sensor to a digit (the value of the monitored parameter), prescribe a reliable/unreliable bit of the conversion process in their control register. This bit must be read together with the monitored parameter and registered in the system.

The data transmission and reception process is considered unreliable if, for example, transmission and reception of the controlled parameter is completed but there were errors in the data transmission protocol or incorrect settings of the data transmission medium, etc.

The SCADA system should automatically generate event logs with protocols of the data reception and transmission process for each data transmission channel. Such an event log should contain the following information: messages about the beginning/end of a communication session, addresses of nodes, communication errors, etc.

The interface between the SCADA system and the DgnMethod software should be carried out using OPC (Open Platform Communications) software technologies, such as OPC HDA (Historical Data Access), OPC A&E (Alarms & Events), OPC XML (Extensible Markup Language), etc.

6 Analysis of the Results and Their Discussion

The DgnMethod software user interface allows us to quickly analyze the performance of the entire SCADA system, clearly illustrating the most vulnerable structural elements of SCADA with the maximum values in the instability coefficients of the structural modules.

Figure 5 shows the diagrams of instability coefficients for the operation of the SCADA structural modules. These demonstrate the dependence of the reliability of a concrete structural module's operation for the corresponding structural modules of the system's lower hierarchy level. These are data sources for the analyzed structural module. In this case, the value of the instability coefficient of the structural module is displayed in a yellow rectangle. When the left mouse button is pressed on the yellow rectangle, the calculated reliability factors of the operation of the corresponding structural modules of the lower hierarchy level appear on the diagram. These are displayed in green rectangles.

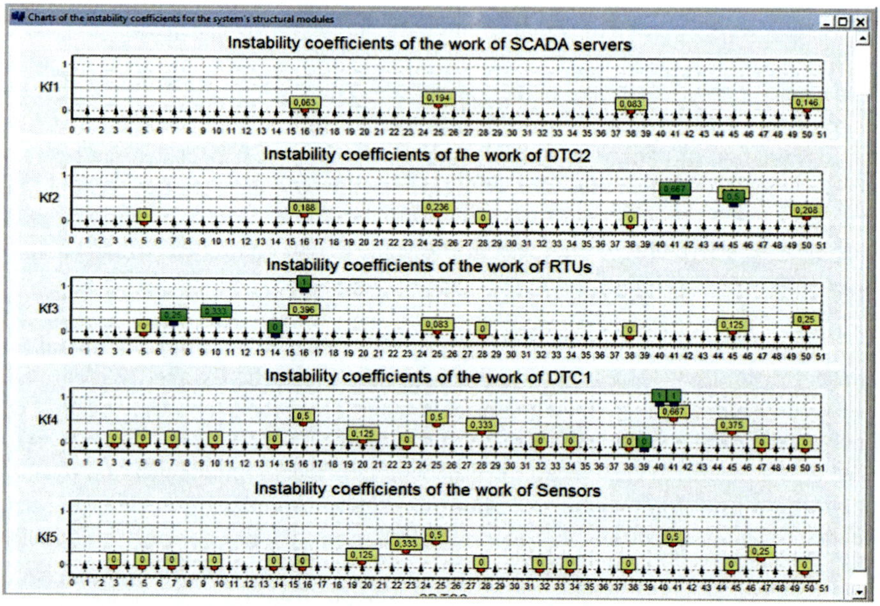

Fig. 5. An instability coefficient diagram showing all system's structural modules.

A retrospective view of changes in the values of the instability coefficients for the operation of structural modules over time is shown in Fig. 6. The current structure of the SCADA system is shown in the upper part of the interface window. When you click the left mouse button on a structural element of the system, a graph with a retrospective change in time of the value of the instability coefficient for the selected structural module appears in the lower part of the interface window.

Experimental studies of the diagnostic model and method for diagnosing SCADA failures are based on determining the reliability of the CP TCO when passing data streams through SCADA structural modules and levels using the DgnMethod software. It shows that this diagnostic model and method allow us the following:

- flexibly configure the SCADA structure;
- monitor the performance of SCADA structural modules in real-time;
- identify the most vulnerable, in terms of operational reliability, structural modules as well as cause-and-effect relationships of their vulnerability.

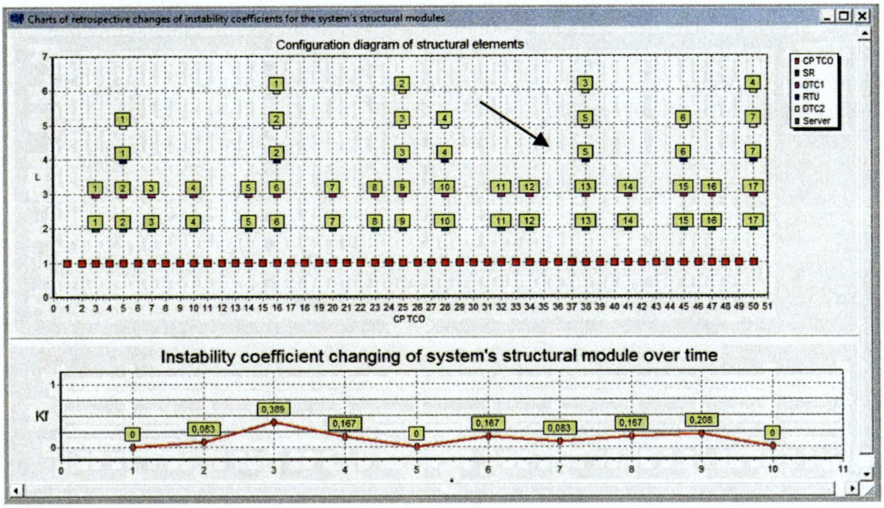

Fig. 6. A diagram showing how the instability coefficients change over time for the system's structural module

7 Conclusions

A distinctive feature of the method proposed for diagnosing the operability of TCO and SCADA is that when identifying points of failure localization in the system, the patterns of changes in information flows in real-time with incomplete data/absence of data in the system's structural elements are taken into account. This ensures the reliability of information technology control of the state of TCO and SCADA in conditions of incomplete information.

The formulas used for calculating the instability coefficients for the SCADA structural module operations are derived taking into account the diagnostic features at each SCADA hierarchy level. They also take into account the data source reliability for each structural module, which allows us to monitor and predict the performance of SCADA structural elements in real-time.

The analysis of the graphs of changes in the values of the instability coefficients for the operation of structural modules in real-time allow the ability to predict the operability of SCADA structural modules. Further, it can perform preventive maintenance of equipment in order to predict and prevent failures.

References

1. Berry, B.: SCADA Tutorial. A Fast Introduction to SCADA Fundamentals and Implementation, pp. 1–12. DPS Telecom, Fresno (2011)
2. Stouffer, K., Falco, J., Kent, K.: Guide to supervisory control and data acquisition (SCADA) and industrial control systems security. Recommendations of the National Institute of Standards and Technology: NIST Special Publication 800-82, pp. 1–170 (2013)
3. MacGregor, J., Cinar, A.: Monitoring, Fault Diagnosis, Fault-tolerant Control and Optimization: Data Driven Methods. Comput. Chem. Eng. **47**, 111–120 (2012)

4. Gaur, A.: Search techniques to contain combinatorial explosion in artificial intelligence. Int. J. Eng. Res. Technol. **1**(7), 1–7 (2012)
5. Giarratano, J., Riley, G.: Expert Systems: Principles And Programming, 4th edn. Thomson Course Technology, Boston (2007)

Comparison of Volume Rendering Methods Using GPU and Specialized Volumetric Accelerator

Sergey I. Vyatkin[1(✉)], Olexandr N. Romanyuk[2(✉)], Roman Y. Chekhmestruk[3(✉)], Sergey O. Romanyuk[4(✉)], and Oksana V. Romanyuk[2(✉)]

[1] Institute of Automation and Electrometry, SB RAS, Ak. Koptyuga, 1, 630090 Novosibirsk, Russia

[2] Vinnytsyia National Technical University, Khmelnytsky Shose, 95, Vinnitsa 21021, Ukraine

[3] 3D GENERATION UA, Pirogova Street 7, Vinnitsa 21021, Ukraine

Rc.ua@3dgeneration.com

[4] Odessa Polytechnic State University, Shevchenko av., 1, Odessa 65044, Ukraine

Abstract. The chapter identities and analysis methods that can be used to solve the problem of volume visualization, using the capabilities of standard GPU. The average computational complexity of volume visualization algorithms is estimated. The computational complexity of volume rendering is analyzed. An overview of alternative implementations of the volume visualization problem and a comparison of the proposed solution with a system based on a highly specialized hardware accelerator for volume visualization is given.

One of the actual applied tasks for any field of science that works with three-dimensional computational or experimental data is their visualization using a computer. Methods for obtaining data can be very different: from modeling to scanning using tomography, ultrasound, and various methods used in medicine, archeology, seismography, geological exploration, etc. In general, volume data is a three-dimensional array of cubic elements (voxels) representing units of 3D space. As a rule, this array contains information about each point in three-dimensional space. Volumetric visualization is the process of obtaining an image of volumetric data on a computer screen to represent an object or phenomenon in a realistic way, with a visual representation of its internal structure. Obviously, we can hardly hope in the near future for a purely software implementation of volume visualization in real time, even in a multiprocessor system, without the cost of creating processing accelerators – optimization algorithms and hardware. Therefore, it is important to further develop the fundamental theory in the field of volumetric visualization, as well as to develop advanced methods of volumetric visualization.

The aim of the work is to research the features of using known and original volumetric visualization methods and evaluation of the quality images and performance.

The object of the research work is to analyze the volumetric visualization methods and comparison of the proposed approaches with well-known volume visualization systems.

The subject of the research is the clut-texture and shaders, combined method of visualization of functionally defined surfaces and 3D textures, as well volume rendering in WebGL.

P. Vorobiyenko et al. (Eds.): IPF 2020, LNNS 212, pp. 359–378, 2021.
https://doi.org/10.1007/978-3-030-76343-5_19

The scientific novelty of the work consists in the development of the volume rendering methods. The problem of volume visualization can be solved using the capabilities of standard GPU. The ability to program the accelerator pipeline allows you to implement specific functionality. At the same time, both volume visualization and polygonal graphics technologies can be used together. A combined method for visualizing four types of data is proposed: analytical description of an object based on perturbation functions; voxel-based terrain; coordinates of vertexes with transformation matrices, the intensity and direction of sunlight, as well as its attenuation and scattering coefficients; and volume clouds based on three-dimensional textures. The mathematical foundations of volumetric rendering and its implementation in WebGL 2 are given to create an interactive volumetric renderer that works completely in the browser.

Keywords: Volume visualization · Volume texture · Perturbation functions · Performance · Pixel shaders · Voxel data arrays · Translucent models · GPU · VP-1000 · WebGl

1 Introduction

The purpose of the visualization is to make objects consisting of arrays of data, or multiple related data sets, visible to the observer. In this case, the media used for visualization does not have to be visual. For example, visual, auditory, tactile, kinesthetic, proprioceptive, olfactory and other sensory organs can be used [1]. One of the most important tasks of 3D rendering is visualization of 3D scalar field samples obtained from such sources as x-ray computed tomography (CT) or ultrasound echographic scanners. Visualization of structural features, such as the boundaries of areas with the same scalar values, as opposed to the scalar values themselves, is also an important direction.

Conceptually, the visualization process converts the contents of the datasets with the three transformations below into images that can be displayed on the display device. These are conceptual steps; different implementations may change the order or combine these steps.

(1) Classification. Maps the original dataset to a new feature dataset that explicitly contains the information we want to see.
(2) Modeling. Parameter maps of feature sets - to visual model objects.
(3) Rendering. Visualization of objects in the model in the form of images.

At the stage of classification, interesting features are highlighted, which can be, for example, the boundaries of regions with identical values. Such boundaries have their-strengths, or-probabilities of existence, and orientations associated with each location in the volume. The classification can be binary or non-binary (non-binary). Binary classification gives the results of the "all or nothing" in relation to the probabilities of existence. On the other hand, a non-binary classification can give any value between 0 and 1. Binary and non-binary classification have their advantages and disadvantages [2].

In the modeling phase, the extracted feature set is mapped to a set of visual media attributes in the visual model. A visual model consists of (1) a modeling space that hosts visual primitive objects that carry information, and (2) visual primitive objects that carry information. The modeling space can be 1D, 2D, or 3D. The modeling space can use a time dimension, although its use is physically limited. In fact, the use of the third spatial dimension is also limited because people cannot see all three dimensions with the same ease. Visual primitive objects can be 0D (point), 1D (line), 2D (surface), or 3D (volume). Several types of visual primitive objects can coexist in the modeling space. For example, a boundary surface can be modeled in 3D modeling space using a point cloud or a polygon mesh at the surface location. A conventional 2D image display has a 2D model space with 0D (dots) regularly positioned visual objects. 0D, 1D, or 2D primitive objects are created by binary classifications. 0D primitives have been popular on vector display devices in the past, and are still useful now, especially when combined with other (e.g. 2D) primitives. Primitive objects are 1D mesh frame and stacks of contours. Polygon meshes can be generated from 3D scalar datasets using binary classification and triangulation algorithms such as [3]. Such polygonal meshes can be easily transformed into surface (2D) primitives [4]. It is also possible to convert directly from a scalar dataset to 1D or 2D features [5].

3D primitive objects are created using non-binary classifications. Primitive modeling object called voxels. Voxel can have various attributes, including opacity and radiance, and radiance can have color components. Models with such 3D primitives are used, for example, in [6].

In the rendering phase, visual models are converted to images that can be displayed on display devices. That is, the rendering process starts. Rendering refers to the process of generating images from visual models. It should be noted that the so-called "volume rendering" algorithms described in the literature often include conceptual rendering steps other than rendering, such as classification and modeling. According to the above definition, "volume rendering" algorithms are more appropriately referred to as volume rendering algorithms.

Understanding information in models depends on both the time resolution, i.e. the frame rate of image generation, and the spatial quality of each image. Given the limited computing resources, most visualization systems make a trade-off between temporal and spatial visualization quality.

A volume dataset can consist of samples of different types of 3D fields, such as scalar or vector fields. However, this paper assumes that the volume dataset is a set of samples of a three-dimensional scalar function that is continuous and bounded. The functions of an object can be obtained from physical objects, in the form of a medical image, or generated by calculation, as in a three-dimensional computational experiment on hydrodynamics. Each sample volume data is called a voxel. Such volume datasets can be classified into two attributes: dynamism and regularity (uniformity), or unevenness.

Data sets of volumes can be divided into three groups depending on their dynamism, that is, on the frequency and degree of their change. In order of increasing dynamics, we list three groups:

1) Static data set. The static dataset remains unchanged throughout the visualization period.
2) Partially dynamic data set. A partially dynamic dataset often changes, but each change is localized to a portion of the dataset.

3) Dynamic data set. The dynamic dataset changes frequently and widely.

The more frequent and extensive the changes to the dataset, the more difficult it is to visualize the dataset. Dynamic data sets have a lower temporal consistency than the less dynamic. Temporal coherence is used for this purpose; it is often used in visualization acceleration algorithms. There are dynamic datasets that change over time, such as the results of calculations in hydrodynamics. However, in the past, visualization algorithms did not take advantage of the consistency that exists for dynamic data. Instead, algorithms developed for static data were used to visualize dynamic datasets.

In the proposed approach, a 3D texture contains only the actual data values as its values at each point. These values are converted to color and transparency values via the CLUT table. However, the difference is that this does not happen immediately, when loading data, but only at the final stage of the fragment-processing pipeline. Thus, the table is stored in the accelerator's texture memory independently of the data.

A combined method of object visualization based on analytical and scalar perturbation functions and three-dimensional textures with the use of graphics processing units is proposed as well. A method of real-time visualization of volumetric clouds is described.

Implementation in WebGL are given to create an interactive volume renderer that works completely in the browser.

2 Estimation of the Average Computational Complexity of Algorithms (in the Case of Software Implementation)

What is the amount of computation when visualizing large data sets? To answer this question, you should identify and evaluate the most time-consuming operations. In both types of algorithms, calculating the illumination and calculating the position of the voxel in the final image are the most time-consuming operations performed for each voxel. Let's assume that the volume is visualized from $N \times N \times N$ voxels. Let's also assume that the forward stroke rendering method (BTF) is used. Then, for each voxel, you need: apply a transformation associated with the observer's position, which requires multiplying the coordinates by an average of 6 matrices:

1. Positions in a voxel space with integer non-negative coordinates, representing the data set as a three-dimensional array;
2. Positions in object space, i.e. relative to some center;
3. Positions in the "world" coordinate system;
4. Positions in the camera's coordinate system;
5. Positions in the coordinate system that cuts off the visible part;
6. The projection matrix on the screen.

Since some multiplications (2, 3, and 4) can be performed once for all points, this requires operations of the order

$$T_2 = 3 \cdot N^3(16 \cdot t_1 + 4 \cdot t_2) \quad (1)$$

where t_1, t_2 and t_3 denote, respectively, the times of addition, multiplication, and division (floating-point operations - FP).

You must also calculate the corresponding color and transparency. In the BTF method, getting color and transparency values is implemented as follows:

$$A = a + A(1 - a) \tag{2}$$

$$C = c \cdot a + C(1 - A) \tag{3}$$

where C and A are the color (3 RGB colors) and transparency of the accumulating buffer, and C and a are the color and opacity of the point. Simple arithmetic gives:

$$T_2 = N^3((1 + 3 \cdot 2)t_1 + (2 + 2 \cdot 3)t_2 + k \cdot t_1 + t_3) \tag{4}$$

where k is the interpolation coefficient for neighboring values to get the final values. Then the total time is:

$$T = T_1 + T_2 = N^3((55 + k)t_1 + 20t_2 + k) \tag{5}$$

Let's give a rough estimate of the algorithm's performance on modern processors, taking into account that for them:

$$t_1 = t_2 = t;\ \ t_3 \approx 40t \tag{6}$$

where t is the time for a floating-point operation, and the number of operations per second for modern processors is about 2^{31}.

Thus, considering only basic operations, for an average resolution and an average quality image (N = 256, K = 1 – "nearest neighbor interpolation"), it will take approximately 2^{31} processor cycles to get a single frame and 2^{36} FP for real-time operation (30 frames per second).

If we try to perform volume visualization with high resolution and trilinear image interpolation (N = 512, K = 8), we will need about 2^{36} processor cycles per frame, or 2^{40} clock cycles for real-time visualization.

Obviously, we can hardly hope in the near future for a purely software implementation of volume visualization in real time, even in a multiprocessor system, without the cost of creating processing accelerators – optimization algorithms and hardware.

3 Possibilities of Using Modern Graphic Accelerators to Solve the Problem of Volume Visualization

The problem of volume visualization can be solved using a polygonal representation of data. In the simplest case, a triangle is created for each volume element, always turned to the observer (the so-called billboard) and colored accordingly. Getting a real picture of the data requires interpolation by neighboring values, which increases the number of triangles – to build a tetrahedron.

It is not difficult to calculate that, with the current speed of processing triangles, in real time, the order of $2^{26}/2^{(1+5)} \approx 2^{19}$ triangles (2^2 is the number of triangles per voxel, and 2^5 is the number of frames per second for real time). Therefore, taking into account only the basic costs, you can draw the order in real time $2^7 \cdot 2^6 \cdot 2^6 (256 \times 128 \times 128)$ voxels containing a single field.

At the same time, we do not take into account the computing costs of the Central processor (CPU).

The texturing speed indicator is crucial for further discussion, because creating a three-dimensional texture as a "stack" of two – dimensional ones is also a possible approach to representing three-dimensional data (usually represented by a series of consecutive "layers"). This value is about 2^{31} texels (texture elements) per second.

Let's estimate the time required to draw a characteristic data cube of the size $512 \times 512 \times 512$ (2^{27} elements). Given the fact that it should be displayed on three sides, because this is the maximum number of sides that we can see both the cube and taking into account the time geometric processing (calculation of the location of the faces, which are placed textures etc.), the amount actually render the array should be multiplied by 4. So, it turns out about 22 frames per second, which is clearly not enough to provide interactive interaction that requires 30 frames per second, but these are numbers of the same order and are achievable at the current speed of performance growth.

Thus, the quantitative indicators of modern graphics accelerators make them quite accessible for solving problems of three-dimensional visualization. At the same time, some basic requirements for the system functionality for solving the problem, despite the abundance of specialized functions, are fundamentally absent in polygonal graphics. Such, for example, the calculation of the gradient at the point (whose role in play set by the outside polygon normals), which is calculated using the correct illumination at the point is a critical requirement for the perception of three-dimensionality of the object person.

4 Application of Standard Graphic Accelerators for Solving the Problem of Volume Visualization

4.1 Alternative Approaches to Solving the Problem of Volumetric Visualization

Until recently, the field of application of "traditional" graphics accelerators was the representation of polygonal data, so such a specific task as volume visualization was solved in other ways:

- Solutions that use the computing power of special graphics stations (such as Onyx from Silicon Graphics) and special servers (such as AquariusNet Server from TeraRecon). The main disadvantages of this solution were (and still are) poor software portability and high price.
- Software implementations of volumetric visualization algorithms. The main drawback of such implementations is that they are either not designed for real-time visualization in principle, making do with the representation of several layers or slices of data, or they do not provide interactivity.
- Specialized accelerators for personal computers that implement volumetric visualization algorithms in hardware, such as VolumePro [7]. Despite the fact that accelerators

of this class provide significant acceleration when working with 3D data, they also have a significant disadvantage in the form of a high price. Another serious inconvenience may be the rigid implementation of the main algorithms "in hardware", so that even a small modification of them becomes impossible. For example, we point to the fact that all accelerators of this class implement the Ray-Casting algorithm [10] in hardware, which excludes the possibility of replacing it with, for example, the Ray-Tracing algorithm. A comparison of the proposed solution and the most powerful specialized accelerator to date, VolumePro 1000, is given later in this chapter.

4.2 Organization of 3D Texture. Basic Approach

A common way to organize 3D texture data is to use the approach where the 3D texture contains color and transparency (RGBA) values at each point. These values are obtained when applying a special table (CLUT - Color LookUp Table) to the source data. The table shows how different data values correspond to certain color and transparency values. The process of applying the CLUT table occurs when filling in the RGBA values of the 3D texture.

3D textures with RGBA values at each point are supported starting with DirectX 8.0 and can run on any graphics accelerator that supports this version of the DirectX API - starting with NVIDIA GeForce3Ti500. OpenGL also supports working with 3D textures.

The only limitation in this case is the amount of texture memory of the accelerator. Let's make a reservation right away that there are restrictions in DirectX 9.0 on the maximum size of a 3D texture on any axis (512 elements). In addition, that this size must be equal to the power of two. They manage with a purely software solution: splitting all volumes into sub-volumes with dimensions of powers of two. Sorting by distance to the observer when traversing the tree under the volume guarantees that the display is correct.

A notable disadvantage of 3D textures organized in this way is the need to completely reload data into the accelerator's texture memory whenever the table for converting data values to colors and transparency is changed.

4.3 Using CLUT Textures and Shaders

In the proposed approach, a 3D texture contains only the actual data values as its values at each point. As in the approach described in the previous Chapter, these values are translated to color and transparency values via the CLUT table. Nevertheless, the difference is that this does not happen immediately, when loading data, but only at the final stage of the fragment-processing pipeline. Thus, the table is stored in the accelerator's texture memory independently of the data.

First, this reduces the amount of memory required for a 3D texture by four times, with the same accuracy of data representation (for example, 8-bit precision data can be stored as 8-bit values, instead of four RGBA values, 8 bits each). Second, to change the representation of bulk data by color and transparency, you do not need to overload the data itself – just change the CLUT table. Reloading the LUT table requires an order of magnitude less time than the frame time, and therefore, in the proposed approach, dynamic changes to the CLUT table are possible in real time.

When implementing 3D textures with RGBA values, the corresponding color and transparency values are interpolated, not the data itself.

To be able to create a CLUT of arbitrary (not just 8-bit, which is clearly not enough for high-resolution data) accuracy, it was decided to store it as a texture. The size of such a texture-in fact, the accuracy of CLUT is limited, as well as the size of any other texture – 2^{27} values, but this is obviously enough for any application. In practice, 12-and 16-bit CLUT are most often used (for 4096 and 65536 values) (see Fig. 1).

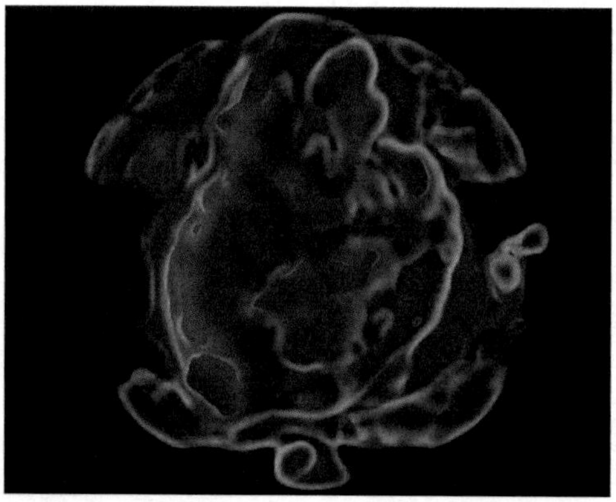

Fig. 1. Rendering a 64 × 64 × 64 volume as a 3D texture using a texture containing a table for converting data values to colors and transparency (CLUT).

Thus, the algorithm for using a three-dimensional texture with a second CLUT texture can be expressed briefly as follows: values from the first texture are used to get color and transparency values from the second one.

In the general case, this approach is called "dependent reading". In addition, despite all the advantages of dependent reading for our task, today it can only be implemented using pixel shaders.

Shaders are small programs that are executed directly by the graphics accelerator. Shaders allow for flexible configuration of the accelerator without reducing the overall level of performance when using algorithms other than the standard ones. In other words, with the help of shaders, it became possible to dynamically program the graphics accelerator pipeline, with the ability to turn on/off and configure its individual nodes.

Therefore, shaders can do the job of rendering a 3D texture with specified tables for converting data values to colors and transparency (CLUT). Another important feature of shaders is flexible control of lighting calculation, and to change the lighting model (for example, to simplify it on a less powerful accelerator), it is enough to load a different shader. Let's just say that to calculate lighting; you need information about the normal at each point (or gradient). This information can also be stored as another three-dimensional texture, and, as experience shows, to ensure correct visualization; an order of magnitude less information is required about the values of normals (data gradients) than the data itself, since the accelerator effectively interpolates the gradient values in the same way as the values of the data itself.

4.4 Working with Large Amounts of Data

A significant optimization is splitting the initial volume into sub-volumes. Sorting by visibility when traversing the tree of these sub-volumes will reduce the number of calculations and guarantee correct rendering. The results of the same sorting will help optimize the use of such a critical resource as texture memory: only necessary data is loaded into the accelerator memory, while unnecessary data is temporarily unloaded. This will avoid numerous problems related to the lack of accelerator resources. To develop an effective algorithm for splitting data, as well as solving problems with their loading, you can use solutions that are well developed in the visualization of the terrain and functionally defined objects, for example, building an octal tree.

4.5 Working with Dynamically Changing Data

The effectiveness of solving these tasks largely depends on how fully the available computing resources are used. In this case, it is advisable to apply General software solutions to improve performance. Plan asynchronous architecture main modules of the system will allow you to upload dynamic data: while the first thread loads the new data to the accelerator, the second waits for the completion of the rendering already loaded the data, the third stream draws on the results screen, accelerator operation etc. A system designed in this way can reduce the downtime of all graphics accelerator pipelines by efficiently distributing tasks.

The bandwidth of thc AGP bus (even for AGP 4X, not AGP 8X) allows, taking into account the overhead, to estimate the data update flow as the volume of the order $128 \times 128 \times 128$ per frame, at 50 FPS. Note that this is about 10% of the bandwidth declared by the manufacturer for the GeForce 6800 graphics accelerator and PCI Express.

4.6 Comparison of VolumePro1000 Performance with the Proposed System

An alternative to implementing a system based on a standard graphics accelerator is to use a highly specialized accelerator. Our laboratory has created a software and hardware complex for volumetric visualization based on the most powerful specialized accelerator to date, VolumePro 1000 [4, 6]. The development of standard graphics accelerators took place in a completely different direction, leading to the appearance of special solutions (for example, vertex cache, compression of 2D textures and z-buffer), which are practically useless for solving the problem of three-dimensional visualization. A major shift in this state of Affairs has been outlined relatively recently, with the advent of support for 3D textures and shaders. So, in addition to schedule time frame VolumePro 1000, schedules of work of the proposed system with different amounts of data on the basis of standard graphics accelerators of different generations (with the use of the index and, therefore, shaders, and without) (see Fig. 2):

- GeForce 3 Ti– the first accelerator (based on the NV20 chip) that supports DirectX8.1 and older;
- GeForce 5200 – the first accelerator that supports DirectX9 and dependent reading-pixel shader's version 2_0;
- GeForce 5900;
- GeForce 6800.

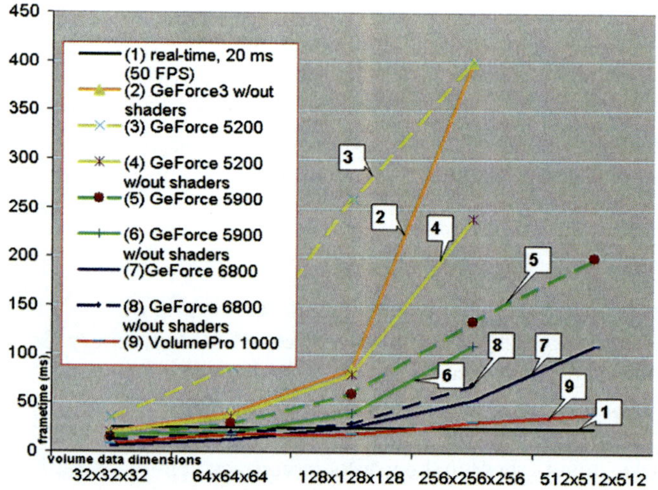

Fig. 2. Graph of the frame time (in milliseconds, on the Y-axis) for VolumePro 1000, compared to the times of the proposed system for different data volumes (X- axis). A bold black line indicates the "real time" limit - 20 MS (50 FPS).

4.7 Comparison of the Functional Characteristics of the Proposed System with the System Based on VolumePro1000

The undeniable advantage of the VolumePro 1000 is a rather weak, linear increase in frame generation time with increasing data volume (see Fig. 1). For standard graphics accelerators of all generations, this growth occurs quadratically (see Fig. 1). It is also important that the VolumePro 1000 allows you to simultaneously work with a volume of up to 8192^3, and standard accelerators no more than 512^3, although this is a solvable problem.

An equally important advantage is the hardware implementation of VolumePro1000 for calculating gradients and lighting (the Phong model). If lighting is implemented on standard accelerators, the necessary information about the normal at each point (gradient) must be stored as another three-dimensional texture. To save texture memory, indexing is also applicable here, and, as experience shows, to ensure correct visualization, an order of magnitude less information is required about the values of normals (data gradients) than the actual data, since the accelerator, similar to the values of the data itself, effectively interpolates the gradient values.

However, calculating lighting on standard accelerators will require additional dependent reading, which, in General, significantly degrades the performance of the system. At the same time, it remains possible to flexibly control the calculation of lighting - using shaders, and to change the lighting model (for example, to simplify it on a less powerful standard accelerator), it is enough to load a different shader. Shaders and a stencil buffer can also be used to implement specific rendering modes such as Maximum, Minimum, or Average Intensity Projection.

Both systems allow changing tables for converting data values to colors and transparency (CLUT) in real time, while the CLUT bit rate of standard accelerators can be arbitrary, in contrast to the fixed 8 or 16 bits for VolumePro 1000.

Moreover, in some respects standard accelerators are superior to VolumePro1000:

- VolumePro1000 provides a fixed set of arithmetic and logical operations, with no more than four volumes. Standard accelerators, using shaders, support any operations on up to 32 volumes (GeForce 6800), in one pass, however, due to an additional performance drop;
- Standard accelerators, unlike VolumePro1000, support perspective rendering.

5 Combined Method of Visualization of Functionally Defined Surfaces and 3D Textures

A combined method of object visualization based on analytical and scalar perturbation functions and three-dimensional textures with the use of graphics processing units is proposed. To display the terrain and the change in levels of detail, the same method as that for color textures is applied, and vertex shaders are used in the case of scattered light. A method of real-time visualization of volumetric clouds is described. For this purpose, it is proposed to form 3D textures by means of pre-processing of the cloud structure and volume-oriented visualization.

In computer graphics, the tremendous computational complexity of algorithms and potentially infinite structural complexity of mapped scenes require specialized hardware support. The high computational cost of the synthesis of realistic images is caused by geometric transformations and by using complex models of illumination, texture mapping, and ambient medium. Many functional capabilities of modern graphics processing units (GPUs) originate from rendering algorithms developed for high-performance graphical workstations and real-time visualization systems for various simulators. The most popular model for visualization of projection images of spatial objects is the polygonal approximation. However, along with many advantages, these models have some drawbacks. Modeling of real objects forms an approximate polygonal model. To ensure a good quality of the image, it is often necessary to increase the number of polygons, which leads to an increase in the visualization time and memory volume. A change in the object scale aggravates the situation because it is impossible to change the number of polygons for the object model rapidly and efficiently.

An adequate three-dimensional image cannot be obtained by scanning a two-dimensional space. Polygonal graphics with scanning of polygons in the image plane is not three-dimensional. Information provided for the user in this technology is incomplete. One of the main drawbacks is the lack of information about the object depth, i.e., information about the beam passing through the object. Moreover, an important aspect for visualization of complex objects is their geometric transformations: unary, binary, and other geometric operations, which are most optimally solved within the framework of solid-state modeling and construction geometry for functional models, e.g., theoretic-multiple operations, which can be used for design of objects and their compositions of unlimited complexity. This is primarily reached by using Boolean operations of integration and intersection. The modeling process becomes significantly more effective if functionally defined primitives and volume-oriented visualization are used.

An alternative visualization technology is proposed, which is based on the method of ray tracing and functional definition of primitives, in contrast to the traditional technology of mapping of polygonal scenes based on triangle rendering. A principal difference of the proposed method from others is a non-polygonal presentation of surfaces and the use of the ray-tracing algorithm for scene rasterization [8]. The rendering mechanism differs from technologies implemented in modern graphics accelerators. The objects are functionally defined, which ensures compact databases. The use of the ray-tracing algorithm eliminates drawbacks typical for systems with rasterization on a plane and problems of large data volumes.

We propose to describe complex geometric forms by defining the (second-order) function of deviation from the basic quadrics in the form

$$\begin{aligned} F(x, y, z) = A_{11}x^2 + A_{22}y^2 + A_{33}z^2 + A_{12}xy + A_{13}xz + \\ A_{23}yz + A_{14}x + A_{24}y + A_{34}z + A_{44} \geq 0, \end{aligned} \tag{7}$$

Based on the quadrics, we construct free forms, which can be considered as a composition consisting of the basic quadrics and perturbations [9]

$$F'(x, y, z) = F(x, y, z) + \sum_{i=1}^{N} f_i R_i(x, y, z) \tag{8}$$

where f_i is the form factor and R(x, y, z) is the perturbation:

$$R_i(x, y, z) = \begin{cases} Q_i^3(x, y, z), & \text{if } Q_i(x, y, z) \geq 0 \\ 0, & \text{if } Q_i(x, y, z) < 0 \end{cases} \tag{9}$$

(Q(x, y, z) is the perturbing quadric).

Rendering of volumetric clouds is a resource-consuming process and can be hardly used in real-time applications. At the same time, there is a need of visualization of volumetric clouds. Thus, developers seek for new tools for increasing game realism. Skyboxes and flat textures work well in 3D scenes, where the camera is far from the clouds and is not displaced to large distances. However, in the open world of the game, where the camera position cannot be predicted in advance, skyboxes yields a static image of the scene, whereas flat textures generates artifacts. This means that the images of the sky, clouds, and scenery are located at far distances around the spectator, generating an illusion of reality. It is impossible to render the scene near or inside the clouds with the use of skyboxes. To improve the situation, the clouds should be located in a 3D scene. This can be done by using a 2D texture, but then the images will look flat and not very realistic. To obtain good results, it is necessary to form volumetric clouds. The formation of clouds and their structure are extremely dynamic processes, and vertical motions due to convection mainly cause these. The clouds consist of small liquid droplets or crystals, which are formed when wet air move upward and expands as the atmospheric pressure becomes lower. This occurs owing to air expansion and reduction of temperature and water vapor condensation in wet air.

To obtain realistic visualization of clouds, the method of beam tracing is used, i.e., tracing of the beams from the camera to the scene. Equidistant samples of cloud visualization are obtained for the beams. The clouds are generated with the use of the noise function and are stored in 3D textures. An improved version of Perlin's noise method is implemented in our method [10]. Tracing of secondary beams is performed at each point inside the cloud. These beams are traced in the direction to the Sun in order to calculate lighting. To obtain the cloud forms from the noise textures, several steps are needed to reach various characteristics, e.g., uprising peaks resulting from convection. Our implementation includes the height distribution function, which controls the cloud form. Three-dimensional textures are limited by free forms.

The last stage of cloud formation is the use of a threshold function, i.e., a simple threshold value. It is used to cut off the "thin" parts of the 3D texture, which allows clouds with a clearer form to be shown. Figure 3 demonstrates objects (aircraft) in volumetric clouds.

The rendering process is divided into two stages and is distributed between the central and graphics processing units. The central processing unit (CPU) divides the object space over the quaternary tree. In searching for object fragments, the cube is divided into four parts in the XY plane, which are tested for intersection with the object. If there is no intersection of the object with the tested volume, then this part of the volume (cube) is eliminated from further considerations. If such intersection may occur in a certain part of the volume, a similar procedure of division is applied. In the general case, this process is finalized when a cell of a certain size corresponds to the part under consideration. This is a function of the CPU. This approach offers the following advantage: large parts of the cube that do not contain the sought object can be rejected already at the early stage. Fragments of the object are used as the initial data at the GPU input. Then the fragment is tested for intersection with the beam directed along the Z-axis, and a binary search for the nearest point of intersection with the object is performed. The task is to find the first point where the function vanishes. After finding this point, one can find the Z coordinate for each beam. Then the normal at the calculated points is determined. When all coordinates and normals at the points are available, the lighting model can be used.

For implementation of the method, we used the DirectX software package (version 12) for vertex and pixel shaders. Vertex shaders perform computations in accordance with Equations for fog and sunlight. The input information for shaders includes the vertex coordinates, transformation matrices, intensity and direction of solar light, and its decay and scattering coefficients.

Fig. 3. Image of an object defined by analytical perturbation functions (plane) in volume clouds and voxel-based terrain.

The images in Fig. 3 were visualized with the use of a computational system containing the Intel Core i7-2700K processor and the NVIDIA GeForce GTX 950 Ti (1024 MHz) GPU with 768 unified shader processors in the real time scale (30 frames per second).

Thus, a large number of processors were involved for simultaneous computation of a large number of beams. The influence of the speed of operation with the memory was taken into account. Registers were involved to the maximum possible extent because it is the high-speed form of available memory.

For GPU operation to be efficient, the maximum possible number of threads performing useful work were involved.

The algorithm of terrain relief mapping and detail level changing is based on the same mechanism as that used for the usual texture. Interpolation of heights is applied for image filtration in the dynamic process by analogy with the color texture. The computation time in the case of terrain relief generation is almost independent of the height map resolution; it is determined only by the screen resolution (quadratic dependence) and depth Z (logarithmic dependence). Landscape does not require relief triangulation. The main point in this technology is the hierarchical mechanism of detail level control, which resembles MipMapping used for the usual color texture. Considering the complete analogy with the texture, all its positive properties are inherited by the technology under discussion. They include the simplicity of surface animation without computational expenses for geometrical transformations of vertices.

6 The Method of Volume Rendering Using Signed Distance Functions

We solve three main tasks: real-time execution, simplicity, and physical correctness. Ray tracing of opaque objects. All objects are primitives with simple intersections with rays.

Phong shading is used to calculate lighting, and the three spherical light sources use a configurable light attenuation coefficient. Shadow rays are not required, because we only illuminate the plane. We will render the entire volume as a separate passage that mixes with the opaque scene. Separately processed opaque and translucent surfaces.

Signed distance functions (SDF) are used for volume modeling. SDF allows you to create organic forms in just a few lines of code. We use smooth conjugation to merge the distance functions of spheres. Smooth pairing is a powerful tool, because you can get something quite interesting just by combining it with a few simple shapes. You need something like a cloud. The great thing about SDF is how easy it is to distort surface, just adding some noise to the SDF. So we add a bit of fractional Brownian motion (fractal noise on top of it Brownian motion, fBM), using the position to index the noise function. Now you need to create the illusion that the volume interacts with the plane of the earth. To do this, add the distance of the signed plane slightly below the ground plane and re-use the smooth conjugation Union with a very large conjugation value (parameter k). The final touch will be to change the xz index of fBM noise over time so that the volume looks like a swirling fog.

A volume is a huge number of particles in a certain area of space. A huge number of particles so, that modeling each of these particles is an impossible task today, even for offline rendering. Good examples of this are fire, fog, and clouds. Strictly speaking, everything is volume, but for the sake of computing speed, simplifications can be made.

We represent a cluster of these particles as density values, usually stored in some 3D grid (or something more complex, such as OpenVDB). When light passes through a volume, a couple of phenomena can occur when light collides with a particle. It can both scatter and go in a different direction, or some of the light can be absorbed by the particle and dissolve. To meet the real-time execution requirement, we will perform what is called single scattering. This means that we will assume that light is scattered only once, when the light collides with a particle and flies towards the camera. So we cannot fake it effects of multiple scattering, such as fog, in which objects at a distance usually look more blurry. However, this is quite enough for our system. Consider the absorption of light in the volume along the camera beam. That is, we will not perform ray marching in the direction of light sources yet.

To do this, we need two actions: perform ray marching inside the volume, calculate the absorption/illumination at each step. To calculate how much light is absorbed in each point, we use the Booger-Lambert-Baer law [11], which describes attenuation of light as it passes through the material. Absorption coefficient is a material parameter. For example, in a transparent volume, such as water, this value will be low, but for something thicker, such as milk, the coefficient will be higher.

To perform volume ray marching, simply take fixed-size steps along the beam and get the absorption at each step, because the density within the volume is not uniform. Not enough self-shadowing yet. Light reaches all parts of the volume equally. Nevertheless, this is not physically correct, depending on the amount of volume between the displayed point and the light source; we will receive a different amount of incoming light.

To do this, you need to do the same thing that we did to calculate the absorption along the camera beam, but only along the light beam. The code for calculating the amount of light that reaches each point will essentially be a repetition of the code, but duplicating it is easier than using HLSL to get the desired recursion.

Objects like clouds are constantly scattered in the surrounding space, so their edges are mixed with empty space space around the volume, which should result in very smooth edges. Therefore, it is necessary to reduce the density depending on how close we are to the edge. In addition, since we work with signed distance functions, it is very easy to implement. You need to add a function that can be used to query the density at any point in the volume. When there is a density function, you can easily add a little noise to the volume to give it extra detail and splendor. In this case, we simply re-use the fBM function that we used to configure the volume shape (see Fig. 4).

Fig. 4. 3D Cloud.

Some light still seeps through the volume. Here we see how the color seeps in where the volume should exactly absorb it. This is because opaque objects are rendered before the volume is rendered. So they do not take into account the shading caused by the volume. This is quite easy to fix—we have a GetLightVisiblity function that can be used to calculate shading, so we just need to call it to light an opaque object.

The following optimizations are implemented:

1) Before performing ray marching in the direction of the light source, you need to check by the amount of light fading whether a significant amount of this light actually reaches the point under consideration. In the implementation, we look at the brightness of the light multiplied by albedo material, and make sure that the value is large enough for the need to perform ray marching.
2) If almost all the light was absorbed by the volume, then you need to stop calculating in advance, without wasting time on ray marching without visible results.
3) Postpone lighting of opaque objects until after performing ray marching of the volume. If the volume absorbed all the light, then we can skip calculating the illumination of opaque objects. However, we will still need to calculate the depth of opaque objects first, so we can pre-opt out of ray marching the volume if we encounter an opaque object.

Testing on GeForce GTX 1660 showed 15 frames per second at a resolution of 800 × 450.

7 Volume Rendering in WebGL

A physical volume rendering is presented that takes into account the absorption and emission of light by the volume, as well as scattering effects.

Volumetric rendering is a conceptually simple method of converting data into images: sampling data along the rays emitted from the eye, and assigning each sample a color and transparency. You can create useful and beautiful images of such scalar fields. In the GPU, such three-dimensional scalar fields are stored as 3D textures; however, 3D textures are not supported in WebGL1, so to emulate them in bulk rendering, additional calculations are required. WebGL2 recently added support for 3D textures, allows you to implement elegant and fast volume rendering in the browser. We will discuss the mathematical basics of volume rendering and how to implement it on WebGL2 to create an interactive volume renderer that works completely in the browser.

To create a physically realistic image from bulk data, we need to model how absorbed, emitted, and light rays are scattered by the medium. Although modeling the propagation of light through the medium at this level produces beautiful and physically correct results, it is too expensive for interactive rendering, which is the goal of visualization. In scientific visualizations, the ultimate goal is to allow scientists to interactively explore their data, as well as ask and answer questions about their research task. Since a fully physical model with scattering will be too expensive for interactive rendering, visualization applications use a simplified emission-absorption model. It also ignores or somehow approximates the cost effects of scattering. We will only consider the emission-absorption model.

In the emission-absorption model, we calculate lighting effects that occur only along the beam, and ignore those that occur from secondary rays. Rays passing through the volume and reaching the eye accumulate the color emitted by the volume and gradually fade until they are completely absorbed by the volume. If we trace the rays from the eye through the volume, we can calculate the light entering the eye by integrating the beam over the volume to accumulate emission and absorption along the beam. Take a ray that enters the volume at a point $\vec{p} = 0$ and exits the volume at a point $\vec{p} = L$. We can calculate the light entering the eye using the following integral:

$$C(r) = \int_{0}^{L} C(p)a(p)e^{-\int_{0}^{p} a(t)dt} dp \tag{10}$$

As the beam passes through the volume, we integrate the emitted color $C(p)$ and absorption $a(p)$ at each point along the beam. The emitted light at each point is attenuated and returns to the eye the volume absorption up to this point, which is calculated by the term $e^{-\int_{0}^{p} a(t)dt}$.

In general, this integral cannot be calculated analytically, so you need to use a numerical approximation. We perform an approximation of the integral by taking a set of samples N along the ray on the interval $p = [0,\ L]$, each of which is located at a distance Δp

from each other, and summing all these samples. The attenuation term at each sampling point becomes the product of the series, accumulating the absorption in the previous samples.

$$C(r) = \sum_{i=0}^{N} C(i\Delta p)a(i\Delta p)\Delta p \prod_{j=0}^{i-1} e^{-a(j\Delta p)\Delta p} \quad (11)$$

To further simplify this sum, we approximate the decay term $e^{-a(j\Delta p)\Delta p}$ of its Taylor series. In addition, for convenience, we introduce alpha $\alpha(i\Delta p) = a(i\Delta p)\Delta p$. This gives us the equation of alpha-compositing performed from front to back:

$$C(r) = \sum_{i=0}^{N} C(i\Delta p)\alpha(i\Delta p) \prod_{j=0}^{i-1} (1 - \alpha(j\Delta p)) \quad (12)$$

The equation shown above reduces to a for loop, in which we step through the ray through the volume and iteratively accumulate color and opacity. This cycle continues until either the ray leaves the volume or the accumulated color becomes opaque $\alpha = 1$. Iterative calculation of the above sum is performed using the familiar front to back compositing equations:

$$\widehat{C}_i = \widehat{C}_{i-1} + (1 - \alpha_{i-1})\widehat{C}(i\Delta p) \quad (13)$$

$$\alpha_i = \alpha_{i-1} + (1 - \alpha_{i-1})\alpha(i\Delta p) \quad (14)$$

These final equations contain the opacity pre-multiplied for proper blending $\widehat{C}(i\Delta p) = C(i\Delta p)\alpha(i\Delta p)$.

To render a volume image, simply trace a ray from the eye to each pixel, and then iterate the above for each ray that intersects the volume. Each ray (or pixel) being processed is independent, so if we want to render an image quickly, we need a way to process a large number of pixels in parallel. This is where the GPU comes in handy. By implementing the ray marching process in the fragment shader, we can use the power of parallel GPU computing to implement a very fast volume renderer.

In order for ray marching to be performed in a fragment shader, you need to force the GPU to execute the fragment shader on the pixels that we want to trace the ray on pixels. However, the OpenGL pipeline works with geometric primitives, and there are no direct ways to execute a fragment shader in a specific area of the screen. To work around this problem, we can render some intermediate geometry to perform a fragment shader on the pixels that we need to render.

The OpenGL pipeline in WebGL consists of two stages of programmable shaders: the vertex shader, which is responsible for converting input vertices to clip space, and the fragment shader, which is responsible for shading triangle-covered pixels.

8 Conclusion

The problem of three-dimensional visualization can be solved using the capabilities of standard graphics accelerators. The ability to program the accelerator pipeline allows you to implement specific functionality.

At the same time, both volume visualization and polygon graphics technologies can be used together. New features are associated with the integration of the three-dimensional visualization package into the virtual reality creation system.

The present chapter describes a combined method of visualization for three types of data: analytical description of the object based on perturbation functions, voxel-based terrain relief, and volumetric clouds based on 3D textures. The transition from rasterization in the image plane to rasterization in the volume combined with the proposed methods for surface and volume definition offers a number of positive issues increasing the realism of scenes being formed. The main positive features of the proposed approach are the high efficiency of the rasterization method and reduction of the amount of data, CPU load, and data flux from the CPU to the GPU.

Using the volume rendering method using signed distance functions three main tasks have been solved: real-time execution, simplicity, and physical correctness.

The OpenGL pipeline is implemented in WebGL, which consists of two stages of programmable shaders: a vertex shader, which is responsible for converting input vertices to clip space, and a fragment shader, which is responsible for shading triangle-covered pixels. A physical volume rendering has been developed that takes into account the absorption and emission of light by the volume, as well as scattering effects.

References

1. Kalawsky, R.S.: The Science of Virtual Reality and Virtual Environments. Addison-Wesley, Boston (1993)
2. Udupa, J., Hung, H.M., Chuang, K.S.: Surface and volume rendering in 3D imaging: a comparison. J. Digit. Imaging **4**(3), 159–168 (1991). https://doi.org/10.1007/bf03168161
3. Bajaja, C.L., Coyleb, E.J., Lin, K.N.: Arbitrary topology shape reconstruction from planar cross sections. J. Graph. Models Image Process. **58**(6), 524–543 (1996). https://doi.org/10.1006/gmip.1996.0044
4. Matheus, C.M., Furuie, S.S.: Automatic coronary wall segmentation in intravascular ultrasound images using binary morphological reconstruction. Ultrasound Med. Biol. **37**(9), 1486–1499 (2011). https://doi.org/10.1016/j.ultrasmedbio.2011.05.018
5. Grevera, G.J., Udupa, J.K.: Shape-based interpolation of multidimensional grey-level images. J. IEEE Trans. Med. **15**(6), 881–892 (1996). https://doi.org/10.1109/42.544506
6. Chen, B., Kaufman, A., Tang, Q.: Image-based rendering of surfaces from volume data. In: Volume Graphics Proceedings of the Joint IEEE TCVG and Eurographics Workshop in Stony Brook, New York, USA, 21–22 June 2001, pp. 279–295 (2001). https://doi.org/10.1007/978-3-7091-6756-4_19
7. Pfister, H., Hardenbergh, J., Knittel, J., Lauer, H., Seiler, L.: The VolumePro real-time ray-casting system. In: Proceedings of SIGGRAPH 1999, Los Angeles, CA, pp. 251–260, August 1999. https://doi.org/10.1145/311535.311563
8. Vyatkin, S., Romaniuk, O., Dudnyk, O.: GPU-based rendering for ray casting of multiple geometric data. In: International Conference ACIT – Advanced Computer Information Technologies, Ceske Budejovice, Czech Republic, 1–3 June 2018, pp.195–198 (2018). eLIBRARY ID: 39134676. https://acit.tneu.edu.ua

9. Vyatkin, S.I., Romanyuk, O.N., Pavlov, S.V., Romanyuk, O.V., Snigur, A.V., Komada, P., Smailova, S., Yeraliyeva, B.Y.: A function-based approach to real-time visualization using graphics processing units. In: Proceedings of the SPIE 11581, Photonics Applications in Astronomy, Communications, Industry, and High Energy Physics Experiments, Wilga, Poland (2020). https://doi.org/10.1117/12.2580212
10. Perlin, K.: Improved noise. ACM Trans. Graph. (TOG) **21**(3), 681–682 (2002). https://doi.org/10.1145/566654.566636
11. Fox, M.: Optical Properties of Solids, 2nd edn., p. 3. Oxford University Press, Oxford (2010). ISBN 978-0199573370

Individual Training Technology in Distributed Virtual University

Zoya Dudar(✉), Igor Shubin(✉), and Andrii Kozyriev(✉)

Department of Software Engineering, Kharkiv National University of Radioelectronics, Kharkiv, Ukraine

{zoia.dudar,igor.shubin,andrii.kozyriev}@nure.ua

Abstract. In the paper the current state of distributed virtual learning systems, their key tasks, advantages and disadvantages are analyzed. The analysis of building local informational and educational environments allowed formulating the basic principles of professional distance education: universality, invariance; scalability; openness; processability of creation, storage and use of resources; continuity etc. The method for building an individual learning path is analyzed. The model of an educational object is constructed. The logical model for structuring the content of text educational materials is improved by introducing predicates of equivalence of document content and equivalence of concepts which allows obtaining a logically justified sequence of representation of educational materials on an individual learning path.

Keywords: Virtual university · Education · E-learning · Predicate algebra · Logical networks

1 Analysis of Problems Educational Environments of Virtual University

E-learning is becoming increasingly relevant and requested. The reasons that led to the emergence of virtual universities and electronic forms of education include the growing amount of information and knowledge. The only form that intelligent content can generate is electronic resources. In the current situation, E-learning acts as a new technological environment for delivering open source knowledge. The increasing demand for higher education has led to the emergence of a market for educational services, in which universities have become competing entities in the field of distance delivery of educational content. These factors have led to the emergence and rapid expansion of a new model of learning in the world – cross-border e-learning or online learning.

It is known that in the most advanced industries in 1.5–2 years half of the knowledge ceases to provide competitive advantages. To create them, we need to generate new knowledge and new ways of acquiring it. If you do not update your knowledge, you will be better off for 5 years, after which you can find yourself outside the competitive market. Creating an individual learning trajectory is possible only in the case of multilevel testing and subsequent decision making about the complexity or simplification of tasks

P. Vorobiyenko et al. (Eds.): IPF 2020, LNNS 212, pp. 379–399, 2021.
https://doi.org/10.1007/978-3-030-76343-5_20

for a particular student, which requires appropriate methods and models of adaptation of educational resources to the individual needs of the user.

1.1 Problems of Creation of Information and Educational Environments of Virtual University

Today, e-learning is the main means of distance education. At the same time, information technologies in e-learning should be aimed primarily at solving such key tasks [1, p. 5]:

- Improving the quality of teaching by providing students with access to a large amount of knowledge, better materials and better communication with teachers, meeting a wider range of needs of different categories of students.
- Removing territorial barriers to access to education by giving students the opportunity to study off-campus at a time and place appropriate for each student.
- Reducing the cost of training and, consequently, the cost of training by automating all educational activities that previously required significant labor costs.

Researchers denote the pros and cons of distance learning as follows.

Pros:

- Advantages of efficiency (the teacher, the head of research activity may be in another city or country, but at the same time can very quickly coordinate the process of program development, course work, educational process, etc.).
- Information advantages (ease in remote learning when receiving large information streams, ie fast access to the Internet).
- Communication advantages (quick contact of participants of training).
- Pedagogical advantages (they include the integrity of learning and the pursuit of knowledge).
- Psychological advantages (these pluses are that the psycho-logical stress that arises during face-to-face communication is removed).
- Financial advantages (here it is necessary to calculate the economic value of the issue, namely the reduction of road costs, rent of premises, reduction of bureaucracy, etc.).
- Ergonomic advantages (drawing your own schedule) [2].

Cons:

- Disadvantage of direct contact (absence of direct contact, verbal contact between student and supervisor).
- Disadvantage the availability of equipment (availability of personal computer and Internet access. Need for constant access to information sources).
- Disadvantage the identification of the student's personality.
- Low motivation of students, manifested by insufficiently active participation in discussions in forums, lack of prior knowledge of the subject area, unclear goals and project expectations, etc. [3].
- Skeptical attitude and uncertainty of professors in the quality of distance learning, which corresponds to the university level of education.

1.2 Concept of Distributed University

There is a concept of an electronic university, with some authors using the term "distributed university" [4, p. 32]. However, the concept of e-university is much broader than simply providing distance education, as it involves the management of a multidisciplinary scientific and educational complex, which is a modern university. The difficulties of managing such a complex are due to the sophisticated structure of the organization of educational and scientific processes at the university, the variety of forms of education and sources of their funding, a large nomenclature of specialties and specializations, a large number of students and entrants. In these conditions, the management's effective decisions can be made only on the basis of complete, accurate and up-to-date data on the student and staff composition, financial status of the university, income and expenses, other indicators of the university's activity [5, p. 87]. Such data can be obtained and processed within the framework of a single integrated university management system, called "Electronic University", the basis of which are software complexes and databases containing multifaceted information on all aspects of its activity and consisting of separate subsystems. Thus, the system "Dean's Office" provides management of the curriculum and contingent of students, allows you to keep track of academic performance, including on the basis of score-rating system, generates various-net reporting and background information. Information systems of accounting, planning, personnel management, and public procurement management provide automation of the processes of administrative and economic life of the university. Management systems of scientific research provide information support for scientific activities of universities.

Analysis of the creation of local information and educational environments, software tools for educational purposes allow us to formulate the basic requirements for the principles of building information technologies (methods, models and software implementation for the task of e-learning), which could ensure the effective formation of network structures of education:

- Universality, i.e. integration into a single whole of different educational institutions from schools to universities.
- Invariance of environment and technology to the level and profile of education.
- Scalability, which allows to increase the resources of educational institutions at different levels.
- Openness, that is, ensuring compatibility with international educational standards and technical specifications.
- Technological process of creation, storage and use of educational-methodological and other resources that provide for conducting the educational process through the Internet.
- Continuity, that is, the ability to use the formats and forms used in the education system (in particular, compatibility with library and workflow systems).
- Equality and independence of all educational institutions and regions in the educational environment.
- The economic interest of all participants in the educational process (individuals and legal entities) and the self-sufficiency of the structures included in the educational process.

The stated principles show that the information and educational environment of open and transboundary professional distance education can represent a set of equal regional (branch) information and educational environments and in some sense be the development of the idea on which the Internet was created, where system-wide agreements (protocols) multiple networks are connected with their own administrative, technical and economic policies [4, p. 166].

The World Wide Web is becoming more widely used in education. Web education offers students unprecedented access to learning resources, far exceeding the reach of the traditional classroom, and making interaction possible to a much greater extent than traditional distance education. The education sector, in their view, will lose its power as a key element in providing education and thus will have to reinvent itself as the education sector.

Many studies have focused on teleconferencing as a completely new phenomenon in education. Conducting teleconferences, which can take place in real time, can be the main form of interaction between the teacher and students, and between students themselves. Teleconferencing is characterized as "distance learning of the new generation", based on the adequate formation of knowledge and judgments in the process of group classes or as a "new area" of learning, which is qualitatively different from classroom learning and from regular distance learning or distance.

Analysis of the creation of local information and educational environments, software tools for educational purposes allow to formulate the basic principles of information and educational environment of professional distance education: universality, invariance; scalability; openness; manufacturability of processes of creation, storage and use of resources; heredity, etc.

Types of e-learning are considered: synchronous, asynchronous and mixed, as a combination of classical classroom learning with computer-mediated activities. Studies have shown the effectiveness of blended learning compared to non-network traditional counterparts. This model increases the flexibility and individualization of student learning. Among the main advantages: the ability to customize learning and automatic personalized data collection for learning and further evaluation, can be used to reallocate resources to improve user performance.

2 Analysis of Current E-learning Systems

E-learning is a promising model of learning based on the use of new multimedia technologies and the Internet to improve the quality of learning, by facilitating access to resources and services, as well as sharing them at a distance. A virtual university is a form of organizing a learning environment by combining e-learning resources to improve the quality of learning. Analysis of the creation of local information and educational environments allowed to formulate the basic principles of professional distance education: universality, invariance; scalability; openness; manufacturability of processes of creation, storage and use of resources; heredity, etc. A review of hypermedia adaptive learning tools has shown that this is a new research area in the field of distance learning, the task of which is to include in distance learning systems the possibility of individualization. Analytical review of e-learning problems, such as intensification, individualization, sequence of materials, adaptability and flexibility – allowed to identify the main approaches to their solution: the formation of individual learning trajectories, development of learning models and learners, as well as methods of adaptive learning.

At the center of the educational process implemented by networked educational institutions is the consumer – a student of a distributed university, who has the opportunity to access a quality educational process, regardless of the time and place of their stay. Common to all students in a distributed university, regardless of their geographical location, are the main components of the educational process – a single information educational environment, information base, electronic information system for accounting of educational activities, teachers-consultants who have passed certification.

Hypermedia tools are actively used in teaching to implement almost all components of computer information technology. The combination of the advantages of hypertext technology and a multi-layered multimedia environment has led to the creation of a special class of information systems that are most adequate to reality. Nonlinear organization of information, inherent in hypertext, which sets semantic associations and dynamic relationships between the elements identified in it, allow you to build a system of alternative models of individual learning (including the implementation of different strategies or teaching methods for one subject area).

After analyzing the systems that are most used and widespread, and based on standard approaches that have become classic in distance learning in today's conditions, such as Moodle, BlackBoard, ANGEL, Desire2Learn, ILIAS, Lotus LearningSpace, WebCT, proved that often, a training course, which is presented in the field of computerized learning tools, is a compilation of static hypertext documents. Those who want to learn can use the same type of material for learning without taking into account their personal characteristics. A student cannot receive prompt assistance from a teacher or other person, as is usually the case in the classroom. There are also other problems associated with insufficient differentiation of the learning process in automated learning environments.

Insufficient standardization and lack of a unified approach to the construction of automated learning systems in Ukraine affects not only the emergence of such software products, but also the overall level of distance education. The main way to solve the problem of the level of education is to create separate pedagogical computerized software for each discipline. But this process is very costly in terms of both material and human resources. Thus, there is a question of developing software that would meet the requirements for the organization of the educational process based on the methodology of e-learning.

As a result of the analysis of the formation, development and current state of distributed learning systems, we can conclude that the world has developed and successfully used distance learning systems that offer services for the study of various curricula and courses. Analysis of both domestic and foreign systems of online distance learning has identified a number of basic or key elements that determine the quality of distance learning. The widespread introduction of information technology in the electronic industry is constrained by the lack of qualitatively new educational and methodological support and its software support. The need to effectively use a large amount of information, which is dynamically changing, adapted to the needs of a particular user, is an urgent task today.

On the basis of the analysis of creation of local information and educational environments, software of educational purpose it is possible to formulate the basic principles of the information and educational environment of professional distance education: universality, invariance; scalability; openness; manufacturability of processes of creation, storage and use of resources; heredity, etc. Types of e-learning are considered: synchronous, asynchronous and mixed, as a combination of classical classroom learning with computer-mediated activities. Studies have shown the effectiveness of blended learning compared to non-network traditional counterparts. This model increases the flexibility and individualization of student learning. Among the main advantages: the ability to customize learning and automatic personalized data collection for learning and further evaluation, can be used to reallocate resources to improve user performance.

The analysis of features of existing software products, allowed to allocate the modular object-oriented dynamic educational environment Moodle. The advantages of the system are due to flexible settings and rich functionality, including placement of educational material in various formats; knowledge test; interactive interaction of users, choice of individual forms of training. A review of adapted hypermedia learning tools within the virtual university showed that this is a new research area in the field of distance learning on the Web-platform, the task of which is to include in distance learning systems the possibility of individualization. With the help of adaptive intellectual technologies, the educational system takes into account the individual abilities of the student, his previous knowledge and skills. Based on these data about the student, the learning process is the optimal way for him.

The above review and analysis showed that, despite the diversity of existing conceptual models of hypertext systems, they are characterized by a lack of connection between the structural component and meta-information and users. If the AHAM model takes into account the presence of the user in its concept, DFHM - the presence of metainformation, and formalized and conditional-typical models determine the structure of hypertext in general, the presence of all three components is absent in existing models.

3 Different Solutions for Building of Virtual University

3.1 Adapted Hypermedia Training Tools Within a Virtual University

The special training tools most adapted for use in telecommunication (network and satellite) technologies of distance learning include:

- Electronic multimedia textbooks
- Multimedia lectures and virtual laboratory workshops
- Computer training and test systems
- Video lectures and a number of other electronic educational resources

Organizational forms of training when using telecommunication (network and satellite information) technologies of distance learning can be:

- Group consultations in the form of teleconferencing with remote audience
- One-on-one consultations and tests using telecommunication tools online and offline

- Live lectures for distributed remote audiences in teleconference mode
- Telecommunication offline video broadcasts of lectures for distributed groups of students
- Consulting workshops in the mode of television conferences before students perform virtual laboratory work [4, p. 135]

There are three basic types of distance education technologies used in pedagogical practice:

1. Case technology, when the teaching and learning materials are clearly structured and appropriately assembled into a special set ("case", " portfolio"), which are forwarded to the student for independent study; with periodic consultations with specialist consultant tutors or instructors in remote (regional) training centers or points established for these purposes
2. TV technology based on the use of television lectures with consultations with teachers
3. Network technology based on the use of the Internet transport environment, both to provide students with teaching materials and interactive interaction between teachers and students and students. [4, p. 142]

The virtual space of knowledge is a system of interdependent facts, definitions, mathematical dependencies, etc. These elements can be presented to the student in the form of tutorials, lecture fragments, presentations and more. Organizing the storage and processing of such information, presented in electronic form, is a very difficult task for developers of electronic systems of training. Knowledge space, like any complex system, has the property of decomposition. This allows tutors to organize the representation of knowledge as a collection of some fragments, such as definitions, theorems, formulas, drawings, elements of sound, animation, etc., which, unlike forms of tutorial type, can be used to store and further work with database technology information. In [6], an object model for storing educational and methodological information in a relational database is considered, taking into account the individual needs and characteristics of the student.

One of the new research areas in the field of distance learning on Web-platforms is adaptive intelligent technologies in the process of providing learning services. The task of this field of research is to include personalization opportunities in distance learning systems. With the help of adaptive intellectual technologies, the educational system takes into account the individual characteristics of the student, his previous skills and abilities. Based on these indicators about the student, the learning process is built in an optimal way for him.

Thus, analyzing the current state of affairs in the field of e-learning systems, researchers have formulated a number of requirements that make it possible to increase the efficiency of the learning process using an e-system, as well as facilitate and enable many users to develop training courses for such a system.

The virtual space of knowledge is a system of interdependent facts, definitions, mathematical dependencies, etc. These elements can be presented to the student in the form of textbooks, fragments of lectures, presentations, etc. The organization of storage and processing of such information presented in electronic form is a very difficult task for developers of electronic learning systems. The space of knowledge, like any complex system, has the property of decomposition. This allows you to organize the presentation of knowledge as a set of fragments, such as definitions, theorems, formulas, figures, sound elements, animations, etc., which, in contrast to forms such as "tutorial" makes it possible to use for storage and further work with database technology. In an object model for storing educational and methodical information in a relational database taking into account the individual needs and characteristics of the student is proposed.

Cloud services are an important part of today's information technology. One of the most famous are Google Docs, Amazon Cloud Drive, Microsoft OneDrive. To work with cloud services, Web 2.0 technology is used as the technological basis. However, in the distance from the World Wide Web, which is a key element of the Web page, cloud technology involves the use of software as a service (SaaS – Software as a Service). SaaS involves the deployment and use of software that requires only a Web browser. In addition to SaaS, there are also IaaS (Infrastructure as a Service), PaaS (Platform as a Service) and DaaS (Desktop as a Service) models. In the IaaS model, users are provided with access to cloud infrastructure resources for their own use (storage, information processing, etc.). This model uses virtualization technologies. The PaaS model allows you to install additional (native) software, unlike IaaS, where this is not possible. An example of this model is Microsoft Azure, Google App Engine, VMWare Cloud Foundry, Oracle PaaS Platform [7]. In the DaaS model, the customer receives not a certain set of applications, but a fully prepared workplace with the required set of programs.

As an example, researchers propose a hybrid model of communications in the cloud environment, which is based on the integration of the Learning Management System (Moodle) with the cloud service package Google Apps. The Moodle distance learning management system provides: tools for the development of author's distance learning courses; placement of educational materials in different formats (*.doc, *.pdf, *.djvu, *.html, etc.); availability of educational materials for students and means of control measures 24/7; receiving statistics/reports (user activity, events, test results and distance learning by a student, etc.); wide opportunities for communication and group work (file sharing, forum, chat, internal mail, seminar, wiki); use of the evaluation system; students download files with completed tasks; use of the calendar (schedule of online consultations, display of terms of control measures, etc.); involvement of students in the active educational process; creating conditions for active interaction between the subjects of distance learning (students, teachers, methodologists, etc.); the ability to add a variety of plugins to the course, which allows you to use third-party software.

Recently, adapted hypermedia systems have become increasingly popular in distance education, providing means of access to user-controlled information. Adaptive hypermedia systems bring together the ideas of hypermedia systems and intelligent learning systems and allow personalized access to information. The problems of distance learning support are considered, special attention is paid to the analysis of methods and means of adaptive hypermedia used by modern adaptive learning web systems. The main goal identified by researchers is to develop adaptive hypermedia, which is based on a simplified format of rules for the movement of learning materials by students. Navigation rules are functions of a finite algebra of predicate and predicate operations that allow us to describe arbitrary models of artificial intelligence. The format of the rules is also an object of development and should allow developers to describe the navigation rules for their application in automated learning systems. Adaptive hypermedia systems use different types of user models to adapt the content of automated learning systems (ALS) and links within it according to the level of knowledge and interests of the user. Education has always been one of the main applications of adaptive hypermedia. Most adaptive hypermedia learning systems use techniques that allow developers to describe navigation rules for students to move ALS content. Thus, in the technology of automatic course sequence establishment on the basis of the developed method of designing the function of adaptation to individual features of students in the interactive mode is offered.

3.2 Mathematical Tools for Modeling Adaptive Hypermedia in a Virtual Learning System

The issues of constructing a complex of mathematical models, organize the learning process, occupy a central place in creating various distance education management systems. Such an approach to the design of an integrated environment, the researchers note, made it possible to implement the basic ideas of teaching automation as a process of managing the cognitive activity of a study, specifying the content of the main issues of didactics [8]. To construct fragments of the resulting hyper structures, it is necessary to classify the selected information, divide it into separate logical elementary groups, logical units. The introduced predicates act as equivalences, therefore, factorize the output days of the set of considered and express the patterns of information structure in the process of constructing hypermedia structures through a predicate that is objectively determined by the classifier.

Adaptive hypermedia technologies use different types of user models to adapt to their individual content and hyperlink page content characteristics. When organizing personal training, the input enters information about the subject of learning from training resources distributed on the network. At the output, we get many learning materials that form one logical sequence in an individualized learning approach. The mechanism of this function is an adaptation organized in the form of navigation rules. In this case, the user is able to independently study the material under the control of adaptive hypermedia technology, which offers and controls the receipt of materials through the nodes for the knowledge of the subject of learning.

Basic hypertext models were developed and proposed in the 1990s to provide an unambiguous and systematic description of the structure of hypermedia systems. Depending on the purpose, all models can be divided into two main groups: hyperbase models; layered hypermedia models.

The first hyperbase model is considered to be HAM [18]. Using graphs to formalize the vertex-link structure, this model has addressed a number of key issues, such as the specification of the browsing history and the implementation of access control. But despite a number of advantages and simplicity of the proposed model, graph theory had some shortcomings, for example, in the case of distribution of information between several nodes and in the description of complex, composite objects. As a result, alternative models were later proposed. Among them is Tompa's formalization [19] based on hypergraphs. The most important improvement, compared to HAM, was the logical separation of hypertext network information from its structure. Another important feature is the formalization of links, in which the start and end points of the view can be a set of information fragments (nodes). The Trellis model is similarly constructed [20, 21], which also proposes the separation of the information component from the structural one together with the extension of the navigation semantics specification, which provides for the use of Petri nets to model parallel ways of viewing information. The considered models became the basis for such models as Hypertext Design Model (HDM) [22] and Object Oriented HDM [23].

Layered hypermedia models are reference models for developing hypertext management systems. The layers in this model correspond to different dimensions of modeling, namely: storage of structure and presentation of data. The Dexter model [9] is one of the most famous and influential in this class. This model offers a reference architecture by introducing a common vocabulary to compare different hypertext systems and models (Fig. 1).

The interface between the runtime layer and the storage layer is called presentation specification. The interface between the storage layer and the internal component layer is called anchoring specification, which is associated with identifying the start and end locations of the links in the content part of the nodes. Other models developed in this direction largely extend the Dexter model. For example, the Amsterdam hypertext model [10] has expanded Dexter's ability to represent and integrate hypertext with a variety of dynamic multimedia objects.

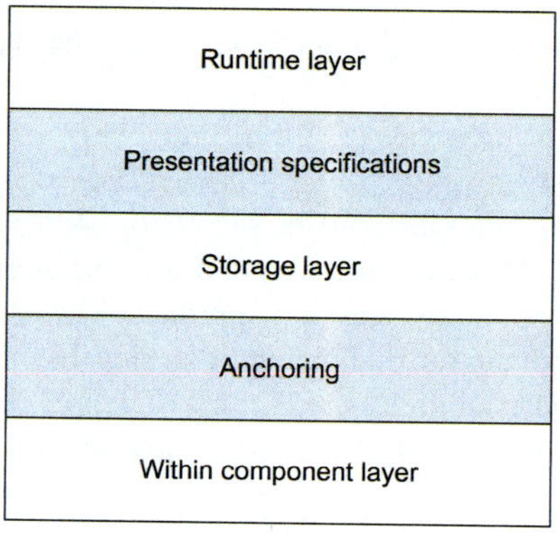

Fig. 1. Layered architecture by Dexter's model

Adaptive hypermedia application model (AHAM) [11] has extended the Dexter data storage layer to the specifications of teacher and user models. The teacher model works as a basis for adaptation and provides a mechanism for modeling the structure of a user's characteristics such as knowledge and goals. The teacher model defines the rules of information adaptation. Tochtermann and Dittrich [12] proposed the so-called Dortmund Family of Hypermedia Models (DFHM), which aims to summarize the architectural features of the most common hypermedia models. The purpose of these models is flexibility and support for various variations of data types. This family can be defined as a family of interrelated models, but not as a single model with a fixed set of data types.

As shown in Fig. 2, DFHM supports logical constructs for components, nodes, links, composite nodes, folders, and hyperdocuments. The component in this model family is an abstraction of a particular type of data.

A node is an autonomous, self-contained piece of information that can contain components and other nodes. The links connect the nodes to each other and anchor them to the components. Links can be grouped into structures that define different access contexts. Views combine and group nodes, links, components, information objects, thus forming contexts for user access. Folders have semantics similar to the file system. All the above-mentioned components together form a hypertext document.

According to Bashmakov's classification [13], the following types of hypertext models are considered: a formalized model; conventional model. As mentioned in [13, 14], the concept of information-reference article (IRA), which acts as an information unit of hypertext, is at the heart of the models of hypertext. In the formalized model of hypertext, the IRA corresponds to an information object whose content is characterized by semantic unity and logical integrity. In specific technologies, IRA is called differently: page, article, topic, etc. [14, 15]. It can contain information submitted in various forms: text, tables, code snippets (macros, scripts), integrated interactive objects, links to similar objects that connect to the IRA when it is downloaded.

For identifying resources on the Internet, a URL is used that specifies the addresses of the corresponding resources [16]. The hyperlinks pointing to fragments of the current resource are local, the ones pointing to external sources are called global.

As shown in [17], in a formalized model, the IRA is described by a tuple:

$$(x_0, x_1, \ldots, x_{11}) \quad (1)$$

where x_0 – the name of the IRA; x_1 – heading of the IRA; x_2 – abstract of the IRA; x_3 – entry point to the IRA; x_4 – multiple text fragments included in the IRA; x_5 – multiple interactive objects included in the IRA (graphic images, videos, etc.); x_6 – multiple software objects included in the IRA; x_7 – IRA certificate; x_8 – a sign of accelerated review of IRA; x_9 – a sign of detailed revision of the IRA; x_{10} – a list of hyperlinks inside the IRA; x_{11} – a list of hyperlinks between IRAs.

The name is a formal ID of the IRA and is used to address it by software. Within the hypertext, all IRA should have unique names. It is headed by the meaningful name of the IRA [17]. If hyperlinks from other IRA do not point to the IRA, then it becomes the main topic and is included in the list of the main topics of the hypertext. If the IRA has no external links, then the IRA completes one or many hypertext navigations paths. Accelerated browsing helps the user to quickly get acquainted with the IRA [15, 17]. Often, the IRA Quick Line is formed by elements x_1 i x_2 (caption and annotation, reflecting the main ideas outlined in IRA).

Activating the detail view feature ensures that all IRA content is represented. In this mode, the user can go any way that contains the elements $x_4 \ldots x_7$. As the scope of the IRA is not limited, there is a reference that provides additional information related to the content of the IRA. The elements $x_7 \ldots x_{11}$ are implemented through the interactive components of the user interface that provide navigation on the hypertext.

One of the drawbacks of a formalized hypertext model is the lack of explicit definition of hyperlink types. In the model all hyperlinks have explicitly specified type. This

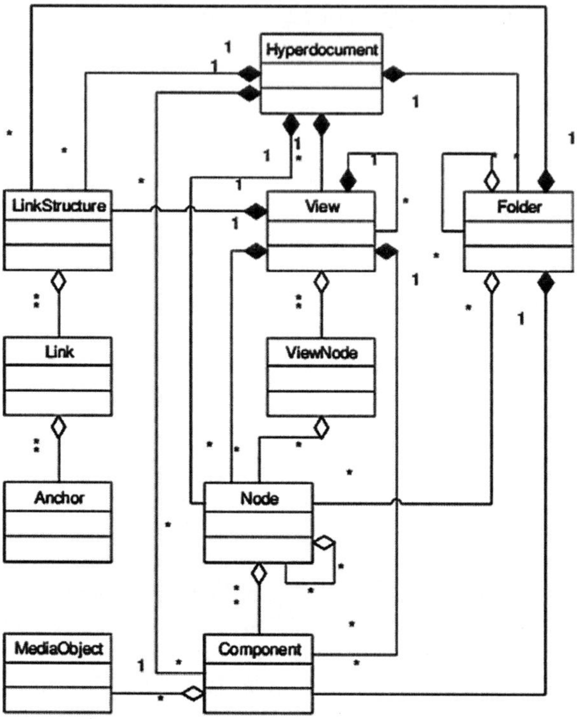

Fig. 2. DFHM basic model

hypertext model contains a thesaurus, a list of major topics and a collection of links. An indispensable component is the domain thesaurus to which the information system belongs. As noted in [13], a thesaurus is an orderly list of terms that reflect semantic relations between them.

Thus, the above review showed that, despite the diversity of existing conceptual models of hypertext systems, they are characterized by a lack of linkage of the structural component with meta-information and users in their structure. If the AHAM model takes into account the concept of user presence, DFHM means the presence of meta-information, and formalized and conditional models define the structure of the hypertext as a whole The presence of all three of these components is absent in the existing models. Intellectual information processing tools are required to describe the process of choosing an individual education path. Intelligence theory, modeling tools for human intellectual functions based on the method of comparative identification and the toolkit of finite predicate algebra make this possible.

4 Mathematical Models of Learning Individualization in Adaptive Hypermedia Learning Systems

Modern society is becoming increasingly informatized, in which the main products of production are information and knowledge. The information society requires a new level of education and new methods of providing it. A sustainable information society is characterized by the following features: increasing the role of information and knowledge in society; increasing the share of information communications, products and services in the gross domestic product; creating a global information space that provides effective information interaction of people, their access to leading information resources and meeting their needs for information data and services.

With the advent of modern information technology there have been radical changes, the essence of which is that the processes of innovation are becoming increasingly important. They characterize not only the technical sphere but are connected with all spheres of human life. There is a new independent type of public resource – information and innovation resource, which saves most other resources. In addition, the viability and progress of society have largely depended on the development and improvement of information and innovation infrastructures, the formation, location and use of information resources.

Due to dynamic growth, modern production needs more educated people than in the past. The experience of leading universities in the United States and the United Kingdom shows that the faculty, as a link and educational principle of the university, is becoming redundant, and the university is no longer tied to the area (new universities have abandoned geographical names). In addition, the concept of cyclical learning disappears you can enter some universities at any time. Instead of a faculty, there is now a list of courses needed to obtain a qualification, the university becomes a global organization, and at the center of the direct educational process is no longer a professor who gathers an audience, but a student who is served by professors.

The project of the world's first university of distance education was launched in 1960 in the UK and was named Open University. The Open University was founded personally by the Queen of Great Britain, and the Speaker of the House of Commons was appointed Chancellor, that is, everything was done to make the University both massive and prestigious. The name of the university was chosen to show its affordability due to the low price and the possibility of not attending classes often. As of today, the university has about 200 000 students a year.

Consider the main objectives of the e-education system in educational institutions of higher education:

- creation of a single network information and communication;
- educational space of higher education organizations to provide continuous online support for the educational process;
- organization of training of scientific and pedagogical staff and teaching and support staff in the field of e-education technologies;
- formation of electronic content available to all participants;
- network interaction, which protects the copyright of its developers and provides for monitoring the quality of electronic content;
- creation of an interuniversity electronic document management system;
- (To ensure networking in the field of e-education: curricula, individual curricula, basic educational programs implemented by partners, the agreement on networking, forms of documents on education);
- providing information protection of the e-education system;

- constant development and updating are provided by the system of electronic education services.

An important area of research today and the design of mathematical models of knowledge representation and processing of heterogeneous information data - are models of application and adaptation in intelligent hypermedia systems for distance or distance learning. It will be necessary to conduct a comparative analysis of the application of modernized methods for modeling the processes of knowledge transfer in order to increase the efficiency of knowledge acquisition. In addition, the analyst of the state of solving the problem speaks of the need to develop a mathematical apparatus for modeling the intellectual functions of society to formalize the learning process and adapt learning tools in the construction of intelligent information systems.

Given the results of the analysis of existing challenges and issues in the field of automation of intelligent adaptive learning hypermedia systems, the need for research of models and methods, development of modernized learning systems on the artificial intelligence platform for their use in automating the development of intelligent distance learning, namely in the algebra of predicate operations, elaboration of the base of logical navigation summaries of the rules of content adaptation and adaptive hiding of hyperlinks. In accordance with a certain set of methods and technologies of intelligent hypermedia systems, the prospects of their use for the need for adaptive learning are outlined. This paper proves the advantages of using a combination of methods and technologies to reproduce an intelligent adaptive learning hypermedia system that can be applied and comprehend all aspects of learning and help the consumer to fully master the learning material, ensuring full knowledge of a particular discipline.

A prerequisite for the study of the mathematical apparatus for modeling data processing processes, as a basis for the design of modern information technology used algebra of finite predicates, which is recognized as a universal language of delineation. The analysis of the basic models of knowledge representation is carried out, the main methods of knowledge identification are studied (definition, classification, method of comparative analysis). When designing a general complex model of intelligent adaptive educational hypermedia information system, a mathematical tool based on algebra of finite predicates for depicting knowledge and designing teaching methods in intelligent hypermedia systems with elements of adaptation to the consumer model is outlined.

4.1 Expected Results of Implementation of E-learning

It is advisable to describe the expected results from the implementation and realization of the concept of e-learning and distance learning technologies:

- equal opportunities for education regardless of place of residence, health and financial security of the student;
- the possibility of organizing the educational process that meets the requirements and goals of modern education and reflects the basic principles of a person-centered approach;
- maximum individualization of the educational process;
- dissemination and focus on self-education;

- flexibility of the organizational structure of the learning process;
- diversity of the content of the educational resource;
- development of effective pedagogical technologies of e-learning, distance educational technologies of realization of educational programs;
- simultaneous access to sources of educational information (Training courses, electronic libraries, information systems, etc.) of a large number of students, communication via the Internet with each other and with teachers;
- concentrated presentation of educational information and multimedia access to it;
- use of the latest achievements of information and telecommunication technologies in the educational process;
- use of new generation educational materials that contain digital educational resources;
- combination of different didactic models of educational classes with the use of e-learning and distance learning technologies, increasing the share of independent work of students;
- modularity of studying educational material, laboratory workshops, etc.;
- flexibility, wide variability and differentiation of access to interactive tests for control and self-control of students;
- increasing the number of students;
- reduction of training costs;
- overcoming the staff shortage in the labor market;
- increasing the efficiency of research and teaching staff;
- efficient use of material and technical resources;
- improving the quality of education.

In the process of classical training in the format of "human-human" an experienced teacher can identify the strengths and weaknesses of the student, assess his general erudition and level of knowledge of a particular discipline – and, based on these data, determine the scope and optimal form of presentation. This property is implemented in adaptive learning systems. The construction of individual learning trajectories is considered for the tasks of lifelong learning (Long Life Learning), the solution of which is especially relevant for the training of employees. A distinctive feature of information processing tasks, in particular the learning process, is that they are extremely easy to solve with the human mind and very difficult to formalize, with which it is possible to automate their solution. Research in the field of modeling of human intellectual functions has shown the universal nature of finite predicate algebra for modeling intelligent data processing processes. The general result of the research was the awareness of the need for a declarative approach in modeling learning tasks. Unlike the traditional information technology algorithmic approach, the declarative model is not an algorithm. The model of the problem, which is presented in declarative form, is usually a set of facts, knowledge of facts and rules, according to which actions are taken on facts that allow to obtain new knowledge on the basis of existing ones.

4.2 Predicate Algebra in Adaptive Hypermedia Computer Systems

The algebra of predicates and predicate operations in the theory of intelligence can be used to describe databases and knowledge bases and model learning and describe the rules for constructing its strategy. Thus, using the algebra of predicates and predicate operations, it is possible to create an integrated model of learning based on traditional models of knowledge representation, as well as models of knowledge representation in natural language.

Using the algebra of finite predicates as a mathematical apparatus, a multilevel model of hypermedia space organization is substantiated and constructed, the concept of educational predicates is introduced, methods of teaching and restoring knowledge are developed, which allow to increase the intensity and efficiency of computerized learning.

Existing means of formalizing and evaluating the hypertext structure provide a number of prerequisites necessary for their application. First, the necessary measure is to remove the subjective component, which involves taking into account in addition to the formal, subjective characteristics that have weight for a particular person. In this case, the equivalence of all information blocks and the links between them is accepted. That is, it is not the semantic content of the hypertext that is studied, but only the structure of its connections. Second, because real hypertext snippets may contain references to themselves (or "anchors" in terms of hypertext markup language) and more than one reference to the same snippets of other hypertext documents, in the case of duplication of navigation elements, these links in no way affect the logical structure and, as a consequence, the reachability. Therefore, these relationships are not taken into account in the analysis of the structure. Third, a restriction is introduced on the set of documents that can be analyzed. Documents and snippets in a given set can be accessed through other, non-set, connections. Since such pathways may be non-obvious and confusing, and their existence is not excluded, existing tools impose restrictions on the consideration of reachability in the superset of the set of documents under consideration.

The analysis of mathematical means of representation of objects of subject area of construction of an individual trajectory of training is carried out. The use of the mathematical apparatus of the algebra of logic and the theory of intelligence is substantiated. The basic concepts are introduced and the use of the method of logical networks for the presentation of knowledge and determination of the parameters of the problem of constructing an individual learning trajectory in a distributed virtual environment is proposed. The language of predicate algebra is universal, it can formally describe the structure of an arbitrary object. The result of the formal description of any object in the language of a predicate algebra is a certain predicate $P(x_1, x_2, \ldots, x_m)$. It expresses the relation P, which is the set of all sets $x_1, x_2, \ldots, x_m$, which satisfy the equation $P(x_1, x_2, \ldots, x_m) = 1$. This relation expresses the structure of the object being described. Getting a predicate P from predicates $P_1, P_2, \ldots, P_n$ of the system $\{P_1, P_2, \ldots, P_n\}$ is called composition. If the predicate P is represented as a conjunction of the predicates $P_1, P_2, \ldots, P_n$, then it is called a conjunctive composition. A logical network is a graphical representation of the result of a binary conjunctive decomposition of a multisite predicate. An arbitrary logical network can be created from poles and branches. Each pole of the logical network is matched by a model object variable named the attribute of that pole. Each pole has a designation of its own subject variable. Each domain is associated with its domain, in other words, the domain of the attribute change of that pole. Each pole of the logical network at any time carries knowledge of the meaning of its attribute. This knowledge is called the pole state. By specifying the state of all poles of a network at a given time, we obtain the state of the network at the same time. The poles of the network are divided into two classes – external and internal. Each outer pole is connected to only one branch, each inner pole to more than one.

Each branch of the logical network has some correspondence to a strictly defined binary relation of the model. Each branch connects the two poles corresponding to the subject variables, which are related by the corresponding relation. Modeling of logical networks consists of several stages. Initially, it is necessary to identify the object variables $x_1, x_2, \ldots, x_m$ of the modeling object and the relationship between them. In the second stage, the recorded relations in the language of the predicate algebra in the form of logical equations are recorded. In the third stage, the decomposition and minimization of the formula description of the object is carried out. The fourth step is to build the logical network directly. The final step is to test the correctness and effectiveness of the model for which the machine experiments are conducted.

Defined on the set X and Y, binary ratio R, has a definition – if xRy_1 and xRy_2 $x \in X, y_1, y_2 \in Y$ then $y_1 = y_2$, that is for every $x \in X$ exists a single value $y \in Y$. Every n-local ratio R_n defined on the set $X_1, X_2, \ldots, X_{n-1}, Y$ is called functional ratio if $(x_1, x_2, \ldots, x_{n-1}, y_1) \in R_n$ and $(x_1, x_2, \ldots, x_{n-1}, y_2) \in R_n$, then $y_1 = y_2$. It means that for every fixed set $(x_1, x_2, \ldots, x_{n-1})$ exists a single value y such, that $(x_1, x_2, \ldots, x_{n-1}, y) \in R$.

Formally, the predicate ratio replacement operation for modeling any intellectual processes, including processes for acquiring new knowledge (training), and modeling the process of adaptation to user needs, is proposed to be described as follows:

$$P(x_1, x_2, \ldots, x_m) = \begin{cases} 1, \textit{if } x_1, x_2, \ldots, x_m \in P \\ 0, \textit{if } x_1, x_2, \ldots, x_m \notin P \end{cases}$$

Based on the proposed general model, an arbitrary knowledge of the fact can be given by the ratio:

$$\{(a_{11}, a_{21}, \ldots, a_{m1}), (a_{12}, a_{22}, \ldots, a_{m2}), \ldots, (a_{1r}, a_{2r}, \ldots, a_{mr})\}$$

where r is a number of sets in the ratio.

This definition is expressed in:

$$x = a_{11} \text{ and } x_1 = a_{21} \text{and} \ldots \text{and } x_1 = a_{m1} \text{ or } x_1 = a_{12} \text{ and } x_1 = a_{22} \text{ and} \ldots \text{and } x_1 = a_{m2} \text{ or} \ldots \text{or } x_1 = a_{1r} \text{ and } x_1 = a_{2r} \text{and} \ldots \text{and } x_1 = a_{mr}$$

In the algebra of predicates, this definition is written in the form of the formula:

$$x_1^{a_{11}}, x_2^{a_{21}}, \ldots, x_m^{a_{m1}} \vee x_1^{a_{12}}, x_2^{a_{22}}, \ldots, x_m^{a_{m1}} \vee \ldots \vee x_1^{a_{1r}}, x_2^{a_{2r}}, \ldots, x_m^{a_{mr}}$$

which is commonly known as Disjunctive Normal Form (DNF) of the predicate.

Predicate algebra describes knowledge of facts only. Algebra of predicate operations formalizes operations on knowledge, represented as relations in some subject space. Predicate algebra describes the declarative component of knowledge, and predicate operations algebra describes the procedural component of learning adaptation. The algebra of predicate operations with constants and variables is introduced, that is, a kind of algebra of predicate operations with basic elements consisting of *identical predicate operations* and *constant predicate operations.*

Identical predicate operation by the variable $X_j (j = 1, \ldots, n)$ is an operation $F(X_1, X_2, \ldots, X_j, \ldots, X_n) = X_j$ at any $X_1, X_2, \ldots, X_j, \ldots, X_n \in M$. This operation is called an *argument selection operation.* There are n of such operations.

Constant predicate operation is an operation $F(X_1, X_2, \ldots, X_j, \ldots, X_n) = P$ at any $X_1, X_2, \ldots, X_j, \ldots, X_n \in M$, where P is a predicate on the set M. Each predicate $P \in M$ has its own constant predicate operation.

The study has shown that the learning process can be described using *disjunctive conjunctive algebra of predicate operations.* The disjunctive conjunctive algebra of predicate operations is called the algebra of predicate operations, whose basic operations are disjunction operations $X \vee Y$, conjunction operations $X \wedge Y$ for any predicate operations X and Y, and the basic elements are constants $P \in M$ and predicates of recognition of predicates $X_j^P (j = 1, \ldots, n), P \in M$. With any carrier N, the algebra of predicate operations is complete. Disjunctive conjunctive algebra of predicates on a carrier M is a subalgebra of algebra of predicate operations on a carrier N. The algebra of predicate operations together with its subalgebra, which is the algebra of predicates, is called the algebra of predicates and predicate operations.

When considering the tasks of creating the knowledge base, the concepts of *educational material* and *methodical material* are combined into a single concept of *educational and methodical material.* The concept of educational predicate (architectural construct) is introduced. It is a didactic complete fragment of educational hypermedia of space, which has a clearly stated purpose of study, theoretical material, tasks for fixing theoretical material and achievement of necessary practical skills, control questions tasks for current and final knowledge control.

For each element of information presentation in the hyperstructure the ratio of belonging Δ of selected educational documents t_a, t_b to one construct is introduced as follows:

$$t_a \Delta t_b \Leftrightarrow (\forall p \in P)(R(t_a, p) = R(t_b, p))$$

Ratio Δ defines the minimal distributed definition of integrity. Ratio Δ has the properties of reflectivity $t_a \Delta t_b$; symmetry $(\forall t_a, t_b \in T)(t_a \Delta t_b) \Rightarrow t_a \Delta t_b$; transitivity $(\forall t_a, t_b, t_c \in T)(t_a \Delta t_b)(t_b \Delta t_c) \Rightarrow t_a \Delta t_c$, so that the ratio Δ is an equivalence.

The ratio Π of definitions p_i, p_j is used for understanding the subject, which lays in the basis of hyper structure constructing:

$$p_i \Pi p_j \Leftrightarrow (\forall p \in T)\big(R(t, p_i) = R\big(t, p_j\big)\big)$$

The ratio Π determines the patterns of structuring integrity. In the development of elements of the algebra of predicates and predicate operations, a number of predicates is constructed and their functionality is proved.

Predicate E_Δ is suggested to use for the determination of didactic proximity of documents t_a and t_b of many similar forms.

If $E_\Delta(t_a, t_b) = 1$, then $R(t_a, p) = R(t_b, p)$ for any p from the set P. That means that all the properties of documents t_a and t_b, which match the definitions of the set Ψ coincide.

If $E_{\Delta}(t_a, t_b) = 0$, then definition $p \in P$, for which $R(t_a, p) \neq R(t_b, p)$ proves didactic variety of t_a and t_b.

Similarly predicate $E_{\Pi}(p_i, p_j) = 0$ can be used for defining of the functional equality of definitions p_i and p_j from the set P.

If $E_{\Delta}(t_a, t_b) = 1$, then $R(t_a, p) = R(t_b, p)$ for any document t from the set T, so these definitions correspond to the document $t \in T$, or do not correspond at the same time.

Both predicates E_{Δ} and E_{Π} are equal, so they factorize sets of different forms and definitions. Predicate E_{Δ} defines the distribution of the set T into layers S of didactically similar documents. Predicate E_{Π} defines the distribution of the set P into layers of L functionally equal definitions. Definitions from different layers L are not functionally equal.

To formally represent the educational hypermedia space, we choose a hierarchical nature of construction. Formally, such information is displayed by a logical network at the top of which is a didactic predicate. In constructing the organization of the educational space, the following basic levels of the model of representation of the knowledge base in the educational system are highlighted:

$$M = \langle VP, VPR, P \rangle$$

where $VP = \{vp_j\}, j = 1, \ldots, k$ is a set of topics, i.e. part of a logical network, which correspond to the topic of a subject; $VPR = \{vpr_z\}, z = 1, \ldots, m$ is a set of subtopics; $P = \{p_i\}, i = 1, \ldots, n$ is a set of predicates.

$$P = P(MI, MP, (MKT, MKR))$$

where MI is an educational material for presenting knowledge about subjects; MP is a training material used to replicate knowledge previously acquired in the classroom; (MKT, MKR) is a block representing the knowledge base for the organization of test control of the educational process: MKT is a knowledge of the material for the control on topics; MKR is a knowledge of the material for testing in sections.

Thus, a general multilevel model of adaptive organization of educational hypermedia space (disciplines) has been developed, which is described in terms of algebra of predicate operations with constants and variables, which makes it possible to automate the decision of the task of creating training programs with elements of artificial intelligence.

5 Conclusion

The existing information technologies, models and methods of individualization of education in the distributed virtual university are analyzed, the basic requirements to development of information technology of intellectual support of electronic form of training are formulated. Analysis of e-learning problems, such as intensification, individualization, sequence of materials, adaptability and flexibility – allowed to identify the main approaches to their solution: the formation of individual educational trajectories, development of learning models and learners, as well as methods of adaptive learning.

The use of mathematical means of intelligence theory for modeling of distributed virtual educational systems is substantiated. Algebraic logical models of individualization of learning for adaptive hypermedia computer systems based on the use of TSA tools and logical networks have been further developed, which allows to build a semantic database of educational resources to solve the problem of modeling an individual educational trajectory.

An approach to the formation of an individual learning trajectory based on the construction of mathematical models of distributed learning systems is formed. The model of structuring the content of hypermedia educational materials by introducing predicates of content equivalence and structuring the concept of educational object is improved, which allows to use several dream characteristics and get a logically substantiated sequence of presentation of educational materials on an individual learning trajectory.

References

1. Tikhomirova, N.: Educational process in an electronic university: conditions and directions of transformation. In: Tikhomirova, N.V., Minashkin, V., Dubeykovskaya, L. (eds.) Higher Education in Russia, 2011, №. 2, pp. 3–11 (2011)
2. Grigorash, O.: Improving the efficiency of quality management of the educational process. In: Grigorash, O. (ed.) Higher Education in Russia, 2013, №. 1, pp. 72–78 (2013)
3. Bazanova, E.: Mass online courses in academic writing: managing student learning motivation. In: Bazanova, E., Sokolova, E. (eds.) Higher Education in Russia, 2017, no. 2 (209), pp. 99–109 (2017)
4. Bochkov, V.: The State, Tendencies, Problems and Role of Distance Learning in Cross-Border Education: Textbook. Allowance, 405 p. RUDN, Moscow (2008). V. Bochkov, G. Krasnov, V. Filippov
5. Mikhailov, S.: Electronic university. In: Mikhailov, S., Degtyarev, A., Tregubov, M. (eds.) Higher Education in Russia, №. 5, pp. 87–91 (2009)
6. Yu., S.K.: Models and algorithms for content formation in the virtual space of knowledge of e-learning systems. In: Kabalnov, Yu.S., Grigoriev, A.I., Ufa, Sh. (eds.) UGATU, vol. 13, no. 2(35), pp. 109–118 (2009)
7. Cloud technologies and education / [Z. S. Seydametova et al.]; under common. ed. Z.S Seydametova. - Simferopol: DIIPI Publishing House, 2012. - 204 p. (2012)
8. Shubin, I.Yu.: Methods and models for the construction of intelligent adaptive hypermedia systems. In: Ya, V., Svyatkin, I., Shubin, Yu. (eds.) Eastern European Journal of Advanced Technologies – Kharkov, 2012, №. 3/11(57), pp. 11–13 (2012)
9. Gorbach T.: Calculation of minimal forms of finite predicate algebra. In: Shubin, I.Yu., Gorbach, T.V., Snisar, S.M. (eds.) Computational Intelligence (results, problems, perspectives): Mater. IV International. Research Practice Conference - KNU them. T. Shevchenko. - Kiev, 2017, pp. 158–159 (2017)
10. Gorbach, T.V.: Models of finite predicate algebra in the construction of software adaptive systems. Shubin, I.Yu., Gorbach, T.V., Goncharov, P.V. (eds.) Information Systems and Technologies (IST-2017): Mater. 6 International. Scientific-Technical Conference, Koblevo, Kharkiv, pp. 289–290 (2017)
11. Gorbach, T.V.: Information technology of constructing an individual trajectory of learning in a distributed virtual university. In: Gorbach, T.V., Shubin, I.Yu., Slavgorodsky, V.Yu. (eds.) Information Technologies: Science, Technology, Education, Health ‘ i: Abstracts of XXVI International. Research Practice Conference - Kharkiv. - 2018. - Part 1, p. 18 (2018)

12. Gorbach, TV.: Building individual trajectory of learning in a distributed learning environment. In: Gorbach, T.V. (ed.) Information Systems and Technologies (IST-2017): Mater. 7 International. Scientific-Technical Conference, Koblevo, Kharkiv, 2018, pp. 253–255 (2018)
13. Gell, M.: Future shocks. In: Gell, M., Cohrane, P. (eds.) Training Tomorrow, 1994/95, pp. 27–29 (1994/95)
14. Barket, R.: Interactive distance learning: perspective and thoughts. In: Barket, R., Holly, C. (eds.) Business Communication Quarterly, 1996, p. 88, December 1996
15. Dmitrieva, V.: Distance learning: essence, problems of implementation. In: Dmitrieva, V., Prokofiev, V., Samoilenko, P. (eds.) Specialist, 1996, vol. 11, pp. 37–40 (1996)
16. Ostar, A.: Partnership between the interactive university and its constituencies. Econ. Dev. Rev. **9**(1), 56–57 (1991)
17. Bochkov, V.E.: Branch features and characteristics of economics of the open education system: features of the regulatory-legislative regulation of the application of distant educational technologies. V.E. All-Russian Scientific Pract. Conference, 26–27 February 2004: B2 Part 1, Moscow-Kazan: IEIL and Taglimat, 2004, pp. 147–156 (2004)
18. The technique of adaptive interactive lectures for the «Multimedia Systems» course. In: Gorbach, T., Scherbak, A., Shubin, I., Svyatkin, Y. (eds.) Sb. scientific Works of the 13th International. Conference Ukrainian Association of Distance Education (VIRT 2011), Kharkiv-Yalta: UADO, 2011, pp. 181–186 (2011)
19. Gorbach, T.: Methods of classification of unstructured multimedia data in educational systems. In: Gorbach, T., Kirichenko, I., Karmanenko, O. (eds.) Modern (e-) Learning: Proceedings of the IX-th International Conference, 2014. - Kyiv, Ukraine, Sofia, Bulgaria, 2014, pp. 29–30 (2014)
20. The methods of adaptation in computer-based training systems. In: Shubin, I., Karmanenko, O., Gorbach, T., Umyarov, K. (eds.) Information Technologies in Innovation Business (ITIB): Proceedings of the International Scientific and Practical Conference, 2015, Kharkiv, Ukraine (Scopus), pp. 64–67 (2015)
21. Gorbach, T.: Computing in the algebra of finite predicates for the construction of adaptive systems in education. In: Shubin, I., Gorbach, T., Karmanenko, O. (eds.) Computational Intelligence (results, problems, prospects): Mater. III International. Scientific-Practical Conference; KNU named after T. Shevchenko. - Kyiv, 2015, pp. 116–117 (2015)
22. Gorbach, T.: Information technologies for modeling adaptive systems. In: Shubin, I., Gorbach, T., Karmanenko, O.A. (eds.) Information Systems and Technologies (IST-2016): mater. 5 International. Scientific and Technical konf., Koblevo, Kharkiv, 2016 - Kharkiv: MONU, KhNURE, 2016, pp. 244–245 (2016)
23. Gorbach, T.: Construction of algorithms for minimization of finite predicates of arbitrary orders in computerized learning systems. In: Gorbach, T., Snisar, S., Shubin, I. (eds.) Quality Strategy in Industry and Education: mater. XIII International. Conference Tech. University of Varna, Bulgaria, 2017, pp. 457–463 (2017)

An Approach to Designing the Educational Intelligent Analytical System

Larysa Globa[1,2](✉), Rina Novogrudska[1,2](✉), and Maryna Popova[1,2](✉)

[1] Igor Sikorsky Kyiv Polytechnic Institute, 37 Peremohy Avenue, Kyiv, Ukraine
lgloba@its.kpi.ua, rinan@ukr.net, pm@man.gov.ua

[2] National Center "Junior Academy of Sciences of Ukraine", 38-44 Dehtyarivska Street, Kyiv, Ukraine

Abstract. Nowadays, the evolution of the scientific communication system, the development of internet resources openness, the filling of electronic knowledge banks and the transformation of scientific and educational content into the general public domain are critical for the entire world community.

The analysis of available world and domestic informational Internet-resources shows that there is a problem of inconsistency while choosing standards for information resources metadata creation. The urgent task is not the implementation of uniform requirements for the description of information resources reflecting the current state of science and technology, but the development of software and information tools that ensure the integration of existing disparate resources and data as well as their metadescriptions developed in accordance with different standards and in different formats. The paper deals with the approach to design the built-in language-invariant tools of semantic analysis and program modules for the dynamic structuring of various informational resources. Such software tools make it possible to link informational resources involved in educational and research process, as well as allow solving a number of important tasks, such as eliminating information inequality, organizing educational, scientific and research processes, promoting information and knowledge, intensifying the communication processes of the scientific community, popularizing science among young people, creating conditions for the formation of high-quality scientific personnel.

Keywords: Information-analytical system · Transdisciplinary · Student youth · Ontology · Module

1 Introduction

In the modern world knowledge and information generate new knowledge. Thus their volumes and influence on the productive development of society are growing. This challenge requires new ways and means of global knowledge dissemination and usage for the purpose of further progress, and is the main property of a knowledge and information society.

At the same time, information and communication technologies, according to modern development conditions, are not just a technical infrastructure related to computerization,

P. Vorobiyenko et al. (Eds.): IPF 2020, LNNS 212, pp. 400–414, 2021.
https://doi.org/10.1007/978-3-030-76343-5_21

but also a global tool for disseminating the ideas of the scientific and technological revolution and the latest technologies. Therefore, the symbiosis of information, knowledge and infocommunication technologies allows to use the information society resources for solving practical problems of science and education.

Nowadays, open access to science is becoming the property of the whole world, not just a single group of scientists whose results of long-term work remain buried in various universities or research organizations archives. The evolution of the scientific communication system, the development of internet resources openness, the filling of electronic knowledge banks and the transformation of scientific and educational content into the general public domain is critical for the entire world community. An important aspect of this process is the growing number of people involved in scientific and practical activities, as well as the attraction of student youth to research activities. However, for student youth effective work within the framework of their research activities, it is necessary to provide convenient mechanisms for accessing to available arrays of scientific information, such as reports on scientific and technical activities, information on scientific and technical products, research results, etc.

Modern informational systems development is aimed at providing open access to information resources. The analysis of the available world and domestic informational Internet-resources shows that there is a problem of inconsistency while choosing standards for information resources metadata creation though. Domestic libraries, archives, electronic journals and file repositories while providing access to their own resources still do not use all the opportunities and achievements of modern information technologies to spread their content to a wider range of users. Such features make the unified standards implementation a very complex and time-consuming task that requires considerable resources. At the same time, the modern user wants to receive information immediately and in full, rather than search through the scattered sites of different institutions. Thus, for end-user, the origin of the information, its format, standard and technology of creation is not important. This is especially important for student youth that may use such information in their research activities but may not be able to find information in distributed sources. Therefore, the urgent task is not to implement unified requirements for the description of information resources that reflect the current state of science and technology, but to develop software and informational tools to ensure integration of existing disparate resources and data, as well as their metadescriptions developed according to different standards and in various formats.

The paper describes the approach to transdisciplinary information-analytical system development that allows supporting student youth educational and research activities.

The structure of the paper is the following: Sect. 2 gives analyses of related works and backgrounds for the research. Section 3 depicts characteristic features of educational intelligent analytical system for student youth educational and research activities support. Section 4 describes the structure of such a system. In Sect. 5, the main features of "POLYHEDRON" software that is proposed to be used as the basis for educational intelligent analytical system development is given. Section 6 presents conclusions and plans for future work.

2 State of Art and Related Works

The practice of computer technologies in education usage starts at the beginning of the 1960s. This process has transformed from the computation of learning stages to the educational services provision based on the use of information and communication technologies. During this time, more than 2,000 different system solutions have been developed, implemented, and are now used. Some of LMS (learning management systems) and CMS (Content management systems) have been implemented. The most popular of them are: Moodle (LMS), ILIAS (LMS), ATutor (LMS), IBM LearningSpace (CMS), Apple Classroom (CMS), eLearning Server 3000 (LMS), LAMS (LMS), Google Classroom (CMS), Microsoft Teams (CMS), Claronline (LMS), SharePointLMS (LMS + CMS), Live@EDU (LMS + CMS), eFront (LMS + CMS), Coursera (LMS + CMS), Canvas (LMS).

Educational information content development process based on these platforms was implemented using certain standards, namely: AICC, SCORM, xAPI (TinCan API), cmi5. Taking them into account during creating educational content provides certain flexibility of the educational content management due to the content hierarchical organization.

Ontological models based on the usage of OWL and Protégé tools (quite popular ontology systems) are also used for the flexibility of teaching content. To increase the level of intellectualization of content management in these solutions, software components of neural network technologies can be used to improve the content management procedures of training courses. This significantly increases the level of management and learning interaction and content quality. Such approaches are implemented when it is necessary to support the research activities of pupils and students.

Another form of organization of digital education services is the use of virtual digital laboratories. They are implemented based on virtual STEM-educational centers [1], that in terms of methodology, are designed with the involvement of certain ontological solutions and social technologies. With the help of such online resources, pupils and students can participate in real and virtual educational research, as well as in international research projects, such as ICE Cubes Service, EDU-ARCTIC, BIOTALENT, and others.

However, it worth taking into account that nowadays educational program are significantly transformed every 4–6 years. During this period, a large amount of educational and methodological literature is accumulated. Their content must be taken into account during using and creating educational digital content. Therefore, in the process of digital educational resources development and usage, there is a problem with Big Data processing. However, there are some solutions to solve it. This requires the inclusion in their operating environment cognitive services (services that have the nature of cognitive computing), such as analysis, structuring, selection, synthesis, etc.

Moreover, modern digital services in education should take into account the following two categories - the ability of the educational environment to provide effective educational services and digital educational logistics, intellectual tools of which form a complex chain of transactions for each student based on analysis and use of capabilities. Capabilities themselves should represent the full range of resources that provide the learning process. None of the modern systems such as LMS, CMS solve these problems.

Without above mentioned, it is practically impossible to provide efficiency of educational and cognitive activity of both pupils and students.

Innovative global educational programs are aimed at developing global skills of XXI century. The content and structure of them are defined in the documents of the educational alliance "The Partnership for 21st Century Learning (Skills)" (P21), which aims to prepare young people for employment and self-realization. Based on extensive research, collaboration with teachers, education experts and business leaders from around the world, the community has developed a map of key skills for successful living and working in the 21st century, with key definitions, examples, interdisciplinary themes, forms of communication for each of the selected competencies. The activity of P21 is a catalyst to provide an education system for the XXI century, promotes the creation of partnerships between education, business, the public, and the government. Digital educational assets should be created with a focus on providing modern challenges for the organization and provision of educational and cognitive activities of pupils and students.

One of the promising areas of the integrated use of distributed information resources is the study of network-centric transdisciplinary knowledge systems as heterogeneous ordered information environments. Computer ontologies are selected by the software and information tools for working with knowledge in such systems, which make it possible to distinguish basic concepts from a certain knowledge domain and establish connections between them. Ontology allows, on generally accepted and generally accessible principles, to conceptually expose and represent the interdisciplinary connections of certain knowledge domains [2]. Ontology, as an analogue of the "model", serves as a means of communication between transdisciplinary distributed information resources and systems through a combination of descriptions of resource content based on an object-oriented formalization procedure with descriptions of interpretation functions that control the process of supplying and filling an information resource. The formal structure of the ontology simplifies computer processing, which allows you to represent the specific tasks of the information environment [3]. The use of ontological tools is also effective in the aggregation of information resources from various sources and environments, their presentation and interpretation in the process of scientific research [4].

Ontologies play an important role in various aspects of education, such as search and web access to learning material [5], computer-aided processes to increase the efficiency of educational activities [6], exchange of information between different departments or educational institutions [7] and organization of the studying process [8].

Nowadays, the ways in which knowledge is acquired, perceived and transferred are constantly changing due to the exponential growth of information and communication technologies. Article 12 of the World Declaration on Higher Education in the 21st Century [9] determines that higher education institutions should be an example of new information and communication technologies benefits and potential usage. Therefore, the importance of ontologies usage is obvious, especially as a technological tool used for adequate information exchange and knowledge access.

Studies on the use of ontologies in the education system can be divided into groups according to the purpose of their application:

- ontologies usage to develop curricula, create descriptions of their contents in the form of courses and information resources, manage learning outcomes and simulate curriculum management [10, 11];
- ontologies usage in the process of e-learning [12, 13];
- ontologies usage to improve the mechanisms of recommendations of academic sources and resources or a detailed description (metadata) of interests corresponding to disciplines [14, 15];
- ontologies application for academic assessment, the implementation of which requires the integration of relevant data, usually distributed in separate systems [16, 17];
- ontologies usage for educational institution management [18, 19];
- ontologies application for software development in order to improve information retrieval [20].

A large number of research papers are dedicated to ontologies development for students' characteristics analysis. For example, the authors of [21] developed an ontological model of the student profile, taking into account the process of thesaurus formation. Using such an ontological model it is possible to increase the student's competence and objectively evaluate his knowledge.

Based on the results of the domestic and foreign experience analysis, the paper authors considered the advantages of ontologies usage in education and science, as well as best developments for educational, scientific and research ontological information systems. Such analysis is taken into account to develop an approach to designing the educational intelligent analytical system (EIAS).

3 Characteristic Features of Educational Intelligent Analytical System

In the research it is proposed to designing the educational intelligent analytical system for student youth educational and research activities support. Such system software and information tools will provide integrated access to information arrays reflecting the results of fundamental and applied research held in different organizations.

Architectural and technological features of the EIAS for student youth educational and research activities support are:

- model-driven architecture is characterized by a high level of formalization of the domain ontology and the mechanisms of ontology-driven;
- high level of transdisciplinary knowledge integration;
- ontological software and information tools for the computer-aided construction of domain ontologies (methodology, technology and software implementation).

  The ontology-driven architecture of the EIAS will allow to realize all the advantages of a network-centric information educational and research environment and to solve the problems of knowledge bases openness, unification of the conceptual structures presentation, and computer-aided construction of new knowledge bases taking into account already existed ones.

  EIAS includes the following software and information tools:

- identification of the basic categories of scientific and technical products representation, their contextual descriptions and semantic relationships, computer-aided classification and cataloging based on the procedures of semantic-linguistic analysis in the form of ontological interactive documents;
- formation of contextual connectivity of information arrays based on the identification of contextual indeterminacies and their inclusion in an existing or being created network ontological representation by computer-aided classification and additional identification;
- interactive interoperable semantic and contextual connectivity of the results of fundamental and applied research general categories based on dynamically generated thematic catalogs;
- formation of ontological knowledge capsules, which include a network ontological presentation of educational research projects of the Junior Academy of Sciences of Ukraine students, semantically connected network ontological representations of the achievements of National Academy of Sciences of Ukraine in the field of science and technology and connected indexes of additional information sources, if necessary, as well as contextual connectedness indexes with thematically related scientific information, educational programs of the Ministry of Education and Science of Ukraine, educational and methodological materials;
- aggregated presentation of physically and thematically distributed network information resources created in various formats, according to various standards and technologies, in the form of transdisciplinary knowledge systems;
- formation of ontological descriptions of educational research projects of the Junior Academy of Sciences of Ukraine students and the achievements of the NAS of Ukraine in the form of interactive documents, the transdisciplinary contexts of which provide the formation of an interactive knowledge base;
- supporting the users interaction of transdisciplinary information and analytical system in the process of implementing joint collective projects in various networked environments according to thematic profiles.

In the field of scientific and innovative research the formation of such a EIAS is one of the important practical applications of ontological engineering. With the help of models, methods and technologies that will be created, qualitatively improved results of fulfilling urgent tasks in the field of education can be implemented:

- computer-aided development of transdisciplinary knowledge systems for educational and research purposes;
- significant reduction in labor, time and financial costs for the knowledge systems formation.

In addition, ontological means of forming transdisciplinary knowledge systems are able to ensure the functioning of a network-centric information-analytical environment:

- effective computer-aided processing of general language and subject knowledge will significantly accelerate the processes of designing knowledge systems and scale the services of their use;
- unlike the subjective, system-ontological approach involves strict (as far as possible at this stage of the development of science) structuring of domain terms and concepts.

The categorical level is represented by the top-level ontology, which design should be included in the general algorithm for the development of modern knowledge bases.

4 Educational Intelligent Analytical System Structure

Technological basis of EIAS is the built-in language-invariant tools of semantic analysis and program modules for dynamic structuring of various information resources that deals with educational and research process.

Basic components of the educational intelligent analytical system are:

- repository of ontologies and their metadata,
- interface modules for ontologies displaying (including the knowledge prism presentation module),
- index zones for the information resources full-text index formation,
- specialized modules for specialized types of information processing and displaying (for example geospatial information),
- administration modules for ontology repositories, index zones and system management.

Let us give a short description of some main constituent elements of the information-analytical system (see Fig. 1).

Ontology Viewer Module. The ontology viewer module provides an ontology representation and search through index zone. Anonymous users can only view public ontologies (from public ontology sites or ontologies for which link access is enabled), while authorized users can additionally view ontologies from their personal account and ontologies accessed by the administrator.

Ontology Libraries Subsystem. The ontology libraries subsystem is used to represent the set of user-owned ontologies. Using ontology libraries, it is possible to select one or more ontologies that the user will work with. The system includes three types of libraries: public library presents the contents of the selected ontological site, personal library presents the contents of the selected user personal account, and application library displays a list of ontologies from all available IAS applications.

Multicriteria Optimization Module. The multicriteria optimization module is de-signed to rank alternatives by specified criteria. The module automatically creates a list of criteria for optimization, using ontology, then the expert should rank the criteria by importance according to one of the ranking scales. As the result, the module displays the ontology object list, ranked according to the selected parameters.

The Subsystem of Natural-Language Linguistic Information Operations. The subsystem of natural-language linguistic information operations is intended for processing and analysis of natural-language texts. The subsystem can be used to determine the semantic closeness or distance between alternative documents. It also easily determines the scope and content of changes made to the document (the work done to improve or

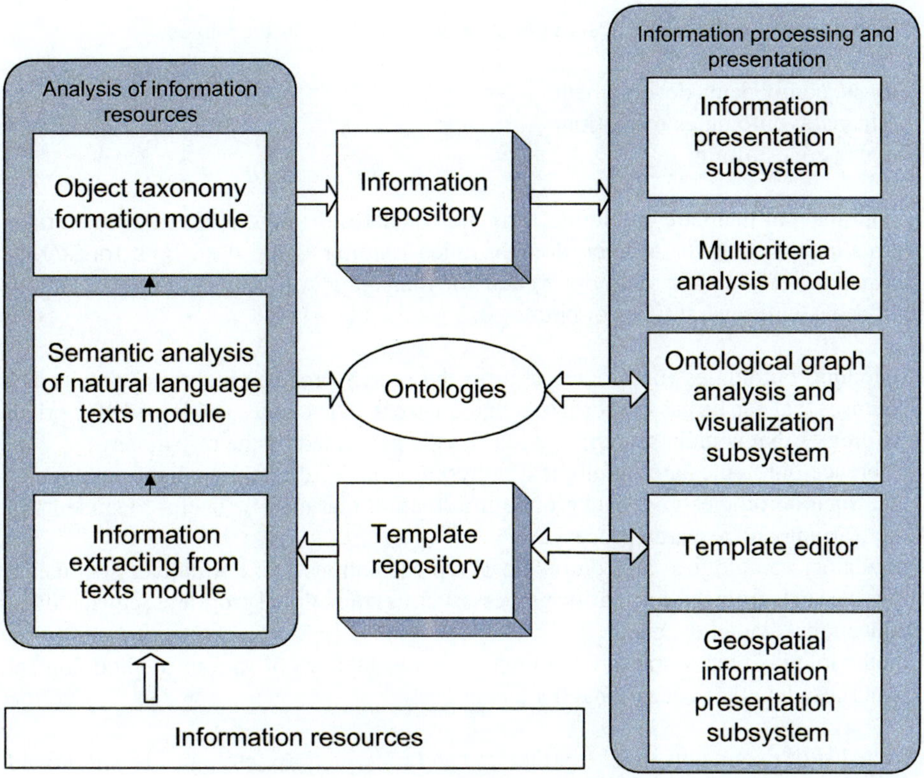

Fig. 1. Structural diagram of the EIAS.

distort documents). The subsystem allows performing pre-analysis, structural, morphological analysis and probabilistic-statistical analysis of a given text, while highlighting word forms, collocations and their linguistic characteristics. On this basis, the text is marked and then the linguo-stylistic portrait of the author is formed. Then, with the help of the text analysis and comparison methods, the text marked on the previous step is compared with the texts templates that were previously stored in the lexicographic base. The result of the subsystem's operation is the determination of the overall similarity coefficient for each pair of texts that wsere analyzed. Using such coefficient similar texts are sorted by coefficient meaning descending.

The Knowledge Prism Module. Such a module is designed to display tree-like data structures in a multidimensional view. It allows to represent ontologies available from IAS in the form of multidimensional cube taking into account all relations between ontologies data.

The process of the EIAS design consists of three consecutive stages:

1. Basic components development.
2. Thematic ontologies formation.
3. Index zones filling.

The stage of thematic ontologies formation declares the ontologies design and their informational filling for a given domain. Such ontologies are used both for storing information (information ontologies) and for managing software modules execution. EIAS deals with several types of ontologies:

- searching ontologies (ontologies used in the search process) includes domain thesauruses that can be used to form search requests in index zones and knowledge prism ontologies that contain search requests already generated by the researcher;
- reference ontologies are ontological representations of directories, lists, catalogues, etc. Such ontologies give background information that may be useful to students in the learning and research process;
- regulatory ontologies are ontological representations of normative documents involved in learning and scientific process, such as educational programs, curriculums, educational standards, etc.;
- methodological ontologies are ontological representations of various methodological materials and other educational documents.

The list of documents that are structured and presented as ontologies is determined in the process of the system designing.

An important step in IAS development is to fill the index zones with information resources, both structured (created on the second stage of system development) and unstructured (articles, monographs, dissertations and other documents not included in the list intended for structuring).

The following latest scientific and technical products within EIAS will also be created:

- A network-centric transdisciplinary educational cluster of support for students educational and cognitive activities on the technological basis of verbally active cognitive services.
- Library of educational programs transdisciplinary ontologies on thematic profiles of subject disciplines.
- Interactive educational knowledge systems in the form of transdisciplinary ontologies of information and educational resources (textbooks, lectures, and teaching materials, monographs, popular publications in basic sciences, and other educational materials recommended by the Ministry of education for the educational process).
- Transdisciplinary means of integration of EIAS network-centric environment with electronic repositories of libraries and other educational institutions based on two-level conceptual indexing.
- Library of cognitive tools for personal network-centric web-oriented ontological platforms creation for students and teachers.

- Library of cognitive procedures for implementing the learning interaction of students with educational interactive knowledge systems in the format of narrative discourse [22].

EIAS for student youth educational and research activities support developed in such a way realize the most complete informational network resource for knowledge representation. This resource will contain an ontological description of the fundamental and applied researches results from Ukrainian NAS institutions and will provide their semantic linking with information from Junior Academy of Sciences of Ukraine, MES curricula and general educational and methodological materials.

EIAS allow to solve a number of important issues while developing modern information and communication systems:

- information inequality liquidation (access to information is provided regardless of time, space, social affiliation),
- organization of educational, scientific and research processes promotion,
- information and knowledge processes intensification for the scientific communications of the society,
- science promotion among young people,
- creating the conditions for the formation of high-quality scientific personnel, dissemination of technical, economic, social and political innovation.

5 Software POLYHEDRON for Educational Intelligent Analytical System Implementation

It is proposed to use software platform "POLYHEDRON" [23] for educational intelligent analytical system implementation. Cognitive information technology "POLYHEDRON" is designed to support the processes of linguistic and semantic analysis of large volumes of spatially distributed unstructured information (Big Data), their structuring, establishing contextual links between processed documents, forecasting and support of rational choice processes with the subsequent formation of information-analytical WEB-oriented solutions.

"POLYHEDRON" tools provide realization of the following information processes:

- Linguistic and semantic analysis of network information resources, created based on the use of various information technologies and standards, which have a significant number of interdisciplinary relationships;
- Transdisciplinary integration with other network information systems and WEB-oriented applications;
- Taxonomization of narratives of arbitrary documents and reflection of their structure, including intertextual relations;
- Creation of ontological interactive documents;

- Detection of latent information in the analyzed information resources;
- Deep Learning (Machine Learning);
- Support for Semantic Web formats and protocols;
- Big Data processing.

Computer ontologies are used as software tools for knowledge management in "POLYHEDRON". The ontology allows to outline and reflect the interdisciplinary linkages of particular knowledge using generally accepted principles. The ontology, as an analogue of the term "model", serves as a means of communication between transdisciplinary distributed information resources and systems by combining descriptions of resource content based on an object-oriented formalization procedure. The ontology model provides interpretive functions that control the process of data extraction from an information resource. Also, its formal structure simplifies computer processing to reflect the specific tasks of different subject domains.

"POLYHEDRON" operational capabilities and features are following:

- Means of context-semantic and linguistic-structural analysis of natural language text and construction of documents taxonomy;
- Means of classification and domain ontologies generation;
- Ontology of the choice task for information-analytical support of decision-making and provision of multicriteria analysis and rating processes;
- Means of semantic search of lexical structures on the basis of linguistic processing of a large number of distributed network text arrays (Big Data);
- Linguistic cluster (an electronic library with means of associative search of semantically connected information arrays).

By means of "POLYHEDRON" it is possible to get ontological representation of educational and research activity results that are semantically linked to other information, such as:

- fundamental and applied research results from Ukrainian NAS institutions,
- MES curricula and training programs,
- general educational and methodological materials.

Thus, "POLYHEDRON" can be used as software for the development of educational intelligent analytical system for student youth educational and research activities support (Fig. 2).

6 Benefits of the Ontological Approach to Designing the Educational Intelligent Analytical System

The ontological approach to designing the educational intelligent analytical system for student youth educational and research activities support allows to create the modern class of artificial intelligence systems based on cognitive computing. Interactive services of the EIAS will provide interaction of educational processes participants with

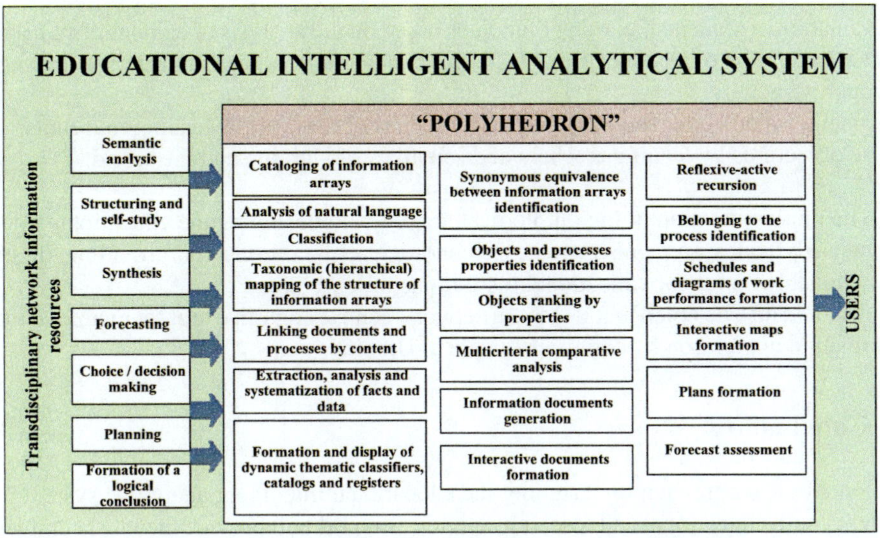

Fig. 2. Components of software platform "POLYHEDRON"

information-educational and corresponding educational-methodical materials in the format of ontological interactive document [24, 25]. This is the world's first such scientific and technical solution in the field of creating interactive knowledge bases. The practical value of the EIAS is its ability to ensure optimal use in the educational process of the necessary information and training materials in accordance with the curriculum. EIAS provides the formation of individual ontological learning scenarios. EIAS cognitive services computer-aided find the necessary contexts corresponding to the part of the curriculum, which make up the content of science-based informational and educational materials (textbooks, lectures, and teaching materials, monographs, popular publications) recommended for use in the educational process by the Ministry of Education and Science of Ukraine.

Verbally active EIAS services provide the formation of single information and educational space for all educational institutions. The conceptual and technological basis of a single educational space is transdisciplinary ontologies of all informational and educational resources. Electronic repositories of libraries and other educational institutions with different levels of educational accreditation can also be computer-aided encapsulated in the environment of single information and educational space.

EIAS can fully provide the educational process with appropriate information-educational and educational-methodical materials in terms of quarantine of all participants in the educational process at the national level by:

- creation of individual curricula and educational scenarios;
- formation of educational knowledge bases, which include educational programs and their taxonomies, video lectures, textbooks, information, and consulting materials, etc.;

- evaluation of students' learning outcomes based on the analysis of an integrated narrative description of all educational materials and documents that characterize students' cognitive activity;
- computer-aided connection to the virtual scenes of the world's leading museums;
- use of lectures by leading teachers and scientists of Ukraine and the world.

The implementation of the ontological approach to designing the educational intelligent analytical system also creates technological conditions for providing digital educational logistics for educational institutions of Ukraine.

It should also be noted that EIAS is a network-centric cognitive tool for counteracting information aggression by countries hostile to Ukraine.

7 Conclusions

The ontological approach to designing the educational intelligent analytical system for student youth educational and research activities support will provide the most complete information-oriented network resource containing a description of scientific and technical products in the form of interactive documents and will provide semantic connectivity with the research projects of the Junior Academy of Sciences of Ukraine, the curricula of the Ministry of Education of Ukraine and educational materials on the basis of the use of mechanisms and tools of ontological management of cognitive and research activities of users to form a network-centric transdisciplinary knowledge base.

The systemological structure of such an interactive knowledge base will correspond to the competencies conceptual system of subjects of education, the structure and content of educational processes declared in the educational programs of the Ministry of Education and Science. All categories and concepts reflecting various disciplinary processes in the form of information resources and digital assets and using in the processes of education and scientific research will have deep semantic connections.

In the short term of the work results implementation, it is expected to improve the quality of the implementation of educational research projects by providing access to transdisciplinary information resources describing the achievements of the NAS of Ukraine, methodological support and tools and services of a network-centric ontology-driven interactive knowledge base.

In the medium term, an increase in the level of awareness and interest in science and technology is expected, which will significantly expand and enrich the content of the network-centric interactive knowledge base with the development of young researchers and scientists, specialists, experts who want to share their own research experience.

In the long term, the results of work are transformed into an information channel for representatives of state bodies of science, education and business in the areas of introducing modern scientific and innovative results into the educational sector.

References

1. Virtual STEM-centers: ontological approach: collective monograph. Kyiv: Institute of Gifted Child NAPSU (2019). 164 p. ISBN 978-617-7734-18-4

2. Novogrudska, R.L., Globa, L.S., Koval, O.V., Senchenko, V.R.: Ontology model of intelligent modeling system for marine facilities identification. In: International Conference on Radio Electronics & Info Communication (UkrMiCo), pp. 1–5 (2017). https://doi.org/10.1109/UkrMiCo.2017.8095426
3. Globa, L.S., Popova, M.A., Yushko, N.A.: Improved approach to quality control of telecommunication service providers. In: Proceedings of the 8th International Conference on Applied Innovations in IT (ICAIIT), pp. 29–34 (2020). https://doi.org/10.25673/32745
4. Globa, L.S., Novogrudska, R.L., Koval, A.V.: Ontology model of telecom operator big data. In: Proceedings of IEEE International Black Sea Conference on Communication and Netwoorking (BlackSeaCom), pp. 1–5 (2018). https://doi.org/10.1109/BlackSeaCom.2018.8433710
5. Piedra, N., Chicaiza, J., Lopez-Vargas, J., Tovar, E.: Seeking open educational resources to compose massive open online courses in engineering education. An approach based on Linked Open Data. J. Univ. Comput. Sci. **21**(5), 679–711 (2015)
6. Alomari, J.: Ontology for academic program accreditation ontology of accreditation board of engineering and technology (ABET). Int. J. Adv. Comput. Sci. Appl. **7**(7), 123–127 (2016)
7. Muñoz, A., Lagos, K., Vera-Lucio, N., Vergara-Lozano, V.: Ontological model of knowledge management for research and innovation. In: Technologies and Innovation: Second International Conference (CITI), pp. 51–62 (2016)
8. Valaski, J., Malucelli, A., Reinehr, S.: Ontologies application in organizational learning: a literature review. Exp. Syst. Appl. **39**(8), 7555–7561 (2012)
9. World Declaration on Higher Education for the Twenty-First Century: Vision and Action. https://www.unesco.org/education/educprog/wche/declaration_spa.htm
10. Sarmiento, C., Duarte, O., Barrera, M., Soto R.: Semi-automated academic tutor for the selection of learning paths in a curriculum: an ontology-based approach. In: IEEE 8th International Conference on Engineering Education (ICEED), pp. 223–228. IEEE Press, Malaysia (2016). https://doi.org/10.1109/ICEED.2016.7856077
11. Berestneva, O.G., Marukhina, O.V., Bahvalov, S.V., Fisochenko, O.N., Berestneva, E.V.: Ontological simulation for educational process organisation in a higher educational institution. J. Phys.: Conf. Ser. **803**(1), 1–6 (2017)
12. Kalou, A., Solomou, G., Pierrakeas, C., Kameas, A.: An ontology model for building, classifying and using learning outcomes. In: 12th IEEE International Conference on Advanced Learning Technologies (ICALT), pp. 61–65. IEEE Press, Italy (2012). https://doi.org/10.1109/ICALT.2012.45
13. Oktavia, T., Prabowo, H., Kosala, R., Supangkat, S.H.: A conceptual social learning ontology for higher education in e-learning 2.0. In: Information Management and Technology (ICIMTech), pp. 165–170. IEEE Press, Bandung (2016). https://doi.org/10.1109/ICIMTech.2016.7930323
14. Lahoud, I., Chamekh, F.: Towards an architecture for universities management assisting students in the choice of their specialization. In: Ciancarini, P., Poggi, F., Horridge, M., Zhao, J., Groza, T. (eds.) Knowledge Engineering and Knowledge Management, EKM 2016, The Second International Workshop on Educational Knowledge Management, pp. 69–81. Springer, Cham (2016)
15. Mosharraf, M., Taghiyareh, F., Soleimani, A., Orooji, F.: ADAM: advisory agents modeling system to enhance student-supervisor decision making. In: International Conference on Technology Enhanced Education (ICTEE), pp. 1–7. IEEE Press, Kerala (2012). https://doi.org/10.1109/ICTEE.2012.6477236
16. Izhar, T.A.T., Torabi, T., Bhatti, M.I.: An application of the ontology based GOAL framework in a higher education institution in Australia: a case study. Int. J. Comput. Inf. Syst. Ind. Manage. Appl. **9**, 43–59 (2017)

17. Ternai, K., Szabó, I., Varga, K.: Ontology-based compliance checking on higher education processes. In: Kő, A., Leitner, C., Leitold, H., Prosser, A. (eds.) Technology-Enabled Innovation for Democracy, Government and Governance, EGOVIS/EDEM 2013, Lecture Notes in Computer Science, vol. 8061, pp. 58–71. Springer, Berlin (2013). https://doi.org/10.1007/978-3-642-40160-2_6
18. Zemmouchi-Ghomari, L., Ghomari, A.R.: Process of building reference ontology for higher education. In: Proceedings of the World Congress on Engineering, London, 3–5 July 2013
19. Zemmouchi-Ghomari, L., Ghomari, A.R.: Towards a reference ontology for higher education knowledge domain. Inter. Rev. Comput. Softw. **8**(2), 474–488 (2013)
20. Abed, H.N., Tang, A.Y., Cob, Z.C.: An ontology-based search engine for postgraduate students information at the ministry of higher education portal of Iraq. In: 13th International Conference on Intelligent Systems Design and Applications (ISDA), pp. 69–73. IEEE Press, Bangi (2013). https://doi.org/10.1109/ISDA.2013.6920710
21. Guryanov, I.S.: Problems of adaptive management of educational content in modern learning management system. Sci. World **7**, 38–40 (2015)
22. Stryzhak, O.Ye.: Taxonomic principles of narrative discourse. Med. Inform. Eng. **2** (2020).https://doi.org/10.11603/mie.1996-1960.2020.2.11186
23. Strizhak, O.V., Globa, L.S., Velychko, V.Y., et al.: Certificate of copyright for the work №. 96078 dated 17.02.2020. Computer program "Cognitive IT platform POLYEDRON" ("POLYHEDRON") Official bulletin №. 57 (31.03.2020) pp. 402–403 (2020)
24. Mintser, O., Pryhodnyuk, V., Stryzhak, O., Shevtsova, O.: Transdisciplinary presentation of information using interactive documents. Med. Inf. Eng. (2018). https://doi.org/10.11603/mie.1996-1960.2018.1.8891
25. Stryzhak, O.Ye., Prykhodniuk, V.V., Gayko, S.I., Shapovalov, V.B.: Display of network information in the form of interactive documents. Transdiscip. Approach. Math. Model. Econ. **5**(3), 87–100 (2018)

Principles of Creating an Integrated Development Environment for Educational Computer Systems

Zoya Dudar, Igor Shubin, and Andrii Kozyriev

Department of Software Engineering, Kharkiv National University of Radioelectronics, Kharkiv, Ukraine
{zoia.dudar,igor.shubin,andrii.kozyriev}@nure.ua

Abstract. The various questions of creation of integrated development environment for computer training systems are considered in the given paper. The information technologies that can be used for creation of the integrated development environment are described. The different didactic aspects of realization of such systems are analyzed. The ways to improve the efficiency and quality of learning process with computer training systems for distance education are pointed.

Keywords: Distance learning · E-learning · Learning course model · Computer-based training system · Hypertext · CMS · LMS · LMCS · Multimedia

1 Introduction

The continuous scientific and technical advance in all areas of human activity requires constantly possessing by new knowledge, and the traditional methods of the stationary learning frequently appear ineffective. There has been a need for creation new technological system, which would allow transmitting to a great number of people arbitrary information content and special knowledge. One of the most perspective directions in this respect is the introduction of distance learning on the base of computer and telecommunication technologies [1].

1.1 The Idea of Distant Learning System

Today, the informatization of education is primarily associated with the formation of an electronic educational environment, which is the subject of many scientific studies. Thus, scientists [9] propose the following formulation of a fundamental problem: how to identify and support the targeted development of e-learning in an imbalance between limited resources for infrastructure upgrades, on the one hand, and the rapid emergence of new technological opportunities that require continuous improvement of ICT skills participants of the educational space – on the other.

The idea of e-learning (e-learning) can be defined as a set of hardware and software, methods and organizational measures that ensure the delivery of educational information

P. Vorobiyenko et al. (Eds.): IPF 2020, LNNS 212, pp. 415–435, 2021.
https://doi.org/10.1007/978-3-030-76343-5_22

through the network, as well as testing the knowledge gained in the courses. Finally, perhaps the most important advantage is the lower cost of training, which, according to Cedar Group, is lower by an average of 32–45%.

Talking about e-learning in a broad sense, it is assumed that the training material will be placed in the e-learning system – Learning Management System (LMS), which is a software product. Access to the training system is carried out via the Internet or local network, in free mode or with user authorization – by login and password [10–12].

The first ideas in the field of e-learning (E-learning) were expressed in 1984. These ideas of distant learning have reduced to derivation of National Technological University (NTU) in USA, which by 1991 has been transformed to a consortium from 40 university engineering schools. Further experience NTU was investigated and was recommended as model for international electronic university. Today NTU is completely financed with fee, which is usually brought by the corporations (employers of learners). Six years later, the total income NTU was $13,5 million. Over millions of students are taught on the program of distant learning. Since 1989 there has been the translation of educational courses through the System of a Public Television. Moreover, the program of training and conversion training of the adults exist. It interacts with more than 1500 colleges and universities. The distance learning in Europe has developed intensively in the last seventy years. Now in every European country there is a considerable group of educational institutions, in which has implemented distance learning programs.

In the simplified systems of distance learning which is not operating with principle of continuous feedback, the information necessary for giving the lectures, seminar and other sorts of class, is usually fixed on video cassettes and on video discs, and then it is forwarded to the students. Audio-records and data record on magnetic disks are used too. The materials are transferred immediately in educational institutions. Such method is applied by National center of today distance learning Francium providing distance learning more than 350000 users in 120 countries of a world. In this program, that includes 2500 learning courses take part about 5000 teachers.

Together with programs intended for large audience, set of the lectures and works, which allow to the learners after final examinations to receive the appropriate diploma or the certificate, has received a wide distribution. Examples of such direction of distance learning are the television courses of the Baltic University of Sweden, which combines more than fifty universities of the Baltic locale. Using systems of a satellite television, students and the science officers have possibility to have scientific and educational contacts on subjects representing common interest.

However, distance learning presents the new requirements to techniques of learning and testing of knowledge. First, these requirements concern system's representation of interdependent knowledge from different application areas [1]. Now there is a necessity in cultivation new systems of creation and organization of active information resources. Such information system is the integrated development environment for computer training systems (IDECTS). The given paper is devoted to various aspects of development and use of it.

1.2 Existing E-learning Systems and Their Scope in the Educational Process

In its most general form, e-learning is learning through electronic media. It can be defined by the following types:

- computer training (CBL) – presented by courses that are distributed on independent media;
- web-e-learning – courses and training that use communication networks to communicate with the training system and supply training materials;
- virtual classes and collaboration;
- non-formal learning within social networks (common interests, common practice, etc.) [13].

Open learning is currently one of the most effective learning formats in various educational institutions. This approach allows using modern interactive technologies to teach in one classroom and under the guidance of one teacher students of different courses in a similar field. This technology, successfully developed in Eastern Europe and the United States over the last decade, is actively developing in Ukraine. In today's society, open learning is a way to optimally plan your personal time to gain professional knowledge quickly and in full. Studies of the effectiveness of the method are impressive: this format is twice as effective as traditional full-time education, and the average score on the results of final exams is almost 30% higher. Because thanks to such an approach and system of teaching, the teacher can pay the student several times more attention than in a regular classroom [14].

The philosophy of open learning centers for student learning shows that the teacher becomes an assistant in the learning process. Teachers should observe, guide and provide the necessary materials for students, should facilitate, not complicate methodologies, the learning process.

But with all the advantages of e-learning, there are limitations and disadvantages that need to be considered. First, it requires from students serious self-discipline, independence and consciousness, the ability to allocate their time. Secondly, there is no personal contact between teacher and student, and given the individual characteristics, not everyone is ready to accept a virtual instructor. Third, there is a lack of communication with fellow students to share experiences, there is no possibility of immediate practical application of the acquired knowledge with further discussion of emerging issues. Fourth, students cannot always provide themselves with sufficient technical equipment – to have a computer and constant access to the Internet (especially in the province where distance learning is most appropriate).

In today's conditions, the emergence of a large number of international organizations that develop modern information technology and design industry standards of network information technology: ADL, AICC, ALIC, ARIADNE, CEN/ISSS, EdNA, DCMI, GEM, IEEE, IMS, ISO, PROMETEUS [15]. One of the most important and actively developing international associations is the Global Learning Consortium (IMS). IMS Global is the world's leading nonprofit collaboration aimed at interoperability, innovation and learning.

IMS provides a plug-and-play architecture and ecosystem that provides a foundation on which innovative products can be quickly deployed and work together seamlessly. IMS member providers are innovation market leaders.

Currently, there are two main branches of e-learning systems: commercial LMS\LCMS; freely distributed LMS\LCMS.

Freely distributed LMS\LCMS: ATutor, Claroline, Dokeos, LAMS, Moodle, OLAT, OpenACS, Sakai.

ATutor is an open source LMS used to develop and manage online courses, as well as to create and distribute interactive electronic content. IMS/ISO AccessForAll support allows students to customize the environment and content to their specific needs.

Claroline is a platform for sharing e-Learning and eWorking (learning management system), released under the open source GPL license. It allows hundreds of organizations around the world, from universities to schools and from companies to associations, to create and administer online courses and collaboration spaces. The platform is used in more than 100 countries and is available in 35 languages.

Dokeos is an open source distance learning system that is freely distributed. The system has all the standard functionality of similar software: forums, chats, html-documents, etc. The Dokeos system was created thanks to the Claroline system. However, the system is not widespread in Ukraine, although it has a Ukrainian-language interface.

LAMS is a web application for designing, managing and delivering online learning through collaborative learning activities. It provides teachers with an intuitive visual authoring environment to create sequences of learning activities. It has been based since 2003 on works from the Open University of the Netherlands as the Educational Modeling Language (EML), which describes the "meta-model" of building the educational process.

Moodle is a free online learning management system that allows teachers to create their own private website, filled with dynamic courses that continue learning anytime, anywhere [16]. Designed to be useful and accessible, the Moodle interface is easy to navigate on both desktop and mobile devices.

Moodle is suitable for using more classic learning styles, in particular, hybrid learning, which transforms the system in addition to presentation learning. Moodle works on all computers where you can install PHP and run a MySQL or PostgreSQL database. The Moodle website provides high-quality support to users of the platform free of charge. With the help of Moodle in thousands of countries around the world, including Ukraine, thousands of LMS training systems have been created with training courses hosted on them. The system was designed in 1999 by Martin Dugiamos, MD, who still supports it. The main goal of the author was to create a tool for interaction between teacher and students, which would allow the first to transfer their pedagogical skills online. As a result of hard work, a full-fledged representative of distance learning systems with open source texts was developed, which gained worldwide popularity not only in the academic but also in the corporate sphere of education [17, 19]. Thus, Moodle is distributed free of charge as an Open Source project and has the following key characteristics:

- based on the philosophy of pedagogy of social constructivism and progressive theories of pedagogical measurements;
- suitable for both distance and stationary training;
- has a "light", efficient, compatible web interface;

- simple installation on any platform that supports PHP, only one database is required;
- supports databases such as MySQL, MSSQL, PostgreSQL, Oracle, etc.

2 The Role of Multimedia Technology in Computer Training Systems

In the information society, the development of open educational resources is of particular importance, and both educational content and tools for its development and maintenance, which in combination with the use of service approach and modern distributed information processing technologies significantly reduces the threshold for entering the modern electronic educational environment. organizations.

2.1 Modern Approach of Learning Content Exchange

The main feature of the modern approach to mass provision is the transition from hierarchical systems of distribution and management of learning to a network model of content exchange and provision of educational services, which includes:

- formation of a new worldview of users using the motivation to use new technologies for their own development in order to increase the comfort of existence in the new information society;
- development of Internet infrastructures aimed at supporting mass learning, which will provide support and sustainable development of the educational services industry;
- further development of process and structural approaches to solving the problems of organizing effective interaction of users with the learning environment;
- creation and implementation on the basis of reference models in the environment of open systems of new types of multilevel interaction of applications, general and applied services, and also infrastructures;
- development of IT standards and recommendations to ensure interoperability and scalability as a basis for creating and supporting the operation of reliable e-learning platforms of the new generation;
- transformation of technological platforms of e-learning from didactically neutral to didactically purposeful, formation of an expanded cross-platform portfolio of didactic services;
- creation of multicomponent educational and technological environments with built-in functions of social learning;
- creation of a theoretical and methodological platform for the implementation of universal approaches to targeted actualization, generation and reuse of heterogeneous information resources and knowledge presented in the form of electronic content;
- creation of technologies to support the distributed work with information, resources and knowledge that support the pyramid of learning strategies, as well as intelligent technologies to individualize the learning process to increase the motivation and efficiency of the learning process;
- intensification of research and implementation of new approaches in the field of information security and protection, including personal data, development of big data analysis tools [9].

2.2 Multimedia in Education

The modern computer training systems are developed with the help of the multimedia technology. The given technology has appeared at the butting of many branches of knowledge. Development multimedia CTS is the long-term and expensive process; therefore, it is important to imagine all main stages of courses creation and possible solutions, which can be used at each stage [2].

The courses for representation in multimedia environment are chosen at a preliminary stage. It is one of the most responsible stages of development. The choice is carried out based on results of marketing research. Already existing courses on the given problematic should be detected, must be defined the guessed expenditures and time for creation of this course and, audience, who are addressed these courses and its probable number of copies. The type of an audience allows defining the common requirements to multimedia course.

It is possible to choose groups of courses: for children of preschool age and low school age, general educational courses of a different level and courses of special derivation for beginner and experts.

The training systems for children of preschool and low school age are intended for forming basic skills and knowledge (color, the form, sound, music, speech, reading, letter, natural sciences, foreign language, the computer skills etc.) and have simple for children's perception interface, which are implemented as the fascinating game, travel and so on.

The general educational courses should take into account features of education that connected with a different level of learners schooling (estimation of learner knowledge and adaptation of the system for the optimal presentation) and with different level of computer knowledge (simple and clear the interface). It is important to provide the matching of teach subjects, that the learning of one subject did not go to the detriment to other and did not overtake learning of parallel subjects.

The courses of special derivation should consider a level of knowledge, give a possibility to not repeat already known subjects, to provide the latest information in the given data domain. When course and audience are selected, the purposes of learning and degree of complexity of a material, which will be represented in a training system, should be also defined.

3 The Methodology of Creation of Computer Training Systems

The creation of most intelligent learning systems is based on the algorithm-heuristic theory of learning, developed in. In this theory, it is believed that any cognitive activity can be analyzed in terms of operations of algorithmic, semi-algorithmic, heuristic, semi-heuristic nature. An essential point of this approach is the construction of an adequate model of the listener.

We can assume that the intellectualization of the educational process of software systems to support distance education is to develop effective methods of presenting knowledge of the subject area and the learning environment by dividing them into thematic, functional, semantic, procedural, operational and creating on this basis authorized hypermedia distributed databases knowledge.

An alternative approach used in systems that implement various mechanisms in the field of adaptive navigation is the use of "smart" links. Examples of this approach include hiding links or applying traffic light metaphors to them. This also includes the dynamic construction of links, when you specify not the object itself, but only the criteria for its search. The system selects the most relevant element according to the specified criteria, considering the characteristics of the user and his level of knowledge. Such criteria must be developed by an expert because the efficiency of information retrieval is higher the more precisely the search query is formulated. Poorly structured information often remains unsuitable for learning purposes because the student requires help in building a model of the field of knowledge before, he can construct fairly accurate search queries.

In the theory and practice of automated learning, methods of artificial intelligence and means of modern information technology are spreading. Info communication hypermedia learning environment differs in communication technologies based on the Internet. For example: computer audio and video conferencing or Internet telephony of different types "for each other", "one for many", "many for many"; group cooperative work of students on a single project; use of a single real-time network application. In all generations of virtual learning systems, the most effective form of e-resources is a computer training program.

In most modern developments of intelligent and adaptive systems, mathematical modeling is presented in the form of traditional mathematical equations and functions (eg, differential calculus), as well as statistical and stochastic models, which do not provide sufficient flexibility to describe mental processes and knowledge. This study is based on algebraic-logical functions that allow the most effective formalization of human mental activity the learning process. The theoretical basis of these studies is the theory of intelligence, the tools of which are the algebra of finite predicates. This algebra can be used to describe finite, discrete and deterministic objects, which can be considered textual constructions of natural language that describe knowledge.

When it comes to learning management, certain definitions need to be introduced. For example, the definition of "adaptation" is a complex ontological (conceptual) structure consisting of a set of subjects and relationships. The interaction between its elements, determined by business logic and a set of business rules, is the company's activity. The information system "reflects" logic and rules, organizes and transforms information flows, automates workflows with dates and information, and visualizes results in the form of sets of reporting forms. Therefore, first you need to develop a model of learning, which is a reflection of the information management system. After the development of the model, a "language of communication" of consultations, developers and future users is formed, which allows to reach a common opinion on the development of a learning management system.

Distributed hypermedia systems form a model of goals, preferences and knowledge of each individual user and use this model in the process of interaction with the user to adapt to the needs of this user. The user model is used to gather information about his knowledge, goals, experience and so on, to adapt the content and structure of navigation. Here is an example. For a user with a low level of knowledge, it will initially be useful to learn general background information, but the same information will not be of interest to the expert. Selecting the right information at the right time is the task of creating a user model. The model studies the elements of learning based on the student's ability to learn on the basis that the student takes responsibility for the learning process.

For intelligent technologies of adaptation of group of network educational systems it is necessary to carry also the technology which is called selection of educational models (or a choice of models). Its essence is to analyze and select models for many students simultaneously, while existing adaptive and intelligent educational systems work with one intern (and one intern model) at once.

Models that contain targeted selected information, familiar to the call, information models. The most important factor that ensures the adequacy of the model in accordance with the object under study is the ability to distinguish the significance for the considered information about the object and organize it in a convenient way for learning. A formal description of this model can be a predicate called a stamp: stamp (surname, subject, sign). The terminology of IDEF1X is almost identical to the terminology of IDEF1, but there are a number of fundamental contrasts in the theoretical concepts of these methodologies. Graphically, the IDEF1X data model is represented by a combination of blocks (entities), connecting linear blocks (relationships between entities) and attribute names within blocks.

3.1 Groups of Learning Environments

Adaptive learning environments can be divided into three groups according to the principle used to link the concepts of the subject area and the objects of the content space:

- indexing pages with concepts related to the content of the page. However, links between concepts may not be established;
- breaking pages into fragments and indexing individual fragments. Thus, at rather small partition each fragment can be indexed by exactly one concept;
- direct connection between the content space and the subject area: each page corresponds to a concept, each link – inter-conceptual relations. This approach, unlike the previous two, requires a model of the subject area of the second or third level.

In addition, it should be noted that the last decade has been characterized by the development of two basic trends in IT – the growing use of cloud technologies and social networks in most areas of human activity. Scientists believe that the use of cloud technologies to build an e-learning system is the optimal solution because it provides a solution to a number of important problems, namely:

- Building a reliable distributed information system infrastructure.
- Processing and storing large arrays of information: for example, you can transfer a high-resolution array of videos for a total of several hours from the information system site to Youtube video hosting, and then place this video on the site using embedding.
- Providing simultaneous access for a large number of users to information resources. In particular, when streaming video, the solution of such a problem without the use of cloud technologies requires significant investments in the hardware component of the information system infrastructure.

- Solve format compatibility issues. When multimedia resources are placed on the site of the information system in a single format, some users have problems with their reproduction due to the incompatibility of some part of the client software. Cloud services in most cases convert formats automatically.
- Mobile access support. Adapting information system interfaces for access from mobile devices has become a non-trivial task, while the use of cloud services allows to solve it automatically, as cloud services either provide a suitable adapted interface or offer their specialized mobile applications for access to content.

Similarly, in the case of social networks:

- Communication function. Traditional solutions, such as a forum or mailing list, are less effective than mechanisms for disseminating information and engaging users in communication on social networks.
- Creating a user-friendly information environment.
- Increasing the availability of information resources. Many information resources, including multimedia, can be placed directly on the pages of social networks through the embedding mechanism.
- Attracting the target audience. The use of social networks significantly increases the dissemination of information within virtual professional communities.
- Providing effective feedback to the organizers. Commenting and communicating on a social network is more organic for the modern user than e-mail correspondence or discussion in a specialized forum.

3.2 Development Environments in Training Systems

With the creation of new educational modern computer technologies, it is possible to achieve a high level of intelligence using computer learning environments. To build learning technologies based on computer learning environments, it is possible to use automatic models of student and teacher, which interact with each other through the space of educational material, formalized in the form of marked information trees. In this approach, the teacher and the student are interpreted as adaptive automata, and the learning process consists in their interactive interaction.

Since IDECTS is intended for design automation of CTS it is necessary to analyze the methodology of their constructing for highlighting of main development stages. The experience of using a computer in training had shown that creation of the effective program for learning process is the composite and laborious work, requiring collective operations of the experts of the different profile. Therefore modern methodologies of constructing programs for learning process need a creation a collective of the developers, which includes: the expert in an investigated subject; Methodist is the expert in a method of training, which is realized in the program; the psychologist is the expert in diagnostics of errors of learners and psychology of perception of an learning material; the artist is the expert in aesthetic designing of an learning material; the composer is the expert in musical designing of an learning material; the manager is the organizer of collective activities of programmers and experts. The structure of collective developers is like structure of creative group on creation of learning films with replacement operator group by the programmers.

The life cycle of CTS designing consists of the following stages:

- Decision making about program design.
- Conceptual designing.
- Detailed designing.
- Implementation that consists of programming, debugging, approbation, finishing and estimation of quality of the program.
- Implantation in the learning process and maintenance.

The analysis stage of a problem consists of exploration of learning process of some discipline, selection of fragments of this process accepting effective using of the computer, tentative estimation of expediency of using computers, decision making about a program design of CTS. A role of the CTS in the learning process here is determined.

A subject and title of the project, structure of designer's collective, age and amount of the learners, who will be trained to the given subject, ways of usage in CTS of techniques of training, methods of a data representation and knowledge, requirement to the configuration of technical means, and other items of information, that characterize CTS of particular sort, are usually defined at a stage of conceptual designing. Result of this stage is the conceptual project (the document containing basic performances of the future CTS). The conceptual project is reviewed and if it is necessary it will be corrected.

At a stage of the detailed designing a development of structure and contents of the CTS occurs. Splitting the CTS into units and functions of each unit here is defined. For each type of training system there is an individual technique of designing, which contains one or several learning scripts for programs of this type. On top levels of designing, according to selected models, there is a script of the programs consisting of a sequence of concrete subjects (purposes) and a learning material, appropriate to them, (definitions, tasks, help etc.). On lower layers of designing the algorithms of output of a learning material, analysis of learner answers and control of the learning process are specified. During designing the major notice is given to how a learning material is looked at the screen of the computer. At present time many techniques contain a lot of recommendations about external designing of the text and graphics of learning information. The main requirements that presented to CTS interface are:

1. Psychological adequacy. The ways of interaction of learner with the system should correspond to its guessed sort of activity.
2. Ergonomics is a choice of type and layouts of peripherals as well as methods of operation with them to allow optimally organizing activity of learners.
3. Aesthetically beauty intends availability of some concept (style) of representation of the information and ways of interaction with learner.
4. The minimum complexity of interaction intends that in each instant learner receives only the necessary information.
5. Harmony psychological, ergonomically, and aesthetic solutions at creation of the CTS interface.

The given requirements are not equivalent or independent. Most of them are derived from requirements of psychological adequacy [3].

Result of the given stage is the project of CTS containing the thumbnails of external representation of the learning information and logical organization of the program. The project is reviewed and, if errors are found, it will be changed.

At a stage of realization, the programming languages (general-purpose or authoring) and tools for creation of learning courses are chosen. At the stage, the CTS coded and debugged. Here also the quality of CTS is estimated, and it is approved. That is considered as necessary steps for estimation of CTS efficiency. The expert in learning technologies produces the estimation of CTS quality. The approbation consists in holding experimental training on the designed program by restricted learner's group, with the subsequent interrogation and data processing about learning process. Result of the given stage is the full-functional CTS, the reviews of the experts about its quality and results of approbation.

For successful work of the CTS it is very important to prepare correctly the environment of its operation (create motivation and required level of learner knowledge to provide trouble-free operation of hardware, to conduct preventive measures of "computer phobia" etc.) and to train serving staff (teachers, laboratory assistants) rendering pedagogical and technical help to learners during their work with the CTS.

3.3 Agent Technologies in Education Systems

Modern research uses various agents that form multiagent or multiagent systems (MAC) [20]. Such a system is built as a system of agents that can carry out information or intellectual interaction with each other using the information language ACL (Agent Communication Language). The organizational structure of the IAC is determined by the role functions of agents and the norms of their interaction. The MAC architecture specifies the interaction of agents in the system. According to the organizational structure of the MAC, the following types of agents are distinguished:

- agents-executors and agents-managers -the first obey the second;
- coordinating agents responsible for organizing the interaction of agents;
- interface agents, employees for communication with the external environment;
- channel agents that provide information exchange in the system.

  Classification of agents. According to the type of connections, the following types of agents are distinguished:
- syntagmatic – connections between equal agents (Two-link connection);
- paradigmatic – connections of subordination to the superior in the link (Two-link);
- hierarchical – connections of subordination to the superior coordinating agent (multi-link connections);
- subsidiary – coordinated actions of peripheral agents authorized by the control center (multi-link).

The agent in the general case has the following characteristics: autonomy, goal-oriented, collaborative, reactivity, temporary continuity, personification, communicativeness, flexibility, mobility, adaptability, capacity for reasoning, Trustworthiness.

The current level of development of the theory and technology of design of deliberative agents is far from practical scope. An alternative way of agent intellectualization can be based on situational methodology and principles of computational intelligence. Agents have "mental" characteristics that are expressed through the following categories:

- perceptions – perception of the situation;
- beliefs – a plausible part of the agent's knowledge of the external environment;
- goals – the desired result of influences on the object;
- intentions – action plan.

The organization of an agent system of e-learning provides advantages in terms of individualization and adaptation of the educational process to the needs of the user. In general, the following components of the e-learning system can be distinguished: search of educational content, delivery of educational content, storage, control, user interface and more.

Users of the e-learning system seek to use materials from numerous sources. To ensure this, there must be mechanisms that allow you to find and use the necessary materials in both local archives and the Internet.

Finding educational content for integration and reuse is one of the important tasks of embedded intelligent information agents. One of the main problems of search engines on the Internet is the inefficiency of ranking algorithms for found documents. This is largely due to the fact that search queries on average consist of only two or three words, ie there is simply not enough source information to effectively rank the issue.

Modern researchers divide the problem of finding educational objects into two types of tasks:

1. search for educational resources in a single local archive;
2. distributed search of educational objects on the Internet.

The solution of the first problem has led to certain theoretical and practical results, which allows to implement their application in e-learning support systems. The second problem does not yet have an effective solution based on a generalized theoretical approach. This leads to the need to study the processes of data retrieval on the Internet, as well as to develop theoretical foundations for solving the problem of finding initial objects.

An educational object is a self-sufficient information unit that can be used for educational purposes and has tags as internal links. Creating a training course requires comparing a predetermined sequence of learning objects. Various educational repositories (archives, information portals) are used to store training courses. Features of educational objects include the ability to be reusable, the ability to aggregate or decompose in accordance with educational objectives and meta-description, compliance with standards.

4 The Structure and Functioning of IDECTS

As already was noted above, the IDECTS is intended for automation of development of computer learning and testing systems of distance learning for any domain. Output data of the IDECTS is the computer training system that is oriented to concrete data domain, program environment and hardware. An input data of the system is the learning script, which help to set functionality of a training system that is creating. In other words, the adjustment of the IDECTS to data domain is realized with help of learning script. The learning script should have a module structure. It is obligatory the presence of following units: the unit of registration, the unit of precheck testing, the unit of the hypertext textbook, the unit of on-line learning and the unit of final testing.

The unit of registration is intended for registration of learner in the system, creation the model of learner and support of the electronic journal. The unit of precheck testing is intended for definition learner's degree of readiness for learning of the given subject and for creation and filling learner's model. The unit of the hypertext textbook is intended for presentation of large theoretical information, which consists of blocks both text's and graphic's information joining by hyperlinks. The unit of on-line learning implements learning process by means of involving learner in active dialogue with system. The unit of final testing realizes the function of testing of the learner for checking his knowledge in material and for determining the missed units of knowledge.

The expediency of inclusion of each unit in the learning script depends on the purposes, which are put at creation of a training system by its author. Thus, the process of creation a training system in IDECTS consists of selection of necessary units, their logical coordination and appropriate set-up of their parameters. Each unit should have the set of structured elements, which represent the next level of detail of the learning script. The insert of structured element in the learning script should be done by means of dragging it with the mouse from the toolbar to a particular place in the scripts or choosing them from the appropriate menu. For the unit of registration such unit is the registration form. Registration form should allow to set biographical data, which will be interrogated from the learners and, if it is necessary to assign to them the special password for operation with the system. The creation of registration form should be made in the on-line mode with the built-in IDECTS editor of the screen forms. For the unit of precheck testing a structural unit is the test task.

After structure of learning script is created, the following stage of development of CTS in IDECTS is its filling, i.e. creation and editing of each structural element of learning script. This process should be carried out in the built-in IDECTS editor of the screen forms. The editor of the screen forms should consist of the editor of the screen forms; the editor of properties and bar of the menu with the graphic palette of components. There should be a possibility of creation a new form with the help of special prepared templates. The additional components both decorative and interactive should be there too. The logic analyzer, which allows analyzing the formulas of logic-algebra, and blocks of the text of input fields, is an example of such components. Any training system that was created with the help of the IDECTS should include subsystems of support of the electronic journal and learning protocol.

A problem of integration of the IDECTS in management systems of the learning process is actual now. It has reduced to appearance in the market of the software a great number of tools to automation and creation Web based training courses (WBT-courses). Therefore, it is necessary to have the open structure of storage of learning scripts in CTS and all related materials.

There is a problem of performance of practical operations by the learners on the computer in organization of distance learning. Most effective is possibility of access to server of an educational institution, when the learner receives on the computer only screens, while programs are executed on the server. It considerably simplifies work with software, as it is completely allocated on server of an educational institution. Besides, it is simplified the involvement of the learners in the collective projects. The learners and teachers also will have access to network resources of an educational institution (for example, to high-speed computers or expensive equipment).

In learning process, the additional materials (for example, literature that is not included in the electronic library of institute) are often necessary for the learner. Then the organization of a bookshop (sending the books to the learner by post) or bridging with existing bookshops for sale this literature is necessary.

During organization of distance learning the possibility of messaging by the e-mail between the learners and teachers is important. For this purpose, it is expedient to include in the IDECTS the tools for work with the e-mail. Besides, the control of the mail service is simplified. Thus, the learners should have possibility to realize access to mail by means of the Web-interface (through a browser). Such manner will allow the learner do not immix mail concerning the learning process with the personal mail and will facilitate the search of necessary information. At organization of the mail service it is expedient to give to a learner the possibility to create folders for storage of different types of the messages.

One of attributes of WBT-course is the computer conference (is usually implemented as Web-forum) with participation of learners and teachers. Such conferences are usually conducted not in real-time mode. Due to that the variance in time's zones between the learners and teachers is overcame and the necessity to be present at certain time is eliminated.

The CTS created with IDECTS should provide the learners a possibility to search in forum messages. For protection against the unfair users it is necessary to provide possibility of reviewing of forums. Thus, there will be a person (moderator), who will check all messages and publishes them, only if they perform the requirements of the given forum.

In the IDECTS it is necessary to provide tools of working with the timetable-calendar within the framework of institute and within the limits of separate course. This tool should give an opportunity to publish events by both learners and teachers. It is necessary that the calendar automatically remind of coming events. The possibility for the learner to store in such calendar the events, which would be available only for him, is very useful. It is known that reading from a paper more conveniently, than from the screen monitor, therefore the learner should have possibility to receive paper variant of course. It is very important that IDECTS supported possibilities of compilation materials of courses for printing with adjustment of page setups, size and typeface of the font, and also variants of layout of the text on page.

The velocity of transmission of information grows promptly. Now if the learners have good connection to the Internet the organization of videoconferences is possible. Therefore, at designing the program system like IDECTS it is necessary to provide possibility of a videoconferencing.

The materials of course should be represented with hierarchies of sections. For this purpose, the appropriate tool in CTS should be included. Due to such approach, the learner permanently has the information about level of hierarchy in which he is in the present moment, and about which sections precede the given section in hierarchy [4].

The realization of search subsystem over contents of course is very important for the learners. And the search not only text's, but also graphics, audio- and video data may be useful. For this purpose, it is necessary to provide in the CTS the advanced search engine with possibility of calling it from any place of course.

For simplification of information retrieval by the learners, the IDECTS should contain tools for organization of a glossary. Thus, the titles of concepts that are covered in a glossary should automatically become the links in the text of course. Besides, it should be realized the function of search in glossary too.

During work with materials of course learner may have need to find fast the detailed description of this or that concept. The subject index is necessary for this purpose. It is easier to make it as the list of terms with hyperlinks to section of course, where this term is described in detail.

To support lessons or conversations in real-time mode the chat tool should be included in the IDECTS. With this tool the teacher can create subject's groups for dialogue of the learners in the chat mode. The discussion of the learners with the teacher should be stored in the learning journal with the purpose of further analysis and correction materials of course on base of results of this analysis.

For simplification of creation of new courses, the IDECTS should contain a set of templates of learning courses. The teacher should have possibility to select subject design from a set of templates. The presence of templates of learning courses will allow the teachers faster to study possibilities and tips of work with the system [5].

The presence of drawing boards of share use (whiteboard), in which learners and the teachers can jointly draw in real-time mode, will allow considerably to approximate the process of discussion or interrogation to personal contact between the teacher and learners. It is used with chat or conferences during subject discussions.

It is necessary for the learners and teachers to have possibility to turn on and off the hints in CTS. Apparently, the user cannot memorize all functions at once. The user should have possibility to disable the hints after certain period of using of IDECTS.

For convenience of learner work he should have possibility to add the comments for itself into sections of learning materials (the learner can add the comment, which will be stored on the learning server). Besides, it is necessary to give learners the opportunity to collect all these comments and to move through connected with them sections.

The important moment at organization of distance learning is the ensuring of information safety. Therefore, teacher should have possibility to limit access of learners to all courses and to the separate parts. Using only Web-interfaces for access of learners to all parts of course (the e-mail, conference, chat etc.) simplifies ensuring of safety at a level of network and transport protocols (TCP/IP).

For realization of the learning process it is necessary to group learners into classes. According to a specialty, the certain courses and tests will be anchored to the class. It is necessary to assign one or several teachers for the class, which will have access to this class for work with the learners.

For registration of the learners the questionnaire will be rather convenient. After filling the blank the learner will enroll automatically (this way is especially convenient at enrolling on free courses). For organization of such registration system the automatic activation and termination of the learning process are necessary. In this case the following limitations are indicated:

- The date and time of a registration beginning;
- The date and time of a registration termination;
- The maximal quantity of the learners in the class.

Nevertheless, for each specialty or course it is necessary to have possibility to establish three ways of registration of the learners. The learner is registered, when he filled the questionnaire. The learner fills the questionnaire but is registered only after confirmation of this questionnaire by the manager or teacher. The manager or teacher registers the learner without his involvement.

In the IDECTS it is necessary to provide tool of batch processing to reduce time of processing of the large number of questionnaires, when it is needed to register the group, instead of one learner.

The organization of document circulation, which relates to registration, teaching and graduation of the learners, is imperative. From the moment of registration (from the moment of confirmation of the questionnaire) the learner should be automatically inserted in the database, in which his activities will be subsequently stored (education fee, the marks for the different tests and examinations, the amount of visiting pages with courses, participation in conferences, chats and forums etc.). And it is necessary to have possibility to limit access of the learners to these data, for example only to read.

Inside of IDECTS there should be a system for timetable formation. At distance learning it is necessary to make timetable for on-line consultations, as one teacher can teach in several classes, besides, the consultations can be conducted not only within one class, but also within course, subject of course, course project etc.

For simplification of organization of the learning process it is very convenient to keep account of the teachers and their schedule. The information about classes and degree of teacher work is stored in the system database. Thus, it is easy to adjust current schedule according to the timetable.

Among other functions of IDECTS the central place at designing belongs to the functions of storing data into knowledge base. There are all learning materials of published courses in the knowledge base. Also, the dependences of learning materials from each other are reflected here. The learning material is structured so that it will be visible, in what sequence it will be presented to the learner. Thus, it should be clear, what knowledge and from which sources the learner should receive before he starts learning a concrete learning material.

From all document circulation it is necessary to distinguish the information on progress of the learner. For each learner, the marks for tests and examinations are stored in the CTS database. Also, there should be following information: how much time learner spent on each question, date of test, how much attempt he had to receive this mark, the marks for the previous attempts. Based on this data it is possible easily to define progress of the learners, to build a graphics and diagrams reflecting their degree of knowledge.

It is expedient to create common database of questions. The test questions, which are created for separate course, are stored in the common database. So, different questions can be used in other courses for making total examinations at this course or at all courses, without necessity to include all question databases to developed CTS. It is the way to increase efficiency of checking learners' knowledge by using major variety of types of questions [6].

Types of the test tasks, which should be supported by IDECTS during designing CTS testing subsystem, are featured below:

- The choice of one variant from several (it is necessary to provide possibility to represent variants of the answers both as text, and as graphics images);
- The choice of several variants from given;
- The short answer (one line, which can be parsed by the CTS).
- The full answer (text, which will be checked by the teacher). It works as previous, besides there should be a possibility to send files;
- The choice of accordance. Two lists will be given. It is necessary to set up a correspondence of units of the first list to the units of the second list;
- Input of the formulas with checking their correctness and comparison on equivalence;
- Indicating the coordinates of a position on the screen with the cursor;
- Dragging the particular object inside given area of the screen;
- Layout of texts' or graphics' units in an exact sequence. For example, on the screen the units of the computer network (workstations, servers, routers, concentrators, different types of a cable) are drawn. The learner should arrange them in the exact order with mouse;
- Calculation of value. The set of arguments of the function is offered to learner. He should define value of the function.

5 Development of Adaptation Models in Training Schemes

We introduce the concept of "Learning Scheme" because of generating a specific path for a student to be taught, which is included in the training sequence of the course. At the end of the sequence, user interaction is evaluated by testing for acquired knowledge. According to the test results, the user can proceed to the next section of the training materials or take the same course, if it is necessary to modify the student's profile, taking into account the re-submitted materials.

The passage from the training course to another is not carried out in an arbitrary or automatic form, but according to the exact navigation rules. These navigation rules are necessary to accurately describe transitions in the form of a learning passage network. In fact, the course is based on a set of transitions between other teaching sequences of the course, in developing the path until the last sequence of the course is reached. The passing network displays the dependence of training courses, i.e. each training course may be dependent on one or more educational processes. This implies that in order to start training on a course, you need to have knowledge of previous courses dependent on it. The learning passage network represents all the navigation rules that interpret various training sequences to achieve an educational goal. The second stage is designed to test the theoretical knowledge of the student. For this purpose, one of the tests is used. If a student has errors, he needs to return and take a theoretical course again. In general, during the first and second stages, the adaptation of the hypermedia teaching system is student-oriented and is supported by the selection of important training materials and tests.

During the third stage, the student solves problems in the subject under adaptive control. In each measure of studying this stage, according to the results of solving the problem, a decision is made on:

- Continuation of the study (the student was given a course on the knowledge he needs);
- Successful achievement of the goal (the goal of the study is achieved);
- The end of the stage (the student is sent to study a new theory).

It is necessary to properly emphasize and describe everything about the paths of these components and the order of the components that are important for the student: the material's training course, test or internal path, dynamically formatted problem solving results [7].

Adaptation methods are divided into two types: adaptation of content (content) and adaptation of relationships.

Adaptation of content adapts the content of the site to the characteristics of the learner. Link adaptation accommodates node communications (hyperlinks to other nodes). In this case, the adaptation of connections will be more appropriate, because we need to guide students through hyperspace formed by CTS content.

The adaptation of links is divided into four types according to how the links are adapted: "direct guidance" involves an explanation of the link that the user must follow or the creation of a button to guide the user; "Adaptive order" is based on determining the degree of suitability of each communication to the user; "Adaptive link hiding technology" narrows down available hyperspace, hiding inadequate learner communications; "Adaptive annotation" - involves additional artistic design, such as images and color ties. The method of concealment of connections does not show connections that are not adequate to the current characteristics of the student. Although this forces the user to follow the navigation paths created by the CTS developer, narrowing his freedom of movement through the content, this method allows the developer to direct him along the most optimal path.

To develop intelligent CTS of the proposed architecture, the authors chose to use a model that stores both long-term and short-term information about the student. In order for the student's model to be simpler, we simplify the format of the navigation rule. As a parameter of the learner, we will use the interests and level of knowledge of the user. This is necessary to simulate a learning strategy over time.

We will also use the history of the student's interaction with CTS, which will be a sequence of classes of nodes of the students studied, to simulate short-term information about him. Each node is associated with an alphabetic character that characterizes its class (the alphabetical name of the class). The concept of a class is necessary for formalization and description, navigation rules, and for the logical conclusion to them. The class is used as a component to represent the user model and navigation rules.

Node classes are determined by the following criteria:

- Does the system show content for a specific kind of user?
- Does the site offer users an explanation, asks a question, offers to complete a test task, or does something else for the educational purpose of the didactic plan;
- Which of the CTS content categories the node belongs to (in cases where the content information of the node may belong to more than one category).

The history of the student's interaction with AOS is presented as a sequence of classes of nodes that he visited and a lot of information from which he looked [8].

The navigation method is based on a decision system that manages the connections of nodes and decides which nodes, which are based on a navigation rule and can be connected with the current node, hide. It seems appropriate to implement the following four types of navigation rules:

- Node path rules;
- General rules of the way;
- User node rules;
- General user rules.

We will call the navigation rule, which uses the set of student's parameters from its model, the path rule, and the navigation rule, which uses user-defined parameters, called the user rule. A navigation rule can also be divided into two types: a node rule and a general rule. A node rule is defined and applies only to a specific node. A general rule is to describe the most common navigation paths in hyperspace and the often used segmentation of the learner's parameter range.

In the navigation rule, the CTS developer describes the links that should be shown in accordance with the node identifier or in accordance with the class of the node that is the target of the link. The system hides all links that are not referenced in the navigation rule.

6 Conclusion

In the given paper the theoretical generalization and new solution of the scientific task of designing the integrated development environment for computer training systems are obtained. The designing of learning course in a concrete subject is extremely laborious task. It can handle only the collective of experts (the teacher training similar course, psychologist, stylist, editor etc.). But for achievement of the goal this collective of the experts of the different profile requires the uniform tool, which would allow them cooperate effectively. Such solution is the integrated development environment of computer training systems.

The program implementation of IDECTS that is basing only on one information technology is not possible. An optimal combination is the following technologies: Macromedia Shockwave + Borland Delphi + SQL + Real Audio + Real Video. Also we shall notice that the writers developing CTS with the help of IDECTS will have problems with a relation of possibilities of hypermedia in learning course. As the practice shows, using units of hypermedia in learning considerably improves the process of obtaining knowledge and skills. For example, it is known from the theory of learning, that if at learning with usage the text information the intensity of learning is about 25%, if the texts is supported by a sound it increases up to 50%, and if it is used text, sound and video the intensity of learning is approximately 75% .

Using the approach that is offered in the given paper, it is possible to develop the environment, which automates the process of creation of CTS for all three stages of learning: obtaining of knowledge, acquisition of skills and mastering skills . The application of such environment together with the technology of distance learning will allow providing high quality of learning and increase its efficiency at the expense of involving in the CTS development of team of highly skilled experts and correction of learning styles depending on level of learner knowledge and his psychological characteristics.

References

1. Bilous, N., Shubin, I., Vyrodov, O.: The conception of interactive training system design. J. New Solutions Mod. Technol. Kharkiv Ukraine **79**, 68–70 (2000)
2. Shubin, I., Vyrodov, O., Bilous, N.: Design of distance learning system on basis of modern information technologies. J. Syst. Anal. Control Inf. Technol. Kharkiv Ukraine (93) (2006)
3. Werner, E.: Artificial Intelligence and Tutoring Systems. Computational Approaches to the Communication of Knowledge. Morgan Kaufmann, Los Altos (1987)
4. Bloom, B.S.: Taxonomy of Educational Objectives: The Classification of Educational Goals, by Committee of College and University Examiners. D. McKay Co., New York (1956–1964)
5. Bilous, N., Shubin, I., Vyrodov, O.: Integrated environment for distance learning courses developer: In: Proceedings of 5-th International conference on Internet is a New Technology Environment in Information Society, Veliko Trnovo, Bulgaria (2002)
6. Bondarenko, M., Bilous, N., Shubin, I.: The Ukrainian e-learning region: In: Proceedings of 10-th International LLinE Conference New Partnerships and Lifelong Learning, Helsinki, Finland (2008)
7. Ya, V.: Svyatkin Learning model using the navigation rules for generating and dynamically modifying learning strategies in the basis of finite predicate algebra. Vestnik NTU KhPI National Technical University KhPI, Kharkov, Ukraine (2012)
8. Yu, I., Shubin, Y.V.: Svyatkin methods and models of building intelligent adaptive hypermedia systems. East Euro. J. Adv. Technol. **3/11**(57) (2012). Kharkov
9. Manako, A.: Information technology in education. In: Manako A., Sinitsa, E. (eds.) Control Systems and Machines, № 2, pp. 46–57 (2017)
10. Yangolenko, O.: Analysis of the state of information technologies in the system of higher education. In: Yangolenko, O., Lyutenko, I., Yakovleva, O. (eds.) Bulletin of the National Technical University "Kharkiv Polytechnic Institute". - Kharkiv: NTU «KhPI, 2012, №. 30, pp. 105–109 (2012)
11. University Management System - UMS [Electronic resource]. https://www.techxact.com/university-management-system-ums.html

12. Gorokhovatskyi, V., Vechirska, I., Chetverikov, G.: Method for building logical data transform in the problem of establishing links between objects in intellectual telecommunication systems. Telecommun. Radio Eng. **75**(18), 1645–1655 (2016)
13. Popov VD E-learning in small and medium enterprises. In: Popov, V.D. (ed.) Economic Bulletin of Donbass, 2013, №. 2 (32), pp. 36–41 (2013)
14. Weiss, G.: Multiagent Systems, 867 p. MIT Press (2013)
15. Yakushin, A.: The use of LMS in the educational process of a pedagogical university. In: Yakushin, A.V. (ed.) Modern Information Technologies and IT Education: a Collection of Reports of the Scientific-Practical Conference, pp. 189–193. MAKS press, Moscow (2006). Ed. VA Sukhomlina, resp. ed. EN Nikelina
16. Tetsky, A.: Investigation of the structure of the MOODLE educational content management system using reverse engineering elements. In: Tetsky, A.G., Uzun, D.D. (eds.) Open Information and Computer Integrated Technologies, №. 65, pp. 206–210 (2014)
17. Lingefjärd, T.: Mathematical modeling and technology in teacher education–Visions and reality. University of Gothenburg, 2001, pp. 205–215 (2001). T. Lingefjärd, M. Holmquist
18. Buteau, C.: Evolving technologies integrated into undergraduate mathematics education. In: Buteau, C., Muller, E., Seventeenth, D.C. (eds.) Technology Revisited: ICMI Study Conference, 2006, pp. 74–81 (2006)
19. Roa, H.: A Model for teaching mathematics via problem-solving supported by technology. In: Roa, H., Seventeenth, D.C. (eds.) Technology Revisited ICMI Study Conference, 2006, pp. 66–73 (2006)
20. Kravtsov, G.: The structure of the quality management system of electronic learning resources. In: Kravtsov, G. (ed.) Information Technology in Education, 2011, №. 10, pp. 94–101 (2011)

Author Index

P. Vorobiyenko et al. (Eds.): IPF 2020, LNNS 212, pp. 437–438, 2021.
https://doi.org/10.1007/978-3-030-76343-5